汉英水电工程常用词汇

Chinese-English Common Vocabulary for Hydropower Engineering

中国水电工程顾问集团有限公司 编

中国水利水电出版社
www.waterpub.com.cn

内 容 提 要

本书由中国水电工程顾问集团有限公司组织编写，北京、华东、西北、中南、成都、贵阳、昆明等勘测设计研究院及咨询公司的几十位专家参与此项工作。本书内容涵盖水电工程生命周期内规划、环保与移民、勘测、设计、施工、科研、总承包、项目管理等，按水电工程专业分章汇编。

本书旨在规范水电工程勘测、设计、施工、管理英文版文件的用词，也可作为涉外水电工程人员的简易工具书，可供水电工程勘测、设计、施工相关人员及大专院校有关专业师生参考。

图书在版编目（CIP）数据

汉英水电工程常用词汇 / 中国水电工程顾问集团有限公司编. -- 北京 : 中国水利水电出版社, 2015.6
ISBN 978-7-5170-3487-2

Ⅰ. ①汉… Ⅱ. ①中… Ⅲ. ①水利水电工程－词汇－汉、英 Ⅳ. ①TV-61

中国版本图书馆CIP数据核字(2015)第182863号

书　　名	汉英水电工程常用词汇
作　　者	中国水电工程顾问集团有限公司　编
出版发行	中国水利水电出版社 （北京市海淀区玉渊潭南路 1 号 D 座　100038） 网址：www.waterpub.com.cn E-mail：sales@waterpub.com.cn 电话：（010）68367658（发行部）
经　　售	北京科水图书销售中心（零售） 电话：（010）88383994、63202643、68545874 全国各地新华书店和相关出版物销售网点
排　　版	中国水利水电出版社微机排版中心
印　　刷	北京纪元彩艺印刷有限公司
规　　格	184mm×260mm　16 开本　24.5 印张　902 千字
版　　次	2015 年 6 月第 1 版　2015 年 6 月第 1 次印刷
印　　数	0001—2000 册
定　　价	**156.00** 元

《汉英水电工程常用词汇》
编　委　会

序

在经济全球化的大背景下，随着我国企业实力的增强和视野的日渐开阔，在国家大力支持中国企业“走出去”的宏观环境下，中国企业以积极主动的姿态融入到全球经济一体化的热潮中，极大地推动了我国企业国际化的进程。近年来，中国对外承包工程签订合同额和营业额大幅增长，业务规模持续扩大，合作模式呈现多样化。以EPC为代表的总承包模式、BOT融资模式越来越多。随着国家经济实力的日益加强，开展对外承包工程的宏观环境得到进一步改善，大量的国际项目在政府合作框架下得以集群式、成批量开发。

水电顾问集团一直从事国内国际河流水电水利规划、抽水蓄能电站选点规划、风电和太阳能发电规划，水电、抽水蓄能电站、风电和太阳能发电项目的勘测、设计、科研、施工监理和工程总承包业务，拥有了国内领先、国际一流的水电风电技术优势。水电顾问集团在2014年美国《工程新闻纪录》全球150强设计企业（国际、国内业务）排名中，位居第12位；国际工程设计公司225强排名第38位；在历年中国《建筑时报》“中国承包商和工程设计企业双60强”评比中，一直位列前2名以内。

水电顾问集团坚持“高端切入、规划先行、技术领先、融资推动”的国际优先发展策略，在国际市场上持续发挥水电、风电等可再生能源的技术和管理优势，从规划设计入手，带动我国的工程技术、标准、设备和工程总承包等走向国际市场。为大力支持中国水电风电企业国际营销和工程总承包的顺利执行，近年来，水电顾问集团开展了国际标准培训、中外技术标准对照研究、中国水电风电标准英文版编译等工作。为使中国规程规范和设计报告图纸编译进一步规范化、标准化，提高编译工作的质量和效率，迫切需要一本适用于水利水电勘察设计的英文专业词汇。

《汉英水电工程常用词汇》由中国水电工程顾问集团有限公司组织编写，水电顾问国际公司负责具体实施，北京、华东、西北、中南、成都、贵阳、昆明等勘测设计研究院及咨询公司的几十位专家参与此项工作。希望本词汇的出版能够为水电工程勘测、设计、施工、管理英文版文件的编写提供参考，能够有助于水电工程英文版文件用词的统一和规范，能够为从事海外工程的人员提供便利和帮助。

中国水电工程顾问集团有限公司总经理

前　言

在全球经济一体化发展、越来越多的中国企业走向国际市场的大背景下，水电顾问集团坚持“国际优先”发展战略，在国际市场上崭露头角。为使国外同行更好地了解中国标准，中国水电工程顾问集团有限公司组织了中国水电勘测设计行业标准的编译工作，为提高中国水电标准编译工作的质量和效率，为规范化、标准化设计报告图纸的编译提供参考，编写了《汉英水电工程常用词汇》（以下简称《词汇》）。

本《词汇》涵盖水电工程生命周期内规划、环保与移民、勘测、设计、施工、科研、总承包、项目管理等内容，按水电工程专业分为12章，收录词汇近2万条。各章的选词主要由专业技术人员承担，以现行水电技术标准中的专业词条为基础，参考了数十部英文原版标准、技术文献及几十册国内相关术语规范。

本《词汇》主要特点：

(1) 注重实用，本书除常规的按字母顺序编排外，另采用按专业分章节的编排方式，以方便专业技术人员查找词汇；在专业章节中，对于涉及多专业的大量词条，大多只在一处出现，避免过多重复；考虑工程设计的实际需要，收录了各专业的主要图表、报告名称。

(2) 立足常用，已正式出版的水利水电和电力技术方面的综合词典很多，本书选词则主要基于水电建设常用词条，并且只选取其在水电工程方面的词义，并收录了工程总承包及反映水电建设领域新技术、新工艺的词汇。

(3) 力求准确，同一术语经常有不同译法，尽量选用国外标准或技术文献的用词，例如机电方面的单词，基本出自IEC相关标准；水工、施工词汇大量参考了美国垦务局和陆军工程兵团的标准及手册。为了尽量统一用法，相同词只选取两个较普遍的译法。

编者

2015年6月

体例说明

（1）为查找方便，本书主要内容分三种方式编排：①按专业分类；②按汉语（字母）顺序；③英文索引（只收录部分技术名词）。

（2）考虑实际查找需要，专业分类中收录的主要图表、报告名称及少部分专业短语只在专业分类部分出现，汉英、英汉排序中不列入。

（3）一个词条有多种译法的，只选取两个水电工程中较普遍的用法。

（4）词汇中圆括号（ ）内的词表示可以替代前词，或另一种称谓，或缩写，例如：锚筋束（桩）、permanent action (load)；方括号［ ］内的词是说明性的，或表示该词适用的特定语境，例如：弯管段损失［水头］、贴角［拱坝］。

目　录

序
前言
体例说明

专业分类部分 ······ 1

第1章　综合　Common ······ 3

第2章　工程勘测　Engineering investigation ······ 6
2.1　区域地质与地震　Regional geology and seismicity ······ 6
2.2　工程地质条件　Engineering geological conditions ······ 8
2.3　岩土力学　Rock and soil mechanics ······ 19
2.4　主要工程地质问题　Main engineering geological considerations ······ 21
2.5　天然建筑材料与岩土试验　Natural construction materials and geotechnical test ······ 23
2.6　工程地质勘察　Engineering geological investigation ······ 25
2.7　测量　Engineering survey ······ 28
2.8　主要报告及图表名称　Titles of reports，figures and tables ······ 30

第3章　工程规划　Planning ······ 32
3.1　水文泥沙　Hydrology and sediment ······ 32
3.2　水能及水资源利用　Water power and water resources utilization ······ 36
3.3　经济评价　Economic evaluation ······ 39
3.4　主要报告及图表名称　Titles of reports，figures and tables ······ 41

第4章　环保与移民　Environmental protection and resettlement ······ 43
4.1　环境保护　Environmental protection ······ 43
4.2　水土保持　Water and soil conservation ······ 46
4.3　移民安置　Resettlement ······ 48
4.4　主要报告及图表名称　Titles of reports，figures and tables ······ 51

第5章　水工建筑物　Hydraulic structures ······ 52
5.1　通用　General ······ 52
5.2　挡水建筑物　Water retaining structure ······ 59
5.3　泄水建筑物　Water release structures（outlet structures） ······ 62
5.4　输水建筑物　Conveyance structures ······ 64
5.5　发电厂房　Powerhouse ······ 65
5.6　基础及边坡处理　Treatment of foundation and slope ······ 66
5.7　观测（监测）　Observation（Monitoring） ······ 67
5.8　其他　Miscellaneous ······ 68
5.9　常用计算准则与方法　Common calculation criteria and methods ······ 70
5.10　主要报告及图表名称　Titles of reports，figures and tables ······ 70

第 6 章　水力机械　Hydro-machinery …… 72
6.1　水轮机及附属设备　Turbine and auxiliary equipment …… 72
6.2　辅机系统　Auxiliary machinery …… 80
6.3　安装和调试　Installation and trial running …… 87
6.4　主要图表名称　Titles of figures and tables …… 88
第 7 章　金属结构及安装　Hydraulic steel structures and installation …… 89
7.1　闸门　Gate …… 89
7.2　启闭机　Hoist …… 92
7.3　升船机　Shiplift …… 95
7.4　主要图表名称　Titles of figures and tables …… 96
第 8 章　电工　Electricity …… 97
8.1　通用　General …… 97
8.2　电力系统及电气接线　Electric power system and connection …… 100
8.3　发电机与发电电动机　Generator and generator-motor …… 104
8.4　电气设备　Electric equipment …… 107
8.5　过电压保护、防雷、接地与电气安全　Overvoltage protection，lightning protection，earthing and electric safety …… 115
8.6　电力电子（含 SFC 与励磁设备）　Electronics，including SFC and excitation equipment …… 119
8.7　监控系统　Supervision and control system …… 121
8.8　继电保护系统　Relay protection system …… 126
8.9　控制电源　Control power sources …… 129
8.10　通信、工业电视与门禁系统　Communication，industrial TV and access control system …… 130
8.11　电气设备布置　Electric equipment layout …… 132
8.12　主要图表名称　Titles of figures and tables …… 132
第 9 章　土建工程施工　Construction of civil works …… 134
9.1　施工布置与交通运输　Construction layout，access and transportation …… 134
9.2　建筑材料　Construction materials …… 135
9.3　施工导流　Diversion during construction …… 138
9.4　主体工程施工　Construction of main structures …… 139
9.5　施工工厂设施及施工机械　Construction plants and machinery …… 148
9.6　施工进度　Construction schedule …… 150
9.7　主要报告及图表名称　Titles of reports，figures and tables …… 150
第 10 章　工业与民用建筑　Industrial and civil architecture …… 152
10.1　建筑与结构　Architectonics and structure …… 152
10.2　供暖、通风与空调　Heating，ventilation and air conditioning（HVAC） …… 155
10.3　给排水　Water and wastewater system …… 158
10.4　消防　Fire protection …… 159
10.5　主要图表名称　Titles of figures and tables …… 162
第 11 章　工程投资　Project investment …… 163
11.1　基础价格　Base price …… 163
11.2　建筑及安装工程单价　Price of civil works and installation works …… 163
11.3　设备费　Equipment cost …… 164

11.4 投资估算 Investment estimation …… 165
11.5 主要图表名称 Titles of figures and tables …… 166
第12章 项目管理 Project management …… 168
12.1 招投标 Tender and bid …… 168
12.2 合同管理 Contract management …… 168
12.3 现场管理 Management of construction …… 169
12.4 档案管理 Archives management …… 172
附录1 技术标准编制常用词汇 …… 174
附录2 部分国外机构、团体名称 …… 176
汉语排序部分 …… 179
英文索引（部分技术名词） …… 329

专业分类部分

第1章 综　　合
Common

水资源　water resources

水能资源　waterpower resources (hydropower resources)

水利工程（水资源工程）　water resources project

水电工程　hydropower project (hydroelectric project)

发电系统　power generation system

输变电工程　power transmission and distribution project

水力发电设施　hydroelectric installations

工程规模　project scale (project size)

工程开发模式　project development mode

公私合作模式　public-privated-partnership (PPP)

工程开发任务　project development purposes

发电　power generation

防洪　flood control

供水　water supply

通航　navigation

灌溉　irrigation

防凌　ice control

减淤　desilting

渔业　fishery

旅游　tourism

休闲　recreation

工程勘测　engineering investigation

水工建筑物　hydraulic structure

安全监测　safety monitoring

水工金属结构　hydraulic steel structure

水力机械设备　hydraulic machinery

机电设备　electromechanical equipment

设备总成套　system equipment design and supply

工程施工　engineering construction

建设征地与移民安置　land requisition and resettlement

工程投资　project investment

建设管理　construction management

工程档案　project archives

供暖通风与空调　heating, ventilation and air conditioning (HVAC)

工业与民用建筑　industrial and civil architecture

消防　fire protection, fire control

给排水　water supply and sewerage system

工程造价　construction cost

水电工程等别　rank (scale) of hydropower project

水工建筑物级别　grade of hydraulic structure

设计安全标准　design safety standard

结构可靠度　structural reliability

挡水建筑物　water retaining structure

泄水建筑物　water release structure

输水系统　water conveyance system

通航建筑物　navigation structure

过鱼建筑物　fish pass structure

过木建筑物　log pass structure

开发方式　development scheme

坝式开发　dam type development

引水式开发　conduit type development

混合式开发　dam and conduit type development

河流梯级开发　cascade development

抽水蓄能开发　pumped storage development

跨流域开发　transbasin development, interbasin development

潮汐发电　tidal power

水电站　hydropower station

坝式水电站　dam type hydropower station

引水式水电站（引水道式水电站）　conduit type hydropower station

坝后式水电站　hydropower station at dam-toe

坝内式水电站　hydropower station in dam

河床式水电站　water retaining type hydropower station, hydropower station in river channel

厂顶溢流式水电站　overflow hydropower station

地下式水电站　underground hydropower station

半地下式水电站　semi-underground hydropower station

径流式水电站　run-of-river hydropower station

短期调节水电站　pondage power station

蓄水式水电站 reservoir power station
梯级水电站 cascade hydropower station
潮汐水电站 tidal hydropower station
抽水蓄能电站 pumped storage power station
水电厂 hydropower plant
永久性建筑物 permanent structure
临时性建筑物 temporary structure
主要建筑物 main structure
次要建筑物 secondary structure
新建、扩建、改建 construction, extension and renovation
改造与增容 rehabilitation and upgrading
水库特征水位 characteristic levels of reservoir
校核洪水位 check flood level
设计洪水位 design flood level
防洪高水位 top level of flood control
防洪限制水位 low limit level during flood season
正常蓄水位 normal pool (storage) level
死水位 minimum operating level, dead water level
尾水位 tailwater level
水土保持 water and soil conservation
环境保护 environmental protection
自然保护区 nature reserve
文物保护遗址 cultural relics site
历史文化遗址 historical and cultural site
基础设施 infrastructure
饮用水水源保护区 drinking water source reserve area
劳动安全与工业卫生 labor safety and industrial hygiene
生活饮用水卫生标准 sanitary standard for drinking water
污水综合排放 integrated wastewater discharge
工业企业厂界噪声 industrial enterprises noise at boundary
大气污染物综合排放标准 integrated emission standard of air pollutants
职业病 occupational disease
职业健康与安全 occupational health and safety
风险分析 risk analysis
风险评估 risk assessment (evaluation)
风险控制 risk control
反馈控制 feedback control
反馈信息 feedback information
江河流域综合利用规划 multipurpose utilization plan of the river basin
河流(河段)水电规划 hydropower development planning of river (river reach)
设计程序 design procedure
比选方案 alternative
预可行性研究 pre-feasibility study
可行性研究 feasibility study
招标设计 tender design
施工详图设计 detailed design
运行期 operation period
清洁可再生能源 clean and renewable energy resources
资源普查 resources investigation
规划 planning
设计 design
工程咨询 engineering consultation
融资 financing
设备采购 equipment procurement
工程总承包 EPC contracting
安全鉴定 safety appraisal
施工监理 construction supervision
竣工验收 final acceptance
运行与维护 operation and maintenance
项目法人 project legal person, project entity
初步勘察 Preliminary Investigation❶
初步设计勘察 Initial Design Investigation❶
最终设计勘察 Final Design Investigation❶
评估勘察 Appraisal Investigation❷
可行性设计 Feasibility Design❷
最终设计 Specifications (Final) Design❷
能源革命 energy revolution
互联互通 connectivity
区域经济一体化 regional economic integration
丝绸之路经济带 Silk Road Economic Belt
21世纪海上丝绸之路 21st Century Maritime Silk Road
一带一路 One Belt One Road
优势互补、互利共赢 complement each other's advantages and achieve win-win outcomes

---

❶ 美国能源署勘察阶段分类
❷ 美国垦务局设计阶段分类

图例 legend
符号 symbol
平面布置图 plan，layout
立视图 elevation view
上游视图 upstream view
下游视图 downstream view
透视图 prospective view
立体图 stereogram
鸟瞰图 aerial view
俯视图 top view，plan view
示意图（略图） sketch（schematic drawing）
外形图 outline dimension drawing
目测草图 visual sketch
轮廓图 outline drawing
剖面图 sectional drawing
纵剖面图 longitudinal section（profile）
横剖面图 transversal section，cross-section
展视图（展开图） outspread view
绝对高程 absolute elevation
相对高程 relative elevation
开挖线 excavation line
原地面线 natural ground line
建筑物轮廓线 outline of structure
加工图（车间图） fabrication drawing，shop drawing
装配图（组装图） assembly drawing
体型图 shape drawing
配筋图 reinforcement drawing
细部详图（大样图） detail drawing
施工图 construction drawing
施工详图 detailed drawing
竣工图 as-built drawing
蓝图 blueprint
套用图 drawing use indiscriminately
框图 block diagram
标题栏 title block
签字栏 authentication block
制图 drawn by
校核 checked by
审查 reviewed by
核定 examined by
批准 approved by
项目建议书 Project proposal（recommendation）
项目申请报告 Project application report
预可行性研究报告 Pre-feasibility study report
可行性研究报告 Feasibility study report
招标设计报告 Tender design report
流域地理位置示意图 Sketch of geographic location of the river basin
工程地理位置图 Geographic location of the proposed project
河流（河段）梯级开发示意图 Sketch of cascade developments of the river（reaches）
电站接入系统地理位置图 Geographic location of hydropower plant interconnection
枢纽总布置图 Project layout
枢纽效果图 Project design sketch
工程特性表 Project features
工程量汇总表 Summary bill of quantities
计算书 Calculation sheet
设计说明 Design description

第2章 工 程 勘 测
Engineering investigation

2.1 区域地质与地震
Regional geology and seismicity

(1) 地质年代 Geological time scale, geological age

代（界） Era（Erathem）
纪（系） Period（System）
世（统） Epoch（Series）
期（阶） Stage（formation）
群组 Group
新生代（界） Cenozoic era（erathem）
晚新生代地层 Late Cenozoic Stratum
第四纪（系） Quaternary period（system）
全新世（统） Holocene epoch（series）
更新世（统） Pleistocene epoch（series）
第三纪（系） Tertiary period（system）
晚第三纪（系） Neogene period（system）
上新世（统） Pliocene epoch（series）
中新世（统） Miocene epoch（series）
早第三纪（系） Eogene period（system），Paleogene period（system）
渐新世（统） Oligocene epoch（series）
始新世（统） Eocene epoch（series）
古新世（统） Paleocene epoch（series）
中生代（界） Mesozoic era（erathem）
白垩纪（系） Cretaceous period（system）
侏罗纪（系） Jurassic period（system）
三叠纪（系） Triassic period（system）
古生代（界） Palaeozoic era（erathem）
二叠纪（系） Permian period（system）
石灰纪（系） Carboniferous period（system）
泥盆纪（系） Devonian period（system）
志留纪（系） Silurian period（system）
奥陶纪（系） Ordovician period（system）
寒武纪（系） Cambrian period（system）
元古代（界） Proterozoic era（erathem）
前寒武纪（系） Precambrian period（system）
震旦纪（系） Sinian period（system）
太古代（界） Archaeozoic era（erathem）
前震旦纪（系） Presinian period（system）
冰川时期 Glacial epoch（period）
冰期 ice age
主要大地构造运动，造山运动 main tectonic movement，orogeny
侵入岩分期 intrusive rock period
造山旋回 orogenic cycle
多旋回性 multiple cyclicities
喜马拉雅运动 Himalayan movement
喜马拉雅运动的第Ⅱ幕 Second episode of Himalayan movement
燕山运动 Yanshan movement
印支运动 Indosinian movement
华力西运动 Variscian movement
加里东运动 Caledonian movement
晋宁运动 Jinniing movement
吕梁运动 Luliang movement
年代测定 age dating
放射性测年 radioactive dating，radioactive age determination
同位素年龄 isotopic age

(2) 区域地质 Regional geology

大地构造 geotectonics
大地构造单元 geotectonic element
大地构造体系 geotectonic system
板块 plate
板块构造环境 plate geotectonic setting
板块缝合带 plate suture zone
拼合地带 collage belt
地缝合带 suture belt
蛇绿岩带 ophiolite belt
板块碰撞 plate collision
岛弧带 island arc belt
造山带 orogenic belt
海进 marine ingression

海退　marine regression
地槽　geosyncline
地台　continental platform
地盾　shield
构造线　tectonic line
构造格局　tectonic framework
构造体系　tectonic system
山字形构造　epsilon-type structure
旋扭构造　rotational shear structure
帚状构造　brush structure
环带构造，带状构造　zonal structure
雁行构造，斜列构造　echelon structure
斜接　diagonal connection，juxtaposition
地垒　horst
地堑　graben
褶断山　fault-folded mountain
断块山地　fault-block mountain
桌状山　table mountain
断陷盆地　faulted basin
拗陷盆地　depression basin
超岩石圈断裂　trans-lithospheric fault
深大断裂　deep fault
活断层　active fault
可能活动断层　capable fault
裂谷　rift valley
区域性主要断裂　regional major fault
活动带　active belt
断裂活动强度　intensity of faulting activity
侧向挤压　lateral compression
拖曳运动　drag movement
叠瓦构造　decken structure
褶皱基底　fold basement
沉积建造　sedimentary formation
磨拉石建造　molasse formation
复理石建造　flysch formation
类复理石建造　flyschoid formation
碳酸盐岩建造　carbonatite formation
地台型建造　platform type formation
岩浆活动　magmatic activity
褶皱系　fold system
褶皱束　fold bundle
隆升　uplift

(3) 地震　Seismicity

地震区　seismic area，earthquake area
地震带　seismic belt
地震构造区　seismic structure zone
震源　seismic source，seismic focus
震源深度　focus depth
深源地震　deep-focus earthquake
浅源地震　shallow-focus earthquake
震中　epicenter
震源机制　focal mechanism
发震断层　seismogenic fault
发震构造　seismogenic structure，seismogenic tectonics
发震机制　seismogenic mechanism
潜在震源区　potential seismic source zone
地震活动　seismic activity
地震前兆　premonitory symptom
前震　foreshock
主震　main shock
余震　aftershock
破坏性地震　destructive earthquake
活动断层探测　surveying and prospecting of active fault
地震活动断层　seismo-active fault
断层活动段　active fault segment
地震构造　seismic structure
活动构造　active structure
构造类比　structure analog
古地震　paleo-earthquake
起算震级　lower limit earthquake
震级档　magnitude interval
地震震级　earthquake magnitude
里氏震级　Richter magnitude
地震烈度　seismic intensity，earthquake intensity
麦加利烈度　Mercalli intensity
基本地震烈度　basic seismic intensity
设计地震烈度　design seismic intensity
地震波　seismic wave
体波　body wave
纵波　longitudinal wave，compression wave
横波　transverse wave，shear wave
面波　surface wave
瑞利波　Rayleigh wave
勒夫波　Love wave
中国地震动参数区划图　Earthquake motion parameter zoning map of China

中国地震动峰值加速度区划图 Ground motion peak acceleration zoning map of China
地震动反应谱特征周期 characteristic period of the seismic response spectrum
中国地震动反应谱特征周期区划图 Eigenperiod zoning map of earthquake response spectrum of China
区域地震构造综合分析 comprehensive analysis on regional seismotectonics
地震环境综合评价 comprehensive evaluation on seismic settings
地震安全性评价 seismic safety evaluation
地震危险性分析 seismic hazard analysis
年平均发生率 annual incidence rate
震级-频度关系 magnitude-frequency relation
中等强度和频度的地震活动性 seismic activity of moderate intensity and frequency
本底地震 background earthquake
地震反应谱 seismic response spectrum
反应谱特征周期 eigenperiod of response spectrum
地震动参数 ground motion parameter
地震动峰值加速度 seismic peak ground acceleration
地震加速度 seismic acceleration
水平加速度 horizontal acceleration
垂直加速度 vertical acceleration
峰值加速度 peak acceleration
相应烈度 equivalent intensity
强震 strong earthquake
弱震 weak earthquake
微震 microquake
等震线 isoseismals
极震区 meizoseismal area
地震断层 earthquake fault
地震地质灾害 earthquake induced geological disaster
地裂缝 ground crack
震陷 earthquake subsidence
锯齿状连续延伸 successively zigzag spread
冒水喷沙 water spraying and sand emitting
砂土液化 sand liquefaction
液化指数 liquefaction index
地震观测 seismological observation
国家地震台网 national network of seismograph
地震台 seismostation
子地震台 seismo-substation
地震台网 seismic network
地震仪 seismograph, seismometer
强震仪 strong-motion seismograph
微震仪 micro-seismograph
新构造运动 neotectonics
新构造运动分区 neotectonic zoning
空间继承性 succession in space
新构造形迹 traces of neotectonics
断裂槽谷 fault trough valley
断层三角面 fault triangular facet
断层崖 fault scarp
坡中谷 valley in slope
地形坡折 terrain slope break
水系错动 dislocated water system
掀斜式隆升 tilting uplift
间歇性隆升 intermittent uplift
差异性隆升 differential uplift
隆升速率 uplift amplitude
构造稳定区 tectonically stable zone
安全岛 safety island
相对稳定地块 relatively stable block

2.2 工程地质条件 Engineering geological conditions

(1) 地形、地貌 Topography and geomorphy

地形 topography
地貌 geomorphy, landform
地貌类型 geomorphic type
大陆 continent
山系 mountain system
山脉 mountain range
山地 mountain region (area)
山坡 mountain slope, hillside
山麓 mountain foot
垭口 saddle back
山口 mountain pass
山谷 valley
丘陵 hill
高原 plateau
高地 highland
平原 plain

准平原　peneplain
残丘　inselberg
盆地　basin
洼地　depression
山间盆地　intermontane basin
河谷　river valley
峡谷　canyon，gorge
纵谷　longitudinal valley
横谷　transverse valley
斜向谷　insequent valley
平直河道　straight channel
蜿蜒河道　meandering channel
多叉河道　braided channel
U形谷　U-shaped valley
V形谷　V-shaped valley
凸岸　convex bank
凹岸　concave bank
河湾　river bend
开阔地　open terrain，open ground
坡度　gradient，slope
缓坡　gentle slope，flat slope
陡坡　steep slope，abrupt slope
悬崖　cliff，scarp
凹槽　trough
凹坡　concave slope
凹凸形　concavo-convex
凹陷　hollow
分水岭　water divide，watershed
阶地，台地　terrace
侵蚀阶地　erosional terrace
堆积阶地　constructional terrace
基座阶地　bedrock seated terrace
埋藏阶地　buried terrace
新近纪湖积台地　neogene lacustrine deposit platform
湖积平原　lacustrine plain
牛轭湖　ox-bow lake
三角洲　delta
洪泛平原　floodplain
堆积平原　accumulation plain
剥蚀平原　denudation plain
陆相剥蚀平原　plain of subaerial denudation
侵蚀平原　erosion plain
冲积平原　alluvial plain
冲积扇　alluvial fan
洪积扇　diluvial fan，proluvial fan
坡积裙　talus fan
冲沟　gully
坳沟　shallow flat ravine
冰川地形　glacial landform
冰斗　glacial cirque
冰川谷　glacier valley
悬谷　hanging valley
冰碛阶地　glacial drift terrace
冰冻隆起　frost upheaval
翻浆　frost boiling
磨蚀　abrasion
磨圆碎屑　rounded fragment
风成地貌　eolian landform
黄土高原　loess plateau
黄土梁　loess ridge
黄土峁　loess hill
黄土塬　loess tableland
黄土坪　loess terrace
落水洞　sinkhole
雅丹地貌　Yardang landform
风蚀残丘　eolian monadnock
风蚀壁龛　tafone
蘑菇石　mushroom stone
沙漠　desert
戈壁　gobi，stone desert
侵蚀及剥蚀地貌单元　erosional and denudational geomorphic unit
剥夷面（夷平面）　planation surface
剥蚀　abrasion，denudation
剥蚀面　denudation plane
侵蚀基准面　erosion basis，base level of erosion
古地理学　paleogeography
古地下水　fossil groundwater，paleo-groundwater
古河槽　old channel，buried channel
古河道　ancient river course，paleochannel
古滑坡　ancient landslide，old landslide
古岩溶　ancient karst
古代冰川作用　ancient glaciation

(2) 岩土与矿物　Rock，soil and minerals

岩性学　lithology
第四系沉积物　Quaternary sediment (deposit)
成土母岩　soil-forming rock
成岩作用　diagenesis

成因分类 genetic classification
成因分析 genetic analysis
内因 internal cause
内营力 endogenetic agent
外营力 exogenic agent
成因类型 genetic type
堆积 accumulation
冲积 alluvium
洪积 diluvium
坡积 slope wash
残积 eluvium
冰碛 glacial drift
冰水堆积体 outwash deposit
冰川沉积，冰积层 glacial deposit
风积物 eolian deposit
湖积物 lacustrine deposit
滑坡堆积 landslide deposit
碎屑沉积 clastic sediment
人工堆积 artificial deposit
堆积体 accumulation body
土石混合体 soil-rock mixture
表土 regolith，top soil
孤石 lonestone
漂石（漂砾） erratic boulder
块石 rock block
卵石 cobble，pebble
砾石 gravel
粗砾 coarse gravel
中砾 medium gravel
细砾 fine gravel
砂 sand
粗砂 coarse sand
中砂 medium sand
细砂 fine sand
粉砂（粉土） silt
碎石土 debris soil
砾质土 gravelly soil
粗砾质土，卵石土 cobble soil
壤土 loam
砂壤土 sandy loam
砂性土 sandy soil
黏土 clay
黏性土 cohesive soil，clayey soil
无黏性土 cohesionless soil
砂质黏土 sandy clay
分散性黏土 dispersive clay
红黏土 laterite
有机质土 organic soil
淤泥 mud
淤泥质黏土 muddy clay
冰碛土 moraine soil
冻土 frozen soil
黄土 loess
湿陷性黄土 collapsible loess
自重湿陷性黄土 self weight collapsible locss
非自重湿陷性黄土 non-weight collapsible loess
非湿陷性黄土 non collapsible loess
膨胀土 swelling soil，expansive soil
敏感性土 sensitive soil
人工填土 backfilled soil
颗粒形状 particle shape
棱角状 angular
次棱角状 subangular
次圆状 subrounded
圆状 rounded
扁平状 flat
长条状 elongated
沉积岩 sedimentary rock
岩相 lithofacies
陆相 continental facies
陆相红层 red bed of continental facies
海相 marine facies
海相碎屑岩 clastic rocks of marine facies
浅海相 neritic facies，shallow sea facies
滨海相 littoral facies
湖相 lacustrine facies，lake facies
泻湖相 lagoon facies
河相 fluvial facies
冰水沉积 glaciofluvial deposit
煤层 coal seam
蒸发岩 evaporite
砾岩 conglomerate
角砾岩 breccia
底砾岩 basal conglomerate
砂砾岩 sandy conglomerate
砂岩 sandstone
杂砂岩，硬砂岩 greywacke
石英砂岩 quartz sandstone

长石砂岩 arkose, feldspathic sandstone
铁质砂岩 ferruginous sandstone
岩屑砂岩 lithite
砾砂岩 conglomeratic sandstone
粉砂岩 siltstone
黏土质砂岩 argillaceous sandstone
黏土岩 clay stone (rock)
泥岩 mudstone
粉砂质泥岩 silt mudstone
页岩 shale
黏土页岩 clay shale
炭质页岩 carbonaceous shale
油页岩 kerogen shale (oil shale)
铝土页岩 bauxitic shale
灰岩，石灰岩 limestone
鲕状灰岩 oolitic limestone
粒状灰岩 granular limestone
板状灰岩 platy limestone
泥质灰岩 argillaceous limestone
结晶灰岩 crystalline limestone
致密灰岩 compact limestone
生物碎屑灰岩 bioclastic limestone
白云质灰岩 dolomitic limestone
竹叶状灰岩 wormkalk
泥灰岩 marl
斑脱土（岩），膨润土 bentonite
胶结 cement
硅质胶结 siliceous cement
钙质胶结 calcareous cement
泥质胶结 argillaceous cement
钙质结核 caliche nodule, calcareous concretion
扁豆体，透镜体 lenticle
变质岩 metamorphic rock
浅变质岩 epimetamorphic rock
深变质岩 katametamorphic rock
正变质岩 orthometamorphite, ortho-rock
副变质岩 parametamorphite
叶理化变质岩 foliated metamorphic rock
板岩 slate
千枚岩 phyllite
片岩 schist
云母片岩 mica schist
滑石片岩 talc schist
角闪石片岩 amphibole schist
绿岩（绿闪石片岩） noricite
片麻岩 gneiss
流纹片麻岩 rhyolite-gneiss
花岗片麻岩 granite-gneiss
斜长片麻岩 plagio-gneiss
混合岩 migmatite
混合岩化 migmatization
块状变质岩 massive metamorphic rock
大理岩 marble
钙质白云石大理岩 calc-dolomite marble
石英岩 quartzite
变质砂岩 metasandstone
变质砾岩 metaconglomerate
变质泥岩 metapelite
角岩 hornfels
麻粒岩 granulite
角闪岩 amphibolite
粗面岩 trachyte
粗玄岩 dolerite
构造岩 tectonite
构造角砾岩 tectonic breccia
破碎角砾岩 cataclastic breccia
碎裂岩 cataclasite
挤压片状岩 sliced rock
糜棱岩 mylonite
千糜岩 phyllonite
断层泥 fault gouge, fault clay
火成岩，岩浆岩 igneous rock
岩浆 magma
酸性岩 acid rock
伟晶岩 pegmatite
花岗岩 granite
花岗伟晶岩 granite-pegmatite
花岗斑岩 granite-porphyry
细晶岩 aplite
流纹岩 rhyolite, liparite
中性岩 intermediate rock
闪长岩 diorite
花岗闪长岩 granodiorite
煌斑岩 lamprophyre
正长岩 syenite (sinaite)
二长岩 monzonite
玢岩 porphyrite
安山岩 andesite
英安岩 dacite, quartz andesite
基性岩 basic rock

辉长岩 gabbro
辉绿岩 diabase, dolerite
玄武岩 basalt
响岩 phonolite
超基性岩 ultra-basic rock
辉岩 pyroxenite
橄榄岩 peridotite, olivinite
蛇纹岩 serpenite
深成岩 plutonic
岩床 sill
岩盘 laccolith
岩株 stock
岩脉 vein, dike
石英脉 quartz vein
伟晶岩脉 pegmatite vein
细晶岩脉 aplite vein
绿帘石脉 epidote vein
方解石脉 calcite vein
煌斑岩脉 lamprophyre vein
基质 matrix
母岩 host rock
脉岩 vein rock, dyke rock
捕虏体 xenolith
斑岩 porphyry
火山岩 volcanic rock
喷出岩 extrusive rock, eruptive rock
中酸性喷发岩 intermediate acidic eruptive rock
火山角砾岩 volcanic breccia
火山碎屑岩 pyroclastic rock
凝灰岩 tuff
暗色矿物 dark mineral
亮色矿物 light mineral
伴生矿物 associated mineral
次生矿物 secondary mineral
软弱矿物 soft mineral
蚀变矿物 altered mineral
暗褐黄色 dark brownish yellow
亮灰绿色 light grayish green
玻璃光泽 vitreous luster
珍珠光泽 pearly luster
油脂光泽 greasy luster
金属光泽 metallic luster
滑石 talc
叶蜡石 pyrophyllite
蒙脱石 montmorillonite
蛭石 vermiculite
高岭石 kaolinitic
水云母 hydromica
石墨 graphite
褐铁矿 limonite
石膏 gypsum
绿泥石 chlorite
岩盐 halite
黑云母 biotite
白云母 muscovite
蛇纹石 serpentine
方解石 calcite
硬石膏 anhydrite
重晶石 barytes
白云石（岩） dolomite
黄铜矿 chalcopyrite
萤石 fluorite
磷灰石 apatite
角闪石 hornblende
辉石 augite
蛋白石 opal
阳起石 actinolite
白榴石 leucite
磁铁矿 magnetite
赤铁矿 hematite
透长石 sinidine
正长石 orthoclase
斜长石 plagioclase
黄铁矿 pyrite
绿帘石 epidote
石英 quartz
电气石 tourmaline
红柱石 andalusite
黄玉 topaz
刚玉 corundum
金刚石 diamond
伊利石 illite glimmerton

(3) 岩石结构与构造 Texture and structure of rock

结晶质 crystalline substance
非晶质 amorphous substance
硬度 hardness
解理 cleavage
断口 fracture

贝壳状断口　conchoidal fracture
结构　texture
斑状结构　porphyritic texture
带状结构　banded texture
等粒结构　granulitic texture
多孔结构　hiatal texture
非均匀结构　heterogeneous texture
角砾结构　brecciated texture
碎屑结构　clastic texture
砂质结构　arenaceous texture
黏土质结构　argillaceous texture
泥质结构　pelitic texture
生物结构　biogenetic texture
粒状结构　granular texture
流状结构　fluidal texture
糜棱结构　mylonitic texture
微晶结构　microlitic texture
粗晶结构　coarsely-crystalline texture
粗粒结构　coarse grained texture
斑晶结构　phenocryst texture
晶孔　vugular pore
显晶质结构　phanerocrystalline texture
隐晶质结构　cryptocrystalline texture
玻璃质结构　vitreous texture
包体结构　inclusion texture
变余结构　palimpsest texture
变晶结构　crystalloblastic texture
板状构造　tabular structure，platy structure
千枚状构造　phyllitic structure
片状构造　schistose structure
片麻状构造　gneissic structure
带状构造　banded structure
眼球状构造　augen structure
块状构造　massive structure
流纹状构造　rhyotaxitic structure
气孔状构造　vesicular structure
杏仁状构造　amygdaloidal structure
柱状构造　columnar structure
沉积层理构造　sedimentary bedding structure
相变　facies transformation
岩层　rock formation，rock stratum
岩层产状　stratum orientation，attitude of bed
岩层倾角　stratum dip
岩层走向　stratum strike
岩层间断　stratum gap
岩层交互　beds alternation
标志层　marker bed
标准层　index bed，guide bed
标准化石　guide fossil，index fossil
层面（层理面）　bedding plane
层理　bedding，stratification
交错层理　cross bedding
水平层理　horizontal bedding
波状层理　wave bedding
透镜体　lens
尖灭　wedge out
错动层面　faulted bedding plane
整合　conformity
不整合　unconformity
角度不整合　angular unconformity
假整合　disconformity
夹层　interlayer，intercalation
泥化夹层　argillized seam
软弱夹层　weak interlayer
黏土夹层　clay seam
互层　alternating layers，interbedded layers

(4) 地质构造　Geological structure

褶皱　fold
褶皱作用　folding
褶皱带　fold zone，fold belt
褶皱脊　fold hinge
褶皱轴　fold axis
褶皱翼　fold limb
平卧褶皱　recumbent fold
倒转褶皱　overturned fold
倾覆褶皱　plunging fold
直立褶皱　upright fold，erect fold
锯齿状褶皱　zigzag fold
短轴褶皱　brachy-axis fold
弯曲褶皱　buckled fold
扭曲褶皱　contorted fold
拖曳褶皱　drag fold
闭合褶皱　closed fold
向斜　syncline
背斜　anticline
复背斜　anticlinorium
复向斜　synclinorium
倒转背斜　overturned anticline
倾伏背斜　plunging anticline

单斜构造　uniclinal structure
构造穹窿　structural dome
构造盆地　structural basin
挠曲　bending fold
揉皱　crumple
断层　fault
正断层，下落断层　downthrown fault，normal fault
张性断层　tension fault
逆断层，上冲断层　upthrown fault，thrust fault
逆掩断层　overthrust fault
剪切断层　shear fault
剪切带　shear zone
韧性剪切带　ductile shear zone
平移（走滑）断层　displacement fault，strike-slip fault
左旋扭动　left lateral wrench
左旋走滑活动　left lateral strike-slip activity
右旋走滑位错　right lateral slip offset
层间错动带　interlayer shear zone
共轭断层　conjugated fault
雁列断层　en echelon fault
叠瓦断层　imbricate fault
顺河断层　fault stretching along river
隐伏断层　buried fault
断层擦痕　slickenside，fault striation
断层擦沟　fault striae
断层破碎带　fault fractured zone
断层影响带　fault influenced zone
挤压片理和劈理带　compression schistosity and cleavage zone
断层交会带　fault intersection zone
层间挤压面（带）　crushed bedding plane（zone）
构造挤压带　crushed zone of structure
擦痕面　slickenside
脆性断裂　brittle fracture
脆性破坏　brittle failure
错位　dislocation，malposition
节理，裂隙　joint
原生节理　primary joint
纵节理　longitudinal joint
横节理　cross joint
层节理　bedded joint
柱状节理　columnar joint
构造裂隙（节理）　tectonic joint
张裂隙（节理）　tension joint
剪裂隙（节理）　shear joint
共轭裂隙（节理）　conjugate joints
层面裂隙，顺层节理　bedding joint
层间（层内）节理　intraformational joint
羽状节理　feather joint
羽状张节理　pinnate tension joint
羽状剪节理　pinnate shear joint
次生节理　secondary joint
风化裂隙　weathering fissure
微裂缝，微裂隙　microfissure
张裂缝　tension crack
张开的节理　fissure
节理系　joint system
节理组　joint set
节理密集带　densely jointed belt，joint-concentrated zone
节理连通率　joint persistence ratio
节理频数　joint frequency
裂缝充填物　fissure filling，crack filling
裂缝愈合　crack healing
裂缝开度　crack aperture
产状　orientation，attitude
方位角　azimuth angle
方向角　direction angle
走向　strike
倾向　dip direction
倾角　dip
真倾角　actual dip
视倾角（假倾角）　apparent dip
缓倾角　low-angle dip
交线　line of intersection
倾伏角　plunge
倾伏方向　trend
开度　aperture，separation
闭合　closed
微张　slightly open
张开　open
壁面　wall surface
钙质薄膜　calcium coated
铁质浸染　iron disseminated，ferruginous imbueing
铁锰质浸染　iron-manganese disseminated
起伏度　fluctuation，waviness
粗糙度　roughness，asperity

平直的 planar
起伏的 undulating
有台坎的 stepped
有擦痕的 slickensided
镜面的 mirror，polished
光滑的 smooth
粗糙的 rough
充填物 infilling，filling
节理充填物 joint filling
石英脉充填 quartz vein infilled
方解石充填 calcite infilled
黏土充填 clay infilled
泥质物充填 argillaceous infilled
斜距 oblique distance
劈理 cleavage
流劈理 flow cleavage
板劈理 slaty cleavage
破劈理 fracture cleavage
折劈理 crenulation cleavage
滑劈理 slip cleavage
错动劈理，应变滑劈理 strain-slip cleavage
层面劈理 bedding cleavage
轴面劈理 axial plane cleavage
线理 lineation

(5) 岩体结构 Rock mass structure

节理岩体 jointed rock mass
整体结构 integral structure
完整岩体 intact rock
层状岩体 stratified rock
节理中等发育岩体 moderately jointed rock
块状破裂岩体 blocky and seamy rock
受挤压岩体 squeezing rock
膨胀岩体 swelling rock
块状结构 massive structure，blocky structure
次块状结构 sub-massive structure，sub-blocky structure
层状结构 stratified structure，bedding structure
巨厚层状结构 giant-thick layer structure
厚层状结构 thick layer structure
中厚层状结构 moderately thick layer structure
互层状结构 alternately bedded structure
薄层状结构 thin layer structure
镶嵌结构 interlocked structure，mosaic structure
块裂结构 blocky-fractured structure
碎裂结构 cataclastic structure，disintegrated structure
散体结构 loose granular structure，cohesionless granular structure
碎块状结构 clastic structure
碎屑状结构 fragmental structure
结构面 structural plane，discontinuity
硬性结构面 rigid structural plane
软弱结构面 soft（weak）structural plane
平直结构面 planar structural plane
波状结构面 undulating structural plane
台阶结构面 stepped structural plane
贯通结构面 penetrating（through）structural plane
压性结构面 compressive structural plane
张性结构面 tensile structural plane
扭性结构面 torsion structural plane
拉裂面 pulling apart plane
切割面 cutting plane
临空面 free face
结构体 structural mass，structural body
结构面分级 grading of structural plane

(6) 水文地质 Hydrogeology

地下水 ground water
透水层 permeable layer
含水层 aquifer
隔水层 aquifuge
相对隔水层 relative watertight layer
阻水层 aquiclude
包气带 aeration zone
毛细水 capillary water
毛细管带 zone of capillarity
上层滞水 perch groundwater
潜水 phreatic water
孔隙潜水 pore groundwater
孔隙水 pore water
裂隙水 fissure water
脉状水 veinwater
裂隙潜水 fissure groundwater
潜水位，地下水位 water table，groundwater table
承压水 confined water，artesian water
承压地层 confining stratum
隔水底层 confining underlying bed

隔水顶层 confining overlying bed
裂隙承压水 fissure artesian groundwater
承压水流 confined flow
承压水头 confined water head，artesian head
自流盆地 artesian basin
地下热水 hot groundwater
温泉 hot spring
钙华 calc-sinter
硫磺华 sulphur-sinter
硫化氢 hydrothion
矿化水 mineral water
岩溶水 karst water
泉 spring
下降泉 depression spring，gravity spring
上升泉 ascending spring
自流泉 artesian spring
喷泉 fountain
间歇喷泉 geyser
地下水排泄区 discharge area of ground water
地下水位坡降 gradient of water table
水质 water quality
水化学分析 hydro-chemical analysis
水样 water sample
水样简分析 simplified analysis of water sample
水样全分析 total analysis of water sample
阴离子 anion
阳离子 cation
离子毫克数 milligram value of ion
离子毫克当量 milligram equivalent of ion
矿化度 mineralization，salinity
总矿化度 total salinity
总硬度 total hardness
溶解固形物 dissolved solid matter
不溶解固形物 indissolved solid matter
酸度 acidity
酸碱度（pH） acidity-alkalinity，power of hydrogen（pH value）
侵蚀二氧化碳 corrosive carbon dioxide
游离二氧化碳 free carbon dioxide
游离氧化钙 free calcium oxide
酸性水 acidic water
碱性水 alkaline water
可饮用水 potable water
微矿化水 slightly mineralized water
侵蚀性水 aggressive water
腐蚀性水 corrosive water
无腐蚀 non-corrosive
弱腐蚀 weakly corrosive
中等腐蚀 moderately corrosive
强腐蚀 highly corrosive
腐蚀类型 corrosion types
分解型腐蚀 decomposing corrosion
溶出型腐蚀 dissolving corrosion
酸型腐蚀 acid corrosion
碳酸型腐蚀 carbonic acid corrosion
分解结晶复合型腐蚀 decomposing-crystalline compound corrosion
硫酸镁型腐蚀 magnesium sulfate corrosion
结晶型腐蚀 crystalline corrosion
硫酸盐腐蚀 sulfates corrosion
微矿化水型腐蚀 corrosion by slightly mineralized water
渗流 seepage
导水系数 transmissivity
等时水位线 isochrone
等水压线 isopiestic line
渗径 seepage path
防渗，渗流控制 seepage control，seepage prevention
渗流模型 seepage model
渗流区 seepage area，vadose region
渗流速度 seepage velocity
渗流网 seepage flow net
渗流压力 seepage pressure
渗漏 leakage
渗漏量 leakage quantity
渗漏流量 leakage discharge
渗滤 percolation
入渗 infiltration
渗入 influent seepage
渗透 seepage，permeation
渗透量 seepage quantity
渗流量 seepage flow
渗透坡降 seepage gradient
渗透扬压力 seepage uplift
渗透剖面 seepage profile，permeability profile
渗透速度 seepage velocity
渗透性 permeability
渗透系数 permeability coefficient
透水率 permeable rate

吕荣值　Lugeon value
单位吸水量　specific water absorption
岩土渗透性分级　classification of permeability of rocks (soils)
极微透水　very slightly pervious
微透水　slightly pervious
弱透水　weakly pervious
中等透水　moderately pervious
强透水　strongly pervious
极强透水　very strongly pervious
压力传导系数　pressure conductivity coefficient
渗透力　seepage force
渗透变形　seepage deformation
临界水力坡降　critical hydraulic gradient
出逸坡降　exit gradient
渗出物　exudation
反滤层　filter
饱和土　saturated soil
不饱和土　unsaturated soil
孔隙水压力　pore pressure
超孔隙水压力　excess pore pressure
水文地质试验　hydrogeological test
渗透试验　permeability test
压水试验　water pressure test，Lugeon test
高压压水试验　high pressure water test
稳定水位　steady water level
管路压力损失　pressure loss of tube
压力-时间曲线　plot of pressure versus time
层流　laminar flow
紊流　turbulent flow
裂隙膨胀与压缩　joint dilation and contraction
注水试验　injection test
抽水试验　pumping test
单孔抽水试验　single well pumping test
多孔抽水试验　multiple well pumping test
稳定流抽水试验　steady-flow pumping test
非稳定流抽水试验　unsteady-flow pumping test
变水头渗透试验　variable head permeability test
完整孔　completely penetrating well
非完整孔　partially penetrating well
动水位　dynamic water level
抽水孔　pumping well
观测孔　observation well
降落漏斗　depression cone
影响半径　influence radius
水位降深　drawdown
导水率　transmissibility
释水系数　storage coefficient
给水度　specific yield
示踪试验　tracing test
示踪同位素　tracer isotope
荧光示踪剂　fluorescent tracer
放射性示踪剂　radioactive tracer，radioindicator
示踪元素　tracer element
指示剂法　tracer method
连通试验　hydraulic connectivity test
水温观测　observation of water temperature
水污染　water pollution
水压计　hydraulic gauge

(7) 物理地质现象　Physicogeological phenomenon

风化与蚀变　weathering and alteration
风化　weathering
风化壳　weathering crust
囊状风化　scrotiform weathering
蜂窝状风化　honeycomb weathering
间层风化　interstratified weathering
球状风化　spheroidal weathering
物理风化　physical weathering
化学风化　chemical weathering
风化带　zone of weathering，weathered zone
风化程度　weathering degree
残积土　residual soil
全风化　completely weathered
强风化　highly weathered
弱风化，中风化　moderately weathered
微风化　slightly weathered
新鲜　fresh
崩解（崩解性）　disintegration
风化分解　decomposition weathering
风化破碎　disintegration weathering
松散岩体　loosed rock mass，broken rock mass
蚀变　alteration
物理蚀变　physical alteration
化学蚀变　chemical alteration
热液蚀变　hydrothermal alteration
蚀变带　altered zone
可溶性　solubility，dissolubility
可压缩层　compressible stratum

卸荷，应力释放 stress release (relief)
卸荷裂隙 stress release fissure
卸荷岩体 relaxed rock mass
强卸荷 highly relaxed
弱卸荷 weakly relaxed
深卸荷带 deep relaxed zone
松动岩体 loosened rock mass
蠕变 creep
蠕变岩体 creeping rock mass
危岩 dangerous rock, unstable rock
地面沉降 ground subsidence
地质灾害 geological disaster
崩塌 collapse, avalanche
剥落 spalling, exfoliation
散落 ravelling
岩崩 rockfall
冰崩 ice avalanche
塌陷 caving
倾倒 toppling
溃屈 buckling
坐落 slumping
滑坡 landslide
平面滑动 planar slide
圆弧滑动 circular slide
楔体滑动 wedge slide
蠕滑 creeping slide
顺层滑动 slide along bedding plane
切层滑动 slide cutting bedding plane
推移型滑坡 landslide of pushing type
牵引型滑坡 landslide of dragging type
平移型滑坡 horizontal-moving landslide
顺层滑坡 consequent landslide
切层滑坡 insequent landslide
滑坡舌 slip tongue
滑坡体 slip mass
滑坡基座 slip foundation
滑坡壁 slip cliff, landslide main scarp
滑坡堆积体 landslide deposit
滑坡洼地 landslide graben
滑坡鼓丘 landslide bulge
滑坡剪出口 toe of sliding surface
滑坡裂缝 landslide fracture
滑坡前缘 landslide deposit toe
滑坡后缘 landslide crown
滑坡侧缘 landslide flank
滑坡台阶 landslide terrance
滑坡坝 landslide dam
堰塞湖 landslide lake, dammed lake
滑坡涌浪模拟 landslide-generated waves simulation
泥石流 debris flow
稀性泥石流 diluted debris flow
黏性泥石流 viscous debris flow
泥石流形成区 debris flow forming region
泥石流流通区 debris flow transporting region
泥石流堆积区 debris flow accumulating region
泥流阶地 mudflow terrace
冰川泥石流 glacier debris flow
降雨型泥石流 debris flow of precipitation pattern
共生型泥石流 debris flow of integration pattern
岩溶 karst
岩溶过程 karst process
岩溶率 rate of karstification
岩溶充填率 rate of karst filling
连通率 continuity
裸露型岩溶 bare karst
覆盖性岩溶 covered karst
埋葬型岩溶 buried karst
岩溶地形 karst topography
岩溶（侵蚀）基准 karst base level
岩溶景观 karst landscape
岩溶盆地 karst basin
岩溶洼地 karst depression
岩溶槽谷 karst valley
溶洞 karst cave
钟乳石 stalactite
石笋 stalagmite
石柱 stalacto-stalagmite, column
泉华 sinter
溶斗 doline, funnel
溶沟 karren
溶隙 solution crack
岩溶井 karst well
暗河 underground river
岩溶洞穴网 karstic network
岩溶含水层 karst aquifer
岩溶侵蚀 karst erosion
岩溶通道 karst channel
岩溶突水 karst declogging

2.3 岩土力学 Rock and soil mechanics

（1）岩土物理特性 Rock（soil）physical properties

粒度 granularity
粒组 fraction
比表面积 specific surface area
颗粒级配 gradation of grain
限制粒径 constrained diameter
有效粒径 effective diameter
平均粒径 mean diameter，mean particle size
不均匀系数 coefficient of uniformity
比重 specific gravity
容重 unit weigth
密度 density
天然密度 natural density
饱和密度 saturated density
浮密度 submerged density
干密度 dry density
质量密度 mass density
重力密度 force（weight）density
相对密度 relative density
孔隙率 porosity
孔隙比 void ratio
紧密密度 compaction density
堆积密度 piling density
很松 very loose
松 loose
中密 medium dense
密 dense
很密 very dense
含水量（含水率） water content，moisture content
最优含水量 optimum water content
饱和度 degree of saturation
阿太堡限度 Atterberg limit
液限 liquid limit
塑限 plastic limit
缩限 shrinkage limit
塑性指数 plasticity index
液性指数 liquidity index
土的固结 consolidation of soil
先期固结压力 preconsolidation pressure
超固结比 over-consolidation ratio
正常固结土 normally consolidated soil
欠固结土 underconsolidated soil
次固结 secondary consolidation
固结度 consolidation degree
固结系数 consolidation coefficient
稠度状态 consistency state
固态 solid state
半固态 semi-solid state
塑态 plastic state
液态 liquid state
稠度试验 consistency test
土体的沉降 settlement of soil mass
初始压缩曲线 initial compression curve，virgin compression curve
现场初始压缩曲线 field initial compression curve
密实度 compactness
容许（允许）变形 allowable deformation
湿陷性黄土特性 characteristics of collapsible loess
湿陷系数 coefficient of collapsibility
自重湿陷系数 coefficient of self weight collapsibility
湿陷起始压力 initial collapse pressure
膨胀 expansion，swelling
膨胀系数 coefficient of expansion（swelling）
膨胀率 expansion ratio，specific expansion
膨胀速率 rate of swelling
饱和湿胀应力 saturation swelling stress
动力黏度 dynamic viscosity
运动黏度 kinematic viscosity

（2）岩土力学参数 Rock（soil）mechanical properties

岩石力学 rock mechanics
土力学 soil mechanics
各向同性 isotropy
各向异性 anisotropy
各向异性岩体 anisotropic rock mass
岩石强度 rock strength
抗压强度 compressive strength
单轴抗压强度 uniaxial compressive strength
无侧限抗压强度 unconfined compressive strength
干抗压强度 dry compressive strength
饱和抗压强度 saturated compressive strength
软化系数 softening coefficient
点荷载强度 point load strength

抗拉强度 tensile strength
单轴抗拉强度 uniaxial tensile strength
劈裂强度 splitting strength
抗弯强度（抗折强度） bending strength
岩体强度 rock mass strength
结构面强度 discontinuity strength
抗剪强度 shear strength
抗剪强度（纯摩强度） shear strength，shear-friction strength
抗剪断强度（剪摩强度） shear-rupture strength
摩擦系数 coefficient of friction
外摩擦角 angle of external friction
内摩擦角 angle of internal friction
黏聚力（凝聚力） cohesion
总应力强度 total stress strength
有效应力强度 effective stress strength
峰值强度 peak strength
残余强度 residual strength
长期强度 long term strength
极限强度 ultimate strength
破坏包线 failure envelope
岩石完整性指数 integrity index of rock
普氏坚固系数 M. M. Protogiakonov's coefficient
岩石质量指标 rock quality designation（RQD）
最小二乘法 least square method
优定斜率法 optimum slope method
点群中心法 point group center method
折减系数 reduction factor，coefficient of reduction
应力包络线 stress envelope

(3) 岩土变形特性参数 Rock（soil）deformation properties

弹性模量 elasticity modulus
变形模量 deformation modulus
剪切模量 shear modulus
黏弹性模量 viscoelastic modulus
压缩模量 compression modulus
弹性变形 elastic deformation
塑性变形 plastic deformation
弹塑性变形 elastic-plastic deformation
黏弹性变形 viscoelastic deformation
黏滞塑性变形 visco-plastic deformation
均匀变形 homogeneous deformation，uniform deformation
不均匀变形 inhomogeneous deformation，non-uniform deformation
不可恢复变形（永久变形） irreversible deformation
屈服变形 yield deformation
应变软化 strain softening
应变硬化 strain hardening
屈服强度 yield strength
屈服标准 yield criterion
屈服极限 yield limit
突变段 sharp transition
蠕变变形 creep deformation
蠕变破坏 creep rupture
蠕变曲线 creep curve
蠕变试验 creep test
流变 rheology，flowing deformation
流变特性 rheological behavior

(4) 工程岩体分类 Classification of engineering rock mass

坝基岩体工程地质分类 engineering geological classification of rock mass for dam foundation
坚硬岩 hard rock
中硬岩 moderately hard rock
软质岩 soft rock
抗滑 sliding resistance
抗变形 deformation resistance
地下硐室围岩工程地质分类 engineering geological classification of surrounding rocks for underground chambers
岩体分类Q系统 Q-system of rock mass classification
围岩 surrounding rock
极好围岩 excellent surrounding rock
好围岩 good surrounding rock
一般围岩 fair surrounding rock
差围岩 poor surrounding rock
很差围岩 very poor surrounding rock
岩体评分系统 rock mass rating（RMR）
围岩稳定 surrounding rock stability
基本稳定 basically stable
局部不稳定 locally unstable
不稳定 unstable
非常不稳定 very unstable
长期稳定性 long-term stability
短期稳定性 short-term stability
整体稳定性 overall stability

岩体完整程度 rock mass integrity
完整 integral
较完整 relatively integral
完整性差 poorly integral
较破碎 relatively crushed
破碎 crushed
结构面发育程度 development degree of discontinuity
不发育 undeveloped
轻微发育 slightly developed
中等发育 moderately developed
较发育 relatively developed
发育 developed
很发育 very developed
岩体完整程度评分 rating of rock mass integrity
结构面状态评分 rating of discontinuity conditions
地下水条件评分 rating of ground water conditions
干燥到渗水 dry to seep
滴水 drip
线状流水 linear running water
涌水 water gushing
主要结构面产状评分 rating of orientation of main discontinuity
裂隙产状修正评分 rating adjustment for joint orientation
围岩总评分 total rating of surrounding rock
围岩强度应力比 strength-stress ratio of surrounding rock
岩体完整性 integrity of rock mass
岩体完整性系数 intactness coefficient of rock mass

(5) 地应力 Geostatic stress

初始应力 initial stress
自重应力 self-weight stress
残余应力 residual stress
构造应力 tectonic stress
感生应力 induced stress
附加应力 superimposed stress
地应力场 ground stress field，geostatic stress field
主应力 principal stress
最大主应力 maximum principal stress
中间主应力 intermediate principal stress
最小主应力 minimum principal stress
应力水平 stress level
应力历史 stress history
应力状态 stress state
侧压力 lateral pressure
应力重分布 stress redistribution
周边应力 circumferential stress
应力集中 stress concentration
饼状岩芯 disk-shaped rock core
岩芯饼化 core disking
围岩应力 surrounding rock stress
岩石压力 rock pressure
应力松弛 stress relaxation
松弛区 relaxed zone
松动区 loosened zone
围岩偏压 non-uniform rock pressure

2.4 主要工程地质问题 Main engineering geological considerations

(1) 水库区 Reservoir area

水库调查 reservoir investigation（survey）
人类活动 human activity
人为污染 man-made pollution
人文因素 human factor
古文化遗址 historic culture site，archeological area
自然保护区 nature conservation area（nature reverse）
自然景观 natural landscape
矿产资源 mineral resources
蓄水 impounding，impoundment
初次蓄水 initial impounding，initial filling
壅高水位 backwater level
壅水长度 backwater length
水库淹没 reservoir inundation，reservoir submergence
水库浸没 reservoir immersion
耕种土 agricultural soil，cultivated soil
耕作层 plough horizon
根系层 root zone
毛细上升高度 capillary rise
临界地下水埋深 critical depth of groundwater table
漂凌 drift ice
水库渗漏 reservoir leakage

永久渗漏　permanent leakage
暂时渗漏　temporary leakage
邻谷渗漏　leakage to adjacent valley
河间地块渗漏　leakage through interfluve
水库坍岸　reservoir bank-collapse，reservoir bank ruin
安息角（休止角）　angle of repose
岸边侵蚀　bank erosion
岸边淤积　bank deposit，inwash
水库诱发地震　reservoir induced earthquake
触发因素　triggering factor

（2）坝址区　Dam site area

坝基　dam foundation
坝肩　abutment
基坑涌水　water gushing in foundation pit
坝基渗流　seepage of dam foundation
坝基渗漏　leakage through dam foundation
绕坝渗漏　bypassing leakage，leakage around dam abutment
坝下淘刷　scour beneath dam
抗滑稳定　stability against sliding
深层抗滑稳定　sliding stability in deep layer
浅层抗滑稳定　sliding stability in shallow layer
软弱夹层抗滑稳定　sliding stability along weak intercalation
下游消能区　downstream energy dissipation area
泄流水雾　water spray created by flood discharge
冲坑淘刷　pool scouring
溯源淘刷　retrogressive scouring，head ward scouring
地基　foundation，subgrade
地基沉降　foundation settlement
不均匀沉降　differential settlement，unequal settlement
初始（瞬时）沉降　initial settlement，immediate settlement
固结沉降　consolidation settlement
次固结沉降　secondary consolidation settlement
最终沉降量　ultimate settlement
沉降计算经验系数　empirical coefficient of settlement calculation
地基隆起　foundation upthrow，foundation upheaval
压缩变形　compressive deformation
剪切变形　shear deformation
冲剪　punching shear
边坡　slope
自然边坡　natural slope
水平坡　horizontal slope
顺向坡　dip slope，consequent slope
反向坡　adverse slope，reversal slope
坡体结构类型　type of slope structure
层状同向结构　consequent bedding structure
层状反向结构　reverse bedding structure
层状横向结构　transverse bedding structure
层状斜向结构　oblique bedding structure
层状平叠结构　horizontal bedding structure
稳定未变形边坡　stable and undeformed slope
变形边坡　deforming slope
边坡开裂　slope cracking
边坡坍塌（坍坡）　slope failure，slope collapse
边坡地质模型　slope structure model
底切岸坡，淘蚀岸坡　undercut slope
坍岸，岸坡表层坍落　bank sloughing
潜在不稳定体　potential instable rock mass
局部失稳　local instability
渐进破坏　progressive failure
蠕滑-拉裂　creep sliding-tensile fracture
滑移-压致拉裂　sliding-compressive generated tensile fracture
滑移-弯曲　sliding-buckling
弯曲-拉裂　bending-tensile fracture
塑流-拉裂　plastic flow-tensile fracture
高速滑坡　high speed landslide
自我润滑机制　self-lubricating mechanism
碎屑流机制　clastic flow mechanism
气垫机制　air-cusion mechanism
流土　soil flow
管涌　piping
砂沸　sand boiling
接触流土　soil flow on contact surface
接触管涌　piping on contact surface
接触冲刷　contact scouring
潜蚀　internal scour，underground corrosion
淘刷　scouring
淤堵　clogging
隧洞　tunnel
洞室群　cavern complex，underground complex
傍山隧洞　mountainside tunnel

傍山深路堑　deep cut in hillside
越岭隧洞　mountain tunnel
深埋隧洞　deep buried tunnel
地下洞室　underground cavern
超前导洞　pilot heading
超前孔　guide hole
洞顶　crown，roof
边墙　side wall
洞脸塌方　portal collapse
拱顶塌落　crown collapse
边墙塌落　side wall collapse
松动圈　relaxation zone
冒顶　roof fall
劈裂剥落，片帮　splitting and peeling off，scaling
掉块　fallouts
临时悬吊松动岩石　false hanging loose rock
径缩变形　radial contracting deformation
围岩膨胀内鼓　swelling out of surrounding rock
底板隆起　bottom heaving
地面下沉　land subsidence
塑性挤出　plastic squeezing out
岩爆　rock burst
岩石崩出　blowout
渗水　water seeps
进水　water influx
涌水　water gushing，inrush of water
突水　water bursting
突泥　mud eruption
瓦斯突出　gas burst
有害气体　harmful gas
有毒气体　poisonous gas，toxic gas

2.5 天然建筑材料与岩土试验 Natural construction material and geotechnical test

(1) 天然土石材料　Natural rock and earth materials

石料　rock material
土料　earth (soil) material
风化土料　weathered soil
碎石，岩屑　debris
团粒状土　crumbling soil
砂砾石　sand and gravel
分散性土　dispersive soil
骨料　aggregate
天然骨料　natural aggregate
针状颗粒　needle-shaped particle
片状颗粒　flake-shaped particle
软弱颗粒　soft particle
颗粒分析　grading analysis，particle size analysis
颗粒尺寸　grain (particle) size
颗粒大小分布曲线　particle size distribution curve
颗粒大小频率图　grain size frequency diagram
颗分曲线　gradation curve
细度模数　fineness modulus (FM)
粒度模数　grain modulus (GM)
曲率系数　coefficient of curvature
表观密度　apparent density
堆积密度　bulk density
采石场　quarry
土料场　borrow area
有用层　available layer，useful layer
无用层　unavailable layer，unuseful layer
剥离层　stripping layer
剥离量　stripped volume
剥采比　stripping ratio
平均厚度法　average thickness method
平行断面法　parallel section method
三角形法　triangular method
初查　preliminary investigation
详查　detailed investigation

(2) 现场试验　Field test

现场调查　field investigation，site survey
原位试验　in-situ test
现场检验　in-situ inspection
现场监测　field (in-situ) monitoring
岩体原位测试　in-situ measurement of rock mass
地应力测试　geostress measurement
应力解除法　stress relief method
应力恢复法　stress restoration method
应力路径　stress path
有效应力路径　effective stress path
反复荷载　repetitive load
分级加荷　stage loading
扁千斤顶试验　flat jack test
狭缝法，刻槽法　narrow slot method
水压致裂法　hydraulic fracturing method
尺寸效应　scale effect

时间效应　time effect
标准贯入试验　standard penetration test（SPT）
圆锥贯入试验　cone penetration test（CPT）
标准贯入试验击数　SPT blow count
动力触探试验　dynamic cone penetration test
静力触探试验　static cone penetration test
比贯入阻力　specific penetration resistance
旁压试验　pressure-meter test（PMT）
平板载荷试验　plate bearing test
十字板剪切试验　vane shear test
普氏击实试验　Proctor compaction test
点荷载试验　point-loading test
回弹仪测试　rebond hammar test
孔隙压力消散试验　test of pore pressure dissipation
直剪试验　direct shear test
承压板试验　bearing plate test
抗剪试验　shear tests
快剪试验　quick shear test
慢剪试验　slow shear test
三轴剪试验　triaxial shear test，triaxial compression test
周期加荷三轴试验　cyclic triaxial test
动三轴试验　dynamic triaxial test
共振柱试验　resonant column test
固结试验　consolidation test
固结排水剪试验　consolidated drained shear test（CD）
固结不排水剪试验　consolidated undrained shear test（CU）
不固结不排水剪试验　unconsolidated undrained shear test（UU）
固结排水三轴试验　consolidated drained triaxial test
固结不排水三轴试验　consolidated undrained triaxial test
等速加荷固结试验　consolidation test under constant loading rate
等应变率固结试验　consolidation test under constant rate of strain
控制比降固结试验　consolidation test under controlled gradient
固结比　consolidation ratio
固结曲线　consolidation curve
固结压力　consolidation pressure
崩解试验　slake test，disintegration test
膨胀试验　expansion test，swelling test
阿太堡试验（土的界限含水量试验）　Atterberg test
含水率试验　water content test
烘干法　drying method
酒精燃烧法　alcohol burning method
炒干法　fried dry method
密度试验　density test
蜡封法　wax-sealing method
比重试验　specific gravity test
比重瓶法　pycnometer method
浮称法　buoyancy method
虹吸筒法　siphon cylinder method
颗粒分析试验　grain size analysis test
筛析法　test sieving method
密度计法　densimeter method
移液管法　pipette method
界限含水率试验　side water content test
湿化试验　slaking test
收缩试验　shrinkage test
击实试验　compaction test
钻孔试样　borehole sample
岩芯试样　core sample
有闭合节理的试样　specimen with healed joints
固定式活塞取样器　fixed-piston sampler
环刀　cutting ring
环刀取样器　ring sampler

(3) 试验室试验　Laboratory test

平行试验　parallel test
验证试验　proof test
布氏（硬度）试验　Brinell test
恰贝冲击试验　Charpy test
标准击实试验　standard compaction test（SCT）
物理力学性试验　physical and mechanical property test
物理化学分离制备法　physical and chemical separation method
湿研磨分离制备法　wet-grind separation method
有机质试验　organic matter test
烧失量试验　ignition loss test
游离氧化铁试验　free iron oxide test
阳离子交换量试验　cation exchange capability test
比表面积试验　specific surface area test
薄膜加热试验　thin-film oven test

坩埚　crucible
古氏坩埚　Gooch crucible
烧杯　beaker
锥形烧瓶　Erlenmeyer flask
比重瓶（密度瓶）　pycnometer
克利夫兰开口杯　Cleveland open cup
隔离剂　release agent
环球法　ring-and-ball apparatus
针入度仪　penetrometer
延度仪　ductilimeter
恒温水槽　thermostatic water bath
回馏　reflux distillation
重复性试验　repeatability test
再现性试验　reproducibility test
平行试验　comparative test
平行测定　parallel measure
抽检试验　random test
土样分析　soil sample analysis
颗粒组成　grain composition
颗粒结构　grain structure
颗粒比表面　specific grain surface
颗粒间空隙　intergranular space
颗粒胶结　granular cementation
取样　sampling
试样　sample，specimen
原状土样　undisturbed soil sample
袋装扰动土样　disturbed bag sample
完整试样　intact specimen
蜡封试样　wax-sealed sample
风干试样　air-dried sample
扰动土样，非原状土样　disturbed soil sample
渥太华砂　Ottawa sand
四分法　quartering method
圆柱体试样　cylinder specimen
同组试样　companion specimen
有侧限试样　laterally confined specimen
保留样品　retention of sample
筛余量　sample retained on sieve
量程　range
刻度　division
仪器最小刻度　graduated in divisions，less and accurate to within
精确到［仪器读数］　reading to
精准至［计算结果］　accurate to，to the nearest
精确到±X%以内　accurate to within plus or minus X percent
精确到小数点第X位　accurate to the X-th decimal place
标准值，特征值　characteristic value
平均值　mean value
大值平均值　average of the higher half values
小值平均值　average of the lower half values
加权平均值　weighted average
修正值　corrected value
修正系数　coefficient of correction
方差　variance
标准偏差　standard deviation
均值系数　coefficient of mean value
离差系数　coefficient of skew
变异系数　coefficient of variation
名义值　nominal value
分位数　quantile，fractile

2.6 工程地质勘察 Engineering geological investigation

(1) 勘测　Investigation，survey

勘测阶段　investigation stage
规划阶段　planning stage
预可行性研究阶段　prefeasibility study stage
可行性研究阶段　feasibility study stage
招标设计阶段　tender design stage
施工详图设计阶段　construction detailed design stage
勘测大纲　outline of investigation program
踏勘　reconnaissance
勘探工作量　quantity of exploration work
勘探项目　exploration item
勘探工作布置　arrangement of exploration work
了解　learn about
初步查明　preliminarily identify
查明　ascertain
复查　review，check
复核　reexamine
核定　verify
安全鉴定　safety appraisal
主勘探剖面　major exploration profile（section）
场地复杂程度　complexity of site
地形制约　topographical constrain

地质填图 geological mapping
可利用资料 available data (information)
资料收集方法 data acquisition method
野外查勘 field reconnaissance
施工地质工作 geological works during construction
施工编录 construction mapping
坝基编录 dam foundation mapping
钻孔编录 drilling log
硐室编录 tunnel mapping (log)
竖井编录 shaft log
数字地质编录系统 digital geological logging system
施工地质日志 geological daily record during construction
施工地质通知 geological notification during construction
施工地质备忘录 geological memorandum during construction
地质灾害调查 geologic hazards inquiry
地面勘探 ground surface exploration
探槽 trial trench
探坑 trial pit
边侧支撑 side shoring
框架支撑 shored with cribbing

(2) 地下勘探 Subsurface exploration

钻探 drilling exploration
硐探 adit exploration
井探 shaft exploration
坑(槽)探 pit (trench) exploration
竖井 vertical shaft
斜井 inclined shaft
勘探钻孔 exploration borehole
可钻性 drillability
岩芯 rock core
岩芯钻探 core drilling (boring)
岩芯采取率 core recovery rate
孔斜 borehole deviation
孔深 borehole depth
孔口高程 collar elevation
孔底高程 borehole bottom elevation
孔斜计 clinograph
倾斜仪 clinometer, inclinometer
扩孔 enlarge boring
取岩芯 core extraction
钻孔孔径 borehole size (borehole diameter)
N 型孔 [孔径约 75 mm/3 in.] N size hole
B 型孔 [孔径约 60 mm/2.3 in.] B size hole
H 型孔 [孔径约 99 mm/4 in.] H size hole
大口径孔 [孔径>0.6m] large-diameter hole, calyx hole
两瓣式岩芯管 double split barrel
三瓣式岩芯管 triple split barrel
钻孔定向 borehole orientation
垂直钻进 vertical boring
斜孔钻进 inclined boring
水平钻进 horizontal boring
定向钻进 oriented boring
钻孔顶角 borehole vertex angle
钻孔倾角 borehole inclination
土钻 soil auger
麻花钻 helical
螺旋钻 auger drill
勺钻 bucket auger
潜孔锤 hammer down the hole (HDT)
冲击钻进 percussion drilling
反循环钻进 reverse circulation drilling
冲抓锥钻进 churn and grabbing drilling
桩孔扩底钻进 hole bottom reaming drilling
大口径钻进 large well drilling
厚壁冲击管 heavy wall drive barrel
切土管刃 cutting shoe
取土器 soil sampler
取样程序 sampling procedure
连续取样 continuous sampling
定向岩芯 oriented core
穿过钻杆提取 retrieved through the drill rod
循环介质 circulating medium
清水 clear water
泥浆 slurry
膨润土泥浆,皂土液 bentonite slurry
添加剂泥浆 slurry with additive
天然有机聚合物 natural organic polymer
合成有机聚合物 synthetic organic polymer
可生物降解表面活性剂混合物 biodegradable mixture of surfactant
植物无固相冲洗液 vegetable glue drilling fluid without clay
岩芯钻进观察 observation during coring operation

每段进尺记录　record of each run
每钻次的深度、长度和时间　depth, length, and time for each run
孔内残留岩芯　amount of core left in the hole
岩粉与岩屑　rock cuttings and chips
钻压　weight on bit (WOB), bit pressure
孔内掉块　material caving into the borehole
掉钻　rod drop
埋钻　drill rod burying
卡钻　drill rod sticking
烧钻　bit burnt
漏水　water loss
返水颜色　color of the return water
临时护壁，临时套管　temporary casing
下套管　casing placement
洗孔　borehole cleaning
封孔　borehole backfilling
岩芯编录　core logging
钻孔地质员　inspector
岩芯箱　core box
岩芯隔板卡　core record spacer
岩芯照片　core photograph
勘探洞　exploration adit, trial adit
导洞　pilot drift tunnel
河底平洞　adit under river
隧洞倾斜度　inclination of tunnel
洞周地质测绘　peripheral geologic mapping
向平面投影　projecting to plan
向剖面投影　projecting to profile
展开圆周面为平面　unrolling the circumference to form a plan
隧洞定位点（桩）　tunnel station (stake)
参照点　reference point
假设高程　assumed elevation
洞周应力状态　stress conditions around openings
切向主应力　tangential principal stress
径向主应力　radial principal stress
自然拱　natural arching
拉张区　tension zone
塑性区　plastic zone
扰动区　zone of disturbance

(3) 物探　Geophysical prospecting

遥感　remote sensing
地面物探　surface geophysical prospecting
物探剖面　geophysical survey profile
水下地形　underwater topography
弹性波测试　elastic wave survey
微地震波测试　micro-seismic wave survey
自动增益控制　automatic gain control (AGC)
正常时差修正　normal moveout correction
高差曲线　departure curve
折射法勘探　refraction survey
界面波　boundary wave, interface wave
跨孔测试　cross-hole test
声波测试　acoustic wave survey
钻孔物探测试　geophysical survey in borehole
钻孔照相　borehole photography
钻孔电视　borehole TV
物探综合测井　comprehensive downhole geophysical probe
声波测井　acoustic log
声波电视测井　acoustic televiewer log
活化测井　activation log
放射法测井　radioactivity log
跨孔层析摄影　crosshole tomography
跨孔地震探测　crosshole seismic probe
跨孔电阻率探测　crosshole resistivity probe
地质雷达　ground penetrating radar (GPR)
层析成像　computer tomography (CT)
声发射监测　acoustic emission monitoring
电法勘探　electrical prospecting
重力勘探　gravitational prospecting
重力异常　gravity anomaly
磁异常　magnetic anomaly
地热异常　geothermal anomaly
电测深法　electrical sounding
电剖面法　electrical profiling
高密度电法　resistivity imaging
激发极化法　induced polarization
自然电场法　self-potential method
充电法　Mise-a-la-masse method
可控源音频大地电磁测深法　controlled source audio frequency magnetotellurics (CSAMT)
瞬变电磁法　transient electromagnetic method (TEM)
浅层地震折射波法　shallow seismic refraction
浅层地震反射波法　shallow seismic reflection
瑞雷波法　Rayleigh wave method

弹性波测试 elasticity wave testing
水声探测 sonic echo exploration
放射性测量 radioactivity survey
同位素示踪法 isotopes tracer technique

(4) 工程地质分析与计算
Engineering geological analysis and calculation

稳定性分析 stability analysis
可靠度分析 reliability analysis
概率曲线 probability curve
期望值 expected value
目标函数 objective function
试错法 trial-and-error method, cut-and-try method
确定性分析 deterministic analysis
抗滑安全系数 factor of safety against sliding
容许安全系数 allowable factor of safety
安全度 degree of safety
容许变形量 allowable deformation
容许沉降量 allowable settlement
容许承压力 allowable bearing pressure
容许极限 acceptable limit
地质力学模型 geo-mechanical model
边坡地质模型 geological model of slope
边坡结构模型 slope structure model
动态设计法 method of information design
开挖坡比 excavation gradient
设计安全标准 design safety criteria
上限解 upper bound solution
下限解 lower bound solution
总应力法 total stress method
有效应力法 effective stress method
条分法 slice method
整体圆弧滑动法 mass circle sliding method
不平衡推力法 method of non-equilibrium thrust
分块极限平衡法 method of block limit equilibrium
楔形体平衡法 wedge equilibrium method
过程模拟 process simulation
仿真模拟 analogue simulation
相似性 similarity
本构关系 constitutive relation
边界条件 boundary condition
莫尔破裂包络线 Mohr's rupture envelope
插值法 interpolation method
外延法 extrapolation method
反分析 back analysis
图解法 graphical solution
统计回归分析 statistical regression analysis
经验公式 empirical formula
半经验公式 semi-empirical formula
经验数据 empirical data
经验参数 empirical parameter
经验系数 empirical coefficient
特性参数 characteristic parameter
建议值 recommended value

2.7 测量
Engineering survey

全球定位系统 global positioning system (GPS)
地理信息系统 geographical information system (GIS)
国家大地测量网 national geodetic net
国家控制测量网 national control survey net
国家坐标系统 state coordinate system
1954 年北京坐标系 Beijing coordinate system 1954
1980 年国家大地坐标系 National geodetic coordinate system 1980
假定坐标系 assumed coordinate system
独立坐标系 independent coordinate
平面控制测量 horizontal control survey
平面控制网 horizontal control network
高程控制网 vertical control network
控制点 control point
控制网 control network
控制网原点 origin of control network
一等控制网 primary control network
三角网 triangulation network
三角网平差 triangulation net adjustment
一等导线 first order traverse
一等三角测量 first order (primary) triangulation survey
一等三角点 primary triangulation point
1956 年黄海高程系 Huanghai elevation system 1956
1985 年国家高程基准 national elevation system 1985
一等水准 first order leveling
三等水准测量 third-order leveling

三边测量 trilateration
三角测量 triangulation
三角高程测量 trigonometric leveling
电磁波测距 electromagnetic distance measurement
GPS 拟合高程测量 GPS fitting vertical survey
近景摄影测量 close-range photogrammetry
大地水准面 geoid
大地基准点 geodetic datum
大地控制点 geodetic control point
子午面 meridian plane
子午线 meridian
基准点 datum point，control point
基准面 base level，datum plane
基线 base line
基线测量 base measurement
水准测量 leveling，level survey
水准测量平差 leveling adjustment
水准点 leveling point，bench mark
水准网 level net
测角网 angular network
视距导线 tacheometric polygon
地形测量 topographic survey
平面测量 plane survey
导线测量 traverse survey
闭合导线 closed traverse
附合导线 connecting traverse
导线点 traverse point
导线边 traverse leg
视距测量 stadia survey，tacheometric survey
图根点 mapping control point
视准点，照准点 collimating point，sighting point
视准测量 collimation measurement
碎部测量 detail survey
收敛测量 convergence measurement
视差 parallax
起始边 initial side
最弱边 weakest side
测站归心 reduction to station center
照准点归心 reduction to targer center
基本测站 basic survey station
隔点设站法 setting station between two points
施工测量 construction survey
施工控制网 construction control network
施工坐标系 construction coordinate system
基本导线 primary traverse
施工导线 construction traverse
高程控制点 vertical control point
桩号 chainage
定位桩（导桩） guide pile
定线 alignment
放样 setting out，laying out
复测 repetition survey
地质点测量 survey of geological observation point
钻孔位置测量 survey of borehole position
路线测量 route survey
断面测量 section survey
贯通测量 holing through survey
竖井定向测量 shaft orientation survey
安装测量 erection survey，installation survey
竣工测量 finial construction survey，as-built survey
测回 observation set
测量方法 survey method
前方交会 forward intersection method
引张线法 tension wire alignment method
视准线法 collimation line method
激光准直法 laser alignment method
收敛检验 test for convergence
测量觇标 observation target，observation tower
测量标志 survey mark
标石 markstone，monument
标记（标志） sign，mark
测量资料（数据） survey data
内业 office work，indoor work
外业 field work
地物 surface feature
等高线 contour
等高距 contour interval
坐标 coordinate
高差 altitude difference
精度 accuracy，precision
误差 error
测量误差 true error
比例误差 scale error
系统误差 systematic error
测回差 discrepancy between observation sets
全站仪 total station
水准仪 level
经纬仪 theodolite

照准仪 alidate
水平度盘 horizontal circle
三角台 tribrach
视准仪 collimator
平板仪 plane-table
垂线观测仪 coordinatorgraph for plummet observation
卫星照片 satellite photograph
航摄照片 aerial photograph
遥感图像 remote sensing image
数字地形图 digital topographic map
照片判读 photograph interpretation
图像处理 image processing
图像识别 image recognition, pattern recognition
地图分幅 sheet line system

2.8 主要报告及图表名称 Titles of reports, figures and tables

岩矿鉴定报告 Rock and mineral identification
地震危险性分析报告 Earthquake risk analysis and assessment
地震安全性评价报告 Evaluation (Assessment) of seismic safety
地质灾害危险性评估报告 Risk assessment of geological hazard
物探报告 Geophysical prospecting report
岩土试验报告 Geotechnical test report, Report of rock and soil tests
水质分析报告 Water quality analysis
专门性工程地质问题研究报告 Report for investigation of special engineering geological problems
地下厂房洞室群围岩稳定分析及支护设计专题报告 Report on stability analysis and support design of surrounding rocks of underground chambers
工程地质勘察报告 Report for engineering geological investigation
区域综合地质图 Regional comprehensive geological map
区域构造纲要图 Regional structural outline map
地貌与第四纪地质图 Geomorphologic and Quaternary geological map
枢纽区工程地质图 Engineering geologic map of project area
水库区综合地质图 Comprehensive geological map of reservoir area
坝址工程地质图 Engineering geological map of dam site
坝（闸）址基岩地质图（包括等高线图） Geological map of bed rocks at dam (sluice) site (including contour)
坝址平切面图 Horizontal section of dam site
比较坝址工程地质平面图（剖面图） Engineering geological plan (profile) of alternative dam site
推荐坝址地质横剖面图 Geological cross section (plan) of recommended dam site
坝基（防渗线）渗透剖面图 Permeability profile of dam foundation (seepage control line)
工程地质剖面 Engineering geological section
实测地质剖面图 Field-acquired geological profile
主要建筑物工程地质纵（横）剖面图 Longitudinal (Transversal) geological sections of main structures
拱坝的工程地质平切图 Horizontal geologic sections of arch dam
地下厂房的工程地质平切图 Horizontal geologic sections of underground powerhouse
典型地质剖面 Typical geological profile
喀斯特区水文地质图 Hydrogeologic map of karst region
存在重大工程地质问题部位的水文地质及工程地质平面图（剖面图） Plan (Profile) of hydrogeology and engineering geology of areas with major geological problems
专门性问题工程地质图 Engineering geological map for special geological problems
历史地震震中分布图 Epicenter distribution of historical earthquakes
浸没区及防护地质图 Geological plan of immersion area and protection area
施工地质编录图 Construction geological log
筑坝材料料场位置和横剖面图 Location map and typical cross section of borrow area for embankment materials
天然建筑材料分布图 Distribution of natural construction material sources
料场综合成果图 Comprehensive investigation results for quarry and borrow areas
地层柱状图 Stratigraphic column
综合柱状图 Composite columnar section

钻孔柱状图 Borehole log，Geological log of drill hole
平洞展示图 Geological log of adit，Developed plan of adits
坝基编录图 Geologic log of dam foundation
开挖平面图 Excavation plan
竣工地质图 As-built geological map
极点图 point diagram
极点等值线图 contour plot of the poles
极点大圆图 plot of great circles
断裂极点图 pole plot of discontinuities
等密图 contour diagram
列线图（图算法） nomography
诺模图 nomogram
节理玫瑰图 joint rosette
节理极点投影图 joint polar stereonet
赤平极射投影 stereographic projection
投影网 strereonet
上半球投影 upper hemisphere projection
下半球投影 lower hemisphere projection
等角投影 equal-angle projection
等面积投影 equal-area projection
实测界线 surveyed boundary
推测界线 conjectural boundary
示坡线 slope line
坡边线 slope margin line

第3章　工　程　规　划
Planning

3.1　水文泥沙
Hydrology and sediment

(1) 河流及流域特征　Characteristics of rivers and basins

河流　river
常年河　perennial river
间歇河　intermittent river
悬河　elevated river
干流　main stream
支流　tributary
河段　reach
上游　upper reaches
中游　middle reaches
下游　lower reaches
河源　headwaters, river source
河口　estuary, river mouth
河长　river length
深泓线　thalweg
中泓线　thread of channel (stream)
落差　drop
河道比降，坡降　river gradient (slope)
河槽　river channel, stream channel
废河道　abandoned channel
冲积河槽　alluvial channel
淤塞河道　blocked channel
季节性水道　seasonal watercourse
平衡河槽 [冲淤]　regime channel
不平衡河槽 [冲淤]　non-regime channel
游移河槽　shifting channel
游荡河槽　interlaced channel
蜿蜒河道　meandering channel, twisting channel
平缓河道　mild channel
河床　riverbed
不稳定河床（动床）　mobile channel
河谷平原　valley flat
河漫滩　alluvial flat, flood plain
河流阶地　river terrace
水系（河网）　hydrographic net, river system
河网密度　drainage density
流域　river basin, watershed
流域分水线　river basin divide
流域面积（集水面积）　catchment area, drainage area
流域植被率　percentage of vegetations in drainage area
流域湖泊率　percentage of lakes in drainage area
流域沼泽率　percentage of swamps in drainage area
相邻流域　adjacent drainage area

(2) 气象　Meteorology

气象站　meteorological station
气温　air temperature
环境温度　ambient temperature
摄氏温度　Celsius temperature
华氏温度　Fahrenheit temperature
月平均气温　mean monthly temperature
年平均气温　mean annual temperature
极端最高温度　extreme maximum temperature
极端最低温度　extreme minimum temperature
空气湿度　air humidity
绝对湿度　absolute humidity
相对湿度　relative humidity
降水　precipitation, rainfall
有效降雨量　effective rainfall
可能最大降水　probable maximum precipitation (PMP)
实测最大降水量　maximum observed precipitation
雨强　rainfall intensity
降雨历时　duration of rainfall
雨量站　precipitation station
雨量计　rainagauge
蒸发　evaporation
水面蒸发　evaporation from water surface
陆面蒸发　evaporation from land surface

蒸腾 evapotranspiration
风速 wind speed
风力 wind force
风向 wind direction
蒲福风力 Beaufort force
盛行风 prevailing wind
日照时间 sunshine duration
年日照时数 annual solar radiation hours
日照率 percentage of possible sunshine
寒潮 cold wave
最大冻土深度 maximum frozen soil depth
气候区 climatic zone
严寒地区 severe cold region，chilly cold region
寒冷地区 cold region
温和地区 mild region

(3) 水文分析计算 Hydrological analysis and computation

水文年 water year
水文计算 hydrological computation
水文统计 hydrological statistics
累积频率 cumulative frequency
重现期 recurrence period（interval）
水文频率曲线 hydrological frequency curve
设计频率 design frequency
水文频率分析 hydrological frequency analysis
暴雨 rainstorm
大暴雨 heavy rainstorm
特大暴雨 extraordinary rainstorm
季风雨 monsoon rain
降雨面积（雨区） precipitation area（rain area）
暴雨中心 rainstorm center
暴雨强度 rainstorm intensity
暴雨历时 rainstorm duration
点雨量 point precipitation
面雨量 areal precipitation
定点定面关系［暴雨］ precipitation relationship between fixed point and fixed area
动点动面关系［暴雨］ precipitation relationship between center point and variable area
暴雨时、面、深关系 depth-area-duration（DAD） relationship of storm
暴雨强度历时曲线 storm intensity-duration curve
等雨量线图 isohyet chart
曲线拟合 curve fitting
设计暴雨 design rainstorm
设计雨型 design rainfall pattern
暴雨模式 rainstorm model
典型暴雨 typical rainstorm
移置暴雨 transposition storm
暴雨移置法 storm transposition method
组合暴雨 synthetic storm
极值暴雨样本 sample of extreme storms
露点 dewpoint
地面露点 dewpoint at earth surface
代表性露点 representative dewpoint
可能最大露点 probable maximum dewpoint
产流 runoff yield
蓄满产流 runoff yield under saturated storage
超渗产流 runoff yield under excess infiltration
下渗容量曲线 infiltration capacity curve
初损 initial loss
后损 latter loss
雨水径流量 storm water runoff
设计净雨量 design net rainfall
基流 base flow
汇流 concentration of flow
河槽集流（河网集流） concentration of channel flow
时段单位线 unit hydrograph
瞬时单位线 instantaneous unit hydrograph
综合单位线 synthetic unit hydrograph
概化洪水过程线 simplified flood hydrograph
洪水行进 flood routing
洪水波行进时间 flood wave travel times
不稳定流和动态行进法 unsteady flow and dynamic routing method
过程线上升段 hydrograph ascending limb
过程线下降段 hydrograph recession limb
入流过程线的涨水段 rising limb of inflow hydrograph
径流 runoff
地表径流 surface runoff
地下径流 groundwater runoff
融雪径流 snowmelt runoff
冰湖溃决洪水 glacial lake outburst flood
冰川湖 glaciers lake
径流深 runoff depth
径流模数 runoff modulus
径流系数 runoff coefficient

多年平均年径流量 mean annual runoff
设计年径流量 design annual runoff
设计年径流量年内分配 distribution of design annual runoff within a year
平均流量 mean flow
枯水流量 low flow
丰水流量 high flow
极端枯水期 extreme dry period
极端丰水期 extreme wet period
入库流量 reservoir inflow
出库流量 reservoir outflow
降雨径流关系 precipitation-runoff relation
流量历时曲线 flow-duration curve
洪水 flood
设计洪水 design flood
设计洪水标准 standard of design flood
可能最大洪水 probable maximum flood (PMF)
百年一遇洪水 one percent chance flood, a 100-year recurrence flood
入库设计洪水 inflow design flood (IDF)
入库洪峰流量 peak rate of inflow
分期设计洪水 staged design flood
施工设计洪水 design flood during construction
坝址洪水 flood at dam site
入库洪水 inflowing flood
溃坝洪水 dambreak flood
洪水过程线 flood hydrograph
大洪水 major flood
洪水频率 flood frequency
洪水记号 floodmark
洪峰 flood peak
洪峰流量历时 duration of flood flow
洪水量 flood volume
洪水超高 flood surcharge
波浪爬高 wave runup
风壅水 wind setup
风壅水面高度 wave setup
风成波 wind wave
洪水频率曲线 flood frequency curve
设站流域 gaged basin
未设站流域 ungaged basin
水文系列 hydrological series
系列代表性 series representativeness
系列插补 series interpolation
系列延长 series extension
历时曲线 duration curve
累积曲线 mass curve
差积曲线 residual mass curve
滑动平均曲线 moving average curve

(4) 泥沙 Sediment

河流泥沙 river sediment
河流泥沙运动力学 dynamics of river sediment movement
冲刷 scour
淤积 sedimentation, sediment deposition
泥沙输移比 sediment delivery ratio
泥沙冲淤平衡 sediment erosion and deposition balance
含沙量 sediment concentration
输沙率 sediment discharge
平均输沙率 mean sediment discharge
固体径流，输沙量 sediment runoff
平均输沙量 mean sediment load
累计输沙量 cumulative sediment load
泥沙淤积 sediment accumulation
淤积量 siltation volume
水库泥沙淤积速率 rate of reservoir sedimentation
悬移质 suspended load
推移质 bed load
层移质 laminated load
悬移质输沙率 suspended load discharge
推移质输沙率 bed load discharge
推悬比 ratio of bed load discharge to suspended load discharge
床沙质（造床质） bed material load
冲泻质（非造床质） wash load
拦门沙 bar
产沙模量 modulus of sediment yield
泥沙性质 property of sediment
泥沙粒径 silt grain size
等容粒径 volume equivalent diameter
沉降粒径 settling diameter
中值粒径 median diameter
加权平均粒径 weighted mean diameter
泥沙颗粒分析 grain size analysis
泥沙形态系数 shape coefficient of sediment
均匀沙 uniform sediment
非均匀沙 non-uniform sediment
泥沙输移 sediment transport

泥沙起动　incipient motion of sediment
起动流速　incipient velocity
起动拖曳力，临界拖曳力　incipient tractive force
沉速　settling velocity
群体沉速　settling velocity of grains
悬浮指标　suspension index
悬浮功　suspension work
平衡挟沙能力　regime sediment charge
饱和输沙（平衡输沙）　saturated sediment transport
非饱和输沙（不平衡输沙）　non-saturated sediment transport
含沙量沿程变化　longitudinal variation of sediment concentration
河槽阻力　resistance of river channel
沙粒阻力　grain resistance
沙波阻力　form resistance of sand wave
床面形态　bed form
沙波　sand wave
沙纹　sand ripple
沙垄　dunes
静平床　stationary flat bed
动平床　moving flat bed
推移带　transporting belt
挟沙水流　sediment-laden flow
无沙水流　sediment free flow
泥沙流　silt flow
高含沙水流　flow with hyperconcentrated sediment
阵流　intermittent flow
浆河现象　clogging of river
不冲流速　non-scouring velocity
不淤流速　non-silting velocity
异重流　density current
水库异重流　density current in reservoir
入库泥沙　inflowing sediment
出库泥沙　outflowing sediment
水库淤积　reservoir silting，sediment deposition in reservoir
长期使用库容　long-term storage capacity of reservoir
水库年限　service life of reservoir
水库淤积纵剖面　longitudinal profile of deposit in reservoir
三角洲淤积　delta deposit
锥体淤积　cone deposit，tapered deposit
带状淤积　belt deposit
水库淤积上延（翘尾巴）　upward extension of reservoir deposition
水库淤积极限　limit state of sediment deposition in reservoir
水库淤积平衡比降　equilibrium slope of sediment deposition in reservoir
沿程淤积　progressive deposition
溯源淤积　backward deposition
泥沙防治　sediment control
泥沙矿物成分　solid mineral composition
多年平均含沙量　average annual solid content
泥沙粒径　solid grain size
拦泥库　sediment detention reservoir
拦沙堰　sediment detention weir
滞洪排沙　flood retarding and sediment releasing
异重流排沙　sediment releasing by density current
蓄清排浑　impounding clear water and releasing muddy flow
水库泄空排沙　sediment releasing by emptying reservoir
汛期排沙限制水位　low limit level for sediment flushing in flood season
冲刷漏斗　scouring funnel
水库回水　reservoir backwater
回水变动区　fluctuating backwater zone
回水末端　upstream end of backwater
库尾　reservoir head

（5）水文预报　Hydrological forecast

水文站　hydrological station
水文基准站（参证站）　bench mark station
水文基本站　basic data station
水文设计站　design station
水文代表站　representative station
水位计　hydrological guage
自计水位计　recording gauge
永久水位站　permanent gauge
水尺　staff gage
水文测站　gauging station
水文记录　hydrological records
水位流量关系　stage-discharge relationship
预见期　forecast lead time
短期水文预报　short-term hydrological forecast
中期水文预报　mid-term hydrological forecast

长期水文预报 long-term hydrological forecast
降雨径流的预报 rainfall runoff forecast
洪水预报 flood forecast
枯水预报 low flow forecast
泥沙预报 sediment forecast
水质预报 water quality forecast
冰情预报 ice regime forecast
封冻预报 freezing forecast
解冻预报 ice break forecast
水文模型 hydrological model
确定性水文模型 deterministic hydrological model
随机性水文模型（非确定性水文模型）stochastic hydrological model
人类活动水文效应 hydrological response due to human activities
水文遥测系统 hydrological telemetry system
水情自动测报系统 hydrological telemetry and forecasting system
水情遥测站 telemetry station
水情中继站 relay station
水情中心站 center station
水文测验 hydrometry
遥测 telemetry
驻测 stationary gauging
巡测 tour gauging
间测 intermittent gauging

3.2 水能及水资源利用 Water power and water resources utilization

(1) 水能利用 Waterpower utilization

水能 waterpower，hydropower
水能资源（水力资源） waterpower resources，hydropower resources
水电蕴藏量 hydropower potential
理论蕴藏量 theoretical potential
可开发蕴藏量 exploitable potential
水能利用规划 waterpower utilization planning
设计保证率 design dependability
设计水平年 design time horizon
水能计算 hydroenergy computation
旱季 dry season
雨季 rainy season，wet season
枯水期 low-flow period
丰水期 high-flow period
反调节 re-regulation
调节周期 period of regulation
日调节 daily regulation
周调节 weekly regulation
年调节 yearly regulation，annual regulation
多年调节 overyear regulation，carryover regulation
调节流量 regulated flow
调节系数 regulation coefficient
水库起调水位 initial reservoir elevation
库水位年周期变化 annual periodic fluctuation of reservoir level
消落深度 drawdown
水库特征库容 characteristic capacities of reservoir
楔形库容 wedge storage
静库容 stilling storage
动库容 dynamical storage
总库容 gross reservoir capacity
防洪库容 flood control storage
调洪库容 flood regulation storage
兴利库容，有效库容 effective storage，live storage
共用库容，重复利用库容 common storage，shared storage
超高库容 surcharge storage
死库容 inactive storage，dead storage
防洪专用库容上限 top of exclusive flood control capacity
有效蓄水库容上限 top of active conservation capacity
非有效库容上限 top of inactive capacity
库容分配 storage allocation
库容系数 coefficient of storage
水库保证出水量 safe reservoir yield
水头 water head
毛水头 gross head
净水头 net head
设计水头 design head
平均水头 average head
加权平均水头 weighted average head
额定水头 rated head
水电站引用流量 power discharge
机组过水能力 turbine discharge
发电 power generation

装机容量 installed capacity
发电量 energy output
出力 output
保证出力 firm output
预想出力 expected output
技术最小出力 minimum output
出力系数 coefficient of output
可靠电能 firm energy
季节性电能 seasonal energy
多年平均年发电量 average annual energy output
装机容量年利用小时数 annual utilization hours of installed capacity
多年平均流量 long-period average flow
预测长期流量 long-term prediction flow
可靠流量 dependable flow
可用流量 available flow
常年基流 perennial base flow
径流调节 runoff regulation
洪水调节 flood regulation
枯水调节 low flow regulation
水能计算 hydropower computation
补偿调节 compensative regulation
电源组成 constitution of power sources
电力电量平衡 balance of power and energy
负荷特性 load characteristics
能量指标 energy indexes
负荷曲线 load curve
峰荷 peak load
基荷 base load
腰荷 medial load
日平均负荷率（日负荷系数） average daily load factor
电力系统容量 installed capacity of power system
电力系统备用容量 reserve capacity of power system
电力系统动态备用容量 spinning reserve of power system
工作容量 working capacity
负荷备用容量 standby capacity
事故备用容量 reserve capacity for emergency
空闲容量 idle capacity
重复容量 duplicate capacity
受阻容量 restricted capacity
可调容量 adjustable capacity
水库调节 reservoir regulation
水库群调节 multi-reservoir regulation
水库调洪 reservoir flood routing
水库调度 reservoir scheduling
防洪限制线 guide curve for flood control
上基本调配线，保证供水线 upper critical guide curve
下基本调配线，限制出力线 lower critical guide curve
防弃水线 guide curve for reducing abandoned water
水库库容面积曲线 reservoir storage capacity-area curve
水头保证率曲线 head dependability curve
厂房尾水水位-流量关系曲线 powerhouse tail water rating curve
出力保证率曲线和电量累积曲线 output dependability curve and cumulative power curve

(2) 水资源开发利用
Water resources development and utilization

水资源 water resources
水利区划 zoning of water conservancy
水资源分区 water resources regionalization
水资源规划 water resources planning
水资源分配 water resources allocation
地下水资源 groundwater resources
水平年 target year
水资源供需分析 demand and supply analysis of water resources
水资源保护 water resources protection
水资源危机 water resources crises
水资源评价 water resources assessment
水资源决策支持系统 water resources decision support system
水利计算 computation of water conservancy
径流调节 runoff regulation
承压水 artesian water (confined water)
潜水蒸发 phreatic water evaporation
地下水矿化度 mineralization of groundwater
降雨入渗补给 infiltration recharge by rainfall
地下水越层补给 recharge through weak permeable layer
地下水侧向补给 recharge by groundwater
灌溉回归水补给 recharge from return flow of irrigation

地下水人工补给 artificial recharge of groundwater
抽咸换淡 pump out the saline water and recharge the fresh water
地下水动态 groundwater regime
地下水开采 groundwater exhaustion
地下水下降漏斗 exhaustion cone of groundwater
地下水超量开采 excessive exhaustion of groundwater
地下水降深 drawdown of groundwater
自由孔隙率，贮水度 free porosity
水资源项目（水利项目） water resources project
综合利用规划 multipurpose planning
单一目的项目 single-purpose project
综合利用工程 multi-purpose project
水库综合开发利用 multipurpose development of reservoir
跨流域引水工程 trans-basin diversion project
跨流域调水 trans-basin water transfer
分期开发 staged development
供水 water supply
分层取水 water-taking at different levels
有坝取水 intake with dam
生活用水 domestic water
工业用水 industrial water
供水保证率 water supply dependability
回归水的重复使用 reuse of return flows
工程效益 project benefits
效益分摊 amortization of benefits
发电效益 power generation benefit
容量效益 capacity benefit
能量效益 energy benefit
供水效益 water supply benefit
灌溉效益 irrigation benefit
休闲效益 recreation benefit
防洪效益 flood control benefit
航运效益 navigation benefit
防洪规划 flood control planning
洪水 flood
季节性洪水 seasonal flood
汛期 flood season
防洪标准 flood control standard
洪水保险 flood insurance
河道安全泄量 safety discharge in river
淹没损失 flood damage
历史洪水灾害 historical flood damage
难确定的（无形）洪水灾害 intangible flood damage
洪水之前降低库水位 pre-flood drawdown
保证水位 highest safety stage
警戒水位 warning stage
分洪水位 flood diversion stage
防洪措施 Flood control measure
工程措施，结构措施 structural measure
非工程措施 non-structural measure
分洪工程 flood diversion works
分洪道 flood way
分洪区 flood diversion area
滞洪区 flood detention area
洪泛区 flood plain
行洪区 flood flowing zone
防洪水库 flood control reservoir
蓄洪垦殖 flood storage and reclamation
防洪对象 protected objects against flood
度汛方案 flood protection scheme
防洪堤 flood dyke
防洪墙 flood wall
航运 navigation
航道及运河 waterway and canal
主航道 main waterway
副航道 sub-waterway
渠化航道 channelized waterway
运河 canal
船闸 shiplock
船队 ship fleet
航道规划 waterway planning
通航标准 navigation standard
航道等级 grade of waterway
通航期 navigation period
航道通过能力 navigation capacity
通航密度 navigation density
通航保证率 navigation dependability
航道整治线 regulation line of waterway
航道设计水位 design stage of waterway
最高通航水位 maximum stage of waterway
通航水深 navigation depth
通航流量 navigation discharge
通航流速 navigation velocity
航道弯曲半径 curvature radius of waterway
通航净空 navigation clearance
航道断面系数 cross section factor of waterway

土壤水　soil water
土壤含水量（含水率）　soil moisture content
土壤饱和含水量　saturated moisture content of soil
土壤蒸发量　soil evaporation
作物需水量　water demand of crop
田间需水量　water consumption on farmland
耗水强度　intensity of water consumption
需水模数　modulus of water demand
灌溉定额　irrigation duty
灌溉用水量　irrigation water consumption
设计灌水周期　designed interval of irrigation
灌溉制度　irrigation program
灌溉保证率　dependability of irrigation
灌溉典型年　typical design year for irrigation
泡田　steeping field
泡田定额　duty of steeping field
灌溉水源　water source for irrigation
引水灌溉　water diversion irrigation
蓄水灌溉　water storage irrigation
提水灌溉　pumping irrigation
自流灌溉　gravity irrigation
间歇灌溉　intermittent irrigation
串灌串排　irrigation and drainage from one field to another
灌区　irrigation area
灌溉系统　irrigation system
蓄引提结合灌溉系统　irrigation system with water storage, diversion and pumping facilities
渠系规划　planning of canal system
灌溉渠道　irrigation canal
总干渠　trunk channel
输水渠道　water conveyance canal
配水渠道　water distribution canal
退水渠　canal for water release
塘堰　pond
渠道毛流量　gross discharge in canal
渠道净流量　net discharge in canal
渠道输水损失　water conveyance loss in canal
渠道过水能力　carrying capacity of canal
渠道坡降　gradient of canal
渠床糙率　roughness of canal bed
渠道断面宽深比　ratio of bottom width to water depth in canal
渠道允许不冲流速　permissible unscouring velocity in canal
渠道允许不淤流速　permissible unsilting velocity in canal
渠道冲淤平衡　balance of scouring and silting in canal
农田排水　farmland drainage
排水不良　impeded drainage
涝　surface waterlogging
渍　subsurface waterlogging

3.3 经济评价
Economic evaluation

动能经济指标　energy economic indicator
替代法　substitution approach
投入产出法　input to output approach
备选方案　alternative
方案比选　comparison and selection of alternatives
备选方案评估　alternative appraisal
标准换算系数　standard conversion factor（SCF）
补贴　subsidy
不变价格　constant price
财务补贴　financial subsidy
财务分析　financial analysis
财务价格　financial price
财务净现值　financial net present value（FNPV）
财务可持续性　financial sustainability
财务内部收益率　financial internal rate of return（FIRR）
财务效益　financial benefit
残值　residual value
偿债备付率　debt service coverage ratio（DSCR）
成本回收　cost recovery
成本有效性分析　cost effectiveness analysis
等价年度费用　equivalent annual cost
非外贸产出和投入　non-traded output and input
非外贸货物　non-traded goods
成本效果比　cost effectiveness ratio
成本效果分析　cost effectiveness analysis（CEA）
成本效益分析　cost benefit analysis（CBA）
分配分析　distribution analysis
风险分析　risk analysis
福利费用　welfare cost
福利效益　welfare benefit
公共产品　public goods
供给价　supply price
官方汇率　official exchange rate（OER）

规模经济 economies of scale
国民收入平减指数 GDP deflator
耗减补偿 depletion premium
核算单位 unit of account
换算系数 conversion factor
汇率溢价 foreign exchange premium
或有估价法 contingent valuation method (CVM)
货币的时间价值 time value of money
机会成本 opportunity cost
基准收益率 hurdle cut-off rate
计价单位 numeraire
加权平均资金成本 weighted average cost of capital (WACC)
价格扭曲 price distortion
价格指数 price index
交叉补贴 cross subsidization
交易费 transaction costs
接受补偿意愿 willingness to accept (WTA)
进口平价 import parity price
经济补贴 economic subsidy
经济费用 economic cost
经济分析 economic analysis
经济规模 economies scale
经济价格 economic price
经济净现值 economic net present value (ENPV)
经济内部收益率 economic internal rate of return (EIRR)
经济评估 economic appraisal
经济生存能力 economic viability
经济寿命 economic life
经济效率 economic efficiency
经济效益 economic benefit
经济资源 economic resource
净现值 net present value (NPV)
矩阵方法 matrix approach
口岸价 border price
劳动力机会成本 opportunity cost of labor
利息备付率 interest coverage ratio (ICR)
敏感性分析 sensitivity analysis
敏感性指标 sensitivity indicator
名义价格 nominal price
内部收益率 internal rate of return (IRR)
年金化值 annuities value
平均增量财务费用 average incremental financial cost (AIFC)
平均增量成本 average incremental cost (AIC)
平均增量经济费用 average incremental economic cost (AIEC)
权益资本 equity capital
人力资本法 human capital method
融资主体 financing entity
私人融资 private financing
融资前分析 analysis before financing
融资后分析 analysis after financing
时价 current price
时间偏好率 time preference rate
实际汇率 real exchange rate
实际价格 real price
实际价值 real value
实物量 volume terms
使用价值 use value
通货膨胀率 inflation rate
土地的机会成本 opportunity cost of land
外部效果 externality
外贸货物 traded goods
外贸投入和产出 trade input and output
无形资产 intangible assets
息税前利润 earnings before interest and tax (EBIT)
显示偏好 revealed preference
现存价值 existence value
现金流(量) cash flow (CF)
现金流出(量) cash outflow (CO)
现金流入(量) cash inflow (CI)
现值 present value (PV)
相对价格作用 relative price effect
项目备选方案 project option
项目框架 project framework
项目周期 project cycle
消费者剩余 consumer surplus
效果费用比 effectiveness cost ratio
需求价格 demand price
需求曲线 demand curve
要求回报率 required rate of return (RRR)
意愿调查评估法 contingent valuation
隐含价值法 hedonic method
盈亏平衡点 break-even point
影响陈述 impact statement
影子工资率 shadow wage rate (SWR)

影子工资系数　shadow wage rate factor (SWRF)
影子汇率　shadow exchange rate
影子汇率系数　shadow exchange rate factor (SERF)
影子价格　shadow price
增量产出　incremental output
增量投入　incremental input
增量效益　incremental benefit
债务资金　debt capital
折旧　depreciation
折现　discounting
折现率　discount rate
支付能力　ability to pay (ATP)
支付意愿　willingness to pay (WTP)
终值　final value (FV)
转移支付　transfer payment
准股本资金　quasi-equity
资本化价值　capitalized value
资本金　capital
净资产收益率　return on equity (ROE)
资产负债率　liability on asset ratio (LOAR)
资金成本　capital cost
资金的财务机会成本　financial opportunity cost of capital (FOCC)
资金的经济机会成本　economic opportunity cost of capital (EOCC)
资源成本　resource cost
总投资收益率　return on investment (ROI)
最低可接受收益率　minimum acceptable rate of return (MARR)
上网电价　feed-in tariff
销售电价　consumer tariff
标杆电价　benchmark tariff
上网电量　on-grid energy

3.4 主要报告及图表名称
Titles of reports, figures and tables

流域水系图（标明水文、气象和大中型水利水电工程位置）　Hydrographical net of the river basin (indicating the hydrological and meteorological stations and large and medium-size water resources and hydropower projects)
河流（河段）综合利用示意图　Sketch of multipurpose developments in the river (river reaches)
河流梯级开发纵剖面图　Longitudinal profile of the river on cascade developments
推荐开发方案与环境敏感对象关系示意图　Relationship of the proposed construction scheme and environmental sensitive objects
水库淹没范围示意图　Sketch of reservoir inundation area
供电范围电力系统地理接线图（现状及远景）　Geographical connection of the power system in the power recipient area (current and future)
径流、暴雨洪水、暴雨量、泥沙插补延长的主要相关关系图　Major correlations for interpolated and extrapolated runoffs, storm floods, storm volumes and sediment
年径流、枯水期径流频率曲线图　Frequency curves for yearly runoff and runoff in dry season
洪峰、洪量关系图　Relations between flood peak and volume
洪峰和各时段洪量（暴雨量）频率曲线图　Frequency curves for flood peaks and volumes (storm volumes) of different time intervals
典型洪水及设计洪水过程线图　Hydrographs of typical flood and design flood
主要设计断面水位-流量关系图　Stage-discharge relation of principal design cross-section
水库库容面积曲线　Reservoir storage-capacity-area curve
水量分布曲线　Water distribution curve
尾水水位流量关系曲线　Tailwater rating curve
出力保证率曲线　Output dependability curve
电量累积曲线　Cumulative power curve
水头保证率曲线　Head dependability curve
水库调度图　Graph of reservoir operation
电力系统典型日（周）运行方式示意图　Schematic diagram of representative daily (weekly) operating mode of power system
悬移质、推移质颗粒级配曲线图　Gradation curves for suspended load and bed load
泥沙淤积计算图　Sediment deposition calculation diagram
水库回水计算图　Reservoir backwater calculation diagram
主要评价指标敏感性分析图　Sensitivity analysis of key evaluation indicators

水资源论证报告　Water resources demonstration
河流水电规划报告　River hydropower planning
河流规划成果报告　River planning results
防洪评价报告　Flood control assessment
水情自动测报系统设计报告　Design report of hydrological telemetry system
年、月径流（雨量）系列表　Yearly and monthly runoff (rainfall) series
洪峰、洪量系列表　Flood peak and volume series
典型洪水和设计洪水过程线表　Hydrographs of typical flood and design flood
水库回水计算成果表　Calculation results of reservoir backwater
年、月输沙量系列表　Yearly and monthly sediment runoff series
电力系统电力电量平衡表　Balance of electric power and energy of power system
各项工程规模方案技术经济比较表　Technical and economic comparisons of various project scale alternatives
固定资产投资估算表　Fixed assets investment estimate
投资计划与资金筹措表　Investment plan and financing
总成本费用估算表　Total cost estimation
借款还本付息计算表　Principal and interest repayments
资金来源与运用表　Fund sources and investment portfolios
现金流量表（全部投资）　Statement of cash flow (total investment)
现金流量表（资本金）　Statement of cash flow (capital)
经济效益费用流程表　Flow of economic benefits and costs
资产负债表　Balance sheet
损益表　Income statement

第4章 环保与移民
Environmental protection and resettlement

4.1 环境保护
Environmental protection

人类环境 human environment
生态系统 ecological system
生态平衡 ecological balance
人与环境和谐 harmony between human being and environment
生境适宜性 habitat suitability
环境完整性 environmental integrity
环境意识 environmental awareness
环境容量 environment capacity
环境质量 environmental quality
环境因素 environmental factor
环境关注 environmental concerns
环境研究 environmental study
环境调查 environmental investigation
环境生态学 environment ecology
环境工程学 environment engineering
环境监测 environmental monitoring
环境保护 environmental protection
低碳 low carbon
低碳生活方式 low-carbon lifestyle
环境影响评价 environmental impact assessment (EIA)
环境影响后评价 follow-up environmental impact assessment
环境回顾评价 retrospective assessment of environment
环境现状评价 present situation assessment of environment
环境预测评价 prospective assessment of environment
环境综合评价 comprehensive assessment of environment
环境影响识别 identification of environmental impact
环境评估系统 environmental evaluation system
环境本底值 present environmental baseline
环境约束 environmental constraints
固碳功能 carbon sequestration
碳汇 carbon sink
中华人民共和国环境保护法 Environmental Protection Law of the People's Republic of China
国家环境政策 national environmental policy
环境质量改善 environmental quality improvement
环境公约 environmental covenant
环境目标 environmental objective
经济影响评价 economic impact assessment
环境和社会影响评价 environmental and social impact assessment (ESIA)
环境与社会管理计划 environment and social management plan (ESMP)
环境和社会行动计划 environmental and social action plan (ESAP)
环境监测计划 environment monitoring program
生物多样性管理计划 biodiversity management plan
环境考虑 environmental consideration
无污染的 pollution-free
环境友好的 environment-friendly
环保型方式 environmentally sound manner
有利环境影响 favorable environmental impact
不利环境影响 adverse environmental impact
有利环境影响最大化 maximization of favorable environmental effects
不利环境影响最小化 minimization of adverse environmental effects
可避免的不利环境影响 avoidable environmental impact
不可避免的不利环境影响 unavoidable adverse environmental impact
不可逆转环境影响 irreversible environmental impact
可能的不利影响 possible adverse effect
环境恶化 environmental deterioration

环境退化 environmental degradation
生态退化 ecological degradation
自然灾害 natural disaster
评价系统 assessment system
生态系统服务 ecosystem service
河流生态修复 river eco-restoration
水环境 water environment
替代生境 alternative habitat
累计影响 cumulative impact
可能环境因子 possible environmental factor
环境敏感对象 environmental sensitive object
环境影响评价区 environmental impact assessment area
重要环境问题区域 area of critical environmental concern
可耐受水平 tolerable level
可再生资源 renewable resources
不可再生资源 non-renewable resources
地下自然资源 subterranean natural resource
自然景观 natural landscape
自然保护区 nature reserve
风景名胜区 scenic resort
地质公园 geologic park
森林公园 forest park
国家公园 national park
自然遗产 natural heritage
历史文化遗产 historical and cultural heritage
人造环境 built environment
休闲区 recreation zone
湿地 wetland
人工湿地 artificial wetland
滩涂 beach
针叶林 coniferous forest
阔叶林 broad-leaved forest
风景林 landscape forest
观赏植物 decorative plant
森林面积 forest area
森林覆盖率 forest coverage rate
生物圈 biosphere
植物群 flora
动物群 fauna
陆生物种 terrestrial species
水生物种 aquatic species
珍稀物种 rare species
特有物种 endemic species
濒危物种 endangered species
入侵物种 invasive species
迁移物种 migratory species
作物袭击 crop raiding
野生生物保护 wildlife conservation
野生生物物种 wildlife species
自然栖息地 natural habitat
野生动植物栖息地 wildlife habitat
栖息地改变 habitat change
栖息地消失 habitat loss
哺乳动物物种 mammal species
食草动物 grazer
野生生物迁徙 wildlife migration
动物迁徙路线 animal migration route
觅食地 feeding area
动物诱捕 animal entrapment
过度放牧 overstocking
就地保护 in-situ conservation
迁地保护 ex-situ conservation
人与野生动物冲突 human-wildlife conflict
涉及野生动物的犯罪 wildlife-related crime
气候变化 climate change
联合国气候变化框架公约 United Nations Framework Convention on Climate Change (UNFCCC)
世界卫生组织空气质量标准 WHO Air Quality Guidelines
臭氧层 ozone layer
小气候 microclimate
空气污染 air pollution
污染物质 pollutant
颗粒物 particulate matter (PM)
排放和粉尘控制 emission and dust control
二氧化碳 carbon dioxide (CO_2)
二氧化硫 sulfur dioxide (SO_2)
氮氧化物 nitrogen oxide (NO_X)
酸雨 acid rain
温室气体排放 greenhouse gas emission
水生生态 aquatic ecology
水生生态变化 aquatic ecology alteration
水生物栖息地 aquatic habitat
水生物 aquatic creature
水生植物 aquatic plant
底栖生物 benthonic organism
浮游动物 zooplankton

浮游植物 phytoplankton
水生维管束植物 aquatic vascular plant
水藻（藻类） algae
鱼类种群 fish population
鱼类洄游 fish migration
产卵场 spawning ground
鱼类孵化场 fish hatchery site
产卵鱼 spawner
孵卵 hatching
索饵场 feeding ground
越冬场 wintering ground
鱼类增殖放流站 fish breeding and releasing station
湖鱼 lacustrine fish
地表水 surface water
地下水 ground water
水资源管理 water resource management
水质标准 water quality standard
水质评价 water quality assessment
水质参数 water quality parameter
变温层 epilimnion
均温层（冷水层） hypolimnion
水质监测 water quality monitoring
水环境容量 water environment capacity
环境流量 environmental flow
水污染 water pollution
河流污染 stream pollution
点污染源 point source pollution
面污染源 non-point source pollution
总悬浮颗粒物 total suspended solids（TSS）
化学需氧量 chemical oxygen demand（COD）
生化需氧量 biochemical oxygen demand（BOD）
溶解氧 dissolved oxygen（DO）
溶氧浓度 dissolved oxygen concentration
富营养化 eutrophication
水体掺气 water aeration
絮凝 flocculation
蒸发损失 evaporation loss
噪声 noise
噪声测量 noise level measurement
噪声标准 noise standard
噪声级 noise level
本底噪声 background noise
噪声污染 noise pollution
噪声管理 noise management
噪声控制装置 noise control device
土地扰动 land disturbance
土壤污染 soil pollution
土壤潜育化 soil gleization
土地沙化 land desertification
盐碱化 salinization
沼泽化 swamping
河湖淤泥 lake and river silt
取土坑 borrow pit
植被清除 vegetation clearance
水库库底清理 clearing of reservoir zone
建设场地清理 construction land clean-up
应急准备 emergency preparation
应急机构（组织） emergency organization
应急响应 emergency response
恢复计划 restoration plan
缓解方案 mitigation program
减缓措施 mitigation measure
环境污染综合防治 integrated control of environment pollution
水污染控制 water pollution control
大气污染控制 air pollution control
噪声污染控制 noise pollution control
振动污染控制 vibration pollution control
固体废物污染 solid waste pollution control
电磁辐射污染控制 electromagnetic radiation pollution control
卫生设施 sanitation facility
安全供水 safe water supply
污水排放标准 sewage discharge standard
废水处理 wastewater treatment
废物处理 waste disposal
生活垃圾 domestic waste
临时弃物存放 temporary spoil storage
易燃固体 flammable solid
易燃（可燃）液体 flammable liquid
碱度（碱性） alkalinity
地方病 endemic disease
传染性疾病 infectious disease
水传疾病 waterborne disease
虫媒病 vector-borne disease
皮肤病 skin disease
疾病携带动物 disease-carrying animal

4.2 水土保持
Water and soil conservation

(1) 通用 General

水土流失 soil erosion and water loss
水损失 water loss
水土流失类型 type of soil erosion and water loss
水土流失区 region of soil erosion and water loss
水土流失面积 area of soil erosion and water loss
容许土壤流失量 soil loss tolerance
水土保持措施 soil and water conservation measures
水土保持设施 soil and water conservation facilities
水土流失综合治理 comprehensive control of soil erosion and water loss
水土保持生态环境建设 soil and water conservation for ecological environment rehabilitation
小流域 small watershed
小流域综合治理 comprehensive management of small watershed
土壤侵蚀 soil erosion
土壤侵蚀类型 type of soil erosion
土壤侵蚀规律 mechanism of soil erosion
侵蚀营力 erosion force
自然侵蚀 natural erosion
常态侵蚀 normal erosion
人为侵蚀 erosion caused by human activities
水力侵蚀 water erosion
面蚀 surface erosion
沟蚀 gully erosion
淋溶侵蚀 leaching erosion
暴雨侵蚀 storm erosion
波浪侵蚀 wave erosion
溯源侵蚀 headward erosion
风力侵蚀 wind erosion
冻融侵蚀 freeze-thaw erosion
重力侵蚀 gravitational erosion
混合侵蚀 mixed erosion
磨蚀 abrasion
塌坡 slope collapse
沟道密度 gully density
土地沙化 land desertification
草场退化 grassland degradation
荒漠化 desertification
风沙流 sand air current
土壤侵蚀程度 soil erosion degree
土壤侵蚀强度 soil erosion intensity
土壤侵蚀量 amount of soil erosion
土壤流失量 amount of soil loss
土壤侵蚀模数 soil erosion modulus
土壤通气不良 defective soil aeration
流域产沙量 watershed sediment yield
流域输沙量 amount of sediment delivery
冲沙流量 sediment flushing flow

(2) 规划设计与试验研究 Planning and testing research

土壤侵蚀分区 soil erosion zoning
水土保持区划 soil and water conservation regionalization
水土保持规划 soil and water conservation planning
土地利用规划 land use planning
土地适宜性评价 land suitability assessment
可治理面积 erosion area suitable to control
水土流失治理面积 area of water and soil conservation
水土流失治理程度 erosion control ratio
防风固沙 windbreak and sand-shifting control
可耕地 arable land
基本农田 capital farmland
坝地 farmland formed in silt storage dam
源头坝 watershed dam
绿化 afforestation
造林密度 density of plantation
造林存活率 survival rate of afforestation
枯枝落叶层 litter layer
郁闭度 crown density
植被覆盖率 vegetation cover ratio
水土保持效益 soil and water conservation benefit
水土保持生态效益 ecological benefit of soil and water conservation
水土保持经济效益 economic benefit of soil and water conservation
水土保持社会效益 social benefit of soil and water conservation

蓄水保土效益 water detention and soil conservation benefit
水土流失观测 observation of soil erosion and water loss
径流小区观测 observation of runoff plot
模拟降雨试验 simulated rainfall experiment
标准径流小区 standard runoff plot
天然坡面径流场 natural runoff plot
土壤侵蚀模型 soil erosion model
小流域产流模型 runoff-yield model of small watershed
降雨侵蚀指数 rainfall erosion index
土壤抗蚀性 soil erosion resistance
土壤侵蚀速率 soil erosion rate

(3) 预防监督与管理 Prevention, supervision and management

水土流失预防 prevention of soil erosion and water loss
水土保持监督 soil and water conservation supervision
水土流失重点防治区划分 zoning of critical prevention and control areas of soil erosion and water loss
重点预防保护区 important protection zone
重点监督区 important supervision zone
重点治理区 important rehabilitation zone
水土保持方案 soil and water conservation program
禁止开垦坡度 critical cultivation slope
退耕还林还草 grain-for-green project
水土流失监测 monitoring of soil and water conservation
山洪易发区 susceptible area of mountain torrent
泥石流易发区 susceptible area of debris flow
滑坡易发区 hazardous area of landslide
滑坡泥石流监测预警 monitoring and forecasting of landslide and debris flow

(4) 综合治理 Comprehensive control

工程措施 structural measure, engineering measure
坡面治理工程 slope treatment for erosion control
梯田 terrace
水平梯田 bench terrace
坡式梯田 sloping terrace
隔坡梯田 interval terrace
反坡梯田 back-slope terrace
坡面截流沟 water intercepting and drainage ditch on the slope
水平沟 level ditch
水平阶 horizontal stage
鱼鳞坑 fish-scale pit
造林整地 land preparation for afforestation
固坡工程 slope stabilization project
护坡工程 slope protection works
坡面水系工程 slope water works
坡面集雨工程 rainfall harvesting works on the slope
沟道治理工程 gully erosion control works
沟边埂 ridge along gully
沟头防护工程 protective works of gully head
淤地坝 check dam for farmland forming
拦沙坝 sediment trapping dam
水坠坝 sluicing-siltation earth dam
河道治理 river training (improvement)
治沟骨干工程 key project for gully erosion control
沟道蓄水工程 water storage works in gully
蓄水池 water storage pool
沉淀池 sedimentation basin
拦渣工程 tailing hold structure
泥石流防治工程 debris flow control works
植物措施 vegetation measure
水土保持林 forest for soil and water conservation
水源涵养林 forest for water resources conservation
农田防护林 shelter-belt on farmland
农田林网 forest net in farmland
河岸林地 riparian woodland
等高植物篱 contour living hedgerow
挂网喷草 spraying grass seeds with net
封禁治理 closing hillside for erosion control
水土保持耕作措施 agricultural measure for soil and water conservation
等高耕作 contour tillage
沟垄耕作 furrow-ridge tillage
覆盖种植 covering cultivation
带状间作 strip intercropping
草田轮作 grass and crop rotation

4.3　移民安置
Resettlement

社会经济基线　socio-economic baseline
社会经济问题　socio-economic issue
可持续发展框架　sustainable development framework
经济发展总体规划　overall economic development plan
长远经济发展规划　long-term economic development plan
近期规划　short-term plan
远景规划　long-term plan
发展目标　development objective
城市扩张　urban expansion
经济发达地区　developed region
经济不发达地区　undeveloped region
人口稠密区　densely populated area
贫困线　poverty line
贫穷地区　impoverished area
实地访问　field inquiry
实地调查　field investigation
逐户调查　household investigation
抽样调查　sampling investigation
填表调查　tabulating statistic
公众参与　public participation
当地居民参与　engagement of local residents
利益相关者参与　stakeholder engagement
村民小组　village group
透明度与公信力　transparency and credibility
公认　public acceptance
群众支持　public support
住户收支调查　surveys on household income and expenditure
可以自由处置的收入　disposable income
实物收入　income-in-kind
副业收入　sideline income
劳务收入　service income
实物指标公示　soliciting public comments on inventory in-kind
实物指标确认　confirmation of inventory in-kind
物价增长　price inflation
工程影响的居住区　project-affected community
少数民族居住地区　minority-inhabited areas
民族意识　ethnic awareness
少数民族　ethnic minorities
种族歧视　ethnic discrimination
种族　ethnic group
工程影响人口　project affected persons（PAPs）
人口预测　population prediction
人口增长率　population growth rate
人口密度　population density
人口结构　population structure
人口流入　population influx
自然增长率　natural growth rate
公共健康　public health
生育率　fertility rate
婴儿死亡率　infant mortality rate
户籍人口　registered population
常住人口　resident population
非农业人口　non-agricultural population
户主　household head
流动人口　migratory population
通勤人口　commuting population
动态人口　dynamic population
劳动力　labor force
本地人口　indigenous population
本地居民　indigenous inhabitant
本地文化　indigenous culture
受教育程度　educational background
移民　resettler
农村移民　rural resettler
农村家庭　rural household
直接受影响住户　directly affected household
搬迁人口　relocation population
非自愿迁移　involuntary relocation
非自愿移民　involuntary resettler
产业结构　industrial structure
职业结构　occupation structure
职业分类　occupation classification
职业分析　vocational analysis
职业教育　vocational education
当前生计　current livelihood
就业状况　employment status
就业机会　job opportunity
就业率　employment rate
失业率　rate of unemployment
人身保险　life insurance
劳动保险　labor insurance
水库淹没处理　reservoir inundation treatment

水库淹没区　reservoir inundation zone
居民迁移线　border of resident relocation
土地征用　land requisition
土地征用线　property line，border of land requisition
临时用地　temporary land occupation
土地占补平衡　requisition-compensation balance of land
人口迁移线　relocation border
敏感对象　sensitive object
临时淹没区　infrequent flooded zone
主要淹没影响　major inundation impact
正面影响　positive impact
潜在影响　potential impact
淹没影响调查　inundation inventory survey
水库淹没实物指标　reservoir inundation inventory
淹没耕地　inundated cultivated land
淹没农田　inundated farmland
零星林木　scattering trees
水库影响区　reservoir-affected area
影响处理区　affected area to be treated
坍岸区　bank caving area
浸没区　Inundation area
蓄水分期　impoundment stages
移民安置范围　extent of resettlement
农业安置　agricultural resettlement
大规模农业生产　large-scale agricultural production
非农业生产　non-agricultural production
工矿企业　industrial and mining enterprises
工业迁建　industrial relocation
工业部门　industrial sector
劳动力密集的行业　labor intensity industry
集体所有制企业　collective ownership enterprise
乡镇企业　township owned enterprise
劳动力市场　labor market
服务业　service business
副业　sideline activity
停产期　period of closure
停产损失　loss in production stop
拟定移民方针　development of a resettlement concept
移民生活水平评价预测　assessment and prediction of resettlers' living standard
移民安置总体规划　general planning of resettlement
移民安置行动计划　resettlement action plan (RAP)
移民实施组织设计　resettlement implementation and organization design
职业健康和安全计划　occupational health and safety plan
社会行动计划　social action plan
社会及经济方面的重建　social and economic reestablishment
移民政策　resettlement policy
移民权利　entitlement of resettlers
移民安置方式　resettlement scheme
移民办公室　resettlement office
移民搬迁期　resettlement transition period
移民后期扶持　post-resettlement support
生产扶持　production support
生计恢复　livelihood restoration
生活标准　living standard
税收优惠政策　preferential tax policy
补贴政策　subsidy policy
移民补贴　resettlement subsidy
申诉机制　grievance mechanism
申诉救济制度　grievance redress system
移民试点工程　pilot resettlement project
安置区　host area
移民安置区　resettlement area
环境容量　environmental carrying capacity
土地承载能力　land bearing capacity
选址　siting
定居区　settled zone
生活区　living quarter
常住居民　permanent resident
生产安置人口　agricultural population to be job-arranged
招工安置　allocation with job
农村移民安置　rural resettlement
就近后靠　local resettlement to higher elevation
远迁　distant relocation
货币补偿　monetary indemnity (compensation)
土地　land
土地所有权　land ownership，land tenure
土地使用权　land use right
土地开发整理　land reclamation and improvement
土地流转　land circulation
土地开垦　land reclamation

土地征用标准 land requisition criterion
土地收入 land revenue
未利用地 left-over land
未开垦土地 uncultivated land
河滩地 flood plain
荒地 barren land
低洼地 swale
低洼盐碱地 saline-alkali swale
种植农业 plantation agriculture
复合农林业 agro-forestry
农用地 farmland
示范农场 demonstration farm
机械化程度 mechanization level
人均耕地 cultivated land per capita
播种面积 sown area
单位面积产量 yield per unit area
耕地 cultivated land
水田 paddy field
水浇地 irrigated land
旱地 arid land
菜地 vegetable land
园地 garden plot
自留地 plot for private use
复种 multiple cropping
复种面积 multiple cropping area
复种指数 multiple crop index
耕作密度 cultivated density
有机肥 organic fertilizer
无霜期 frost-free period
歉收年 fail year
粮食作物 grain crops
经济作物 commercial crops
油料作物 oiling crops
饲料作物 forage crops
青苗 young crops
蔬菜 vegetable
粮食产量 grain output
粮食补贴 grain allowance
林地 forest land
林场 forest farm
原种场 foundation seed farm
林带 forest belt
用材林 timber forest
经济林 non-wood forest
薪炭林 firewood forest
防护林 shelter forest
生态公益林 non-commercial forest
防风固沙林 sand-shifting control forest
灌木林地 shrub forest land
疏林地 vegetable and forest land
苗圃 nursery garden
苗木 nursery stock
木柴 firewood
禁猎区 sanctuary
禁猎期，禁渔期 fence time
畜牧业 livestock husbandry
饲养制度 feeding regime
牲畜头数 livestock population
畜舍 livestock shed
繁殖场 breeding farm
渔业 fishery
渔业区 fishing zone
鱼饲养 fish rearing
鱼池（鱼塘） fish pond
鱼苗 fry
鱼苗场 fry raising farm
幼鱼 fingerling
房屋 housing
框架式房屋 frame-type house
砖混房 brick-concrete house
砖木房 brick-wood house
木房 wood house
土房 adobe
人均住房面积 floor area per capita
重置价格 replacement price
迁建规划 relocation plan
移民搬迁设计 relocation design
生产安置设计 production resettlement design
城市集镇迁建 urban and town relocation
城市总体规划 urban master plan
城市集镇新址 new town site
供水计划 water supply scheme
污水处理厂 sewage treatment plant
自吸式污水泵 self-priming sewage pump
专项设施 special facility
文物古迹 culturaland historic relic
压覆矿产 covered mineral resource
防护工程 protection works

补偿费用概算 compensation cost estimate
移民安置费估算 resettlement cost estimate
迁建补偿标准 relocation compensation standard
分项投资 breakdown cost

4.4 主要报告及图表名称
Titles of reports, figures and tables

环境质量报告 Environmental quality
环境影响评价报告 Environmental impact assessment (EIA)
水土保持方案报告 Soil and water conservation
实物指标调查报告 Reservoir inundation inventory survey
移民安置规划大纲 Resettlement planning outline
建设征地和移民安置规划设计报告 Planning of land requisition and resettlement
劳动安全与工业卫生预评价报告 Pre-assessment on labor safety and industrial hygiene
水电工程竣工（截流、蓄水）验收建设征地移民安置设计工作报告 Design of land requisition and resettlement work for hydroelectric project acceptance upon completion (river closure or impoundment)
水电工程竣工（截流、蓄水）验收建设征地移民安置实施工作报告 Implementation of land requisition and resettlement work for hydroelectric project acceptance upon completion (river closure or impoundment)
水电工程竣工（截流、蓄水）建设征地移民安置验收报告 Acceptance of land requisition and resettlement work upon hydroelectric project completion (river closure or impoundment)
水电工程竣工（截流、蓄水）验收项目法人建设征地移民安置工作报告 Project owner's land requisition and resettlement work for hydroelectric project acceptance upon completion (river closure and impoundment)
水电工程竣工（截流、蓄水）验收建设征地移民安置综合监理工作报告 Overall supervision of land requisition and resettlement work for hydroelectric project acceptance upon completion (river closure or impoundment)
水电工程竣工（截流、蓄水）验收建设征地移民安置独立评估工作报告 Independent appraisal of land requisition and resettlement work for hydroelectric project acceptance upon completion (river closure and impoundment)
移民安置规划示意图 Sketch of resettlement planning
城市集镇迁建规划总体布置图 General layout of urban and towns resettlement works
环境保护措施总体布置示意图 General layout of environmental protection works
水土保持规划方案图 Schematic drawing for soil and water conservation planning
水库淹没影响示意图 Sketch of reservoir inundation impact
征地范围图 Boundary of land used for project
库区移民安置规划图 Reservoir resettlement planning

第5章 水 工 建 筑 物
Hydraulic structures

5.1 通用
General

(1) 结构可靠性设计 Reliability design

可靠性（可靠度）reliability
可靠性指标 reliability index
水工结构可靠度设计 design of reliability of hydraulic structure
结构目标可靠性指标 structure target reliability index
概率极限状态设计原则 principles of probabilistic theory and limit states design
水工建筑物结构安全级别 safety level of hydraulic structure
水工结构可靠度水平 reliability level of hydrostructure
设计基准期 design reference period
结构抗力 structural resistance
抗力的变异系数 coefficient of variation of resistance
分项系数 partial safety factor
分项系数设计 limit states design with partial coef-ficient
结构重要性系数 coefficient of structure importance，importance factor of structure
设计状况系数 coefficient of design situation
材料性能分项系数 partial factor of material properties
作用分项系数 partial factor for action
作用效应 action effect
结构的权系数 weight coefficient of structure
随机特性 stochastic characteristic
随机变量 stochastic variable
随机变量设计验算点 design checking point of stochastic variable
随机变量当量正态分布 equivalent normal distribution of stochastic variable
抗力分项系数 partial factor for resistance
结构系数 structure partial factor
可靠概率 probability of survival
失效概率 probability of failure
概率分布 probability distribution
概率分布模型 probability distribution model
概率密度函数 probability density function
概率分布函数 probability distribution function
分位值 tantile
标准正态分布概率密度函数 probability mass function of the standard normal distribution
标准正态分布的反函数 inverse function of the standard normal distribution
随机变量的数学期望值 mathematical expectation value of stochastic variable
功能限值 limit value of function
统计参数 statistical parameter
一次二阶矩法 first - order second - moment method
作用的随机特性 stochastic characteristics of actions
结构分析 structural analysis
结构构件 structural member
结构体系 structural system
结构模型 structural model
本构关系 constitutive relation
计算模式不定性 uncertainties of calculation model
长期组合系数 coefficient of long-term combination
正态分布 normal distribution
对数正态分布 logarithmic normal distribution
极值Ⅰ型分布 extreme value type Ⅰ distribution
设计使用年限 design working life，design service life
基本变量 basic variables
极限状态方程 limit state equation
超越概率 exceeding probability
校准法 calibration
定值设计法 deterministic method
容许应力法 permissible (allowable) stress method

单一安全系数法 single safety factor method
安全裕度 safety margin
概率设计法 probabilistic method
反馈设计 feedback design
极限状态 limit states
分项系数极限状态设计 limit states design with partial coefficients
承载能力极限状态 ultimate limit states
正常使用极限状态 serviceability limit states
不可逆正常使用极限状态 irreversible serviceability limit states
可逆正常使用极限状态 reversible serviceability limit states
设计状况 design situations
持久状况 persistent situation
短暂状况 transient situation
偶然状况 accidental situation
地震状况 seismic situation
全生命周期 total life cycle

(2) 荷载及组合 Load and combination

永久作用(荷载) permanent action (load)
可变作用(荷载) variable action (load)
偶然作用 accidental action
可控制的可变荷载 governable variable load
基本组合 basic (fundamental) combination
偶然组合 accidental combination
作用的代表值 representative value of an action
作用的标准值 characteristic value of an action
可变作用的伴随值 accompanying value of a variable action
可变作用的组合值 combination value of a variable action
可变作用的频遇值 frequent value of a variable action
可变作用的准永久值 quasi-permanent value of a variable action
作用的设计值 design value of an action
作用的地震组合 seismic combination
标准组合 characteristic combination, nominal combination
作用的频遇组合 frequent combinations
作用的准永久组合 quasi-permanent combinations
水力系数 hydraulic factor
单荷载系数法 single load factor method
无系数荷载 unfactored load
单个作用 single action
时变效应 time-dependent effect
建筑物自重 dead load of structure
水压力 water pressure
静水压强 hydrostatic pressure intensity
绝对压强 absolute pressure
相对压强 relative pressure
负压 negative pressure
真空度 vacuity
压强水头 pressure head
位置水头 potential head
势能 potential energy
等压面 equi-pressure surface
内水压力 internal water pressure
外水压力 external water pressure
外水压强 external water pressure intensity
外水压力作用水头 acting water head of external water pressure
扬压力 uplift pressure
全水头扬压力 full uplift pressure
浮托力 buoyancy force
渗流压力 seepage pressure
残余扬压力 residual uplift pressure
渗透压力强度系数 coefficient of seepage pressure intensity
扬压力强度系数 coefficient of uplift pressure intensity
残余扬压力强度系数 coefficient of residual uplift pressure intensity
扬压力分布图 distribution diagram of uplift pressure
渗透压力分布图 distribution diagram of seepage pressure
压力梯度 pressure gradient
流场 stream field
流速 flow velocity
流量 flow rate
水力比降(水力坡度) hydraulic gradient (slope)
动水压强 hydrodynamic pressure
时均压力 time-average pressure
脉动压力 fluctuating pressure
水锤压力 water hammer pressure
渐变流时均压力 time-average pressure of gradually varied flow
反弧段水流离心力 flow centrifugal force on

reverse curve section
水流对尾槛的冲击力 impact force of flow on end sill
静水头 hydrostatic head
重力场 gravitational field
垂直均布压力 vertical uniform pressure
水平均布压力 horizontal uniform pressure
土压力 earth pressure
静止土压力 earth pressure at rest，static earth pressure
主动土压力 active earth pressure
被动土压力 passive earth pressure
垂直土压力 vertical earth pressure
侧向土压力 lateral earth pressure
土与结构相互作用 soil-structure interaction（SSI）
围压 confining pressure
持力层 bearing stratum
受力层，压缩层 compression zone
淤沙压力 silt pressure
风荷载 wind load
基本风压 basic wind pressure，reference wind pressure
风振 wind-induced vibration
风振系数 wind vibration coefficient
地面粗糙度 terrain roughness
风压高度变化系数 variation coefficient of wind pressure with height
风荷载体形系数 wind load shape coefficient
雪荷载 snow load
基本雪压 basic snow pressure，reference snow pressure
积雪分布系数 coefficient of snow accumulation distribution
冰压力 ice pressure
冻胀力 frost heaving force（pressure）
静冰压力 static ice pressure
动冰压力 dynamic ice pressure
切线冻胀力 tangential frost heaving force
水平冻胀力 horizontal frost heaving force
竖向冻胀力 vertical frost heaving force
浪压力（波浪力） wave pressure，wave force
剩余浪压力强度 residual wave pressure intensity
波浪反压力强度 wave counter pressure intensity
恒荷载 dead load
活荷载 live load
等效均布活荷载 equivalent uniform live load
可移动的局部荷载 movable partial load
可移动的集中荷载 movable concentrated load
侧向荷载 lateral load
重复荷载 repeated load
双向偏心受压 biaxial eccentric compression
扭转结构 twisted structure
扭转效应 torsional effect
试载扭转分析法 trial-load twist method of analysis
荷载分布 load distribution
荷载工况 load case
运行与检修荷载 operation and maintenance loads
动力系数 dynamic coefficient
线分布力 force per unit length
面分布力 force per unit area
体分布力 force per unit volume
固定作用 fixed action
自由作用 free action
静态作用 static action
动态作用 dynamic action
多次重复作用 repeated action，cyclic action
低频反复作用 low-frequency cyclic action
温度作用 temperature action（effect），thermal action
温度场的变化 variation of temperature field
热膨胀系数 coefficient of thermal expansion
基本气温 reference air temperature
初始温度 initial temperature
边界温度 boundary temperature
环境温度 ambient temperature
地震作用 seismic action
爆炸作用 explosion action
船舶撞击力 ship impact force
船舶挤靠力 ship breasting force
船舶系缆力 mooring force
启门力 lifting force
闭门力 closing force
持住力 holding force
地基反力系数 coefficient of subgrade reaction
浅基础 shallow footing
深基础 deep footing
板式基础 slab footing
筏基础 raft footing
格栅基础 grill footing

箱形基础 box footing
地基承载力的临塑荷载 critical edge load of foundation bearing capacity
地基承载力的临界荷载 critical load of foundation bearing capacity
地基承载力的极限荷载 ultimate load of foundation bearing capacity
地基变形容许值 allowable foundation deformation
地基承载力特征值 characteristic value of foundation bearing capacity
极限承载力 ultimate bearing capacity
允许承载力 allowable bearing capacity
地基刚度 stiffness of foundation

(3) 水工混凝土结构设计 Design of hydraulic concrete structures

混凝土结构 concrete structure
素混凝土结构 plain concrete structure
钢筋混凝土结构 reinforced concrete structure
预应力混凝土结构 prestressed concrete structure
先张法预应力混凝土结构 pretensioned prestressed concrete structure
后张法预应力混凝土结构 post-tensioned prestressed concrete structure
无黏结预应力混凝土结构 unbonded prestressed concrete structure
有黏结预应力混凝土结构 bonded prestressed concrete structure
矩形横截面形状构件 rectangular cross-sectional shape member
最大压缩应变纤维 fiber of maximum compressive strain
水平梁单元 beam element
深梁 deep beam
简支梁 free beam，simple beam
悬臂梁 cantilever
无约束悬臂梁 free cantilever
静定梁 statically determinate beam
超静定梁 statically indeterminate beam
深受弯构件 deep flexural member
相对受压区计算高度 relative calculation depth of compression zone
剪跨比 ratio of shear span to effective depth (shear span to depth ratio)
宽高比 aspect ratio
截面抵抗矩的塑性系数 ratio of plastic moment to elastic moment
弹性模量 elastic modulus
剪切模量 shear modulus
持续弹性模量 sustained modulus of elasticity
瞬时弹性模量 instantaneous modulus of elasticity
混凝土泊松比 Poisson's ratio of concrete
日照辐射热 radiant heat of sunshine
导热系数 thermal conductivity
导温系数，散热系数 thermal diffusivity
线热胀系数 linear expansion coefficient
等效放热系数 equivalent coefficient of heat evolution
混凝土的比热 specific heat of concrete
混凝土拉应力限制系数 coefficient for limiting concrete tensile stress
混凝土局部受压时的强度提高系数 coefficient of strength increase for partially compressed concrete
剪扭构件混凝土受扭承载力降低系数 coefficient of torsion capacity reduction for concrete member in shear and torsion
受拉区混凝土塑性影响系数 plastic influence coefficient of concrete in tension zone
塑性矩 plastic moment
弹性矩 elastic moment
配筋率 reinforcement ratio
体积配筋率 volumetric reinforcement ratio
竖向分布钢筋 vertically distributed bar
水平分布钢筋 horizontally distributed bar
纵向受拉钢筋 longitudinal tensile rebar
轴心受压构件的稳定系数 stability coefficient of axial compressed member
构件承载能力 bearing capacity of member
抗裂 crack resistance
裂缝宽度控制 crack width control
结构力学法 structural mechanics method
弹性力学法 elastic mechanics method
截面内力 sectional internal force
轴心受拉构件 axial tensile member
小偏心受拉构件 tensile member with a small eccentricity
裂缝宽度限值 allowable value of maximum crack width
裂缝控制等级 crack control class
受弯构件的挠度限值 allowable deflection of flexural member

悬臂构件的挠度限值 allowable deflection value of cantilever member
强度设计法 ultimate strength method
工作应力法 working stress method
混凝土强度等级 strength class（grade）of concrete
混凝土轴心抗压强度标准值 characteristic value of concrete axis compressive strength
混凝土轴心抗拉强度标准值 characteristic value of concrete axis tensile strength
混凝土轴心抗压强度设计值 design value of concrete axis compressive strength
混凝土轴心抗拉强度设计值 design value of concrete axis tensile strength
钢筋强度标准值 characteristic value of rebar strength
钢筋强度设计值 design value of rebar strength
线弹性分析 linear-elastic analysis
非线性分析 non-linear analysis
杆件体系 member system
非杆件体系 non-member system
构件刚度 stiffness of structural member
塑性分析 plastic analysis
混凝土受压构件 concrete member in compression
混凝土受弯构件 concrete member in bending
正截面承载力 bearing capacity of normal section
斜截面承载力 bearing capacity of oblique section
受弯承载力 flexural bearing capacity
受压承载力 compression bearing capacity
受扭承载力 torsion bearing capacity
受冲切承载力 punch shear bearing capacity
局部受压承载力 local compression bearing capacity
局部破坏 local damage
正截面抗裂计算 crack-resisting calculation for normal section
正截面裂缝宽度控制 crack width control in normal section
非杆件体系结构裂缝控制 crack control of non-member system
张拉控制应力限值 allowable value of tension stress control
预应力损失 prestress loss
混凝土干缩影响 influence of concrete drying shrinkage
黏结强度 bond strength
握裹力 bond stress
结构缝 structural joint
伸缩缝 expansion joint
沉降缝 settlement joint
施工缝 construction joint
混凝土保护层 concrete cover
钢筋连接 splice reinforcement
预制构件的接头形式 connection type of a precast member
搭接接头 lapped splices
对接接头 butt splices
机械对接 mechanical butt splicing
单向连续板 one-way continuous slab
多跨单向板 one-way slab with multispans
加肋板 ribbed slab
双向平板，四边支承楼板 two-way flat slab
锚定梁 anchor beam
箱梁 box beam
固端梁 built-in beam，fixed end beam
组合梁 built-up beam，compound beam
共轭梁 conjugate beam
框架梁 frame beam
框架横梁 spanning member
暗梁 concealed beam
薄腹梁 thin-webbed beam
轴心受压柱 axial compression column
轴向受力柱 axial loaded column
框架柱 frame column
桁架拱 truss arch
双柱式墩 queen-post supporting pier
排架桩墩 bent pile pier
板中弯起钢筋的弯起角 bent-up angle of bent-up rebars in the slab
分布钢筋 distribution rebar
现浇板 cast-in-situ concrete slab
受扭构件 torsion member
抗扭箍筋 torsion stirrup
偏心受压构件 eccentric compressed member
连续梁中间支座 intermediate support of continuous beam
水平梁单元的支座 abutment of a beam element
框架梁中间节点 intermediate nodes of frame beam
剪力墙 shear wall
叠合式受弯构件 composite flexural member
柱牛腿（独立牛腿） bracket

墙牛腿（壁式牛腿） corbel
弧形闸门支座 support of radial gate
预应力混凝土闸墩 prestressed concrete pier
钢筋混凝土蜗壳 reinforced concrete spiral case
钢筋混凝土尾水管 reinforced concrete draft tube
钢筋混凝土框架 reinforced concrete frame
框架梁柱节点 beam-column nodes in frame
铰接排架柱 hinged bent column
排架 bent frame
柱下独立基础 independent footing under column
桥跨结构 bridge span structure
受拉力控制截面 section controlled by tension
受压力控制截面 section controlled by compression
混凝土收缩 shrinkage of concrete
混凝土徐变 creep of concrete
混凝土碳化 carbonation of concrete
二期混凝土 phase Ⅱ concrete
轴向力 normal force，axial force
正应力 normal stress
主应力 principal stress
次应力 secondary stress
剪应力 shear stress；tangential stress
抗剪强度 shear strength
抗剪阻力 shear resistance
剪胀性 dilatancy
侧向挠度 lateral deflection
侧向收缩 lateral contraction
侧向压力 lateral pressure
侧向应变 lateral strain
重复应力 repeated stress
附加扰力 additional disturbing force
附加应力（叠加应力） superimposed stress
局部应力 local stress
压曲临界荷载 buckling load
压曲稳定性 buckling stability
压实系数 compaction coefficient
永久变形 permanent deformation
整体稳定 overall stability
轴向荷载 axial load
轴心受压承载力 load-carrying capacity in compression
侧向稳定 lateral stability
附加荷载 additional load
循环应力 cycle stress
预应力 prestress
外加变形 imposed deformation
约束变形 restrained deformation
线应变 linear strain
剪应变 shear strain，tangential strain
主应变 principal strain
极限应变 ultimate strain
屈服强度 yield strength
疲劳强度 fatigue strength
侧移刚度 lateral displacement stiffness
结构的整体稳固性 structural integrity，structural robustness
连续倒塌 progressive collapse
极限变形 ultimate deformation
空间工作性能 spatial behavior
脆性破坏 brittle failure
延性破坏 ductile failure
计算高度 effective height
净高 net height
截面面积矩 first moment of area of section
截面惯性矩 second moment of area of section，moment of inertia of section
截面极惯性矩 polar second moment of area of section，polar moment of inertia of section
截面模量 sectional modulus
截面回转半径 radius of gyration of section
偏心距 eccentricity
偏心率 relative eccentricity
净跨度 net span
矢高 rise
长细比 slenderness ratio
钢筋间距 spacing of bars
箍筋肢距 spacing of stirrup legs
环境作用 environmental action
室内正常环境 normal indoor environment
室内潮湿环境 indoor damp environment
露天环境 outdoor environment
长期水下环境 permanently under water environment
弱腐蚀环境 weak corrosive environment
中等腐蚀环境 moderate erosive environment
强腐蚀环境 severe erosive environment
盐雾作用区 salty fog acting zone
海水浪溅区 seawater spraying zone
水位变动区 water level fluctuating zone
劣化 degradation

劣化模型 degradation model
结构耐久性 structure durability
氯离子在混凝土中的扩散系数 chloride diffusion coefficient of concrete
混凝土抗冻耐久性指数 durability factor of concrete (DF)
可修复性 restorability

(4) 水工建筑物抗震设计 Aseismatic design of hydraulic structures

基本烈度 basic seismic intensity
设计烈度 design seismic intensity
抗震设防标准 seismic fortification criterion
地震烈度 seismic intensity
抗震设防要求 seismic precautionary requirement
抗震设防烈度 seismic fortification intensity
抗震设防区 seismic precautionary zone
综合抗震能力 comprehensive aseismic capability
抗震安全性 aseismatic safety
地震动 ground motion
地震动峰值加速度 ground motion peak acceleration
设计地震加速度 design seismic acceleration
地震作用效应 seismic action effect
地震液化 earthquake-induced liquefaction
设计反应谱 design response spectra
地表加速度峰值 peak ground acceleration (PGA)
最大可信地震 maximum credible earthquake (MCE)
运行基本地震 operating basis earthquake (OBE)
最大设计地震 maximum design earthquake (MDE)
震动卓越周期 predominant period of vibration
标准地面运动分析 standard ground motion analysis
地震的方向分量 earthquake directional components
动力法 dynamic method
拟静力法 pseudo-static method
时程分析法 time history method
振型分解法 mode-superposition method
平方和方根法 square root of sum square method
完全二次型方根法 complete quadric combination method
地震动水压力 seismic water pressure
地震动土压力 seismic earth pressure
地震作用的效应折减系数 reduction coefficient of seismic action effect
特征周期 characteristic period
振动 vibration
振幅 amplitude of vibration
振型 mode of vibration
共振 resonance
阻尼 damp
重力加速度 acceleration of gravity
频率 frequency
自振频率，固有频率 natural frequency
自振周期 natural period of vibration
自由度 degree of freedom
土层剪切波速 shear-wave velocity of soil layer
整体剪切破坏 general-shear failure
可液化土层 potential liquefaction soil layer
水平向地震作用 horizontal seismic action
竖向地震作用 vertical seismic action
地震惯性力 seismic inertia force
场地类别 site category
坝面附加质量 added mass at dam face
抗震措施 aseismatic measure
抗震计算 aseismatic checking
防震缝 aseismatic joint
设计地震动 design ground motion
多遇地震 frequently occurred earthquake, low-level earthquake
设防地震 precautionary earthquake
罕遇地震 rare earthquake, high-level earthquake
运行安全地震动 operational safety ground motion
极限安全地震动 ultimate safety ground motion
设计基本地震加速度 design basic acceleration of ground motion
地震影响系数曲线 seismic effect coefficient curve
设计地震动特征周期 design characteristic period of ground motion
基本周期 fundamental period
结构动力特性 dynamic properties of structure
抗震等级 seismic grade
抗震概念设计 seismic concept design
抗震构造措施 structural details of seismic design
结构抗震性能 earthquake resistant behavior of structure
基本振型 fundamental mode of vibration
结构影响系数 influential coefficient of structure
位移放大系数 displacement magnification factor
位移延性系数 displacement ductility ratio
内力调整系数 adjustment coefficient of internal

force
弹性抗震设计　seismic elasticity design
延性抗震设计　seismic ductility design
基于性能的抗震设计　performance-based seismic design
基于位移的抗震设计　displacement-based seismic design
基于能量的抗震设计　energy-based seismic design
抗震结构整体性　integral behavior of aseismatic structure
结构振动控制　structural vibration control
消能减震　energy dissipation and earthquake response reduction
阻尼器　damper
隔震　seismic isolation
隔震装置　isolation device
抗震鉴定　seismic appraisal
抗震加固　seismic strengthening
溃坝分析　dam break study
增量危害评估　incremental hazard evaluation
大坝失事后果　consequences of dam failure
潜在的下游危害　downstream hazard potential
潜在破坏面　potential failure surface
临界破坏面　critical failure surface
抗倾覆稳定　stability against overturning
抗倾覆验算　overturning resistance analysis
倾覆力矩　overturning moment
抵抗力矩　resisting moment
简化楔体　simplified wedge method
疲劳验算　fatigue analysis
稳定计算　stability calculation
滑移验算　slip resistance analysis
变形验算　deformation analysis
工程类比、经验类比　project analogue

(5) 水工建筑物抗冰冻设计　Design of hydraulic structure against ice and freezing

最大冻土深度　maximum frozen soil depth
季节冻土　seasonally frozen ground
季节冻结深度　depth of seasonal freezing
地基土设计冻深　design freezing depth of foundation
封冻　complete freezing
冻结指数　freezing index
冻胀量　amount of frost-heaving
地表冻胀量　amount of frost-heaving of ground surface
冰情　ice regime
冻害　frost damage
冰盖　ice cover
冰坝　ice dam
武开江　ice breakup due to hydraulic and climatic effect

5.2 挡水建筑物
Water retaining structure

(1) 混凝土坝　Concrete dam

大体积混凝土坝　mass concrete dam
重力坝　gravity dam
混凝土实体重力坝　concrete solid gravity dam
混凝土空腹重力坝　concrete hollow gravity dam
混凝土宽缝重力坝　concrete slotted gravity dam
碾压混凝土坝　roller compacted concrete dam (RCC dam)
基础约束系数　restraint coefficient of foundation
混凝土徐变引起的应力松弛　stress relaxation caused by concrete creepage
整体水工模型试验　monolithic hydraulic model test
断面水工模型试验　sectional hydraulic model test
坝体结构　dam structure
坝段　dam section
溢流坝段　overflow section
挡水坝段　non-overflow section，water retaining section
廊道　gallery
防渗和止水系统　seepage control and water stop system
坝顶超高　freeboard
胸墙　breast wall，parapet
防浪墙　wave wall
坝踵　dam heel
坝趾　dam toe
坝肩　abutment
坝块　monolith，dam block
横缝　transverse joint
纵缝　longitudinal joint
斜缝　inclined joint
错缝　staggered joint

(2) 混凝土拱坝 Concrete arch dam

拱坝 arch dam
单曲拱坝 single-curvature arch dam
双曲拱坝 double-curvature arch dam
薄拱坝 thin arch dam
中厚拱坝 medium-thick arch dam
厚拱坝 thick arch dam
空腹重力拱坝 hollow gravity arch dam
重力拱坝 gravity arch dam
三中心变厚度拱坝 three-centered variable thickness arch dam
变中心拱坝 variable centre arch dam
变半径拱坝 variable radius arch dam
定圆心拱坝 constant-center arch dam
定中心角拱坝 constant-angle arch dam
单心拱 single-centered arch
双心拱 two-centered arch
三心拱 three-centered arch
对数螺线型拱坝 logarithmic spiral arch dam
厚高比 ratio of thickness to height
拱坝轴线 axis of arch dam
拱坝体型 arch dam configuration
拱圈线型 arch shape
拱座 abutment pad
拱圈中心角 central angle of an arch
拱坝竖向曲率 vertical curvature of the arch dam
拱圈分块（楔块） voussoir
贴角［拱坝］ fillet
外弧面，拱背 extrados
内弧面，拱腹 intrados
倒悬度 overhanging degree
变厚度拱 variable-thickness arches
拱端局部加厚度拱 locally-thickened arches at both abutments
推力墩 thrust block
重力墩 gravity block
翼坝 aliform dam
垫座 concrete socket
倒悬 overhang
下切 undercutting
恒定中心 constant center
应力控制指标 stress control index
应力等值线 stress contour
封拱 closure of arch
封拱温度 closure temperature
拱冠梁 crown cantilever
拱梁分载法 trial-load method
拱梁网格体系 grid system consisting of a series of arches and cantilever units
拱坝整体安全性 integral safety of the arch dam
拱坝弧高比 ratio of arch crest length to height (crest length-height ratio)
等效线性温差 equivalent linear temperature difference
综合变形模量 comprehensive deformation modulus

(3) 土石坝 Earth-rock fill dam

土坝 earth dam
均质土坝 homogeneous earth dam
非均质土坝 non-homogeneous earth dam，zoned earth dam
堆石坝 rockfill dam
石渣坝 rock debris dam
填筑坝（土石坝） earth-rock fill dam，embankment dam
碾压式土石坝 rolled earth-rock fill dam
土质心墙堆石坝 earth core rockfill dam
黏土心墙土石坝 earth-rock fill dam with clay core
黏土斜墙土石坝 earth-rock fill dam with sloping clay core
直心墙堆石坝 vertical core rock fill dam
土质防渗体分区坝 zoned earth rockfill dam with impervious soil core
非土质材料防渗体分区坝 zoned earth rockfill dam with non-soil impervious core
混凝土面板堆石坝 concrete face rockfill dam
坝顶 dam crest
坝底 dam base
坝坡 dam slope
马道，戗台 berm
坝顶超填 camber
填筑分区 embankment zoning
防渗层 impervious layer
反滤料区 filter zone
堆石区 rockfill zone
任意堆石区 randomfill zone
过渡区 transition zone
排水区 drainage zone
级配反滤料铺盖 weighted graded filter

坝壳　dam shell
防渗斜墙　diaphragm
填筑碾压标准　embankment rolling criteria
不连续级配土　gap graded soil
颗粒级配良好　well-distributed gradation
干容重　dry unit weight
筑坝料的含水量　water content of embankment material
天然含水量（含水率）　natural water content
最优含水量（含水率）　optimum water content
堆石料的孔隙率　porosity of rockfill material
允许渗透坡降　allowable seepage gradient
原始浸润面　original phreatic surface
浸润线　phreatic line，seepage line
流网　flow net
流线　stream line
内部侵蚀　internal erosion
土质防渗体　soil impervious zone
心墙　core wall
铺盖　blanket
截水槽，齿槽　cutoff trench
反滤准则　filter criteria
渗透破坏　seepage failure
坝体排水　drainage of embankment
垂直排水，烟囱式排水　chimney drain
柱状排水［土石坝内］　pillar drain
坝趾排水　toe drain
上游防渗铺盖　upstream impervious blanket
垂直防渗措施　vertical impervious measure
水平排水垫层　horizontal drainage cushion
反滤排水沟　filter drainage ditch
减压井　relief well
下游透水铺盖　downstream permeable weight blanket
渗流计算　seepage calculation
坝坡出逸段　embankment escape area
出逸坡降　escape gradient
渗透稳定计算　seepage stability calculation
防止管涌　protection against piping
堆石坝体　embankment
垫层区　cushion zone
特殊垫层区　special cushion zone
主堆石区　main rockfill zone
上游堆石区　upstream rockfill zone
下游堆石区　downstream rockfill zone
抛石区　riprap zone
盖重区　weighted cover zone
混凝土面板　concrete face slab
趾板　plinth，toe slab
平趾板　flat plinth
趾墙　toe wall
填补块［面板坝］　fillet
下游混凝土防渗板　downstream concrete slab connected with the plinth.
周边缝　periphery joint
垂直缝　vertical joint
面板接缝　face joint
一期面板　first-stage facing
二期面板　second-stage facing
面板脱空　separation between concrete slab and cushion
柔性填料　flexible filler
增模区　increased-modulus zone
护栏　guard rail
路缘石　curb stones
透水性　permeability
低压缩性　low compressibility
水平反滤层　horizontal filter
允许水力梯度　allowable hydraulic gradient
起始水力坡降　beginning hydraulic gradient
贯穿性裂缝　through crack
开级配沥青混凝土　open-graded asphalt concrete
密级配沥青混凝土　dense-graded asphalt concrete
沥青混凝土面板　asphalt concrete facing
沥青混凝土心墙　asphalt concrete core
简式断面　simplified section
复式断面　composite section

(4) 其他坝型　Other dam types

支墩坝　buttress dam
平板坝　flat slab dam
大头坝　solid-head buttress dam
连拱坝　multiple arch dam
圬工坝　masonry dam
毛石圬工坝　rubble masonry dam
橡胶坝　rubber dam，inflatable dam
干砌石坝　loose-rock dam
石笼坝　gabion dam
木笼坝　crib dam
木笼填石坝　rock-crib dam

翻板坝 shutter dam
水力冲填坝 hydraulic fill dam
副坝 secondary dam (saddle dam)

5.3 泄水建筑物
Water release structures (outlet structures)

(1) 溢洪道与泄水隧洞 Spillway and discharge tunnel

正常（常用）溢洪道 service spillway
非常溢洪道 emergency spillway
岸边溢洪道 river-bank spillway
侧槽式溢洪道 side channel spillway
斜（陡）槽式溢洪道 chute spillway
滑雪道式溢洪道 ski-jump spillway
开敞式溢洪道 open spillway
坝顶溢洪道 crest spillway
顶部溢流式溢洪道 overflow spillway
分离式溢洪道 detached spillway
反弧面溢洪道 ogee spillway
竖井跌水式溢洪道 drop-inlet spillway
有闸门控制的溢洪道 gated spillway
无闸门控制的溢洪道 ungated spillway
自溃式非常溢洪道 fuse-plug spillway
竖井式溢洪道 shaft spillway, glory hole spillway
迷宫式溢洪道 labyrinth spillway
台阶式溢洪道 stepped spillway
辅助溢洪道 auxiliary spillway
坝体溢流段 overflow dam section
非溢流坝段 non-overflow dam section
坝顶溢流 crest overflow
坝面泄流 discharging along the downstream dam face
坝后厂房顶溢流 discharging through a spillway on the powerhouse roof at dam toe
泄水闸室 sluice chamber
泄水闸板 sluice board
闸槛 ground sill
中墩 intermed pier
边墩 abutment pier
溢洪道闸墩 spillway pier
悬臂挑台 cantilevered extension
泄槽 chute
泄槽的纵坡 longitudinal slope of the chute
消能防冲设施 energy dissipation facility
出水渠 outlet channel
下泄水流的流态和雾化 discharging flow pattern and atomization
河道的冲淤 scour-and-deposition in the river channel
闸门控制的堰 gated weir
无闸门堰 ungated weir
开敞式堰 open weir
带胸墙的实用堰 practical weir with breast wall
宽顶堰 broad-crested weir
驼峰堰 hump-shaped weir
斜背堰 inclined weir
平顶堰 flat-topped weir
长顶堰 long crested weir
锐缘堰 sharp-crested weir
圆顶堰 round crested weir
泄水堰 sluice weir
出流堰 effluent weir
自由溢流堰 free fall weir
水力设计 hydraulic design
溢流堰的流量系数 discharge coefficient of overflow weir
开敞式堰面曲线 open weir surface curve
溢流坝堰面曲线 spillway surface curve
坝身孔口泄流 flow discharge through dam outlet
坝身泄水孔 outlet through dam
泄水底孔 bottom outlet
泄水深孔 deep outlet
泄水中孔 middle outlet
泄水表孔 surface outlet
排沙孔 sediment flushing outlet
下泄流量 outflow discharge
掺气 aeration
掺气设施 aerating facility
掺气槽 aeration slot
掺气水流 aerated flow
水力要素 hydraulic elements
水面线 water surface profile
水工隧洞 hydraulic tunnel
泄洪隧洞 spillway tunnel
放水隧洞 relief tunnel
龙抬头泄洪洞 inlet-raised discharge tunnel
冲沙隧洞 sluice tunnel, flushing tunnel
导流隧洞，引水隧洞 diversion tunnel

过水隧洞　waterway tunnel
有压隧洞　pressure tunnel
无压隧洞　free-flow tunnel
高压隧洞　high pressure tunnel
放空洞　emptying tunnel
洞式溢洪道（溢洪洞）　tunnel spillway

(2) 消能建筑物（消能工） Dissipator

联合消能　combined energy dissipation
底流消能　bottom flow energy dissipation
面流消能　surface flow energy dissipation, roller bucket type energy dissipation
挑流消能　ski-jump energy dissipation
洞内消能　inside-tunnel energy dissipation
淹没于水下的消能工　submerged energy dissipator
辅助消能工　auxiliary energy dissipator
戽斗式消能工　bucket-type energy dissipator
冲击式消能工　impact-type energy dissipator
滑雪道式消能工　ski-jump energy dissipator
跌水消能工　drop energy dissipator
挑流鼻坎　deflecting bucket, flip bucket
扭曲式挑坎　distorted type flip bucket, torsioned flip bucket
窄缝式挑坎　slit-type flip bucket
上挑鼻坎　upturned deflector
齿槽式挑流鼻坎，差动式鼻坎　slotted spillway bucket
溢流堰反弧鼻坎　upcurved spillway bucket
连续挑坎　continuous flip bucket
异形挑坎　special-shape flip bucket
出口消力戽　outlet bucket
扩散式消力戽　diffusion bucket
宽尾墩　end-flared pier
陡槽消力墩，分流墩　chute block
消力槛　baffle sill
消力墩　baffle block
水舌挑距　trajectory distance of nappe
冲坑水垫厚度　water cushion depth of plunge pool
波动或掺气后水深　water depth in fluctuation or after aeration
淹没系数　submersion coefficient
水垫塘　plunge pool
消力池　stilling basin
水跃消力池　hydraulic jump basin
消力池底板　stilling-basin slab
扩散与束流墙　spray walls and training wall
厂前挑流　spillway with the flip bucket in front of the powerhouse at dam toe
折冲水流　deflected current
水跃　hydraulic jump
共轭水深　conjugate depth
挑流水舌　trajectory nappe
贴附水舌　adhering nappe
掺气水舌　aerated nappe
淹没水舌　drowned nappe
自由水舌　free nappe
跌水　hydraulic drop
空蚀破坏　cavitation damage
高速水流　high-velocity flow
防空蚀措施　cavitation prevention measure
流速场　velocity field
流速脉动　velocity fluctuation
水流空化数　flowing cavitation number
初生空化数　inception number of cavitation
临界流　critical flow
超临界流，急流　supercritical flow, rapid flow
亚临界流，缓流　subcritical flow, tranquil flow
临界水深　critical water depth
河床抗冲能力　riverbed erosion-resistant capacity
下游水位衔接　downstream flow connection
冲坑水垫深度及范围　depth and extent of water cushion at the scour pit
消力池护坦抗浮稳定性　stability against floatation of stilling basin apron
水跃淹没度　submergence degree of hydraulic jump
空化　cavitation
防空蚀设计　cavitation prevention design
高速水流区　high-velocity flow area
泄洪雾化　atomization during flood discharging
闸墩墩头　pier nose
尾槛　end sill
护坦（海漫）　apron
防淘齿墙　scour prevention key wall
二道坝　auxiliary weir
防冲槽　scour-resistant slot
导水墙　guide wall
丁坝　groin
掺气减蚀模型试验　model test for aeration cavitation resistance

5.4 输水建筑物
Conveyance structures

水电站引水渠 headrace canal
前池 forebay
调节池 regulating pond
自动调节渠道 self-regulating canal
非自动调节渠道 non-self-regulating canal
进水渠 approach channel
水电站进水口 intake
开敞式进水口 open inlet
有压式进水口 pressure inlet
坝式进水口 intake integrated with the dam
塔式进水口 intake tower
岸塔式进水口 intake tower against the bank
喇叭形进水口 bell mouth inlet
闸门竖井式进水口 intake with gate shaft
岸坡式进水口 intake with inclined gate slots in the bank
多层进水口（分层取水口） multi-level intake
侧式进/出水口 side intake/outlet
竖井式进/出水口 shaft intake/outlet
独立式进水口结构 free-standing intake structure
倾斜式进水口结构 inclined intake structure
矩形进水口结构 rectangular intake structure
圆形进水口结构 circular intake structure
不规则进水口结构 irregular intake structure
低水位泄洪进水口 low-level drawdown intake
倾斜式进水塔 inclined intake tower
贯通式漏斗漩涡 through funnel vortex
消涡 vortex elimination
淹没深度 inundation depth
通气孔 air vent
回流区 backflow area
进水口底板高程 intake bottom elevation
防沙措施 sediment control measure
防污措施 trash control measure
防冰措施 preventive measure against icing
进水口平台 intake platform
拦污栅 trashrack
淹没式拦污栅 submerged trashrack
粗粒径含沙量 coarse sediment concentration
电站输水道 plant waterway
输水系统 conveyance system
水道 conduit
发电隧洞 power tunnel
引水隧洞 headrace tunnel
输水隧洞 convey tunnel, delivery tunnel
尾水隧洞 tailrace tunnel
尾水渠 tailrace channel
隧洞形状 shape of tunnel
圆形断面 circular cross section
圆拱直墙式，城门洞形 inverted U-shaped section
马蹄形断面 horseshoe section
异形断面 odd-shaped
隧洞上平段 upper horizontal section
隧洞下平段 lower horizontal section
拱冠 crown
拱肩 spandrel
拱脚 springer
坝基廊道 foundation gallery
排水廊道 drainage gallery
闸门廊道 gate gallery
灌浆廊道 grouting gallery
电缆廊道 cable gallery
检查廊道 inspection gallery
调压井 surge shaft
阻抗式调压室 restricted orifice surge chamber
圆筒式调压室 cylindrical surge chamber
双室式调压井 double-chamber surge shaft
差动式调压室 differential surge chamber
气垫式调压室 air cushion surge chamber
溢流式调压室 overflow surge chamber
水室式调压室 water-chamber-type surge chamber
尾水调压室 tailrace surge chamber
最高涌浪水位 highest surge level
最低涌浪水位 lowest surge level
波动稳定断面 cross-section area of oscillating stability
过水断面 wetted section
过流能力 discharge capacity
封堵体 plug
折流板 baffle plate
明渠流 open channel flow
满管流 full conduit flow
稳定非均匀流 steady non-uniform flow
总水头 total head
压力水头 pressure head
水头损失 head loss
沿程水头损失 linear head loss

局部水头损失 local head loss
弯管段损失[水头] bend loss
输水系统水工模型试验 hydraulic model test for water-delivery system
压力钢管 steel penstock
明管 exposed penstock
埋管 buried penstock
自由式压力钢管 free penstock
半固定式压力钢管 semi-fixed penstock
坝内埋管 embedded penstock within dam
坝后背管 penstock laid on downstream face of dam
钢衬钢筋混凝土管 steel lined reinforced concrete penstock
支管 branch pipe
两岔管 bifurcated pipe
分岔管 manifold penstock
三梁岔管 three-girder reinforced bifurcation
内加强月牙肋岔管 crescent-rid reinforced bifurcation
月牙加劲板 crescent stiffener
球形岔管 spherical branch pipe
无梁岔管 shell type branch pipe
贴边岔管 hem reinforced branch pipe
钢管圆度偏差 penstock roundness tolerance
压力钢管伸缩节 penstock expansion joint
波纹伸缩节 bellow expansion joint
套筒伸缩节 sleeve expansion joint
钢管凑合节 adjustor of steel pipe
镇墩 anchorage block
支墩 supporting pier
鞍形支座 saddle support
平面滑动支座 sliding ring girder support
滚动支座 rolling ring girder support
摇摆支座 rocking ring girder support
膜应力 film stress
抗外压稳定临界压力 critical external compressive resistance of buckling
加劲环 stiffener ring
支撑环 supporting ring
阻水环 cut-off collar
止推环 thrust collar
凑合节 compensating joint
管段 pipe segment (section)
管壁等效翼缘宽度 equivalent flange width of pipe shell
软垫层 soft cushion
预留环缝 preformed circumferential welding seam
水压试验 hydrostatic pressure test
防腐蚀措施 corrosion proof measure
分岔角 fork angle
腰线 frieze
转折角 turning angle
偏转角 angle of band, deflection angle
转角 outer corner
相贯线 intersection line
锥管母线 generating line of conic tube

5.5 发电厂房
Powerhouse

水电站厂房 hydroelectric powerhouse
河床式厂房 water retaining powerhouse
坝后式厂房 powerhouse at dam toe
引水式厂房 conduit-type powerhouse
岸边式厂房 powerhouse on river bank
坝内式厂房 powerhouse within dam
地下式厂房 underground powerhouse
半地下式厂房 semi-underground powerhouse
露天式厂房 outdoor powerhouse
半露天式厂房 semi-outdoor powerhouse
窑洞式厂房 cavern powerhouse
闸墩式厂房 pier-head powerhouse
挑流式厂房 flyover type powerhouse
溢流式厂房 overflow type powerhouse
主厂房 main powerhouse, machine hall
副厂房 auxiliary plant, service building
主机间 generator hall
机组段 unit bay
机墩 turbine pier
圆筒式机墩 cylinder pier of turbine
矮式机墩 short pier of turbine
墙式机墩 wall-type pier of turbine
安装间 erection bay
上部结构 superstructure
下部结构 substructure
中央控制室 central control room
发电机层 generator floor
水轮机层 turbine floor
蜗壳层 spiral casing floor
尾水管层 draft tube floor
渗漏排水井 leakage water dewatering pit

尾水平台　tailrace platform

开关站　switchyard

露天开关站　open switchyard

主变压器场　main transformer yard

厂房抗浮稳定计算　stability against floating calculation of powerhouse

厂房地基应力计算　foundation stress calculation of powerhouse

5.6　基础及边坡处理

Treatment of foundation and slope

工程处理措施　engineering measure

应急措施　emergency measure

增稳措施　stabilizing measure

常规处理　conventional treatment

常规检查　routine inspection

建基面　foundation surface (interface)

砂砾石地基　sand and gravel base

坝基开挖　excavation of dam foundation

基岩与覆盖层分界线　boundary of bedrock and overburden

断层破碎带处理　treatment of fault and fracture zone

喀斯特处理　karst treatment

软基处理　treatment of soft foundation

桩基础　pile foundation

沉井基础　open caisson foundation

真空排水预压加固软基　soft foundation reinforced by vacuum drainage preloading

加固　reinforcement

回填　backfill

置换　replacement

振冲　vibroflotation

打桩　piling

灌浆　grouting

锚固　anchorage

衬砌　lining

支撑　supporting

卸载　unloading

防冲措施　erosion control measure

抗冲性能（抗侵蚀性能）　erosion-resisting performance

抗冲刷性　scour resistance

抗腐蚀性　corrosion resistance

坝基面渗透压力　seepage pressure on foundation surface

坝基渗流量　seepage flow in foundation

坝基排水孔　drainage holes in dam foundation

防渗帷幕　impervious curtain

防渗铺盖　impervious blanket

防渗心墙　impervious core

防渗墙　cutoff wall, watertight diaphragm

地下连续墙　underground diaphragm wall

防水层　sealing layer, waterproof layer

保护层　protective layer

工程边坡　cut slope, engineered slope

坝坡　dam slope

库岸　reservoir bank

下游坡（背水坡）　downstream slope

上游坡(迎水坡)　upstream slope

河岸内坡　inside bank slope, landside slope

河岸外坡　outside bank slope, river side slope

底切岸坡，淘蚀岸坡　undercut slope

半挖半填的边坡　cut and fill slope

开挖坡比　excavation gradient, gradient of cutting

护坡　slope protection

堆石护坡　rockfill slope protection

抛石护坡　riprap slope protection

削坡减载　cutting slope and unloading

拦石网　protecting wire mesh

压脚　loading at foot

削头压脚　cutting head and loading at foot

抗剪洞　shear resisting plug

锚固洞　retaining concrete plug

设计锚固力　design anchoring

系统锚固　systematic anchoring, pattern anchoring

预应力锚固　prestressed anchorage

格构锚固　anchored framework

钢筋网　reinforcing mesh

喷混凝土　shotcrete

排水孔　drain hole

岩锚支护　rock bolting

土锚杆支护　soil nailing

土钉墙　soil nail wall

悬臂桩　cantilever sheet-pile (pile)

抗滑桩　slide-resisting pile

拉索桩　cable-stay pile

随机锚杆　spot bolting

系统锚杆　pattern bolting

超前锚杆　forepoling bolt

预应力锚杆 prestressed anchor
永久性预应力锚杆 permanent prestressed anchor
临时性预应力锚杆 temporary prestressed anchor
有黏结预应力锚杆 bonded prestressed anchor
无黏结预应力锚杆 unbounded prestressed anchor
锁定荷载（锁定吨位） lock-in load
永存荷载(永存吨位) eternal tensile load
损失吨位 decreased load
张拉 tensioning
设计张拉力 design tension
超张拉力 extra design tension
预张拉 pretension
补偿张拉 compensatory tension
抗拔力 pull-out force, pull-out resistance
延伸率，伸长率 elongation
回缩量 retraction range
内缩量 drawn-in
张拉段 tensile section
锚固段 anchored section
内锚固段 inner anchored section
预应力钢筋束 prestressing tendon
拉力型锚索 tensioned grout tendon
拉力分散型锚索 tensioned multiple-head tendon
拉压复合型锚索 tension-compression combined tendon
压力集中型锚索 compression-concentration tendon
压力分散型锚索 compression-dispersion tendon

5.7 观测（监测） Observation (Monitoring)

系统观测 systematic observation
安全监测 safety monitoring
原型观测 prototype observation
工况观测 conditional observation
控制观测 control observation
同步观测 simultaneous observation
监测系统 monitoring system
观测网 observation network
观测系列 observation series
表面监测 surface monitoring
内部观测 internal monitoring
外部观测 external monitoring
流动监测 mobile monitoring
结构性能观测 structural behavior measurement
目测 eye observation
水位基点，测站基面 gauge-datum
仪器观测 instrumentation
精密测量观测 precise surveying measurement
长期观测 long-term observation
常规观测 routine observation
首次蓄水期 first impound period
初蓄期 initial impound period
变形监测 deformation observation
位移监测 displacement observation
渗透压力监测 seepage pressure monitoring
渗流量监测 seepage flow monitoring
应力应变监测 stress-strain monitoring
滑坡监测 landslide monitoring
收敛变形 convergent deformation
监测仪表与自动化元件 measuring instrument and automatic element
埋设仪器系统 embedded instrument system
位移传感器 displacement sensor
三向位移计 three-way displacement meter
多点位移计 multipoint displacement meter
基岩变形计 bedrock deformeter
钻孔测斜仪 bore-hole inclinometer
倾角计 inclination transducer
滑动测微计 sliding micrometer
静力水准系统 static leveling system
水管式沉降仪 water level settlement gauge
套筒式沉降仪 telescopic settlement gauge
环式渗透仪 ring infiltrometer
孔隙压力计（盒） pore pressure meter (cell)
测压管 piezometer
伸长计，应变计 extensometer
双向应变计组 two-way extensometer group
三向应变计组 three-way extensometer group
弦式应变计 wire strain gauge
夹式引伸计 clip typed extension meter
无应力计 non-stress meter
“无应力”应变计 “no-stress” strain meter
钢筋计 reinforcement stress meter
锚杆应力计 rock bolt stress meter
锚索（杆）测力计 anchorage cable (rock bolt) dynamometer
土压力计 soil pressure gauge
脉动压力计 pulse pressure gauge

速度计 speed gauge
加速度计 accelerometer
水听器 hydrophone
钢弦式传感器 string transducer
差动电阻式传感器 differential resistive transducer
电感式传感器 inductive transducer
压阻式传感器 piezoresistive transducer
电容式传感器 capacitive transducer
电位器式传感器 potentiometric transducer
热电偶式传感器 thermocouple transducer
光纤光栅传感器 optical fiber grating transducer
磁致伸缩式传感器 magnetostrictive transducer
伺服加速度式测斜仪 servo acceleration inclinometer
电解质式测斜仪 electrolysis inclinometer
测缝计 joint gauge
双向测缝计 two-way joint meter
三向测缝计 three-way joint meter
塞尺 feeler gauge
电阻式温度计 resistance thermometer
热电偶温度计 thermocouples
分布式光纤温度传感器 optical fiber distributed temperature transducer
激光准直系统 laser alignment system
倒铅垂线 inverse plummet
正铅垂线 right plummet
铅垂线竖井 plumbline well
引张线仪 wire alignment transducer
观测墩 instrument pier
殷钢尺 invar tape
活动觇标 movable target
固定觇标 stationary target
基准觇标 reference sighting target
坝外基准点 off-dam reference
观测站 reading station
三角觇标系统 triangulation target system
惠斯登电桥测试装置 Wheatstone bridge test set
流量计 flow meter
压差流量计 differential pressure flow meter
堰顶水位计 weir gauge
量水堰 measuring weir
帕氏量水槽 Parshall flume
监测断面 monitoring section
历时系列 duration series
起始数据 initial data
初始读数 initial reading
本底值 background value
计算基准值 fiducial value
蓄水基准值 fiducial value before first impound
设计警戒值 design threshold
年变化 annual variation
年变幅 annual amplitude
既有结构 existing structure
趋势分析 trend analysis
模型分析 model analysis
观测误差 observation error
偶然误差 accidental error
零点漂移 zero shift
灵敏度漂移 sensitivity shift
分辨率 resolution
阈值 threshold

5.8 其他 Miscellaneous

(1) 抽水蓄能及潮汐电站 Pumped storage power station and tidal power station

抽水蓄能电站 pumped storage power station
混合式抽水蓄能电站 mixed pumped storage power station
纯抽水蓄能电站 pure pumped storage power station
日调节水库 daily regulating reservoir
周调节水库 weekly regulating reservoir
综合循环效率 comprehensive cycle efficiency
上水库 upper reservoir
下水库 lower reservoir
进/出水口 intake/outlet
库盆防渗设计 reservoir basin seepage control design
上水库充水方式 impounding mode for upper reservoir
机组发电工况黑启动 black start of units in generating mode
潮汐电站 tidal power station
潮汐发电开发方式 tidal power generation development mode
单库双向开发 single-lagoon two-way tidal power development
单库单向开发 single-lagoon one-way tidal

power development
双库单向电站 two-lagoon one-way tidal power station

(2) 通航建筑物及过鱼设施 Navigation buildings and fish passage facilities

船闸 navigation lock, ship lock
多线船闸 multi-line (multiple) lock
多级船闸 flight lock, multi-stage lock
闸首 lock head
闸室 lock chamber
船闸输水系统 conveyance system of lock
船闸充水和泄水 filling and emptying
引航道，进水渠 approach channel
导航建筑物 guide structure
靠船建筑物 berthing structure
升船机 ship lift
鱼道 fishway
斜面式鱼道 inclined plane fishway
闸室式鱼道 lock chamber type fishway
鱼梯 fish ladder, fall and fall fishway
过鱼闸 fish lock

(3) 堤防及河道整治 Dike and river training

河道治理 river training
渠系建筑物 canal structure
滞洪坝 detention dam
丁坝 spur dike
围堤 border dike
自溃堤 breaching dike
截水堤 cut-off dike
潜坝 submerged dike
透水堤 permeable dike
顺坝 training dike, parallel dike
埽工 fascine works
沉排 mattress
分水堤 divide dike
渡槽 aqueduct, flume
倒虹吸管 inverted siphon
跌水建筑物 drop structure
排洪槽 over-chute
浅滩整治 shoal training
治导线（整治线） training alignment
护岸工程 bank protection works
控导工程 river control works
裁弯工程 cut off works
控制流路 main current control

(4) 水工挡土墙 Retaining wall

浆砌石挡土墙 mortar-rubble masonry wall
干砌石挡土墙 dry-laid rubble masonry wall
背面台阶式挡土墙 retaining wall with stepped back
重力式挡土墙 gravity retaining wall
半重力式挡土墙 semi-gravity retaining wall
衡重式挡土墙 shelf retaining wall
悬臂式挡土墙 cantilever retaining wall
扶壁式挡土墙 counterfort retaining wall
空箱式挡土墙 chamber retaining wall
板桩式挡土墙 sheet-pile retaining wall
加筋土挡土墙 reinforced earth wall
锚杆式挡土墙 anchor retaining wall
岸墙 side wall
翼墙 wing wall
前趾 fore toe

(5) 排水及冲沙 Draining and flushing

自由排水 free draining
重力排水 gravity drain
竖式排水 vertical drainage
水平排水 horizontal drainage
棱体排水 prism drainage
贴坡排水 slope face drainage
明渠排水 open-channel drain
地面排水 surface drainage
地下排水 subsurface drainage
暗沟排水 blind drainage, buried drain
管道排水 pipe drain
暗管排水 subsurface pipe drain
沟槽式排水 trench drain
人字形排水系统 chevron drain
背面排水 back drain
铺盖排水（褥垫排水） blanket drain
背水面坡脚排水 counter drain
竖向排水 vertical drainage
侧向排水 lateral drainage
初期排水 initial dewatering
经常性排水 regular dewatering
冲沙槽 flushing channel
冲沙闸 desilting sluice

拦沙坎　silt sill
沉沙池　sand trap
跌水池　plunge pool
连续冲洗式沉沙池　continuous scouring sand basin
定期冲洗式沉沙池　intermittent scouring sand basin
条渠沉沙池　desilting channel
导沙丁坎　groin sill
导沙顺坎　silt training sill
过木　log-passing
排冰排漂　ice and floating debris release

5.9　常用计算准则与方法
Common calculation criteria and methods

列维准则　Levy's criteria
米赛斯准则　Von Mises criteria
特雷斯卡准则　Tresca criteria
极限平衡法　limit equilibrium method
瑞典条分法　Swedish slice method
瑞典滑弧法　Swedish slipcircle method
简布法　Janbu slice method
毕肖普法　Bishop method
简化毕肖普法　Simplified Bishop method
莫尔-库仑准则　Mohr-Coulomb criteria
贝叶斯概率法　Bayes probability method
萨尔玛法　Sarma method
斯宾塞程序　Spencer's Procedure
摩根斯顿-普赖斯法　Morgenstern-Price
广义韦氏附加质量法　Generalized Westergaard Added-Mass
辛普森法则　Simpson's rule
达西定律　Darcy's law
达西-韦斯巴赫公式　Darcy-Weisbach equation
曼宁公式　Manning equation
谢才公式　Chezy formula
雷诺相似定律　Reynolds' similarity law
阿基米德原理　Archimendes' principle
伯努利方程式　Bernoulli equation
库朗条件　Courant condition
欧拉平衡方程　Euler's equilibrium equation
胡克定律　Hooke's law
拉格朗日法　Lagrangian method
纳维一斯托克斯方程式　Navier-Stokes equations
帕斯卡定律　Pascal's law of pressure
圣·维南方程　Saint-Venant equation
反滤准则　filter criteria
太沙基固结理论　Terzaghi's consolidation theory
太沙基承载力理论　Terzaghi's bearing capacity theory
计算机辅助设计　computer aided design (CAD)
计算机辅助绘图　computer aided drawing
应用软件　application software
兼容性　compatibility
传输安全性　communication security
计算机系统安全性　computer system security
数据库　data base
关系数据库　relational database
数据库管理系统　data base management system
数据结构　data structure
数据传送　data transfer
数据安全　data security
施工仿真　construction simulation analysis
数字大坝技术　digital technology for dam
数字化大坝监测管理系统　digital dam monitor management system
三维协同设计　computer supported collaborative 3D design
数值分析　numerical analysis
有限元法　finite element method (FEM)
有限元网格　finite element grid (mesh)
有限差分法　finite difference method (FDM)
离散元分析　discrete element method (DEM)
二维模型（平面模型）　two-dimensional model, plane model
三维模型（立体模型）　three-dimensional model, stereo-model
工控机　industrial personal computer (IPC)

5.10　主要报告及图表名称
Titles of reports, figures and tables

枢纽布置图和主要建筑物剖面图　Project layout and profiles of main structures
坝（闸）址、厂址比较平面布置图及剖面图　Project layout and profiles of dam (sluice) area and powerhouse area of alternative scheme
代表性方案的枢纽平面布置图　Project layout of main structures of the representative scheme
主要建筑物布置平面图及剖面图　Plan and profiles of main structures
拱坝坝轴线展开剖面　Developed section of arch dam axis

对称峡谷中单圆心拱坝平面图 Plan of a single-centered arch at a symmetrical canyon

不对称峡谷中双圆心变厚度拱坝平面图 Two-centered arch dam with variable-thickness arches at a nonsymmetrical canyon

对称（不对称）拱坝的平面图 Plan of a symmetrical (nonsymmetrical) arch dam

面板堆石坝趾板展开剖面 Developed section of plinth of concrete faced rockfill dam

发电输水道（包括坝内压力管道）的平面图和剖面图 Plan and profile of power outlet with penstock through dam

隧洞式溢洪道泄水槽剖面 Profile of tunnel spillway channel

廊道与竖井的平面、立视和剖面图 Plan, elevation, and section of gallery and shaft

泄水道拦污栅结构平面图和剖面图 Plan and section of outlet trashrack structure

坝内监测仪器布置剖面图 Locations of instrumentation installed in dam

坝址、厂址方案比较汇总表 summary of alternative comparison of dam area and powerhouse area

枢纽布置方案比较汇总表 comparison summary of main structure layouts

代表性方案的主要工程量汇总表 summary of main working quantities of the representative alternative

挡水工程（泄水工程、发电工程）工程量表 Bill of quantities for retaining structures (release structures, powerhouse)

建筑物安全监测工程量表 Bill of quantities for safety monitoring of buildings

坝址选择专题报告 Report on dam site selection

坝型比选和枢纽布置专题报告 Report on dam type selection and project layout

正常蓄水位选择专题报告 Special report on selection of reservoir normal pool level

高坝设计研究专题报告 Report on high dam design and research

高坝抗震设计专题报告 Report on seismic design of a high dam

水力学模型试验报告 Report on hydraulic model tests

高水头岔管结构分析研究专题报告 Report on structural analysis of high- head bifurcations

新材料、新结构试验研究专题报告 Report on test and research of new materials and new structures

安全监测系统设计专题报告 Report on design of project safety monitoring system

第6章　水　力　机　械
Hydro-machinery

6.1　水轮机及附属设备
Turbine and auxiliary equipment

(1) 水轮机机型　Types of turbines

水轮机　hydraulic turbine
反击式水轮机　reaction turbine
混流式水轮机　Francis turbine，radial-axial flow turbine
轴流式水轮机　axial flow turbine
轴流转桨式水轮机　Kaplan turbine，axial flow adjustable-blade turbine
轴流定桨式水轮机　Nagler turbine，axial flow fixed-blade turbine
斜流式水轮机　diagonal turbine
贯流式水轮机　tubular turbine
灯泡贯流式机组　bulb tubular unit
竖井贯流式机组　pit tubular unit
全贯流式机组　rim-generator tubular unit
轴伸贯流式机组　（S型机组）shaft-extension type tubular unit
冲击式水轮机　impulse turbine
水斗式水轮机　Pelton turbine
斜击式水轮机　Turgo turbine，inclinedjet turbine
抽水蓄能机组　pump storage unit
三机式机组　tandem unit
可逆式水轮机　reversible turbine
水泵水轮机　pump-turbine
单级可逆式水轮机　single stage pump-turbine
多级可逆式水轮机　multi-stage pump-turbine
立轴机组　vertical shaft unit
卧轴机组　horizontal shaft unit
斜轴机组　inclined shaft unit
直驱机组　direct-driven unit
有齿轮增速箱的机组　unit with gear box
有启动装置的机组　unit with starting device

(2) 水轮机性能参数　Turbine performances

比能　specific energy
位置比能　potential energy
压力比能　pressure energy
速度比能　velocity energy
加权平均水头　weighted average head
位置水头　potential head
压力水头　pressure head
速度水头　velocity head
蓄能泵扬程　storage pump head
蓄能泵零流量扬程　no-discharge head of storage pump
水轮机流量　turbine discharge
蓄能泵流量　storage pump discharge
水轮机空载流量　no-load discharge of turbine
转速　rotational speed
额定转速　rated speed
初始转速　initial speed
水轮机飞逸转速　runaway speed of turbine
稳态飞逸转速　steady-state runaway speed
水轮机最大瞬态转速　maximum momentary overspeed of turbine
蓄能泵反向飞逸转速　reverse runaway speed of storage pump
蓄能泵最大瞬态反向转速　maximum momentary counter rotation speed of sto-rage pump
瞬态转速变化率　momentary speed variation ratio
大气压力　atmospheric pressure（ambient pressure）
汽化压力　vapour pressure
绝对压力　absolute pressure
表计压力　gauge pressure
水力功率　hydraulic power
机械功率　mechanical power
转轮的机械功率　mechanical power of runner
机械功率损失　mechanical power losses
水轮机输入功率　turbine input power
水轮机输出功率　turbine output power
水轮机额定输出功率　rated output power of turbine

蓄能泵输出功率　storage pump output power
蓄能泵输入功率　storage pump input power
蓄能泵零流量功率　no discharge input power of storage pump
转轮输出功率　output power of runner
转轮输入功率　input power of runner
叶轮输入功率　input power of impeller
叶轮输出功率　output power of impeller
轴力矩　shaft torque
转轮力矩　runner torque
摩擦力矩　friction torque
水轮机机械效率　mechanical efficiency of turbine
蓄能泵机械效率　mechanical efficiency of storage pump
水轮机水力效率　hydraulic efficiency of turbine
蓄能泵水力效率　hydraulic efficiency of storage pump
最优效率　optimum efficiency
相对效率　relative efficiency
单位飞逸转速　unit runaway speed
水轮机比转速　specific speed of turbine
单位水推力　unit hydraulic thrust
单位水力矩　unit hydraulic torque
空化　cavitation
空蚀　cavitation erosion（cavitation pitting）
临界空化系数　critical cavitation coefficient
初生空化系数　incipient cavitation coefficient
电站空化系数　plant cavitation coefficient
电站吸出高度　static suction head
冲击式机组排出高度　static discharge head of impulse turbine
蓄能泵吸入高度　static suction head of storage pump
蓄能泵吸入扬程损失　suction head loss of storage pump
蓄能泵净吸上扬程（空化余量）　net positive suction head（NPSH）
安装高程　setting elevation
空化裕量　cavitation margin
空化基准面　cavitation reference level
空蚀保证期　cavitation pitting guarantee period
空蚀保证运行时间　cavitation pitting guarantee duration of operation
基准运行时间　reference duration of operation
实际运行时间　actual duration of operation
正常连续运行范围　normal continuous operating range
高负荷短时运行范围　high-load temporary operation range
低负荷短时运行范围　low-load temporary operation range
低水力比能短时运行范围　low specific hydraulic energy temporary operating range
高水力比能短时运行范围　high specific hydraulic energy temporary operating range
过机含沙量　solid content passing through hydro-turbine
泥沙磨损　sand erosion
磨蚀　combined erosion by sand and cavitation
压力脉动　pressure pulsation
振动位移　vibration displacement
振动速度　vibration velocity
振动加速度　vibration acceleration
离散量　discrete quantity
波动量　fluctuation of quantity
脉动量　pulsation of quantity
周期性振动和脉动　periodic vibration and pulsation
基频　basic frequency
转频　rotational frequency
角频率　angular frequency
简谐振动或脉动　simple harmonic vibration or pulsation
相位角　phase angle
复合振动或脉动　compound vibration or pulsation
随机振动或脉动　random vibration or pulsation
应变脉动　strain pulsation
应力脉动　stress pulsation
主轴转矩脉动　shaft torque pulsation
转速脉动　rotational speed pulsation
功率脉动　power pulsation
导叶扭矩脉动　guide vane pulsation
径向载荷脉动　radial load pulsation
轴向载荷脉动　axial load pulsation
测量通道极限频率　limit frequency of measuring channel
功率谱密度　power spectral density
分析仪的恒等百分比带宽　constant relative bandwidth of an analyzer
最大截止频率　max cutoff frequency

压力波的传播速度 propagation velocity of wave
A/D转换器 A/D converter
采样速率 sampling rate
信号记录时间 signal recording time
过渡过程 transient
调节保证 regulating guarantee
水锤 water hammer
初始压力 initial pressure
水轮机瞬态压力 momentary pressure of turbine
瞬态压力变化率 momentary pressure variation ratio

(3) 水轮机试验 Turbine testing

原型水轮机 prototype turbine
模型水轮机 model turbine
装配试验 assembly test
模型试验 model test
性能试验 performance test
特性试验 characteristic test
飞逸试验 runaway speed test
力特性试验 force characteristic test
负载试验 load test
甩负荷试验 load rejection test
耐压试验 pressure test
效率试验 efficiency test
流速仪法 current meter method
压力-时间法 pressure-time method
声学法（超声波法） acoustic method (ultrasonic method)
热力学法 thermodynamic method
指数法 index method
空化试验 cavitation test
压力脉动试验 pressure fluctuation test
补气试验 air admission test
水轮机功率试验 turbine output test
运行工况 operating condition
最优工况 optimum operating condition
飞逸工况 runaway speed operating condition
空载工况 no-load operating condition
相似工况 similar operating condition
协联工况 combined condition
力特性 force character
导叶力特性 guide vane force character
叶片力特性 blade force character
水推力 hydraulic thrust
径向力 radial force
综合特性曲线 combined characteristic curve
性能特性曲线 performance hill diagram
运转特性曲线 performance curve
飞逸特性曲线 runaway speed curve
协联特性曲线 combination curve
水泵水轮机全特性 complete characteristics of pump-turbine
导叶开度 guide vane opening
导叶角度 guide vane angle
喷针行程 needle stroke
转轮叶片角度 runner blade angle
公称直径 nominal diameter
转轮出口开度 runner outlet width
水斗宽度 bucket width
尺寸比 length scale ratio
平面角 plane angle
重力加速度 acceleration of gravity
动力黏性系数 coefficient of dynamic viscosity
运动黏性系数 coefficient of kinematic viscosity
热力学温度 thermodynamic temperature
绝对温度 absolute temperature
表面张力 surface tension
转动惯量 rotational inertia
转速因数 speed factor
流量因数 discharge factor
力矩因数 torque factor
功率因数 power factor
能量系数 energy coefficient
流量系数 discharge coefficient
力矩系数 torque coefficient
相对压力脉动 factor of pressure fluctuation
比转速 specific speed
托马数 Thoma number
初生托马数 incipient Thoma number
电站托马数 plant Thoma number
空化系数 cavitation coefficient
欧拉数 Euler number
韦伯数 Weber number
弗劳德数 Froude number
雷诺数 Reynolds number
损失分布系数 loss distribution coefficient
可换算的损失 relative scalable loss
不可换算的损失 relative non-scalable loss

总的相对损失 relative total loss
水力性能 hydraulic performance
比尺效应 scale effect
圆周速度 peripheral velocity
圆盘摩擦损失 disk friction loss
漏水损失 leakage loss
容积损失 volumetric loss
双调节机械 double regulated machine
协联关系 cam relationship
中立试验室 independent laboratory
模型对比试验 comparative model test
试验台 rig for model test
气体含量 gas content
溶于水的气体 dissolved air
未溶解于水的气体 undissolved air
迁移着的空化 travelling cavitation
汽蚀核 nuclie
涡带 vortices
尾水管涡带（混流） vortex rope
湍流 turbulence
不稳定流 unsteadiness
原位标定 calibrated in situ
可追溯性 traceability
电化学腐蚀 electrochemical corrosion
几何相似 geometric similarity
水力相似 hydraulic similitude
雷诺数相似 Reynolds similitude
弗劳得数相似 Froude similitude
三坐标测量仪 three dimensional coordinate measuring machine
光学测量系统 optical measuring system
样板 template
轴面形状 meridional contour
叶片进口节矩 blade inlet pitch
叶片进口安放角 blade inlet angle
水斗的倾角 bucket inclination
水斗的出水角 bucket discharge angle
分水刃型线 cut-out profile
分水认脊角 angle of face at the back of cut-out
射流分布圆直径 jet circle diameter
射流相对转轮的偏移值 offset of jet to runner
射流相对转轮的对准度 alignment of jet to runner
外部直径处的节矩 bucket pitch at outer diameter
叶片型线 blade profile
一致性偏差 uniformity tolerance
相似性偏差 similarity tolerance
频闪灯 stroboscopic light
内窥镜 endoscope
外推法 extrapolation
内插法 interpolation
原级方法 primary method
次级方法 secondary method
重量法 weighing method
容积法 volumetric method
移屏法 moving screen method
流速仪 current-meter
毕托管 Pitot tube
薄板堰 thin-plate weir
孔板 orifice plate
文丘里管 Venturi tube
涡轮流量计 turbine flow meter
电磁流量计 electromagnetic flow meter
声学流量计 acoustic flow meter
涡街流量计 vortex flow meter
偏流器 diverter
测压头 pressure tap
均压环 ring manifold
阻尼装置 damping device
液柱压力计 liquid column manometer
重力压力计 dead weight manometer
压力重梁 pressure weighbeam
迟滞 hysteresis
零点漂移 zero shift
涡流测功器 eddy-current brake
水力测功器 hydraulic brake
机械测功器 mechanical brake
扭矩仪 torque meter
摩擦力矩 friction torque
粗糙度 roughness
波浪度 waviness
效率修正 efficiency correction
效率比尺效应换算 efficiency scale up
模型到原型换算 model to prototype conversion
时域分析 time-domain analysis
频域分析 frequency-domain analysis
固有频率 natural frequency
自由振荡 free oscillation
主频 dominant frequency
导叶力矩 guide vane torque

转轮叶片力矩　runner blade torque
具有正斜率的水泵特性（驼峰区）　pump characteristic with positive slope
指数试验　index test
风损　windage loss

(4) 混流式机组结构　Structure of Francis turbine

补气系统　air admission system
转轮叶片　runner blade
底环　bottom ring
推拉杆　connecting rod
联轴螺栓　coupling bolt
连接法兰　coupling flange
尾水管　draft tube
尾水管锥管　draft tube cone
尾水管肘管　draft tube elbow
尾水管里衬　draft tube liner
尾水管扩散段　draft tube outlet part
抗磨板　facing plate，wear plate
基础环　foundation ring
导轴承　guide bearing
导叶　guide vane
导叶端面密封　guide vane end seal
导叶限位块　guide vane end stop
导叶臂　guide vane lever
导叶连杆　guide vane link
导叶过载保护　guide vane overload protection
导叶轴　guide vane stem
导叶轴密封　guide vane stem seal
导叶止推轴承　guide vane thrust bearing
顶盖　head cover
迷宫密封　labyrinth seal
下机坑　lower pit
主轴　main shaft
鼻端固定导叶　nose vane
支墩　pier
支墩鼻端里衬　pier nose liner
机坑　pit
机坑里衬　pit liner
压力平衡管　pressure balancing pipe
控制环　regulating ring
转轮　runner
转轮下环　runner band
转轮下环腔　runner band chamber
转轮下环止漏环　runner band seal
转轮腔　runner chamber
转轮泄水锥　runner cone
转轮上冠　runner crown
转轮上冠腔　runner crown chamber
转轮上冠止漏环　runner crown seal
静止止漏环　stationary seal ring
转动止漏环　rotating seal ring
主轴密封　shaft seal
蜗壳　spiral case
座环　stay ring
固定导叶　stay vane
走道盖板　walkway
导叶轴套　guide vane bearing
回转环　return ring
回转环导叶　return ring vane
单独导叶接力器　individual guide vane servomotor

(5) 轴流式机组结构　Structure of axial flow turbine

操作架　cross head
转轮室　discharge chamber
转轮室上环　discharge chamber ring
喉管　throat ring
转轮体　runner hub
转轮叶片转臂　runner blade lever
转轮叶片连杆　runner blade link
转轮叶片密封　runner blade seal
转轮叶片接力器　runner blade servomotor
转轮叶片枢轴　runner blade trunnion

(6) 贯流式机组结构　Structure of tubular turbine

竖井通道　access shaft
灯泡体　bulb
灯泡体支柱　bulb support
反向推力轴承　counter thrust bearing
齿轮增速箱　gear box，speed increaser
发电机进人孔　generator access hatch
内导水环　inner guide ring
外导水环　outer guide ring
受油器　oil head
转子环　rim

转子环密封　rim seal
贯流式座环　stay cone
水轮机进水流道　turbine inlet water passage
拆卸法兰　dismantling flange

(7) 水斗式机组结构　Pelton turbine structure

制动喷嘴　brake nozzle
叉管　branch pipe
水斗　bucket
分流器　cut-in deflector
折向器（偏流器）　deflector
机壳　housing
配水管路　intake pipe
分流管　manifold
喷针　needle
喷针折向器定位装置　needle-deflector positioner
喷针杆　needle rod
喷针接力器　needle servomotor
喷针头　needle tip
喷嘴　nozzle，injector
喷管　nozzle pipe
喷嘴保护罩　nozzle shield
喷嘴口环　nozzle tip ring
转轮轮盘　runner disk
折向器接力器　deflector servomotor
喷针折向器连杆　needle deflector link
转轮转运车　runner cart
转轮运输门　runner transport door

(8) 导轴承　Guide bearing

轴领　guide bearing collar
轴承体　guide bearing housing
轴颈　guide bearing journal
导轴承分块瓦　guide bearing pad
筒式导轴承轴瓦　guide bearing shell
供油系统　oil supply system
筒式导轴承轴瓦支撑装置　shell supporting device
分块瓦导轴承轴瓦支撑装置　pad supporting device
油盆　oil reservoir
油冷却器　oil cooler

(9) 推力轴承　Thrust bearing

轴承高压油顶起系统　bearing oil injection device
推力轴承基础板　thrust bearing base plate
推力轴承油箱　thrust bearing housing
推力轴承支架　thrust bearing supporting cone
推力头　thrust collar
推力瓦　thrust pad
推力瓦支撑　thrust pad support

(10) 调速器　Governor

机械液压调速器　mechanical hydraulic governor
电液调速器　electric-hydraulic governor
微机调速器　micro-computer based governor
双调整调速器　double regulating governor
通流式调速器　governor without pressure tank，through flow type governor
压力罐式调速器　governor with pressure tank
水轮机控制系统　hydraulic turbine control system
被控制系统　controlled system
水轮机调节系统　hydraulic turbine regulating system
有差调节　difference regulation，deviation regulation
无差调节　no-difference regulation，no-deviation regulation
随动系统　servo-system
操作器　position operator，gate operator
电子负荷调节器　electronic load controller
电动机调速器　governor with motor driven gate operator
比例-积分调速器　proportional-integral governor (PI governor)
比例-积分-微分调速器　proportional-integral-derivative governor (PID governor)
串联 PID 调速器　series PID governor
并联 PID 调速器　parallel PID governor
缓冲型调速器　damping type governor
加速度-缓冲型调速器　acceleration-damping type governor
测速装置　speed sensing device
测频单元　frequency module
测速信号源　speed signal source
齿盘　toothed disk
脉冲传感器　impulse transducer
剩磁　residual remanence
测速发电机　tachogenerator
人工死区单元　artificial dead band module
电机转换器　electro-mechanical converter

电液转换器　electro-hydraulic converter
电液伺服阀　electro-hydraulic servo-valve
电液比例阀　electro-hydraulic proportional valve
配压阀　distributing valve
主配压阀　main distributing valve，control valve
引导阀　pilot distributing valve
比例阀　proportional valve
柱塞线圈　plunger coil
阀芯行程　valve spool stroke
减振器　dashpot
电气缓冲单元　electrical damper module
机械开度限制机构　mechanical opening limiter
电气开度限制单元　electrical opening limiting module
转速调整机械　speed adjusting mechanism
功率给定单元　power setting module
频率给定单元　frequency setting module
机械连杆　mechanical linkage
接力器　servomotor
辅助接力器　auxiliary servomotor
中间接力器　pilot servomotor
主接力器　main servomotor
导叶接力器　guide vane servomotor
轮叶接力器　blade servomotor
协联装置　combination device
分段关闭装置　step closing device
先导伺服定位器　pilot servo-positioner
随动的伺服定位器　following main servo-positioner
综合放大单元　summation and amplification module
自动运行　automatic operation
手动运行　manual operation
限负荷运行　limited load operation
孤立运行　isolated operation
并联运行　parallel operation
空载运行　no-load operation，idling operation
带负荷运行　load operation
甩负荷　load rejection (load shedding)
并网运行　grid-connected operation
稳定状态　steady state
波动状态　oscillation condition
瞬变状态　transient condition
型式试验　type test
出厂试验　workshop test
验收试验　acceptance test
现场试验　field test
无水调试　dry test
空载扰动试验　no-load disturbing test
带负荷试验　load test
甩负荷试验　load rejection test
被控参数　controlled variable
与稳态值的相对偏差　relative deviation from a steady-state value
转速相对偏差　relative deviation of speed
给定（指令）信号　command signal
给定信号相对偏差　relative deviation of command signal
调节器输出信号　governor output signal
接力器行程　servomotor stroke
接力器容积　servomotor volume
主配压阀中间位置　neutral position of main distributing valve
水轮机控制系统静态特性　droop graph of turbine control system
测速装置放大系数　speed sensor amplification
永态差值系数　permanent droop
暂态差值系数　temporary droop
差值曲线的斜率　slope of the droop graph
速动时间常数　promptitude time constant
加速时间常数　derivative time constant
微分环节时间常数　time constant of derivative module
水流惯性时间常数　water inertial time constant
机组惯性时间常数　unit inertial time constant
机组加速时间常数　unit acceleration constant
负载惯性时间常数　load inertial time constant
负载加速时间常数　load acceleration constant
暂态作用阶跃　step of transient function
积分时间常数　integral time constant
积分增益　integral gain
积分作用时间　integral action time
缓冲时间　damping time
阶跃响应曲线　step response curve
微分增益　differential gain
微分作用时间　derivative action time
比例增益　proportional gain
比例放大系数　proportional amplification
开环增益　opening loop gain
微分方程式　differential equation

暂态响应　transient response
频率响应　frequency response
传递函数　transfer function
传递系数　transmission coefficient
被控制变量范围　controlled variable range
油伺服系统　oil servo system
转速稳定性指数　speed stability index
功率稳定性指数　power stability index
接力器关闭时间　servomotor closing time
接力器开启时间　servomotor opening time
延缓时间　cushioning time
压力钢管反射时间　penstock reflection time
水轮机自调节系数　turbine self-regulation coefficient
发电机负载调节系数　generator load self-regulation coefficient
电网负载特性系数　network load characteristic coefficient
被控制系统自调节系数　controlled system self-regulation coefficient
并联结构　parallel structure
串联结构　series structure
并联控制　parallel control
随动控制　follow-up control
死区　dead band
不灵敏度　insensitivity
伺服定位器时间常数　time constant of the servo-positioner
放大器死区　amplifier dead band
放大器不准确度　amplifier inaccuracy
控制系统不动时间　control system dead time
驱动能量　actuating energy
储能机构　energy storage mechanism
压力油罐　oil pressure tank
回油箱　sump tank
设计油压　design oil pressure
工作油压　operating oil pressure
紧急停机油压　tripping oil pressure
最低要求压力　minimum required pressure
紧急停机油体积　tripping oil volume
可用油体积　usable oil volume
剩余（不可用）油体积　residual (not usable) oil volume
满负荷关机　full-load shutdown
动态系统品质　dynamic system behavior
阶跃变化　step change
线性系统　linearized system
正弦变化　sinusoidal change
随机变量　arbitrary variation
傅里叶变换　Fourier transformation
拉普拉斯变换　Laplace transformation
控制功能　control function
平稳切换　bump-free switch-over
选择器　selector
综合点　summing point
峰荷电站　peak-load power station
基荷电站　base-load power station
转速调节器　speed governor
功率输出调节器　power output governor
转速信号给定装置　speed signal setter
开度限制器　opening limiter
反馈信号　feedback signal
函数发生器　function generator
双重控制　dual control
运行方式转换　operating mode transition
水力过渡过程　hydraulic transient
负荷惯性　load inertial
转动惯量　moment of inertia
转动惯矩　rotary moment of inertia
飞轮力矩　fly-wheel effect
回转半径　radius of gyration
接力器响应时间　servomotor response time
单独接力器控制　individual servomotor control
波传递速度　wave travel speed
特征方程式　characteristic equation
可压缩性　compressibility
调压室涌浪　excursion of surge tank water level
消能器　energy dissipator
阻尼比　damping ratio
临界阻尼　critical damping
阶跃位移输入　step displacement input
大波动　large perturbation
正向通道　forward path
反馈通道　feedback path
霍尔发生器　Hall generator
电极　electrode
旋转型传感器　rotational transducer
直线型传感器　linear transducer
气囊式储压器　bladder accumulator
活塞式储压器　piston accumulator

惰性气体 inert gas
氮气 nitrogen
密封隔离 hermetic separation
去油雾设备 oil mist exhaust equipment
流量恒定泵 constant displacement pump
变流量泵 variable displacement pump
飞摆 pendulum
转速偏差 speed deviation
接力器缓冲时间 servomotor cushioning time
接力器作用力 servomotor force
接力器容量 servomotor capacity
转矩偏差 torque deviation
瞬时角速度 instantaneous angular speed
角动量矩 angular momentum
保证转矩 guaranteed torque
水轮机控制传递比 turbine control transmission ratio
转速调节图 speed regulation graph
永态转速调节 permanent speed regulation
电网负荷特性 network load characteristic
水锤数（阿列维常数） water hammer number，Allievi constant
最大瞬时转速变化 maximum momentary speed variation
最大瞬时压力变化 maximum momentary pressure variation
控制系统本体 control systems proper
控制系统部件 control systems component
安全关闭 safety shutdown
过速保护 over speed protection
联锁 interlock
蠕动检测 creep detection
电磁兼容性 electromagnetic compatibility (EMC)
电气干扰源 electrical interference source
液压机械部分 hydro-mechanical part
压力增益曲线 pressure gain curve
过调节 overshoot
液压放大级 hydraulic amplifier stage
正搭叠 overlapped
零搭叠 zero-lapped
负搭叠 underlapped
接力器开启/关闭规律 servomotor opening/closing law
转速整定值 speed set-point value
发散性振荡 divergent oscillation
拓扑结构 topology
转轮圆周频率 runner peripheral frequency
压力波动 pressure fluctuation
转矩波动 torque fluctuation
电力系统稳定器 power system stabilizer (PSS)
自动电压调节器 automatic voltage regulator (AVR)
水的可压缩性 water compressibility
管壁弹性 pipe wall flexibility

6.2 辅机系统 Auxiliary machinery

(1) 起重机 Crane

桥式起重机 bridge crane，overhead crane
吊钩起重机 hook crane
抓斗起重机 grabbing crane
手动起重机 manual crane
电动起重机 electric crane
液压起重机 hydraulic crane
司机室操纵起重机 caboperated crane
地面操纵起重机 floor-operated crane
按钮操纵起重机 pendant-operated crane
遥控起重机 remote operated crane
无线遥控起重机 cableless remote operated crane
有线遥控起重机 cable remote operated crane
无线电操纵起重机 radio-operated crane
红外线操纵起重机 infrared rays operated crane
起重力矩 load moment
轮压 wheel load
吊钩极限位置 hook approach
起升高度 load-lifting height
下降深度 load-lowering height
起升范围 lifting range
起重机轨面高度 crane track height
起升速度 load-lifting speed
下降速度 load-lowering speed
微速下降速度 precision load-lowering speed
运行速度 travelling speed
作业周期 operation cycle time
起重机轨距 track center
小车轨道中心距 rail center of crab
横移 traversing
静载试验 static test
动载试验 dynamic test

起升机构 hoisting mechanism
起重机运行机构 crane travel mechanism
起重葫芦 hoist
桥架 bridge
起重小车 crab，trolley
制动器 brake
卷筒制动器 drum brake
鼓式制动器 shoe brake
盘式制动器 disk brake
滑轮 sheave，pulley
绳索滑轮组 reeving system
吊钩滑轮组 hook assembly
取物装置 load-handling device
轨道总成 rail track
限制器 limiting device，limiter
额定起重量限制器 rated capacity limiter
运动限制器 motion limiter
缓冲器 buffer
终端止挡器 end stop
指示器 indicator
工作参数指示器 operating parameter indicator
额定起重量指示器 rated capacity indicator
净起重量 net load
总起重量 gross load
额定起重量 rated capacity
主梁拱度 bridge girder camber
预拱度 built-in camber
车轮轴距 wheel pitch
端梁 end carriage，end truck
导向轮 guide roller
单梁桥架 single-girder bridge
双梁桥架 double-girder bridge
起重机组别 A crane groups
机构组别 M mechanism groups
起重机械利用等级 U class of utilization of lifting appliances
起重机的谱等级 Q sectrum classes for cranes
机构利用等级 T class of utilization of mechanisms
结构件利用等级 B class of utilization of mechanisms
接电持续率 duty factor
车轮上的载荷 load on wheel
由自重引起的载荷 load due to dead weight，constant load
由水平运动引起的载荷 load due to horizontal motion
由工作荷重引起的载荷 load due to working load
由加速或制动引起的载荷 load due to acceleration or braking
由电机最大转矩引起的载荷 load due to maximum motor torque
由摩擦力引起的载荷 load due to friction forces
由缓冲效应引起的载荷 load due to buffer effect
由工作荷重的垂直位移所引起的载荷 load due to vertical displacement of the work- ing load
起重机械总使用持续时间 total duration of use of lifting appliance
循环持续时间 duration of cycle
小车轨道梁腹板 web of trolley rail girder
起重机的额定行走速度 nominal travel speed of appliance
荷重悬挂点 point of suspension of load
起重循环 hoisting cycle
钢丝绳倾角 angle of inclination of rope
起重机跨距偏差 divergence in span of crane
转向架 bogie

(2) 水泵 Pump

回转动力式泵 rotodynamic pump
离心泵 centrifugal pump
混流泵 mixed flow pump
轴流泵 axial flow pump
蜗壳泵 volute pump
导叶泵 diffuser pump
节段式泵 sectional type pump
侧盖式泵 side cover type pump
径向剖分泵 radially split pump
轴向剖分泵 axially split pump
单级泵 single-stage pump
多级泵 multi-stage pump
单吸泵 single-suction pump
双吸泵 double-suction pump
中心支承式 centerline support type
管道式 inline type
共座式 common baseplate type
分座式 separate baseplate type
齿轮传动式 gear driven type
液力耦合器传动式 hydraulic coupling driven type
皮带传动式 belt driven type

共轴式 close coupled type
液下式泵 wet pit type pump
双壁壳式泵 armoured type pump
地坑筒式泵 pit barrel type pump
抽出式泵 pull-out type pump
自吸式泵 self priming type pump
潜液电泵 submergible motor pump
屏蔽电泵 canned motor pump
平衡鼓式 balancing piston type
平衡盘式 balancing disc type
自身平衡式 self-balancing type
平衡孔式 balancing hole type
锅炉给水泵 boiler feed pump
凝结水泵 condensate pump
循环水泵 circulating water pump
杂质泵 liquid-solids handling pump
砂泵 sand pump
渣浆泵 slurry pump
泥浆泵 sludge pump
污水泵 sewage pump
消防泵 fire water pump
增压泵 booster pump
耐腐蚀泵 anti-corrosive pump
工况点 operating point
规定点 specified point
关死扬程 shut off head
静扬程 total static head
理论扬程 theoretical pump head
出口总水头（排出扬程） total discharge head
入口总水头（吸入扬程） total suction head
排出压力 discharge pressure
吸入压力 suction pressure
几何高度 geometric height
泵基准面 reference plane
必需气蚀余量 net positive suction head required (NPSHR)
可用气蚀余量 net positive suction head available (NPSHA)
临界气蚀余量 critical net positive suction head (NPSH)
允许吸上真空度 allowable suction vacuum
汽蚀比转速 suction specific speed
泵输出功率（有效功率） pump effective power
泵轴功率（输入功率） pump shaft power
原动机输入功率 driver input power
泵效率 pump efficiency
机械损失 mechanical loss
机组效率 overall efficiency
保证效率 guaranteed efficiency
性能曲线 performance curve
特性曲线 characteristic curver
全特性 complete characteristics
等效率曲线 iso-efficiency curve
阻力曲线 system head curve
轴面速度 meridian velocity
速度三角形 velocity triangle
冲角 attack angle
失速 stall
喘振 surging
脱流 separation
扬程系数 head coefficient
圆周速度系数 speed constant
轴功率系数 shaft power coefficient
注水 priming
水封 water sealing
暖泵 warming-up
蜗形体 volute casing
双蜗形体 double volute casing
导流壳体 diffuser casing
吐出壳（吐出段） discharge casing
吸入壳（吸入段） suction casing
中壳（中段） stage casing
多级泵中段 conveyor case
多级泵导叶 conveyor vane
水泵导叶 diffuser vane
吐出弯管 discharge elbow
扬水管 lifting pipe
悬吊管 column pipe
吸入弯管 suction elbow
吸入喇叭管 suction bell
内壳 inner casing
外壳 outer casing
泵盖 casing cover
吸入盖 suction cover
平衡室盖 cover of balancing chamber
水套盖 jacket cover
填料压盖 gland cover
叶轮 impeller
闭式叶轮 closed impeller
开式叶轮 open impeller

不堵式叶轮 non-clogging impeller
叶轮密封环（口环） impeller ring
叶轮螺母 impeller cap
叶轮轮毂 impeller hub
叶轮下环腔 impeller chamber
叶轮引水锥 impeller cone
叶轮上冠 impeller crown
叶轮上冠腔 impeller crown chamber
叶轮上止漏环 impeller crown seal
叶轮下环 impeller skirt
叶轮叶片 impeller vane
水泵扩散段 pump diffuser
反导叶 return ring vane
吸入管 suction tube
诱导轮 inducer
轮叶 impeller blade
泵轴 pump shaft
上轴 upper shaft
下轴 lower shaft
中间轴 intermediate shaft
中间联轴器 intermediate shaft coupling
联轴器罩 coupling guard
联轴器使用系数 coupling service factor
填料轴套 packing sleeve
水轴承套 bearing sleeve
挡套 interstage sleeve
轴套螺母 sleeve nut
减压套 pressure reducing sleeve
平衡套 balancing sleeve
调整环 adjust ring
对开挡环 split ring
机械密封 mechanical seal
浮动环密封 floating ring seal
隔板 interstage diaghragm
壳衬 casing liner side plate
壳体密封环 casing ring
填料环 seal cage
水封环 water seal cage
节流套 throat bushing
中间衬套 interstage bushing
减压衬套 pressure reducing bushing
平衡衬套 balancing bushing
平衡管 balancing pipe
穿杠 tie bolt
水轴承 submerged bearing
水轴承体 bearing spider
流体动力轴承 hydrodynamic bearing
流体动力径向轴承 hydrodynamic radial bearing
流体动力止推轴承 hydrodynamic thrust bearing
隔舌 cut-water
前盖板 front shroud
后盖板 back shroud
允许工作范围 allowable operating range
泵壳最大容许作用压力 maximum allowable casing working pressure
最高允许连续转速 maximum allowable continuous speed
基本设计压力 basic design pressure
最大动密封压力 maximum dynamic sealing pressure
最小容许流量 minimum permitted flow
腐蚀余量 corrosion allowance
自停转速 trip speed
第一临界转速 first critical speed
设计径向负荷 design radial load
轴摆度（径向跳动） shaft runout
端面跳动 face runout
轴的挠度 shaft deflection
循环冲洗 circulation flush
注入冲洗 injection flush
阻隔液体 barrier liquid
节流衬套 throttle bush
喉部衬套 throat bush
筒形壳体 barrel casing

(3) 阀门及附件 Valve and accessories

主阀 main shut-off valve
蝶阀 butterfly valve
球阀 ball valve（spherical valve）
闸阀 gate valve
截止阀 globe valve，stop valve
节流阀 throttle valve
圆筒阀（筒形阀） cylindrical valve，ring gate
针形阀 needle valve
旋塞阀 plug valve
止回阀 check valve，non-return valve
隔膜阀 diaphragm valve
安全阀 safety valve
减压阀 pressure reducing valve
泄荷阀 pressure relief valve

空心锥形泄荷阀　hollow-cone valve (Howell-Bunger valve)
空心射流泄荷阀　hollow-jet valve
排污阀　blow-down valve
调节阀（控制阀）　control valve, adjusting valve
分配阀　dividing valve
疏水阀　steam trap
混合阀　mixing valve
水力控制阀　hydraulic control valve
手动阀门　manual operated valve
电动阀门　electrical operated valve
液动阀门　hydraulically operated valve
气动阀门　pneumatically operated valve
高温阀门　high temperature valve
中温阀门　moderate temperature valve
常温阀门　normal temperature valve
低温阀门　sub-zero valve
超低温阀门　cryogenic valve
进出口端面距离　face-to-face dimension
直通式阀　through way type valve
角式阀　angle type valve
直流式阀　Y-type valve
T形三通式阀　T-pattern three way valve
L形三通式阀　L-pattern three way valve
常开式阀　normally open type valve
常闭式阀　normally closed type valve
保温式阀　steam jacket type valve
波纹管式阀　bellows seal type valve
全径阀门　full-port valve
缩径阀门　reduced-port valve
缩口阀门　reduced-bore valve
单向阀门　unidirectional valve
双向阀门　bidirectional valve
双座双向阀门　twin-seat bidirectional valve
双锁紧泄放阀（DBB）　double-block-and-bleed valve
上密封　back seal
压力密封　pressure seal
壳体　shell
阀体　valve body
阀盖　valve bonnet
阀瓣，启闭件　valve disc
阀座　valve seat
密封面　sealing face
阀杆　stem, spindle
阀杆螺母　yoke nut
填料盒　stuffing box
填料垫　packing seat
支架　yoke
撞击手轮　impact hand wheel
明杆闸阀　outside screw rising stem type gate valve
暗杆闸阀　inside screw non-rising stem type gate valve
楔式闸阀　wedge gate valve
平行式闸阀　parallel gate valve (parallel slide valve)
闸板　wedge
单闸板　single gate
双闸板　double gate
楔式双闸板　wedge double gate
平行式双闸板　parallel double gate
刚性闸板　rigid gate
弹性闸板　flexible gate disc
浮动式球阀　floating ball valve
固定式球阀　fixed ball valve
垂直板式蝶阀　vertical disc type butterfly valve
斜板式蝶阀　inclined disc butterfly valve
对夹式蝶阀　wafer type butterfly valve
屋脊式隔膜阀　weir diaphragm valve
截止式隔膜阀　globe diaphragm valve
填料式旋塞阀　gland packing plug valve
油封式旋塞阀　lubricated plug valve
升降立式止回阀　vertical lift check valve
旋启式止回阀　swing check valve
旋启多瓣式止回阀　multi-disc swing check valve
底阀　foot valve, bottom valve
轴流式止回阀　axial flow check valve
蝶式止回阀　butterfly swing check valve
销轴　hinge pin
摇杆　rocker arm
弹簧式安全阀　direct spring loaded safety valve
杠杆式安全阀　lever and weight loaded safety valve
先导式安全阀　pilot operated safety valve
全启式安全阀　full lift safety valve
微启式安全阀　low lift safety valve
波纹管平衡式安全阀　bellows seal balance safety valve
双联弹簧式安全阀　duplex safety valve
直接载荷式安全阀　direct-loaded safety valve

带动力辅助装置的安全阀 assisted safety valve
带补充载荷的安全阀 supplementary loaded safety valve
真空安全阀 vacuum relief valve
敞开式安全阀 openly sealed safety valve
调节螺母 adjusting screw
弹簧座 spring plate
导向套 dise guide
反冲盘 disc holder
调节圈 adjusting ring
压力释放装置 pressure relief device
重闭式压力释放装置 reclosing pressure relief device
爆破片装置 bursting disk device
折断销装置 breaking pin device
弯折销装置 buckling pin device
剪切销装置 shear pin device
易熔塞装置 fusible plug device
排放面积 discharge area
流道面积 flow area
帘面积 curtain area
额定开启高度 rated lift
密封面斜角 seat angle
密封面积 seat area
净流通面积 net flow area
喉径 throat diameter
整定压力 set pressure
超过压力 over pressure
回座压力 reseating pressure
启闭压差 blowdown
冷态试验差压力 cold differential test pressure
排放压力 relieving pressure
排放背压力 built-up back pressure
附加背压力 superimposed back pressure
理论排量 theoretical discharge capacity
额定排量 certified discharge capacity
当量计算排量 equivalent calculated capacity
频跳 chatter
颤振 flutter
动作性能及排量试验 operational characteristics and flow capacity testing
在用试验 in-service testing
工作台上定压试验 bench testing
薄膜式减压阀 diaphragm reducing valve
弹簧薄膜式减压阀 spring diaphragm reducing valve
活塞式减压阀 piston reducing valve
波纹管式减压阀 bellows seal reducing valve
杠杆式减压阀 lever reducing valve
直接作用式减压阀 direct-acting reducing valve
先导式减压阀 pilot operated reducing valve
固有流量特性 inherent flow characteristic
缓闭止回阀 low speed closed check valve
稳压阀 pressure sustaining valve
水泵控制阀 pump control valve
最小开启压力 minimum open pressure
阀操作机构 valve-actuating mechanism
驱动装置 actuator
多回转驱动装置 multi-turn actuator
部分回转驱动装置 part-turn actuator
电动装置 electric actuator
气动装置 pneumatic actuator
直线型气动装置 linear pneumatic actuator
回转型气动装置 rotary pneumatic actuator
气动装置的行程 stroke of pneumatic actuator
液压驱动装置 hydraulic actuator
电磁驱动装置 electromagnetic actuator
蜗轮传动装置 worm gear actuator
正齿轮传动装置 cylindrical gear actuator
锥齿轮传动装置 conical gear actuator
壳体试验 shell test
密封试验 closure test
试验压力 test pressure
试验介质 test fluid
试验介质温度 test fluid temperature
弹性密封副 resilient seats
冷态工作压力 cold working pressure (CWP)
允许工作压差 design differential pressure
上密封试验 backseat test
波纹管式阀杆密封 bellows stem sealing

(4) 消防系统 Fire prevention system

可燃性 combustibility
易引燃性 ease of ignition
易燃性 flammability
难燃的 difficult-flammable
耐电弧性 arc resistance
不燃性 non-combustibility
自热 self-heating

自燃物　pyrophoric material
气化　gasify
惰化　inerting
自燃　spontaneous ignition
阴燃　smouldering
轰燃　flash over
全燃火　fully developed fire
起火（引燃）　ignite
引火源　ignition source
闪点　flash point
燃点　fire point
引燃温度　ignition temperature
引燃时间　ignition time
放热率　heat release rate
热辐射　thermal radiation
试验热值　experimental heat release
实际热值　actual calorific valve
表面燃烧　surface burn
表面闪燃　surface flash
有焰燃烧　flaming
火焰峰　flame front
火焰传播　flame spread
火焰传播速率　flame spread rate
续焰　after flame
过火面积　burned area
烧毁面积　damaged area
阻燃性　fire retardance
阻燃处理　fire retardant treatment
氧指数　oxygen index
自熄性　self extinguishbility
灰烬　ash
炭　char
烟囱效应　chimney effect
熔渣　clinker
熔滴　melt drip
烧焦　scorch
焦味　smell of scorching
耐火等级　fire resistance classification
耐火极限　duration of fire resistance
火灾气流　fire effluent
急性毒性　acute toxicity
慢性毒性　chronic toxicity
潜伏毒性　delayed toxicity
特殊毒性　specific toxicity
毒物　toxicant
毒效　toxic potency
毒害　toxic hazard
毒害危险性　toxic risk
毒物浓度　toxicant concentration
暴露时间　exposure time
暴露剂量　exposure dose
半中毒浓度　effect concentration 50%（ec_{50}）
半致死浓度　lethal concentration 50%（lc_{50}）
丧失能力　incapacitation
消防信息　fire information
消防有线通信系统　fire wired communication system
消防无线通信系统　fire radio communication system
火灾探测和报警　fire detection and alarm
火灾报警　fire alarming
有线报警　wired alarming
无线报警　radio alarming
自动火灾信号　automatic fire signal
火灾自动报警系统　automatic fire alarm system
手动火灾报警按钮　manual fire alarm call point
故障信号　fault signal
火警电话　fire telephone
误报　false alarm
触发器件　trigger device
点型火灾探测器　spot-type fire detector
线型火灾探测器　line-type fire detector
分散接警　scattered receipt of fire alarms
集中接警　centralized receipt of fire alarms
混合接警　combined receipt of fire alarms
接警持续时间　duration of receiving alarms
火警调度专线　fire alarm dispatching proprietary wires
火警调度台　fire alarm dispatching console
车（船）载无线电话机　mobile radiophone
袖珍无线电话机　hand-held radiophone
消防计算机通信系统　fire computer communication system
公安消防队　fire brigade of public security
消防训练塔　fire station training tower
火警瞭望　fire lookout
消防水源　water source for fire fighting
火场指挥部　fire commanding post
火灾分类　fire classification
耐火隔墙　fire resisting partition

耐火管道 fire resisting duct
耐火竖井 fire resisting shaft
防火阀 fire resisting damper
防火墙 fire wall
防火幕 safety curtain
防火堤 fire bund
集流坑 catchpit
消防电梯 fire lift
灭火剂 extinguishing agent
泡沫灭火剂 foam extinguishing agent
干粉灭火剂 powder extinguishing agent
灭火器 fire extinguisher
固定式灭火器 fixed fire extinguisher
移动式灭火器 mobile fire extinguisher
手提式灭火器 portable fire extinguisher
推车式灭火器 transportable fire extinguisher
消防车 fire vehicle
联用消防车 universal fire truck
消防接口 fire coupling
烟气控制 smoke control
自然控烟 natural smoke control
排烟口 smoke vent
排烟风机 smoke extractor exhaust fan
进风口 air inlet
屋顶通风 roof vent
屋顶挡烟隔板 roof smoke screen
防烟阀 smoke damper
逃生滑梯 slide escape
事故照明 emergency lighting

6.3 安装和调试 Installation and trial running

安装场 erection bay
装配场 assembly bay
水轮机廊道 turbine walkway
水轮机机坑 turbine pit
发电机风罩 generator pit
转轮检修道 runner removal access
浆叶检修孔 blade removal opening
安装工艺 installation technique
组装 assembly
拆卸 disassembly
装配记号 assembly mark
装配螺栓 assembling bolt
连接螺栓 coupling bolt
连接法兰 coupling flange
对开法兰 split flange
大轴找正 alignment of shaft
转轮体装配 runner hub assembly
蜗壳进人孔 spiral case access
组件吊运 handling of component
转子翻身 turning over of rotor
厂内装焊转轮 shop fabricated runner
工地装焊转轮 site fabricated runner
正反向旋转 clockwise and anti-clockwise rotation
发电机空气间隙测量 measuring of air gap between generator stator and rotor
检修密封 standstill seal
工作密封 service seal
安装精度 installation accuracy
安装误差 installation error
安装公差 erection tolerance
平行度 parallelism
直线度 straightness
平面度 flatness
同轴度 coaxiality
水平度 levelness
垂直度 perpendicularity
对称度 symmetry
偏斜 declination
极限偏差 limit deviation
机械加工余量公差 machining allowance tolerance
电气设备干燥 drying of electric equipment
电气盘柜安装 erection of electric panel and cabinet
硬母线加工 preparation of rigid busbar
电缆终端制作 preparation of cable terminal
电缆桥架 cable tray
电缆支架 cable bearer
电缆导管 cable ducts
金属套 metallic sheath
铠装层 armour
焊缝代号 welding symbols
帮条焊 article for the welding
坡口焊（斜角焊） bevel welding，groove welding
单面焊接 one side welding
双面焊接 double side welding
管子坡口加工 polish of tube groove

管子对口 aligning of pipes
预埋件 embedded parts
检定 verification
整定试验 set test
定值校验 setting verify
单体调校 adjustment and calibration of individual component
一次性通过 pass at the first attempt
一次性完成 complete at the first attempt
动水启动试验 starting test in dynamic water
动水关闭试验 closing test in dynamic water
空载试验 no load test
空载试运行 trial running without load
满负荷试运行 trial running with full load
72小时带负荷连续试运行 72h commissioning with rated load
30天考核试运行 30d commissioning

6.4 主要图表名称 Titles of figures and tables

技术供水系统图 Principle diagram of cooling water supply system
检修排水系统图 Principle diagram of dewatering system
渗漏排水系统图 Principle diagram of drainage system
压缩空气系统图 Principle diagram of compressed air system
油系统图 Principle diagram of oil treat-ment system
水处理系统图 Principle diagram of water treatment system
水力测量系统图 Principle diagram of hydraulic measuring system
消防供水系统图 Principle diagram of firefighting water supply system
原理图 Schematic diagram
调速器控制原理图 Governor control schematic diagram
厂房平面布置图 Plan view of powerhouse layout
厂房纵剖面布置图 Longitudinal section view of powerhouse layout
厂房横剖面布置图 Cross section view of powerhouse layout
水轮机布置平面（剖面）图 Plan (profile) of turbine layout
水轮机总体布置图 Turbine general arrangement
尾水管里衬基础图 Foundation of draft tube liner
机坑里衬组装图 Assembly of pit liner
技术供水泵安装布置图 Installation and layout of cooling water pump
桥机竣工图 As-built drawing of bridge crane
材料表 Bill of materials
管夹材料表 Bill of material of pipe support
压水系统计算书 Calculation sheet of water depression system
滤水器说明书 Specification of strainer
运行维护手册 Operation and maintenance manual

第7章 金属结构及安装
Hydraulic steel structures and installation

7.1 闸门
Gate

(1) 闸门类型 Types of gates

定轮闸门 fixed wheel gate
滚轮闸门 free roller gate
履带式闸门（链轮闸门） caterpillar gate，roller chain gate
提升式平板闸门（串轮闸门） Stoney gate
工作闸门 service gate
事故闸门 emergency gate
快速闸门 quick-acting shutoff gate，stop gate
检修闸门 bulkhead gate
导流闸门 diversion gate
露顶式闸门 emersed gate
潜孔式闸门 submerged gate
泄水闸门 sluice gate
冲沙闸门 flush gate
防洪闸门 flood gate
泄洪闸门 flood-discharge（flood-relief）gate
尾水（管）闸门 tail gate，draft-tube gate
首部闸门 head gate
进水口闸门 intake gate
出水口闸门 outlet gate
底孔闸门 ground gate
船闸闸门 lock gate
平面闸门 plain gate，vertical-lift gate
横拉闸门 lateral movement gate
浮箱闸门（浮动闸门） floating gate
升卧式闸门 lifting-lie gate
叠梁闸门 stoplog gate
排针闸门 pin gate
双扉闸门 double-leaf gate
多扉闸门 multi-leaf gate
滑动闸门 sliding gate
舌瓣闸门,拍门 flap gate
翻板闸门 hinged crest gate，wicket gate
自动翻板闸门 balanced wicket，flashboard
人字闸门 miter gate
一字闸门 single leaf swing gate
弧形闸门 radial gate，tainter gate
反向弧形闸门 reverse tainter gate
下沉式弧形闸门 submersible taiter gate
立轴式弧形闸门 sector gate，vertical axes tainter gate
扇形闸门 sector gate
鼓形闸门 drum gate
屋顶闸门 roof gate
拱形闸门 arch gate
圆筒闸门 cylindrical gate
环形闸门 ring gate
附环闸门 ring-follower gate
圆辊闸门 roller gate
球形闸门 ball gate
管形阀 tubular valve
空注阀 hollow jet valve
锥形阀 fixed cone valve
蝴蝶阀（蝶阀） butterfly valve
球阀 spherical valve，ball valve
套筒式控制阀 sleeve valve
排气阀 air release valve
拦污栅 trash rack
移动式拦污栅 portable trash rack
固定式拦污栅 fixed trash rack
清污机 screen cleaning device
拦污漂 trash boom
浮式拦漂排 floating boom

(2) 主要零部件及结构件 Main components and elements

门叶结构 gate leaf
闸门面板 skin plate
主横梁（横向主梁） horizontal girder
主纵梁(竖向主梁) vertical girder
单腹板梁 single girder

双腹板梁　double girder
边梁　side beam
隔板　diaphragm
加劲板、加筋板　stiffener
转角水封　corner seal
节间水封　seal between sections
转铰水封　swing seal device
双向水封　bidirectional seal
橡塑复合水封　PTFE-covered seal
P 型橡皮水封　P-type rubber seal
I 型水封　flat type seal
Ω 型水封　center bulb seal
充压式水封　pressure-actuated seal
压紧式水封　clamp seal，compressed seal
压模法　molded
挤压法　extruded
水封座板　base plate for seal
水封压板　keeper plate，clamp bar
前翼缘　upstream flange
后翼缘　downstream flange
腹板　web
支撑　end support
滑动支撑　sliding support
滚动支撑　roller support
滑块（支承滑块）　slide block，bearing block
定轮　fixed wheel
简支轮　simple support wheel
悬臂轮　cantilevered wheel
踏面　rolling face
链条　roller chain
链板　chain plate
辊轮　roller
均衡座，支承走道　roller race
吊耳　lifting eye（lifting hook）
充水阀　filling valve
拉杆　rod，hanger
锁定装置　dogging device（dog）
栅叶　rack
栅架　rack frame
栅条　rack bar
支臂　radial arm，end frame
直支臂　parallel radial arm
斜支臂　inclined radial arm
臂柱　arm column，strut
臂柱加强杆（臂柱连接系）　strut bracing
边柱　side column
支铰　trunnion
固定铰　trunnion yoke
活动铰　trunnion hub
支铰轴　trunnion pin
轴套　bushing，sleeve
球面滑动轴承　spherical plain bearing
自润滑轴承　maintenance free bearing，self-lubricating bearing
滚动轴承　antifriction bearing，rolling bearing
径向轴承　transverse bearing，radial bearing
推力轴承　axial bearing，thrust bearing
顶枢　top anchorage assembly
底枢　pintle assembly
斜接柱　miter end
门轴柱　quoin post
导卡　miter guide
斜接柱支枕垫　miter block
门轴柱支枕垫　quoin block
环氧填料　epoxy filler
蘑菇轴头　pintal
承轴巢　pintle socket
承轴台　pintle shoe
门背连接系，背拉杆　diagonal
角撑板　gusset plate
门槽　gate slot，gate groove
突扩式门槽　sudden-expand gate slot
底槛　bottom sill
门楣　lintle
主轨　track
反轨　converse guide
止水板　seal plate
侧轨　side guide
铁轨　railroad
起重机轨　crane rail
护角　armor angle
支铰大梁　trunnion girder，trunnion beam

(3) 主要参数　Main parameters

孔口型式　opening type
闸门型式　gate type
门槽型式　gate slot type
孔口净宽　opening span（opening width）
孔口净高　opening height
弧门半径　gate radius，skin plate radius

总水压力　total hydrostatic load
支承型式　type of end support
支承跨度　support span
止水间距　span of side seals
止水高度　sealing height
操作条件　operating condition
静水启闭　operation under balanced head
动水启闭　closing and opening in flow
平压方式　method of water pressure balance
节间充水　water filling by gate split
旁通管充水　water filling by pass pipe
动水小开度提门充水　water filling by gate opening in gap
吊点距　distance between lifting eyes
启门力　lifting force
闭门力　closing force
持住力　holding force
水柱压力　water column pressure

(4) 制造和安装　Fabrication and installation

闸门安装平台　working platform for insta-lling gate
碳素结构钢　carbon steel
优质碳素结构钢　quality carbon structural steel
低合金结构钢　low-alloy structural steel
合金结构钢　alloy structural steel
不锈钢　stainless steel
不锈钢复合钢板　composite stainless steel plate
锻件　steel forging
铸钢件　steel casting
铸铁件　cast-iron
结构件　structural component
标准件　standard component
焊接件　weldment
装配件　assembly
化学成分　chemical composition
机械性能　mechanical property
装配焊接　erection welding
焊缝　weld
剖口　groove
堆焊　overlay welding
塞焊　plug weld
熔透焊缝　full-penetration weld
角焊缝　fillet welds
包角焊　seal welds
焊接工艺　welding process
焊工资质　welder's qualification
焊缝分类　clarification of welds
焊缝检验　weld inspection，weld testing
焊接变形　weld deformation
焊接质量　weld quality
热影响区　heat affected zone
预热　preheat
清根　back gouging
碳弧气刨　air carbon arc gouging
消氢处理　de-hydrogenation treatment
外观检测　visual inspection
无损探伤　non-destructive testing
渗透探伤　penetrating testing (PT)
超声波探伤　ultrasonic testing (UT)
磁粉探伤　magnetic particle testing (MT)
射线探伤　radiographic testing (RT)
缺陷　defects
裂纹　crack
夹渣　slag
咬边　undercut
气孔　porosity
未焊透　incomplete fusion，cold weld
缺陷焊缝　poor weld
未熔透焊缝　skin weld
消除应力　stress relief
调质　quenched and tempered
焊前预热　pre-heating of welding piece
焊后热处理　heat treatment after welding
正火　normalizing
退火　annealing
回火　tempering
单个构件　individual member
零件　component
复杂构件　complex assembly
工艺流程　workmanship procedure
下料　cutting
机加工　machining，machine work
加工精度　machining precision
公差　tolerance
错位　dislocation
间隙　gap
扭曲　distortion
螺栓连接　bolted connection

高强螺栓 high-strength bolts
摩擦面 friction surface
抗滑移系数 slip coefficient
预拉力 pretension force
施工扭矩 construction torque
检查扭矩 check torque
初拧 early screw
终拧 final screw
测力扳手 torque wrench
螺母 screw nut
垫圈 washer
弹簧垫圈 spring washer
止动垫圈 lock washer
螺孔 screw hole
工厂预组装 shop pre-assembled
出厂验收 factory acceptance
现场拼装 site assembly
静平衡试验 static balance test
防腐 corrosion prevention
涂漆 painting
表面预处理 surface preparation
表面清洁度 surface cleanness
涂料配套（涂层系统） coating system
热喷涂 thermal spray
涂料 paint；coating
阴极保护 cathodic protection
牺牲阳极 sacrificial anode
强制电流 impressed current
无机改性水泥浆 inorganic modified grout
镀锌钝化 galvanized and passivation
底涂层 primer coat
封闭涂层 barrier coat
中间漆 intermediate coat
面漆 surface coat
干漆膜厚 dry paint film thickness
涂层道数 number of coats
涂层附着力 coating adhesion
划格法［涂层附着力试验］ knife and tape test
拉开法［涂层附着力试验］ pull-off test
备品备件 spare parts

7.2 启闭机
Hoist

(1) 启闭机类型 Types of hoists

门式启闭机（门机） gantry crane
双向门机 trolley-mounted gantry crane
单向门机 gantry crane with fixed hoist
台车式启闭机（台车） trolley
桥式启闭机（桥机） bridge crane
固定式启闭机 fixed hoist
卷扬式启闭机 wire rope hoist，cable hoist
液压启闭机 hydraulic hoist
螺杆启闭机 screw stem hoist
链式启闭机 chain hoist
电动葫芦 electric hoist
单梁电动葫芦 monorail hoist，monorail crane
齿轮齿条机构 rack and pinion mechanism
移动式起重机（汽车吊） mobile crane
悬臂起重机 jib crane
盘香式启闭机 incense coil hoist

(2) 主要机构和零部件 Main mecha-nism and components

起升机构 lifting mechanism
行走机构（运行机构） travelling mechanism
回转机构 slewing mechanism
小车 trolley，truck
大车 railroad car
机架 base frame
门架 gantry
门腿 support column
机房 machine room
司机室 cab
车轮组平衡架 balancing stand of wheels
起升高度指示器 gate position indicator
极限位置装置 limit switch；limiter
编码器 encoder
绝对型编码器 absolute rotary encoder
荷载指示器 load indicator
荷载限制器 load limiter
抓梁 lifting beam
液压自动抓梁 lifting beam with remote hydraulic-control of latching and unlatching
机械式自动抓梁 lifting beam with latching and unlatching mechanisms
自动挂脱梁 hook-release lifting beam
移轴装置 hook-release device
电动机 electric motor
联轴器 coupling
制动器 brake

盘式制动器 disc brake
块式制动器 shoe type brake
减速器（减速箱） speed reducer
三合一减速器 gearmotor
行星齿轮减速器 planetary reduction
开式齿轮 open gearing
限速器 speed governor，overspeed governor
卷筒 wire rope drum
折线式卷筒 lebus groove drum
钢丝绳 wire rope
镀锌钢丝绳 drawn galvanized rope wire
金属芯钢丝绳 independent wire rope core (IWRC) wire rope
纤维芯钢丝绳 fiber core (FC) wire rope
线接触钢丝绳 linear contact lay wire rope
面接触钢丝绳 facial contact lay wire rope
钢丝绳涂油器 rope lubrication device
定滑轮 upper pulley，sheave
动滑轮组 movable block
平衡滑轮 equalizer pulley (sheave)
车轮 wheel
轨道 crane rail，track
夹轨器 rail clamping device
阻进器、缓冲器 buffer
锚定装置 anchor
风速仪 anemograph
避雷器 arrester
集中润滑 centralization lubrication
递进式集中润滑 progressive centralization lubrication
浸油润滑 oil-bath lubrication
喷油润滑 spray lubrication
吊钩 hook
螺栓 screw
蜗杆 worm
有杆腔 rod chamber
无杆腔 rodless chamber
蜗轮 worm gear
滑触线 trolley conductor
电缆卷筒 cable drum
液压系统 hydraulic system
开环控制同步纠偏回路 synchro-control circuit with open-cycle
闭环控制同步纠偏回路 synchro-control circuit with closed-cycle
液压泵 pump
变量泵 variable displacement pump
定量泵 fixed displacement pump
轴向柱塞泵 axial piston pump
叶片泵 vane pump
齿轮泵 gear pump
手动变量轴向柱塞泵 manual variable displacement axial piston pump
液压阀 hydraulic valve
伺服阀 servo valve
比例控制阀 proportional control valve
滑阀 spool valve
插装阀 cartridge valve
换向阀 directional control valve
电磁换向阀 solenoid operated directional valve
电液换向阀 electro-hydraulic directional control valve
电液比例换向阀 electro-hydraulic proportional directional valve
压力阀 pressure valve
顺序阀 sequence valve
减压阀 pressure reducing valve
流量阀 flow valve
节流阀 throttle valve
调速阀 flow control valve
单向阀 check valve
液控单向阀 pilot operated check valve
平衡阀 counterbalance valve
直动式 direct operated
先导式 pilot operated
压力传感器 pressure sensor
油箱 oil reservoir (oil tank)
液压油 hydraulic oil
滤油器 oil filter
滤油车 oil filter vehicle
脱水过滤汽化装置 water-extraction and filtration unit
带滤清器的呼吸器 breather cap with air filter
除湿剂 dehumidizer
液位计 oil level gage；liquidometer
液位传感器 oil level sensor
声波发射器 pinger
注油口 pouring orifice
磁性吸铁装置 magnetic particle unit

加热装置 heater
冷却器 cooler
压力表 pressure gage
压力表开关 pressure meter switch
压力继电器 pressure relay
集成式行程检测装置 integrated travel detector
内置式行程检测装置 built-in travel detector
蓄能器 accumulator
液压缸 cylinder
活塞杆 piston rod
镀铬活塞杆 chromcplate steel rod
不锈钢活塞杆 stainless rod
陶瓷活塞杆 ceramic-coated rod
活塞 piston
端盖 end cover
导向套 guide sleeve (bush)
密封圈（密封件） sealing ring (element)
天然橡胶 natural rubber (NR)
丁腈橡胶 acrylo nitrile butadiene rubber (NBR)
聚氨酯 polyurethane (PUR)
聚四氟乙烯 Polytetrafluoroethene (PTFE)
氟橡胶 fluoroelastomer (FPM)
尼龙 polyaminde (PA), Nylon
防尘圈 wiper seal
刮污环 scraper ring
液压马达 hydraulic motor
液压管道、油管 hydraulic tubing, hydraulic pipeline
吸油管 oil suction pipe
回油管 oil return pipe
软管 flexible hose
管接头 pipe fittings
焊接式管接头 weld fittings

(3) 主要参数 Main parameters

启闭容量（启闭力） hoisting capacity
工作行程 operating stroke
启闭速度 hoist speed
轨距 track gauge
轮距 wheel track
轴距 wheel base
工作级别 classification group
工作寿命 service life
短时工作制 short-time duty
防护等级 degree of protection
多层缠绕 spooling of a wire rope in multilayer
预拉伸 pre-stretching
排量 displacement
正排量 positive displacement
额定压力 rated pressure
工作压力 operating pressure
压力损失 pressure loss
沿程损失 linear pressure loss
局部损失 local pressure loss
泄漏 leak
过滤能力 filter capacity
过滤精度 filter fineness
滤芯强度 filter element strength
耐油 oil resistance
防水（耐水） waterproof
防尘 dust-tight
防震 quakeproof
防锈 rust protection
抗老化 ageing resistance
耐压性 resistance to pressure
抗撕裂强度 tearing strength
清洁度 cleanliness
热备用 hot standby
自动复位 automatic reset

(4) 制造和安装 Fabrication and installation

工程陶瓷涂层 ceramic coating
磷化处理 phosphating
镀镍处理 nickel plating
配管 piping
酸洗 acid pickling
专用工具 special tool
率定（校准） calibration
径向跳动 circular runout
齿轮副侧隙 side clearance of both gears
出厂试验 shop test
空载试验 no-load test
负荷试验 load test
静载试验 static test
动载试验 dynamic test
最低动作压力试验 start-up pressure test
耐压试验 pressure test
内泄漏试验 inner-leakage test

外泄漏试验 outer-leakage test
试验大纲 testing schedule

7.3 升船机
Shiplift

(1) 升船设备 Equipment of ship lifting

升船机 shiplift (ship elevator)
垂直升船机 vertical ship lift
斜面升船机 inclined ship lift
配重式（平衡重式） counterweight type
全平衡式 full balanced type
部分平衡式 partial balanced type
浮筒式 flotation tank type
湿运式升船机 wet shiplift
干运式升船机 dry shiplift
钢丝绳卷扬式 winch，wire rope hoist
齿轮齿条式 rack and pinion
液压顶升式 hydraulic jack

(2) 主要机构和零部件 Main mechanism and components

驱动系统 drive system
安全机构 safety mechanism
承船厢 ship chamber
承船车、承船架 ship carriage，platform
主提升设备 main hoist
牵引绞车 winch
平衡重 counterweight
力矩平衡重 torque counterweight
重力平衡重 gravity counterweight
可控平衡重 gravity counterweight
平衡链 balance chain
对接装置 connection device
对接锁定装置 locking device for connection
顶紧装置 push device
间隙密封装置 clearance seal device
防撞装置 anticollision device
导向装置 guiding device
缓冲装置 buffer device
承船厢调平装置 chamber leveling device
机械同步 mechanical synchronization
同步轴 synchronizing shaft
充泄水系统 water filling and sluicing system
安全制动器 safety brake
扭矩传感器 torque sensor
润滑系统 lubrication system
护舷 fender
系船柱 bollard
高低轮 high-low wheels
高低轨 high-low tracks
钢丝绳张力均衡装置 balancing device of wire rope tension
钢丝绳张力检测设备 inspection device of wire rope tension
托轮 riding wheel
托辊 roller
侧铲 side shovel

(3) 主要参数 Main parameters

通过能力 through capacity
船队吨位 shipping tonnage
客货运量 passenger and freight volume
年单向通过能力 annual through capacity in one-way
通航水头 head of navigation
水位变幅 water level amplitude
水位变率 change rate of water level
提升力 lifting force，hoist capacity
升程（提升高度） lifting range (lift height)
提升速度 lifting speed
驱动功率 drive power
船厢总重 gross weight of ship chamber
承船厢有效尺度 valid size of ship chamber
允许误载水深 permitting water depth misloading
承船厢整体挠度 deflection of ship chamber
卷筒直径与钢丝绳直径比 drum to wire rope diameter ratio
承船厢干舷高 chamber freeboard

(4) 安装 Installation

船厢水平度 chamber levelness
纵向倾覆 lengthways overturn
卷筒直径制造误差 manufacture error of drum diameter
钢丝绳直径制造误差 manufacture error of wire rope diameter
钢丝绳弹性模量偏差 elasticity modulus error of wire rope
承船厢结构制造对称性偏差 symmetry error of

ship chamber
静态调平　leveling in static state
动态调平　leveling in dynamic state
预加力矩　pre-torque
整机组装　assembly of the whole
整机空载试验　no-load test of the whole
分系统调试　subsystem debagging
整机总联调　joint debagging of the whole
试通航　trial navigation

7.4 主要图表名称
Titles of figures and tables

闸门与启闭机布置图　Layout of gates and hoists
闸门总图　General drawing of gates
门叶总图　General drawing of gate leafs
门槽总图　General drawing of gate slots
充水阀总图　General drawing of filling valve assembly
定轮总图　General drawing of wheel assembly
水封总图　General drawing of seal details
启闭机总图　General drawing of hydraulic hoist (gantry crane)
液压系统图　General drawing of hoist hydraulic system
液压缸总图　General drawing of hoist cylinder assembly
升船机总图　General drawing of shiplift
产品安装维护使用说明书　Erection, maintenance and operation manual

第8章 电 工
Electricity

8.1 通用
General

电 electricity
电荷 electric charge
电场 electric field
电场强度 electric field intensity (strength)
静电场 electrostatic field
静电感应 electrostatic induction
电势 electric potential
电位 electric potential
电势差（电位差） electric potential difference
等位线 equipotential line
等位面 equipotential surface
等位体 equipotential volume
地电位 earth potential
电压 voltage
电压降 voltage drop，potential drop
电动势 electromotive force (e. m. f)
电源电压 source voltage
反电动势 counter-electromotive force，back electromotive force
电介质 dielectric
介电常数 dielectric constant，permittivity
电流 electric current
全电流 total current
极化电流 polarization current
磁场 magnetic field
磁场强度 magnetic field strength
磁通量 magnetic flux
磁感应 magnetic induction
磁通密度 magnetic flux density
安匝 Ampere-turn
自感 self-inductance
自感应 self-induction
自感电动势 self-induced e. m. f
互感 mutual-inductance
互感应 mutual induction
互感电动势 mutual induced e. m. f
感应电压 induced voltage
耦合 coupling
耦合系数 coupling coefficient
磁化强度 magnetization intensity
磁化 magnetization
磁化电流 magnetizing current
绝对磁导率 absolute permeability
相对磁导率 relative permeability
磁化率 magnetic susceptibility
磁化曲线 magnetization curve
磁滞 magnetic hysteresis
磁滞回线 magnetic hysteresis loop
磁滞损耗 magnetic hysteresis loss
磁饱和 magnetic saturation
剩磁（剩余磁化强度） remanent magnetization
剩磁通密度 magnetic remanence
退磁 demagnetization
磁体 magnet
磁极 magnetic pole
铁磁性 ferromagnetism
铁磁性物质 ferromagnetic substance
永久磁体 permanent magnet
永磁材料 permanent magnet material
软磁材料 magnetically soft material，soft magnetic material
磁屏蔽 magnetic screen
涡流 eddy current
涡流损耗 eddy current loss
趋肤效应 skin effect
邻近效应 proximity effect
电磁场 electromagnetic field
电磁能 electromagnetic energy
电磁波 electromagnetic wave
电磁力 electromagnetic force
电磁感应 electromagnetic induction
电磁干扰 electromagnetic interference (EMI)
电磁兼容 electromagnetic compatibility (EMC)

电磁辐射　electromagnetic radiation
电磁屏蔽　electromagnetic screen
电磁体　electromagnet
波导　waveguide
楞次定律　Lenz law
法拉第定律　Faraday law
焦耳效应　Joule effect
焦耳定律　Joule law
压电效应　piezoelectric effect
光电效应　photoelectric effect
光电发射　photoelectric emission
电-光效应　electro-optic effect
接触电位差　contact potential difference
霍尔效应　Hall effect
法拉第效应　Faraday effect
直流电流　direct current（DC）
直流电压　direct voltage（DC voltage）
交流电流　alternating current（AC）
交流电压　alternating voltage（AC voltage）
周期　period
频率　frequency
角频率　angular frequency
相位　phase
相量　phasor
相角差　phase difference
相位差角　phase difference angle
相位移角　displacement angle
相位超前　phase lead
相位滞后　phase lag
正交　in quadrature
反相　in opposition
同相　in phase
相量图　phasor diagram
圆图　circle diagram
峰值　peak value
峰-峰值　peak-to-peak value
谷值　valley value
峰-谷值　peak-to-valley value
瞬时值　instantaneous value
方均根值（有效值）　root-mean-square value (r. m . s value)
脉冲　pulse
电路　electric circuit
电路图　circuit diagram
电路元件　circuit element
集中参数电路　lumped circuit
分布参数电路　distributed circuit
线性电路　linear circuit
非线性电路　nonlinear circuit
理想电压源　ideal voltage source
理想电流源　ideal current source
负荷（负载）　load
用电　electric power consumption
电量　electrical energy
导体　conductor
超导体　superconductor
电阻　resistance
电阻器　resistor
电导　conductance
电导率　conductivity
电阻率　resistivity
电感　inductance
电感器　inductor
电抗器　reactor
电抗　reactance
感抗　inductive reactance
电容　capacitance
容抗　capacitive reactance
阻抗　impedance
阻抗的模　modulus of impedance
输入阻抗　input impedance
输出阻抗　output impedance
转移阻抗　transfer impedance
导纳　admittance
电纳　susceptance
感纳　inductive susceptance
容纳　capacitive susceptance
阻抗匹配　impedance matching
串联　series connection
互联　interconnection
星形连接　star connection
三角形连接　delta connection
多边形连接　polygon connection
回路　loop
支路　branch
节点　node
网孔　mesh
网络　network
端子　terminal
端口　port

等效网络　equivalent network
有源网络　active network
无源网络　passive network
网络变换　network transformation（conversion）
网络化简　network simplification
星形-三角形变换　star-delta conversion，star-delta transformation
三角形-星形变换　delta-wye conversion，delta-star transformation
状态变量　state variable
状态方程　state equation
状态矢量　state vector
欧姆定律　Ohm law
基尔霍夫第一定律　Kirchhoff's first law
基尔霍夫第二定律　Kirchhoff's second law
基尔霍夫电流定律　Kirchhoff's current law（KCL）
基尔霍夫电压定律　Kirchhoff voltage law（KVL）
叠加定理　superposition theorem
初始条件　initial condition
稳态　steady state
稳态分量　steady state component
瞬态　transient
瞬态分量　transient component
激励　excitation，stimulus
响应　response
时间常数　time constant
衰减　decay
强制振荡　forced oscillation
阻尼振荡　damped oscillation
自由振荡　free oscillation
功率　power
瞬时功率　instantaneous power
有功功率　active power
无功功率　reactive power
视在功率　apparent power
复功率　complex power
功率因数　power factor
谐振　resonance
铁磁谐振　ferroresonance
电压谐振　voltage resonance
串联谐振　series resonance
电流谐振　current resonance
并联谐振　parallel resonance
谐振频率　resonance frequency
谐振曲线　resonance curve
频率特性　frequency characteristics
固有频率　inherent frequency，natural frequency
频带　frequency band
通带　pass band
带通滤波器　band pass filter
带阻滤波器　band stop filter
磁路　magnetic circuit
磁阻　reluctance
磁导率　permeability
主磁通　main flux
漏磁通　leakage flux
三相制　three phase system
三相四线制　three phase four wire system
对称三相电路　symmetric three phase circuit
多相制　multiphase system，polyphase system
相序　phase sequence
中性导体　neutral conductor
中性点　neutral，neutral point
相电压　line to neutral voltage，phase to neutral voltage
线电压　line-to-line voltage
相电流　phase current
线电流　line current
多相网络的不平衡状态　unbalanced state of a polyphase network
中性点位移　neutral point displacement
对称分量法　method of symmetrical component
正序分量　positive sequence component
负序分量　negative sequence component
零序分量　zero sequence component
非正弦周期量　non-sinusoidal periodic quantity
基波　fundamental，fundamental component
二次谐波　second harmonic component
高次谐波　harmonic，harmonic component
谐波分析　harmonic analysis
直流分量　DC component
基频　fundamental frequency
基波功率　fundamental power
基波因数　fundamental factor
总畸变因数　total distortion factor（TDF）
谐波因数　harmonic factor
谐波含量　harmonic content
谐波次数　harmonic number，harmonic order
总谐波畸变　total harmonic distortion（THD）
脉动因数　pulsation factor

有效纹波因数　r. m. s ripple factor
峰值纹波因数　peak-ripple factor，peak distortion factor
拍　beat
拍频　beat frequency
傅里叶级数　Fourier series
拉普拉斯变换　Laplace transform
拉普拉斯逆变换　inverse Laplace transform
傅里叶变换　Fourier transform
快速傅里叶变换　fast Fourier transform（FFT）
傅里叶逆变换　inverse Fourier transform
标积　scalar product
点积　dot product
向量积　vector product
叉积　cross product
频谱　frequency spectrum
传递函数　transfer function
微分电路　differential circuit
积分电路　integrating circuit
运算放大器　operational amplifier
理想变压器　ideal transformer
阻抗变换器　impedance converter
特性阻抗　characteristic impedance
行波　traveling wave
波长　wave length
入射波　incident wave
反射波　reflected wave
折射波　refracted wave
反射系数　reflection coefficient
折射系数　refraction coefficient
驻波　standing wave
SI基本单位　SI base unit
SI导出单位　SI derived unit
安培　Ampere
牛顿　Newton
焦耳　Joule
瓦特　Watt
伏特　Volt
欧姆　Ohm
库仑　Coulomb
法拉　Farad
亨利　Henry
赫兹　Hertz
西门子　Siemens
韦伯　Weber
特斯拉　Tesla
乏　var
高斯　Gauss
奥斯特　Oersted
麦克斯韦　Maxwell

8.2 电力系统及电气接线
Electric power system and connection

电力系统　electric power system
电网　electric power grid
交流系统　alternating current system（AC system）
直流系统　direct current system（DC system）
一次系统　primary system
二次系统　secondary system
工频　power frequency
系统标称电压　nominal voltage of power system
电压等级　voltage level
低压　low voltage（LV）
中压　medium voltage（MV）
高压　high voltage（HV）
超高压　extra-high voltage（EHV）
特高压　ultra-high voltage（UHV）
高压直流　high voltage direct current（HVDC）
特高压直流　ultra-high voltage direct current（UHVDC）
系统运行电压　operating voltage of power system
发电厂出力　output of power plant
发电计划　generation schedule
基荷机组　base load set
调节负荷机组　controllable set
峰荷机组　peak load set
输电　transmission of electricity，power transmission
交流输电　AC power transmission
直流输电　DC power transmission
输电容量　transmission capacity
输电效率　transmission efficiency
变电　transformation of electricity，power transformation
供电　power supply
间隔　bay
电网结构　network structure，network configuration
智能电网　smart grid
主干电网　main grid
输电线路　transmission line
架空线路　overhead line

联络线　interconnection line
发电厂接入系统　generation interconnection
发电厂送出工程　generation interconnection project
单电源供电　single supply
双电源供电　duplicate supply
馈线　feeder
辐射馈线　radial feeder
支线　branch line
T接线路　tapped line，teed line
中性点接地方式　neutral point treatment
中性点不接地系统　isolated neutral system
中性点直接接地系统　solidly earthed neutral system
中性点阻抗接地系统　impedance earthed neutral system
中性点消弧线圈接地系统　arc-suppression-coil earthed neutral system
中性点谐振接地系统　resonant earthed neutral system
中性点经变压器接地系统　transformer earthed neutral system
系统阻抗　system impedance
故障阻抗　fault impedance
系统稳态　steady state of power system
潮流　load flow
潮流计算　load flow calculation
标幺制　per unit system
标幺值　per unit value
有名制　ohmic system
有名值　actual value
基准值　base value
无限大母线　infinite bus
参考节点　reference node
无源节点　passive bus
负荷节点　load bus
电压控制节点　voltage controlled bus
松弛节点　slack bus
节点导纳矩阵　bus admittance matrix，Y bus matrix
平衡节点　balancing bus
节点阻抗矩阵　bus impedance matrix，Z bus matrix
负荷数学模型　mathematical model of load
负荷电压特性　voltage characteristics of load
负荷频率特性　frequency characteristics of load
动态负荷特性　dynamic characteristics of load
静态负荷特性　steady state characteristics of load
系统短路　short-circuit
短路计算　short-circuit calculation
短路容量　short-circuit capacity
短路电流允许值　short-circuit current capability
远端短路　far-from generator short circuit
近端短路　near-to generator short-circuit
短路电流　short-circuit current
短路点电流　current at the short-circuit point
故障电流　fault current
故障点电流　current at the fault point
短路电流交流分量　AC component of short-circuit current
短路电流周期分量　periodic component of short-circuit current
对称短路电流　symmetrical short-circuit current
对称短路视在功率　symmetrical short-circuit apparent power
短路电流直流分量　DC component of short-circuit current
短路电流非周期分量　aperiodic component of short-circuit current
直流时间常数　DC time constant
短路电流峰值　peak short-circuit current
稳态短路电流　steady state short-circuit current
瞬态短路电流　transient short-circuit current
超瞬态短路电流　subtransient short-circuit current
短路电流的热效应　heat effect of short-circuit current
金属性短路　dead short
绝缘故障　insulation fault
永久性故障　permanent fault（persistent fault）
瞬时故障　transient fault
线路故障　line fault
母线故障　busbar fault
消缺处理　defect elimination
负荷转移　load transfer
三相对称短路　symmetrical short-cicuit，three phase short-cicuit
不对称短路　unsymmetrical short-cicuit
单相接地短路　phase-to-earth fault，single line-to-earth short-cicuit

两相短路 phase-to-phase short-cicuit, line-to-line short-cicuit
两相对地短路 two phase-to-earth short-cicuit, double-line-to-earth short-cicuit
相间短路 interphase short-cicuit
双重故障 double faults
发展性故障 developing fault
故障清除 fault clearance
正序网络 positive sequence network
负序网络 negative sequence network
零序网络 zero sequence network
正序阻抗 positive sequence impedance
负序阻抗 negative sequence impedance
零序阻抗 zero sequence impedance
正序短路电流 positive sequence short-circuit current
负序短路电流 negative sequence short circuit current
零序短路电流 zero sequence short-circuit current
运算曲线法 calculation curve method
系统暂态 transient state of power system
暂态过程 transient process
小扰动 small disturbance
大扰动 large disturbance
负荷稳定性 load stability
电压稳定性 voltage stability
频率稳定性 frequency stability
电力系统静态稳定性 steady state stability of power system
电力系统动态稳定性 dynamic stability of power system
电力系统暂态稳定性 transient stability of power system
稳定裕度 stability margin
快速励磁 high response excitation
交流电机内角 internal angle of an alternator
两电动势间相角差 angle of deviation between two e. m. f. 's
电力系统振荡 power system oscillation
自持振荡 self-sustained oscillation
低频振荡 low frequency oscillation
同步摇摆 synchronous swing
振荡周期 oscillation period
电力系统瓦解 power system collapse
黑启动 black start
电力系统运行 power system operation
正常运行方式 normal operation mode
稳态运行 steady state operation
正常检修运行方式 normal maintenance operation mode
两相运行 two-phase operation
同步时间 synchronous time
系统同步运行 synchronous operation of power system
发电机组计划运行 scheduled operation of a generating set
电机同步运行 synchronous operation of a machine
同步电机异步运行 asynchronous operation of synchronous machine
失步运行 out-of-step operation
再同步 resynchronization
两系统同步 synchronization of two systems
准同步并列 ideal synchronization
电网解列 islanding, network splitting
合环 ring closing
解环 ring opening
孤立系统 isolated power system, island in a power system
孤立运行 isolated operation
互联运行 interconnected operation
分网运行 separate network operation
电力系统异常运行 abnormal operation of power system
不平衡运行 unbalanced operation
异步运行 asynchronous operation
非全相运行 open phase operation
功率调节 power control
频率调节 frequency control
调频方式 mode of frequency regulation
一次调频 primary control of the speed of generating sets
二次调频 secondary control of active power in a system
发电机组二次功率调节 secondary power control operation of a generating set
机组调差系数 droop of a set
电力系统调差系数 droop of power system
电力系统功率调节特性 power characteristics of power system
功率调节范围 controlling power range
发电机组的调节范围 control range of a generating set

功率角　load angle
功角特性　load-angle characteristic
线路过负荷能力　overload capacity of transmission line
联络线输送容量　transmission capacity of a link
调压方式　voltage control method
无功功率电压调节　reactive power voltage control
供电连续性　continuity of power supply
停电　outage，interruption
每千瓦时停电损失　loss of power outage per kWh
负荷备用　load reserve
事故备用　emergency reserve
检修备用　maintenance reserve
旋转备用（热备用）　spinning reserve（hot reserve）
冷备用　cold standby reserve
备用容量　reserve capacity
联络线负荷　connection line load，tie-line load
自动低频减负荷　automatic underfrequency load shedding
负荷的功率调节系数　power regulation coefficient of load
电力系统调峰　peak load regulating of power system
调峰容量　peak load regulating capacity
调峰填谷　peak shaving and valley filling
调频容量　frequency regulating capacity
调相容量　synchronous condensing capacity
满发利用小时数　full output power hours
年利用小时数　annual utilization hours
电力平衡　power balance
电量平衡　energy balance
电力系统有功功率平衡　active power balance of power system
电力系统无功功率平衡　reactive power balance of power system
电力负荷曲线　power load curve
日负荷曲线　daily load curve
周负荷曲线　weekly load curve
有功负荷　active load
无功负荷　reactive load
低谷负荷　valley load
负荷同时率　load coincidence factor
有功电能　active energy
无功电能　reactive energy
冲击负荷　impact load
负荷中心　load center
不对称负荷　unsymmetrical load
功率损耗　power loss
电能损耗　energy loss
输电损耗　transmission loss
电能质量　power quality
公共连接点　point of common coupling
供电质量　quality of supply
用电质量　quality of consumption
频率偏差　deviation of frequency
电压质量　voltage quality
电压偏差　deviation of voltage
电压恢复　voltage recovery
电压消失　loss of voltage
电压调整　voltage regulation
三相电压不平衡　three phase voltage unbalance
三相电流不平衡　three phase current unbalance
电压波动　voltage fluctuation
波形畸变　waveform distortion
基波分量　fundamental component
谐波分量　harmonic component
谐波源　harmonic source
电压暂降　voltage dip，voltage sag
低电压穿越　low voltage ride through（LVRT）
故障穿越　fault ride through（FRT）
日负荷　daily load
日最大负荷利用小时　daily maximum load utilization hours
检修间隔　maintenance interval
机组技术数据　unit technical data
机组开机时间　start-up time of unit
机组停机时间　shutdown time of unit
零起升压　voltage build up from zero
残压起励建压　voltage build up from residual levels
实际有功出力曲线　actual active power output curve
实际无功出力曲线　actual reactive power output curve
备用容量曲线　reserve capacity curve
变电站　substation
智能变电站　smart substation
数字化变电站　digital substation

无人值班变电站 unmanned substation
敞开式变电站 open-type substation
气体绝缘金属封闭变电站 gas insulated metal-enclosed substation
户内变电站 indoor substation
户外变电站 outdoor substation
升压变电站 step-up substation
降压变电站 step-down substation
开关站 switchyard
发电厂 power plant (power station)
基荷电厂 base load power plant
调峰电厂 peak load power plant
孤立电厂 isolated power plant
可靠性 reliability
可用性 availability
可维修性 maintainability
平均无故障工作时间 mean time to failure (MTTF)
平均失效间隔时间 mean time between failures (MTBF)
运行时间 operating duration
备用时间 stand-by duration
可用时间 up duration
不可用时间 down duration (outage duration)
计划停运时间 planned-outage duration, scheduled-outage duration
维修时间 maintenance duration
强迫停运时间 forced-outage duration
修复时间 repair duration
停电持续时间 interruption duration
停运率 outage rate
利用系数 utilization factor
故障停电平均持续时间 average interruption duration
供电可靠性 service reliability
单费率电价 flat rate tariff
分时电价 time-of-day tariff
峰荷电价 peak-load tariff
非峰时电价 off-peak tariff
单母线接线 single-bus configuration
单母线分段接线 sectionalized single-bus configuration
单母线带旁路接线 single-bus with transfer bus configuration (main and transfer bus configuration)
双母线接线 double-bus configuration
双母线分段接线 sectionalized double-bus configuration
双母线带旁路接线 double-bus with transfer bus configuration
桥形接线 bridge configuration
一个半断路器接线 one-and-a-half breaker configuration, breaker-and-a-half configuration
双断路器接线 double-breaker configuration
三分之四断路器接线 4/3 breaker configuration
角形接线 ring bus configuration
变压器-线路组接线 transformer-line configuration
母线 busbar
工作母线 main busbar
备用母线 reserve busbar
旁路母线 transfer busbar

8.3 发电机与发电电动机 Generator and generator-motor

水轮发电机 hydrogenerator
立式水轮发电机 vertical hydrogenerator
卧式水轮发电机 horizontal hydrogenerator
伞式水轮发电机 umbrella hydrogenerator
半伞式水轮发电机 semi-umbrella hydrogenerator
悬式水轮发电机 suspended hydrogenerator
灯泡式水轮发电机 bulb type hydrogenerator
同步调相机 synchronous condenser
定子 stator
定子铁芯 stator core
定子绕组 stator winding
定子线圈 stator coil
定子线棒 stator coil bar
定子机座底板 soleplate of the stator frame
电枢绕组 armature winding
叠绕组 lap winding
波绕组 wave winding
绕组绝缘 winding insulation
上机架 upper bracket
下机架 lower bracket
上盖板 upper cover
上导轴承 upper guide bearing
下导轴承 lower guide bearing
转子 rotor
凸极电机 salient pole machine
励磁绕组 excitation winding, field winding

转子绕组　rotor winding
滑环　collector ring，slipring
集电环　collector ring，slipring
转子风扇　rotor fan
阻尼绕组　damping winding，amortisseur winding
感应电动势　induced electromotive force，induced voltage
转子中心体　rotor hub，spider hub
转子支架　spider
磁轭　magnetic yoke
推力轴承　thrust bearing
推力头　thrust block
推力轴承机架　thrust-bearing bracket
镜板　thrust runner，thrust collar
推力瓦　thrust bearing shoe
导轴瓦　guide bearing shoe，guide bearing segment
巴氏合金瓦　babbitt bearing
金属塑料瓦　metal-PTFE bearing，metal-TEFLON bearing
钨金瓦　babbitt bearing
定子机座　stator frame
主引出线，相线端子　main terminals，phase terminals
中性点端子　neutral terminals，neutral leads
励磁绕组端子　field winding terminals，excitation leads
制动闸　brake，jack
下机架底板　soleplate of the lower bracket
下部主轴　lower shaft
上部主轴　upper shaft
传动端　drive end（DE）
非传动端　non-drive end（NDE）
滑环　slipring
滑环盖　slipring cover
刷握　brush holder
转子磁轭　rotor rim，rotor yoke
转子支架　spider
埋入式测温计　embedded temperature detector
电晕防护　corona shielding
防电晕层　anti-corona coating
轴承绝缘　bearing insulation
旋转磁动势　rotating magnetomotive force
电枢反应　armature reaction
电磁制动转矩　electromagnetic braking torque
电磁负荷　electromagnetic load
电磁功率　electromagnetic power
不平衡电流　unbalanced current
电话谐波因数　telephone harmonic factor（THF）
绝缘的耐热等级　thermal class of insulation
发电机额定容量　rated capacity
最大连续容量　maximum continuous capacity
饱和特性　saturation characteristic
V形曲线特性　V-curve characteristic
铁心及机座振动　stator core and frame vibration
发电机效率　generator efficiency
铜损耗　copper loss
铁损耗　iron loss
杂散负载损耗　stray load loss
摩擦通风损耗　friction and windage loss
总损耗　power losses
负序电流承载能力　negative sequence current withstand capability
发电机冷却　generator cooling
冷却系统　cooling system
空-空冷却发电机　air-to-air cooled machine
空-水冷却发电机　air-to-water cooled machine
水直接冷却发电机　direct water-cooled machine
空气冷却系统　air cooling system
空气冷却器　air cooler
发电机轴承油冷却器　generator bearing oil coolers
推力轴承高压顶起系统　thrust bearing high-pressure lift system
推力轴承顶起油泵　thrust-bearing oil lift pump
灭火装置　fire-extinguishing device
径向通风　radial ventilation
轴向通风　axial ventilation
轴电流　shaft current
接地电刷　earthing brush
电制动　electrical braking
机械制动　mechanical braking
电腐蚀　electro-erosion
定子端部绕组绝缘　stator end winding insulation
铁心松弛　core loosening
并网运行　paralleling operation
电压和频率的偏离值　voltage and frequency variations
空载特性　no-load characteristics，open circuit characteristics
短路特性　short-circuit characteristics
负载特性　load characteristics

调相运行 condensing operation
过励磁运行 overexcitation operation
欠励磁运行 underexcitation operation
进相运行 leading operation, leading power factor operation
迟相运行 lagging operation, lagging power factor operation
发电机异常运行 abnormal operation of generator
发电机特殊运行 special operation of generator
带不平衡负荷运行 unbalanced loading operation
低励及失磁 underexcitation and loss of excitation
带励磁失步运行 out-of-step operation with excitation
发电机故障 generator fault
定子绕组绝缘故障 stator winding insulation fault
定子线棒接头开焊 welded joint breaks of stator bar
定子绕组匝间短路 stator winding inter-turn short circuit
铁心故障 core fault
转子绕组接地 rotor winding earth fault
转子绕组匝间短路 rotor winding inter-turn short circuit
励磁系统故障 excitation system fault
技术供水系统故障 service water system fault
油系统故障 oil system fault
温度监测 temperature monitoring
电气预防性试验 electrical preventive test for generator
绕组直流电阻测量 DC winding resistance measurement
绕组绝缘电阻测量 winding insulation resistance measurement
铁心故障探测 stator core fault detection
定子绕组直流耐压试验 stator winding DC voltage withstand test
泄漏电流试验 leakage current measurement
定子绕组交流耐压试验 stator winding AC voltage withstand test
转子绕组交流阻抗测定 AC impedance test for rotor winding
定子铁芯的损耗发热试验 stator core loss and temperature rise test
转子动平衡及超速试验 dynamic balance and overspeed test for rotor
耐电压试验 withstand voltage test
通过试验确定同步电机的参数 determining synchronous machine quantities from tests
短时电压升高试验 short time voltage rising test
温升试验 temperature rise test
短时过电流试验 short time overcurrent test
交接试验 acceptance test
发电机冷却系统试验 generator cooling system test
状态监测 condition monitoring
同步电机相量图 phasor diagram of synchronous machine
直轴同步电抗 direct-axis synchronous reactance
交轴同步电抗 quadrature-axis synchronous reactance
直轴瞬态电抗 direct-axis transient reactance
交轴瞬态电抗 quadrature-axis transient reactance
直轴超瞬态电抗 direct-axis subtransient reactance
交轴超瞬态电抗 quadrature-axis subtransient reactance
同步电机的时间常数 time constants of synchronous machine
同步发电机瞬态电势 voltage behind transient reactance of synchronous generator
同步发电机超瞬态电势 voltage behind subtransient reactance of synchronous generator
直轴瞬态短路时间常数 direct-axis transient short-circuit time constant
直轴超瞬态短路时间常数 direct-axis subtransient short-circuit time constant
直轴瞬态开路时间常数 direct-axis transient open-circuit time constant
直轴超瞬态开路时间常数 direct-axis subtransient open circuit time constant
交轴瞬态短路时间常数 quadrature-axis transient short-circuit time constant
交轴超瞬态短路时间常数 quadrature-axis subtransient short circuit time constant
交轴瞬态开路时间常数 quadrature-axis transient open-circuit time constant
交轴超瞬态开路时间常数 quadrature-axis subtransient open-circuit time constant
电枢短路时间常数 armature short-circuit time constant

保梯电抗　Potier reactance
定子绕组电阻　stator winding resistance
短路比　short-circuit ratio
同步发电机功率角　power angle of synchronous generator
同步发电机静态稳定　steady state stability of synchronous generator
同步发电机动态稳定　dynamic stability of synchronous generator
同步发电机瞬态稳定　transient stability of synchronous generator
同步电机的振荡　oscillation of synchronous machine
同步电机的突然短路　sudden short-circuit of synchronous machine
同步电机的电枢反应　armature reaction of synchronous machine
可逆式电机　reversible machine
发电电动机　generator-motor
变极式发电电动机　pole changing generator-motor
发电机工况　generator mode（generating mode）
水轮机工况　turbine mode
电动机工况　motor mode
水泵工况　pump mode（pumping mode）
调相工况　condenser mode
水泵启动　pump starting
背靠背启动　back-to-back starting（BTB starting）
变频启动　variable frequency starting
异步启动　asynchronous starting
同步启动　synchronous starting
静止变频器　static frequency convertor（SFC）
飞轮力矩　flywheel moment
机组动平衡试验　dynamic balancing test of unit
盘车　barring（turning）machine for alignment
盘车装置　barring gear
水轮发电机组试运行　trial starting of hydrogenerator set
72小时试运行　72h trial running

8.4 电气设备
Electric equipment

(1) 高压电气设备　High voltage equipment

电容器　capacitor，condenser
电力电容器　power capacitor
电力电子电容器　power electronic capacitor
纸介质电容器　paper capacitor
金属箔电容器　metal foil capacitor
金属化电容器　metalized capacitor
并联电容器　shunt capacitor
电容器成套装置　capacitor installation
串联补偿　series compensation
并联补偿　shunt compensation
串联电容补偿装置　series capacitive compensator
并联电容补偿装置　shunt capacitive compensator
静止同步补偿装置　static synchronous compensator
静止无功补偿器　static var compensator（SVC）
无功功率补偿　reactive power compensation
滤波器　filter
无源滤波器　passive filter
有源滤波器　active power filter
静止移相器　static phase shifter
断路器　circuit-breaker
柱上断路器　pole-mounted circuit-breaker
外壳带电断路器　live tank circuit-breaker
落地罐式断路器　dead tank circuit-breaker
真空断路器　vacuum circuit-breaker
SF_6 断路器　SF_6 circuit-breaker（sulfur hexafluoride circuit-breaker）
磁吹断路器　magnetic blow-out circuit-breaker
隔离开关　disconnector，isolating switch
单极隔离开关　single pole disconnector
三极隔离开关　three pole disconnector
单柱式隔离开关　single-column disconnector
三柱式隔离开关　three-column disconnector
水平旋转式隔离开关　center rotating disconnector
剪刀式隔离开关　pantograph disconnector
半剪刀式隔离开关　semi-pantograph disconnector
双臂伸缩式隔离开关　pantograph disconnector
V形隔离开关　V-type disconnector
单臂伸缩式隔离开关　semi-pantograph disconnector
破冰式隔离开关　ice-breaking disconnector
馈线隔离开关　feeder disconnector
母线分段隔离开关　busbar section disconnector
柱上隔离开关　pole-mounted disconnector
接地开关　earthing switch，earthing disconnector
快速接地开关　high speed earthing switch
母线接地开关　earthing switch for busbar

负荷开关　switch，load break switch
通用负荷开关　general purpose switch
专用负荷开关　limited purpose switch
负荷隔离开关　load-disconnector switch
高压开关装置（高压配电装置）　high voltage switchgear
高压开关设备和控制设备　high voltage switchgear and controlgear
户外开关设备和控制设备　outdoor switchgear and controlgear
户内开关设备和控制设备　indoor switchgear and controlgear
金属封闭开关设备和控制设备　metal enclosed switchgear and controlgear
开关柜　switch cubicle
铠装式金属封闭开关设备和控制设备　metal clad switchgear and controlgear
间隔式金属封闭开关设备和控制设备　compartmented switchgear and controlgear
箱式金属封闭开关设备和控制设备　cubicle switchgear and controlgear
充气式金属封闭开关设备和控制设备　gas filled switchgear and controlgear
绝缘封闭开关设备和控制设备　insulation enclosed switchgear and controlgear
气体绝缘金属封闭组合电器　gas insulated metal enclosed switchgear and controlgear（GIS）
混合式气体绝缘金属封闭开关设备　hybrid gas insulated metal enclosed switchgear（HGIS）
电弧电压　arc voltage
弧后电流　post arc current
截断电流　cut-off current，let through current
电流零点　current zero
电弧电流零区　arc current zero period
复燃　reignition
重击穿　restrike
开关设备的极　pole of switchgear
主电路　main circuit
辅助电路　auxiliary circuit
电接触　electric contact
固定电接触　stationary electric contact
可动电接触　movable electric contact
触头的行程　travel of contacts
超行程　contacting travel，overtravel
时间行程特性　time travel diagram
脱扣器　release
分励脱扣器　shunt release
灭弧管　arc-extinguishing tube
灭弧室　arc-extinguishing chamber
灭弧装置　arc-control device
纵吹灭弧室　axial blast interrupter
横吹灭弧室　cross blast interrupter
纵横吹灭弧室　mixed blast interrupter
外能灭弧室　external energy interrupter
真空灭弧室　vacuum extinction chamber
触头　contact
主触头　main contact
静触头　fixed contact
动触头　moving contact
弧触头　arcing contact
控制触头　control contact
控制开关　control switch
动合触头（点）　make contact
常开触头（点）　normally open contact
动断触头（点）　break contact
常闭触头（点）　normally closed contact
操动机构　operating mechanism
电动机操动机构　motor operating mechanism
气动操动机构　air operating mechanism
液压操动机构　hydraulic operating mechanism
储能操动机构　stored energy operating mechanism
手动储能操动机构　independent-manual operating mechanism
手动操动机构　dependent manual operating mechanism
关合　making
预期电流　prospective current
预期峰值电流　prospective peak current
短路关合能力　short-circuit making capacity
短路开断能力　short-circuit breaking capacity
时间-电流特性　time-current characteristic
短时耐受电流　short-time withstand current
峰值耐受电流　peak withstand current
外施电压　applied voltage
恢复电压　recovery voltage
直流稳态恢复电压　DC steady-state recovery voltage
峰值电弧电压　peak arc voltage
电气间隙　clearance
极间电气间隙　clearance between poles

对地间隙 clearance to earth
触头开距 clearance between open contacts
隔离距离 isolating distance
燃弧时间 arcing time
关合时间 make time
峰值关合电流 peak making current
开断 breaking
开断时间 break time
开断电流 breaking current
关合-开断时间 make-break time
合闸 closing
合闸位置 closed position
合闸时间 closing time
合闸速度 closing speed
分闸 opening
分闸位置 opening position
分闸时间 opening time
分闸速度 opening speed
合-分时间 close-open time
合-分操作 close-open operation
分-合时间 open-close time
重合时间 reclosing time
无电流时间 dead time
操作循环 operating cycle
操作顺序 operating sequence
防跳跃装置 anti-pumping device
密度传感器 density sensor
密度继电器 density relay
断路器合闸电阻 closing resistor of circuit-breaker
断路器恢复电压 recovery voltage of circuit-breaker
断路器瞬态恢复电压 transient recovery voltage of circuit-breaker (TRV)
回路的预期瞬态恢复电压 prospective transient recovery voltage of circuit
断路器工频恢复电压 power frequency recovery voltage of circuit-breaker
起始瞬态恢复电压 initial transient recovery voltage (ITRV)
瞬态恢复电压上升率 rate of rise of TRV，RRRV
断路器首开极因数 first-pole-to-clear factor of circuit-breaker
快速瞬态过电压 very fast transient overvoltage (VFTO)
年漏气率 yearly gas leakage rate
高压开关设备联锁装置 high voltage switchgear interlocking device
带电显示装置 voltage presence indicating device
电流引入回路 current injection circuit
电压引入回路 voltage injection circuit
电寿命试验 electrical endurance test
机械寿命试验 mechanical endurance test
五防 five preventions [prevention of closing or opening of disconnectors with load，prevention of accidental closing or opening of circuit breakers，prevention of closing of earthing switch to live parts，prevention of connection of live parts to earth，prevention of access to live bay]

(2) 变压器与电抗器 Transformer and reactor

电力变压器 power transformer
主变压器 main transformer
备用变压器 standby transformer
联络变压器 system interconnection transformer
升压变压器 step-up transformer
降压变压器 step-down transformer
油浸式变压器 oil immersed transformer
隔离变压器 isolating transformer
干式变压器 dry type transformer
SF_6 绝缘变压器 SF_6 gas insulated transformer
单相变压器 single phase transformer
三相变压器 three phase transformer
双绕组变压器 two winding transformer
三绕组变压器 three winding transformer
自耦变压器 autotransformer
分裂变压器 split-phase transformer
接地变压器 earthing transformer
有载调压变压器 on-load tap-changing transformer
壳式变压器 shell type transformer
心式变压器 core type transformer
户内变压器 indoor type transformer
户外变压器 outdoor type transformer
高压绕组 high voltage winding
低压绕组 low voltage winding
中压绕组 intermediate-voltage winding
附加绕组 auxiliary winding
一次绕组 primary winding
二次绕组 secondary winding
公共绕组 common winding
相绕组 phase winding
绕组的全绝缘 uniform insulation of a winding

绕组的分级绝缘 non-uniform insulation of a winding
散热器 radiator
自然风冷却 air natural cooling (AN)
强迫风冷却 air forced cooling (AF)
油浸自冷却 oil natural air natural cooling (ONAN)
油浸风冷却 oil natural air forced cooling (ONAF)
强迫油循环风冷却 oil forced air forced cooling (OFAF)
强迫油循环水冷却 oil forced water forced cooling (OFWF)
强迫导向油循环风冷 oil directed air forced cooling (ODAF)
强迫导向油循环水冷 oil directed water forced cooling (ODWF)
分接 tapping
额定分接 principal tapping
主分接 principal tapping
变比 transformation ratio
额定电压比 rated voltage ratio
阻抗电压 impedance voltage at rated current
短路阻抗 short-circuit impedance
穿越阻抗 through impedance
空载损耗 no-load loss
空载电流 no-load current
负载损耗 load loss
附加损耗 supplementary load loss
总损耗 total losses
开口三角形连接 open-delta connection, broken-delta connection
Z形连接 zigzag connection
曲折形连接 zigzag connection
变压器相位移 phase difference for transformer
连接组标号 connection symbol
变压器调压装置 voltage regulator of transformer
线路压降补偿 line drop compensation
有载分接开关 on-load tap-changer
无励磁分接开关 off-circuit tap-changer
储油柜 oil conservator
油枕 oil conservator
油枕油位计 conservator oil level indicator
磁油位计 magnetic oil level gauge
瓦斯继电器 Buchholz relay
变压器压力突变 transformer sudden pressure rising
有载调压开关 on load tap changer (OLTC)
有载调压开关及其驱动机构 OLTC with drive mechanism
硅胶呼吸器 silicagel breather
压力释放阀 pressure relief valve
顶部滤油阀 top filter valve
净油器 oil filter
排油阀 drain valve
绕组温度计 winding temperature indicator
接地端子 earthing terminal
油箱 oil tank
泄油池 oil leakage sump
防火墙 firewall
套管 bushing
瓷套管 porcelain bushing
油浸纸套管 oil impregnated paper bushing
电容型套管 condenser bushing
油气套管 oil-SF_6 bushing
电抗器 reactor
串联电抗器 series reactor
并联电抗器 shunt reactor
消弧电抗器 arc suppression reactor
消弧线圈 arc suppression coil, Petersen coil
滤波电抗器 filter reactor, smoothing reactor
油浸式电抗器 oil immersed reactor
干式电抗器 dry type reactor
限流电抗器 current-limiting reactor
空心电抗器 air-core reactor
互感器 instrument transformer
仪用变压器 instrument transformer
组合式互感器 combined instrument transformer
电子式互感器 electronic instrument transformer
光电式互感器 optical instrument transformer
保护用互感器 protective instrument transformer
测量用互感器 measuring instrument transformer
电流互感器 current transformer
母线式电流互感器 bus type current transformer
电缆式电流互感器 cable type current transformer
套管式电流互感器 bushing type current transformer
钳式电流互感器 split core type current transformer
棒式电流互感器 bar primary type current transformer
支柱式电流互感器 support type current transformer
总加电流互感器 summation current transformer
中间式电流互感器 current matching transformer
零序电流互感器 zero sequence current transformer,

residual current transformer
电流误差　current error
准确级　accuracy class
拐点电压　knee point voltage
互感器负荷　burden of instrument transformer
互感器额定负荷　rated burden of instrument transformer
仪表保安系数　instrument security factor
额定准确限值一次电流　rated accuracy limit primary current
扩大电流值　extended rating current
电压互感器　voltage transformer
电容式电压互感器　capacitive voltage transformer
电磁式电压互感器　inductive voltage transformer
接地电压互感器　earthed voltage transformer
匹配用电压互感器　voltage matching transformer
电容分压器　capacitive voltage divider
额定电压比　rated voltage ratio
电压误差　voltage error
额定电压因数　rated voltage factor

(3) 电力线路与电缆　Transmission line and cable

交流线路　AC line
直流线路　DC line
交流线路的相　phase of AC line
架空线路的回路　circuit of an overhead line
气体绝缘管道输电线，气体绝缘管道母线　gas insulated line (GIL)
单回路　single circuit line
双回路　double circuit line
多回路　multiple circuit line
线路自然功率　natural load of a line
架空线路的导线　conductor of overhead line
导线振动　conductor vibration
电晕干扰　corona interference
可听噪声　audible noise
工作荷载　working load
额定荷载　normal load，primary load
专用荷载　special load
试验荷载　test load
破坏荷载　failure load
设计荷载　design load
垂直荷载　vertical load
纵向荷载　longitudinal load
横向荷载　transverse load
风荷载　wind load
冰荷载　ice load
风偏　wind deflection
风偏角　angle of wind deflection
覆冰厚度　radial thickness of ice
覆冰区　ice coverage area
污秽区　polluted area
输电线路走廊　transmission line corridor
跨越　crossing
经济电流密度　economic current density
载流能力　current carrying capacity
允许载流量　permissive carrying current
平均运行张力　everyday tension
综合拉断力　comprehensive breaking strength (UTS)
集中荷载　centralized load
计算拉断力　rated tensile strength (RTS)
相间净距　phase-to-phase clearance
相对地净距　phase-to-earth clearance
导线对塔净空距离　clearance between conductor and structure
杆塔　support (structure of an overhead line)
铁塔　steel tower
塔高　tower height
耐张塔　tension support，angle support
直线杆塔　intermediate support
换位杆塔　transposition support
终端杆塔　terminal support，dead end tower
转角杆塔　angle support
单回路塔　single circuit steel tower
双回路塔　double circuit steel tower
地线支架　earth wire peak
节点　node，panel point
钢筋混凝土杆　steel reinforced concrete pole
档距　span length
临界档距　critical span
重力档距　weight span
弧垂　sag
导线最大弧垂　maximum sag of conductor
弧垂观测　visual of sag
架空线路的耐张段　tension section of overhead line
悬链线　catenary
悬链线常数　catenary constant
纵断面　longitudinal profile
线路转角　line angle

导线排列　conductor configuration
水平排列　horizontal configuration
三角形排列　triangular configuration
倒三角排列　delta configuration
垂直排列　vertical arrangement
换位　transposition
对地净距　earth clearance
对障碍物的净距　clearance to obstacles
相间距离　phase-to-phase spacing
保护角　angle of shade
拉线　guy，stay
拉线棒　anchor rod
拉线盘　anchor
金具　fitting
线夹　clamp
耐张线夹　tension clamp，dead-end clamp
悬垂线夹　suspension clamp
连接金具　link fitting，insulator set clamp
接触金具　contact tension fitting
保护金具　protective fitting
母线金具　busbar fitting
挂环　link，eye
挂板　clevis，tongue
球头　ball
碗头　socket
挂钩　hook
U形挂环　U-shackle
U形螺丝　U-bolt
U形挂板　U-clevis
联板　yoke plate
调整板　adjusting plate
花篮螺丝　turn buckle
跳线线夹　jumper flag，jumper lug
防振锤　vibration damper
均压环　grading ring
屏蔽环　shielding ring
重锤　counterweight，suspension set weight
绝缘子保护金具　insulator protective fitting
铜铝过渡板　copper to aluminium adapter board
母线固定金具　busbar support clamp
母线伸缩节　busbar expansion joint
封端球盖　corona bar，cap
导线的夜间警告灯　night warning light for conductor
航空警告标志　aircraft warning marker
绝缘子　insulator
瓷绝缘子　porcelain insulator
支持绝缘子　supporting insulator
棒式绝缘子　rod insulator
玻璃绝缘子　glass insulator
复合绝缘子　composite insulator
耐污绝缘子　anti-pollution type insulator
绝缘子串　insulator string
悬垂绝缘子串　suspension insulator string
耐张绝缘子串　tension insulator string
V形绝缘子串　V string of insulator
绝缘子串组　insulator set
悬垂绝缘子串组　suspension insulator set
耐张绝缘子串组　tension insulator set
污闪　pollution flashover
雾闪　fog flashover
股线　wire，strand
绞线　stranded conductor
铝绞线　all aluminium conductor (AAC)
钢绞线　steel strand wire
钢芯铝绞线　aluminium conductor steel reinforced (ACSR)
铝合金绞线　all aluminium alloy conductor (AAAC)
钢芯铝合金绞线　aluminium alloy conductor steel reinforced (AACSR)
铝包钢加强铝绞线　aluminium clad steel reinforced aluminium conductor
加强型绞线的芯　core of reinforced conductor
单导线　single conductor
分裂导线　conductor bundle
子导线　sub-conductor
电缆　cable
油浸纸绝缘电缆　oil impregnated paper insulated cable
挤包绝缘电缆　cable with extruded insulation
聚氯乙烯电缆　polyvinyl chloride insulated cable
交联聚乙烯绝缘电缆　cross-linked polyethylene insulated cable (XLPE)
自容式充油电缆　self-contained oil- filled cable
耐火电缆　fire resistant cable
阻燃电缆　non-flame propagating cable，flame retardant cable
铠装电缆　armoured cable
钢丝铠装电缆　steel-wire armoured cable
钢带铠装电缆　steel-tape armoured cable
无铠装电缆　non-armoured cable

直埋电缆　direct-burial cable
控制电缆　control cable
单芯电缆　single core cable
多芯电缆　multicore cable
编织层　braid
护套　sheath，jacket
导体屏蔽　conductor screen
绝缘屏蔽　insulation shielding
电缆故障定位　cable fault locating
电缆接头　cable joint
分支接头　branch joint
过渡接头　transition joint
电缆终端头　cable terminal
连接箱　link box
分配箱　distribution box
直埋敷设　cable direct burial laying
电力电缆间水平净距　horizontal clearance between power cables
电缆架层间垂直净距　vertical clearance between cable racks
架间水平净距　horizontal clearance between cable racks
架与壁间水平净距　horizontal clearance between rack and wall
电缆线路路径选择　cable route selection
电缆线路巡视检查　cable line inspection
电缆试验　testing of power cable
电缆交接试验　acceptance test of cable
电缆路由　cable routing
电缆选择　cable selection，cable sizing
电缆敷设　cabling
电缆管道　cable trunking，cable duct
电缆隧道　cable tunnel
电缆架　cable rack
电缆托架　cable tray
电缆沟　cable trough，cable ditch
电缆竖井　cable shaft
电缆排管　cable duct bank
硬母线　rigid busbar
软母线　flexible busbar
离相封闭母线　isolated-phase bus (IPB)
共箱母线　non-segregated phase busbar
线槽　trunking
母线槽　busway
槽型母线　channel busbar
管型母线　busduct
矩形母线　rectangular busbar
环氧绝缘母线　epoxy insulated busbar

(4) 厂用电、低压电器、电动机与照明 Station service equipment，low voltage equipment，motors and lighting

厂用电　station service (service power)
厂用电接线　station service single line diagram
负荷统计　load estimation
同时系数　coincidence factor
负载系数　load factor
厂用电率　internal consumption rate，service power rate
低压电缆线路　low voltage cable line
厂用变压器　station service transformer
机组自用变压器　auxiliary transformer of unit
照明变压器　lighting transformers
备用电源　stand-by power supply
保安电源　emergency power supply
柴油发电机　diesel generator
厂用配电系统　station service distribution system
配电装置　power distribution unit
双重化馈线　duplicate feeders
厂用电开关设备　station service switchgear
铠装金属封闭开关设备　metal-clad switchgear
SF_6 断路器　SF_6 circuit breaker
真空断路器　vacuum circuit breaker
抽屉式断路器　withdrawable circuit breaker
进线断路器　incoming breaker
母联断路器　bus tie breaker
馈线断路器　feeder breaker
母线转换断路器　switched busbar circuit-breaker
分段断路器　section circuit-breaker
隔室　compartment
隔离开关　switch-disconnector
熔断器式隔离器　fuse-disconnector
隔离开关熔断器　switch-fuse-disconnector
主配电屏　main distribution board
分配电屏　sub-distribution board
端子箱　terminal box，marshalling kiosk
电动机控制中心　motor control centers (MCC)
低压电器　low voltage apparatus
熔断器　fuse
跌落式熔断器　drop-out fuse

真空熔断器 vacuum fuse
限流式熔断器 current-limiting fuse
空气断路器 air circuit breaker
塑壳断路器 moduled case circuit breaker (MCCB)
微型断路器 miniature circuit breaker (MCB)
磁力启动器 magnetic starter
接触器 contactor
控制器 controller
电磁脱扣器 magnetic release
热过载脱扣器 thermal overload release
瞬时脱扣器 instantaneous release
过流脱扣器 over-current release
定时限脱扣器 definite time-delay overcurrent release
反时限脱扣器 inverse time-delay overcurrent release
过载脱扣器 overload release
低压脱扣器 undervoltage release
小电流接地选线装置 feeder earth fault detector in an isolated neutral system
主令开关 master switch
无触点开关 non-contact switch
控制测量设备 control and metering equipment
控制开关 control switch
切换开关 transfer switch
电压不稳定 voltage instability
谐波电压源 source of harmonic voltage
谐波电流源 source of harmonic current
谐波谐振 harmonic resonance
电动机 motor
厂用电动机 station electric motor
感应电动机 induction motor
异步电动机 asynchronous motor
同步电动机 synchronous motor
多转速电动机 multispeed motor
外壳防护等级 degree of protection of enclosure
鼠笼式转子 squirrel cage rotor
绕线式转子 wound rotor
深槽式转子 deep slot type rotor
全电压启动 full voltage starting, direct on line starting
过载能力 overload ability
星-三角启动 star-delta starting
自耦变压器启动 autotransformer starting
降压启动 reduced voltage starting
串联电抗启动 reactor starting
转子串联电阻启动 rotor resistance starting
频敏电阻启动 frequency-sensitive rheostat starting
软启动 soft starting
变频启动 variable frequency starting
启动转矩 start torque
转速调整 speed adjustment
变极调速 pole changing speed control
变频调速 variable frequency speed control
制动转矩 braking torque
固有制动转矩 inherent braking torque
电制动转矩 electrical braking torque
机械制动转矩 mechanical braking torque
电磁制动 electromagnetic braking
电制动 electric braking
能耗制动 dynamic braking
直流制动 DC injection braking
再生制动 regenerative braking
蠕行 crawling
空转 idling
滑差，转差率 slip
失电制动器 power-off brake
热稳定 thermal equilibrium
负载持续率 cyclic duration factor
满载 full load
满载值 full load value
工作制 duty
工作周期 duty cycle
周期工作制 periodic duty
工作制类型 duty type
连续工作制 continuous running duty
短时工作制 short-time duty
断续周期工作制 intermittent periodic duty
直流电机 direct current machine (DC machine)
串励绕组 series winding
并励绕组 shunt winding
电刷 brush
电刷磨损 wear of brush
带集电环感应电动机 slip-ring induction motor
绕线转子感应电动机 wound-rotor induction motor
笼型感应电动机 cage induction motor, squirrel cage induction motor
单相感应电动机 single phase induction motor

伺服电动机 servomotor
伺服机械装置 servomechanism
启动器 starter
备用电源自动投入 automatic bus transfer
剩余电流 residual current
剩余动作电流 residual operating current
剩余电流动作保护装置 residual current operated device (RCD)
电照明 electrical lighting
正常照明 normal lighting
应急照明 emergency lighting
疏散照明 escape lighting
安全照明 safely lighting
备用照明 stand-by lighting
光通量 luminous flux
发光强度 luminous intensity
亮度 luminance
照度 illuminance
维护系数 maintenance factor
频闪效应 stroboscopic effect
光强分布 distribution of luminous intensity
灯具效率 luminarie efficiency
显色指数 colour rendering index
色温 colour temperature
反射比 reflectance
照明功率密度 lighting power density (LPD)
坎德拉 candela
流明 lumen
勒克斯 lux
吸收比 absorptance
反射率 reflectivity
折射率 refractive index
光度计 photometer
亮度计 luminance meter
色度计 colorimeter
灯具 luminaire
对称配光型灯具 symmetrical luminaire
非对称配光型灯具 asymmetrical luminaire
直接型灯具 direct luminaire
漫射型灯具 diffused luminaire
广照型灯具 wide angle luminaire
深照型灯具 narrow angle luminaire
防尘灯具 dust-proof luminaire
防爆灯具 luminaire for explosive atmosphere
可移式灯具 portable luminaire
嵌入式灯具 recessed luminaire
吸顶灯具 ceiling luminaire，surface-mounted luminaire
疏散标志灯 escape sign luminaire
出口标志灯 exit sign luminaire
指向标志灯 direction sign luminaire
道路照明灯具 luminaire for road lighting
荧光灯 fluorescent lamp
紧凑型荧光灯 compact fluorescent lamp
发光二极管灯 light emitting diode lamp，LED lamp
高压钠（蒸汽）灯 high pressure sodium (vapour) lamp
金卤灯 metal halide lamp，metal halogen lamp
荧光灯 fluorescent lamp
三基色 tri-phosphor
太阳能灯 solar lamp
应急灯 emergency lamp
投光灯 spot lamp
泛光灯 flood lamp
格栅灯 grille lamp
防水灯 water proof lamp，under water lamp
电感镇流器 inductive ballast，magnetic ballast
电子镇流器 electronic ballast
启动器 starter

8.5 过电压保护、防雷、接地与电气安全

Overvoltage protection，lightning protection，earthing and electric safety

跨步电压 step voltage
接触电压 touch voltage
危险电压 dangerous voltage
安全电流 safety current
对地泄漏电流 earth leakage current
接地故障电流 earth fault current
电击电流 shock current
避雷器 surge arrester，lightning arrester
阀式避雷器 valve type arrester
磁吹阀式避雷器 magnetic blow-out valve type arrester
电涌保护器 surge protective device (SPD)
碳化硅阀式避雷器 silicon carbide valve type surge arrester

金属氧化物避雷器 metal oxide arrester (MOA)
无间隙金属氧化物避雷器 gapless metal oxide arrester
管式避雷器 tube type arrester
击穿保护器 sparkover protective device
避雷器阀片 valve disc of arrester
避雷器均压环 grading ring of arrester
避雷器压力释放装置 pressure-relief device of arrester
避雷器额定电压 rated voltage of arrester
避雷器工频参考电压 power frequency reference voltage of arrester
避雷器直流参考电压 direct current reference voltage of arrester
避雷器持续运行电压 continuous operating voltage of arrester
避雷器残压 residual voltage of arrester
避雷器工频放电电压 power frequency sparkover voltage of arrester
避雷器标准雷电冲击放电电压 standard lightning impulse sparkover voltage of arrester
避雷器标称放电电流 nominal discharge current of arrester
避雷器的保护特性 protective characteristics of arrester
避雷器电导电流 conducting current of arrester
避雷器泄漏电流 leakage current of arrester
高电压技术 high voltage technology
高电压试验设备 high voltage testing equipment
工频试验变压器 power frequency testing transformer
串级工频试验变压器 cascade power frequency testing transformer
工频谐振试验变压器 power frequency resonant testing transformer
高压整流器 high voltage rectifier
直流高压发生器 high-voltage DC generator
串级直流高压发生器 cascade high-voltage DC generator
冲击电压发生器 impulse voltage generator
冲击电流发生器 impulse current generator
保护电阻器 protective resistor
冲击电流分流器 impulse current shunts
冲击电压分压器 impulse voltage divider
串联谐振试验装置 series resonant testing equipment
直流耐压试验 DC voltage withstand test
交流耐压试验 AC voltage withstand test
冲击耐压试验 impulse voltage withstand test
操作冲击耐压试验 switching impulse voltage withstand test
工频耐压试验 power frequency voltage withstand test
冲击电流试验 impulse current test
雷电冲击截波试验 chopped lightning impulse test
破坏性放电试验 disruptive discharge voltage test
介质损耗试验 dielectric dissipation test, loss tangent test
绝缘老化试验 insulation ageing test
人工污秽试验 artificial pollution test
干试验 dry test
湿试验 wet test
绝缘隔离装置 insulated isolated device
高电压试验测量系统 measuring system for high voltage testing
方波电压发生器 square wave voltage generator
高压电桥 high voltage bridge
高压示波器 high voltage oscilloscope
电阻式分压器 resistive divider
阻容分压器 resistance-capacitance voltage divider
数字记录仪 digital recorder
高压标准电容器 high voltage standard capacitor
高压耦合电容器 high voltage coupling capacitor
局部放电检测仪 partial discharge detector
无线电干扰测试仪器 radio interference meter
标准球隙 standard sphere gap
球间隙 sphere gap
火花检测器 spark tester
电场测量探头 electric-field probe
泄流计 leakage tester
雷暴日 thunderstorm day
雷暴小时 thunderstorm hour
雷电流 lightning current
雷电流峰值 peak value of lightning current
雷电流总电荷 total charge of lightning current
雷电流平均陡度 average steepness of lightning current
地面落雷密度 earth flash density (GFD)
雷电电磁脉冲 lightning electromagnetic pulse

(LEMP)
绕击率 shielding failure rate due to lightning stroke
反击率 risk of flashback
建弧率 arc over rate
雷击跳闸率 lightning outage rate
直接雷击 direct lightning strike
感应雷击 indirect lightning strike
避雷线 overhead earthing wire
避雷针 lightning rod
架空地线 overhead earthing wire
雷电冲击全波 full lightning impulse
雷电冲击波前时间 front time of a lightning impulse
雷电冲击半峰值时间 time to half value of a lightning impulse
标准雷电冲击全波 standard full lightning impulse
标准雷电冲击截波 standard chopped lightning impulse
雷电冲击波保护比 protection ratio against lightning impulse
操作冲击波前时间 time to peak of switching impulse
操作冲击半峰值时间 time to half value of switching impulse
操作冲击波 90%值以上时间 time above 90% of switching impulse
标准操作冲击波 standard switching impulse
冲击电流 impulse current
冲击电流波前时间 front time of impulse current
冲击电流半峰值时间 time to half value of impulse current
方波冲击电流峰值持续时间 duration of peak value of rectangular impulse current
方波冲击电流总持续时间 total duration of a rectangular impulse current
标准冲击电流 standard impulse current
操作冲击截断时间 time to chopping of switching impulse
波头截断冲击波 impulse chopped on the front
波尾截断冲击波 impulse chopped on the tail
破坏性放电 disruptive discharge
破坏性放电电压 disruptive discharge voltage
冲击放电电压 impulse sparkover voltage
冲击波波前放电电压 impulse front discharge voltage
耐受电压 withstand voltage
标准操作冲击耐受电压 standard switching impulse withstand voltage
操作冲击波保护比 protection ratio against switching impulse
标准雷电冲击耐受电压 standard lightning impulse withstand voltage
雷电冲击截波耐受电压 withstand voltage of chopped lightning impulse
工频耐受电压 power frequency withstand voltage
标准短时工频耐受电压 standard short duration power-frequency withstand voltage
保护装置的保护水平 protection level of protective device
标准大气条件 standard reference atmospheric condition
介质损耗因数 dielectric loss factor
尖端放电 point discharge
气体放电 gas discharge
保护火花间隙 protective spark gap
气体介质击穿 gas dielectric breakdown
局部放电 partial discharge
电晕放电 corona discharge
电晕效应 corona effect
电晕损失 corona loss
火花放电 sparkover
电弧放电 arc discharge
空气间距 air clearance
气体绝缘介质 insulating gas
液体绝缘介质 insulating liquid
沿面放电 discharge along dielectric surface
吸收比 absorption ratio
绝缘配合 insulation coordination
绝缘配合因数 insulation coordinating factor
伏-秒特性曲线 volt-time characteristics
安-秒特性 Ampere-time characteristics
耐雷水平 lightning withstand level
基本冲击绝缘水平 basic impulse insulation level (BIL)
操作冲击绝缘水平 basic switching impulse insulation level
标准绝缘水平 standard insulation level
海拔校正系数 altitude correction factor
绝缘故障率 failure rate of insulation
绝缘性能指标 performance criterion of insulation

内绝缘 internal insulation
外绝缘 external insulation
主绝缘 main insulation
自恢复绝缘 self restoring insulation
非自恢复绝缘 non-self restoring insulation
户内外绝缘 indoor external insulation
户外外绝缘 outdoor external insulation
绝缘老化 ageing of insulation
绝缘击穿 insulation breakdown
爬距，泄漏距离 creepage distance
爬电比距 specific creepage distance
耐受概率 probability of withstand
相对地过电压标幺值 per unit value of phase-to-earth overvoltage
相间过电压标幺值 per unit value of phase-to-phase overvoltage
过电压 overvoltage
内过电压 internal overvoltage
大气过电压 lightning overvoltage
操作过电压 switching overvoltage
雷电过电压 lightning overvoltage
合闸过电压 closing overvoltage
分闸过电压 breaking overvoltage
重合闸过电压 reclosing overvoltage
惯用最大操作过电压 conventional maximum switching overvoltage
惯用最大雷电过电压 conventional maximum lightning overvoltage
统计操作过电压 statistical switching overvoltage
统计雷电过电压 statistical lightning overvoltage
感应过电压 induced overvoltage
谐振过电压 resonance overvoltage
暂时过电压 temporary overvoltage
瞬态过电压 transient overvoltage
工频过电压 power frequency overvoltage
振荡解列操作过电压 oscillation overvoltage due to system splitting
土壤电阻率 soil resistivity
接地体 earthing electrode
自然接地体 natural earthing substance
人工接地体 artificial earthing electrode
接地线（接地导体） earthing conductor
接地网 earthing grid
接地引下线 down conductor（down lead）
接地装置 earth-termination system
接地汇流排 main earthing conductor
接地装置对地电位 potential of earthing connection
工频接地电阻 power frequency earthing resistance
冲击接地电阻 impulse earthing resistance
防静电接地 static electricity protection earthing
保护接地 protective earthing
等电位连接 equipotential bonding
局部等电位连接 local equipotential bonding
保护导体 protective conductor
降阻剂 resistance reducing agent
绝缘杆 insulating stick
接地和短路装置 earthing and short-circuiting equipment
短路装置 short-circuiting device
警告牌 warning board
标示牌 signboard
操作票 operation order
紧急操作 emergency operation
误操作 misoperation
电气故障 electrical fault（failure）
基本绝缘 basic insulation
故障点 fault point
故障相 fault phase
故障区段 fault section
电气安全措施 electrical safety measure
电气事故 electrical accident
电气事故报警 electrical fault alarming
电气安全 electrical safety
电气防火间距 electrical interval of fire prevention
电气保护遮栏 electrically protective barrier
电气保护外壳 electrically protective enclosure
电气保护屏蔽体 electrically protective screen
电气防火 electrical fire prevention
消防设施 fire protection equipment
防火墙 fire protection wall
电气消防通道 fire fighting access to electrical equipment
电击 electric shock
外部导电部分 external conductive part
验电 live line detection
作业距离 working distance
最小安全距离 minimum approach distance（minimum working distance）
电气距离 electrical distance

8.6 电力电子（含 SFC 与励磁设备） Electronics，including SFC and excitation equipment

电力电子技术 Power electronics
换流，变流 power conversion
通断 switching
整流 rectification
逆变 inversion
交流/直流变流器 AC/DC converter
电压型交流/直流变流器 voltage stiff AC/DC converter
电流型交流/直流变流器 current stiff AC/DC converter
整流器 rectifier
逆变器 inverter
电压源逆变器 voltage source inverter（voltage fed inverter）
电流源逆变器 current source inverter（current fed inverter）
负载换相逆变器 load commuted inverter（LCI）
有源电力滤波器 active power filter
交流变流器 AC converter
变频器 frequency converter
直流斩波器 DC chopper
半导体变流器 semiconductor converter
电力电子开关 electronic power switch
电子器件 electronic device
电子阀器件 electronic valve device
可控阀器件 controllable valve device
不可控阀器件 non-controllable valve device
整流二极管 rectifier diode
反向阻断阀器件 reverse blocking valve device
开关阀器件 switched valve device
半导体阀器件 semiconductor valve device
换相电抗器 commutation reactor
换相电容器 commutation capacitor
阻尼器 snubber
阻容吸收器 RC snubber
阀臂 valve arm
主臂 principal arm
续流臂 free-wheeling arm
关断臂 turn-off arm
桥式连接 bridge connection
全控连接 fully controllable connection
换相 commutation
换相电压 commutating voltage
换相电路 commutation circuit
负载换相 load commutation
自换相 self-commutation
自动顺序换相 auto-sequential commutation
熄断 quenching
阀器件熄断 valve device quenching
脉冲控制 pulse control
脉宽调制控制 pulse width modulation control（PWM）
触发延迟角 trigger delay angle
触发超前角 trigger advance angle
直流变流器的转换因数 transfer factor of DC converter
导通状态 on state，conducting state
阀电压降 valve voltage drop
断态 off state，blocking state
正向阻断状态 forward blocking state
反向阻断状态 reverse blocking state
导通方向 conducting direction
关断期 hold-off interval
基本周期 elementary period
基本频率 elementary frequency
电路断态工作峰值电压 circuit crest working off-state voltage
电路反向工作峰值电压 circuit crest working reverse voltage
换相失败 commutation failure
穿通 break-through
击穿 breakdown
正向击穿 forward breakdown
反向击穿 reverse breakdown
触发 triggering
开通 firing
误通 false firing
直通 conduction through
失通 firing failure
阀器件闭锁 valve device blocking
脉波数 pulse number
电路角 circuit angle
换相数 commutation number
直流功率 DC power
变流因数 conversion factor
整流因数 rectification factor

逆变因数　inversion factor
交流变流因数　AC conversion factor
直流变流因数　DC conversion factor
过渡电流　transition current
直流电压调整值　direct voltage regulation
电子阀器件的门槛电压　threshold voltage of an electronic valve device
直流侧纹波电压　ripple voltage on DC side
直流波形因数　DC form factor
直流纹波因数　DC ripple factor
稳压特性　stabilized voltage characteristic
稳流特性　stabilized current characteristic
自动开通　automatic switching on
自动关断　automatic switching off
跃变特性　jumping characteristic
稳定电源　stabilized power supply
恒压电源　constant voltage power supply
恒流电源　constant current power supply
允差带　tolerance band
从属运行　slave operation
傅立叶级数基波分量　fundamental component of a Fourier series
基波频率　fundamental frequency
谐波频率　harmonic frequency
间谐波频率　interharmonic frequency
间谐波分量　interharmonic component
分谐波分量（次谐波分量）subharmonic component
总畸变含量　total distortion content
谐波含量　harmonic content
总谐波率　total harmonic ratio
总畸变率　total distortion ratio
总谐波因数　total harmonic factor
总畸变因数　total distortion factor
电力电子设备　electronic power equipment
变流设备　converter equipment（converter assembly）
触发器触发设备　triggering device
平衡温度　equilibrium temperature
冷却媒质　cooling medium
热转移媒质　heat transfer agent
直接冷却　direct cooling
间接冷却　indirect cooling
强迫冷却　forced cooling
输入变压器　incoming transformer
输出变压器　outgoing transformer
网桥　network bridge
机桥　machine bridge
平波电抗器　smoothing reactor
直流电抗器　DC reactor
直流电流互感器　DC current transformer
励磁系统　excitation system
励磁装置　exciter
励磁控制　excitation control
磁场绕组　field winding
额定励磁电流　rated field current
额定励磁电压　rated field voltage
空载励磁电流　no-load field current
空载励磁电压　no-load field voltage
励磁系统额定电流　excitation system rated current
励磁系统额定电压　excitation system rated voltage
励磁系统顶值电流　excitation system ceiling current
励磁系统顶值电压　ecitation system ceiling voltage
励磁系统空载顶值电流　excitation system no-load ceiling current
励磁系统空载顶值电压　excitation system on-load ceiling voltage
励磁系统额定响应　excitation system nominal response
强行励磁　forced exciting
旋转励磁装置　rotating exciter
直流励磁机　DC exciter
交流励磁机　AC exciter
带旋转整流器的交流励磁机　AC exciter with rotating diodes
无刷励磁机　brushless exciter
静止整流励磁装置　static exciter
并励静止整流装置　potential source static exciter
电压调节器　voltage regulator
负荷电流补偿器　load current compensator
过励限制器　overexcitation limiter
欠励限制器　underexcitation limiter
伏赫比限制器　Volts per Hertz limiter（V/F limiter）
续流二极管　freewheel diodes
熔断器监视　fuse monitoring
励磁变压器　excitation transformer
晶闸管整流器　thyristor rectifier
触发装置　trigger set
过压保护　overvoltage protection

灭磁回路 de-excitation circuit, field suppression circuit
直流起励装置 DC field flashing device
晶闸管桥 thyristor bridge (SCR bridge)
磁场断路器 field breaker
磁场放电电阻 field discharge resistor
风扇 ventilation fan
磁场接地监测用的滑环 slip ring for field earth fault detection
缓冲电路 snubber circuit
脉冲放大器 pulse amplifier
逆变灭磁 de-excitation by inversion
跨接电路 crowbar circuit
自动电压调节器 automatic voltage regulator (AVR)
发电机电压设定 generator voltage set point
增减开关 raise/lower switch
磁场电流设定 field current set point
励磁电流调节器 excitation current regulator (ECR), field current regulator (FCR)
调节和控制柜 regulating and control cubicle
晶闸管桥的(n-1)冗余 (n-1) redundancy of thyristor bridges
并联的整流桥数量 number of rectifier bridges in parallel
每个桥臂串联的晶闸管数量 number of thyristors in series per arm
双通道自动电压调节器 dual automatic channel voltage regulator
功率整流器 power converter
响应比 response ratio
功率因数控制器 power factor controller
两个自动通道之间的跟踪 followup function between two automatic channels
线路充电 line charging
手动-自动之间的跟踪 manual-automatic followup function
圆盘形晶闸管 disk type thyristors
脉冲变压器 pulse transformers
快速熔断器 ultra-rapid fuses
五极隔离开关 five-pole isolator
弧电压 arcing voltage
反向并联晶闸管 antiparallel connected thyristors
非线性放电电阻 non-linear discharge resistor
分流器 shunt
磁场电流变送器 field current transducer
磁场电压变送器 field voltage transducer

8.7 监控系统 Supervision and control system

(1) 计算机监控系统 Computer supervision and control systems

控制方式 control mode
无人值班水电厂 unmanned hydropower plant, (unattended hydropower plant)
无人值班(少人值守)水电厂 unmanned (with few watchers) hydropower plant, [unattended (with few watchers) hydropower plant]
有人值班水电厂 attended hydropower plant
中央控制室 central control room
控制盘台 control panel and console
远方控制 remote control
现地控制 local control
本地终端 local terminal
远程终端 remote terminal
移动终端 mobile terminal
闭环控制 closed-loop control
开环控制 open-loop control
自动调节系统 automatic modulating control system
连锁控制 interlock control
顺序控制系统 sequence control system
机组启停控制 automatic control for unit start-up and shutdown
光字信号器 annunciator
工程师工作站 engineer workstation
操作员工作站 operator workstation
通信工作站 communication workstation
培训工作站 training workstation
语音报警工作站 voice alarm workstation
蜂鸣器 buzzer
电铃 electric bell
历史数据库服务器 historical data server
打印机 printer
历史数据存储 historical data memory
定时打印 periodic logging
追忆打印 post-trip logging
扫描速率 scan rate
采样周期 scan period
过程控制级 process control level
监控级 supervision level

厂级信息系统 information system for plant level
厂级监控信息系统 supervisory information system for plant level
GPS接收和授时装置 GPS reception and clock distribution device
备用供电系统 standby supply system
等位机架 equipotential frame
电厂控制级 plant control level
电磁发射 electromagnetic emission
电磁干扰 electromagnetic interference (EMI)
分辨率 resolution
分布式I/O distributed I/O
分层分布的监控系统 hierarchically distributed supervision and control system
以太网 ethernet
总线 bus
环形 ring
星形 star
通信协议 protocol
故障导向安全 fail-safe
容错 fault-tolerance
计算机监控系统 computer supervision and control system (CSCS)
数据采集与监控系统 supervisory control and data acquisition (SCADA)
诊断 diagnostics
消除抖动 debounce
先选择后操作 select before operate
交换机［网络］ switch
交流采样 direct acquisition from CT and VT
开关量，二进制量 binary variable，on/off variable
模拟量 analog variable
数字量 digital variable
人机接口 human-machine interface (HMI)，human-computer interface
事件 event
事件日志 event log
事件顺序记录 sequence of events (SOE)
数据服务器 data server
系统软件 system software
现场总线 fieldbus
现地控制单元 local control unit (LCU)
机组现地控制单元 unit LCU
机组事故停机后备装置 backup unit emergency shutdown device
开关站现地控制单元 switchyard LCU
公用设备现地控制单元 coomon device LCU
坝区现地控制单元 dam LCU
数字量输入模件 digital input (DI) module
模拟量输入模件 analog input (AI) module
数字量输出模件 digital output (DO) module
模拟量输出模件 analog output (AO) module
电压变送器 voltage transducer
电流变送器 current transducer
有功功率变送器 active power transducer
无功功率变送器 reactive power transducer
频率变送器 frequency transducer
多元件变送器 multi-element transducer
组合变送器 multi-section transducer
电压表 voltmeter
电流表 ammeter
有功功率表 wattmeter
无功功率表 varmeter
频率表 frequency meter
功率因数表 power factor meter
电能表 energy meter
有功电能表 watt-hour meter
无功电能表 var-hour meter
多费率电度表 multi-rate energy meter
双向电能表 bidirectional energy meter
通信模件 communication module
电源模件 power supply module
I/O模件 I/O module
棒图 bar graph
定值表 setting list
现地控制级 local control level
响应时间 response time
远程I/O remote I/O
系统软件 system software
应用软件 application software
支持软件 support software
工具软件 tool software
通信软件 communication software
诊断软件 diagnosis software
数据采集软件 data acquisition software
数据处理软件 data treatment software
数据库软件 database software
接口软件 interface software

智能电子设备　intelligent electronic device (IED)
主计算机　main computer
电力调度　power dispatching
电力系统调度信息　information for power system dispatching
电力系统调度管理　dispatching management of power system
电力系统分层控制　hierarchical control of power system
控制中心　control center
电力系统负荷预测　load forecast of power system
调度命令　dispatching command
电力系统经济调度　economic dispatching of power system
模拟屏　mimic board
电力系统调度自动化　automation of power system dispatching
电力系统负荷曲线　load curve of power system
主站（控制站）　master station，controlling station
子站　slave station，controlled station
远动　telecontrol
遥信　teleindication，telesignalization
遥测　telemetering
遥控　telecommand
遥调　teleadjusting
远动配置　telecontrol configuration
远程监视　telemonitoring
远程切换　teleswitching
远程指令　teleinstruction
电力系统远动技术　telecontrol technique for power system
事故追忆　post disturbance review
调度规程　dispatching regulation
调度自动化计算机系统　computer system of dispatching automation
数据通信　data communication
安全隔离通信　security isolation communication
异步远动传输　asynchronous telecontrol transmission
同步远动传输　synchronous telecontrol transmission
模拟通信　analogue communication
监视信息　monitored information
状态信息　state information
双态信息　binary state information
事件信息　event information
返回信息　feedback information
增量信息　incremental information
故障状态信息　fault state information
组合告警　group alarm
总告警　common alarm
脉冲命令　pulse command
保持命令　maintained command
持续命令　persistent command
调节命令　adjusting command
设定命令　set-point command
增量命令　incremental command
选择命令　selection command
组命令　group command
指示命令　instruction command
选择并执行命令　select and execute command
时标　absolute chronology，time tagging
同步对时　synchronous clock
有功功率与频率控制　control of active power and frequency
无功功率与电压控制　control of reactive power and voltage
总响应时间　overall response time
总传送时间　overall transfer time
远动传送时间　telecontrol transfer time
平均传送时间　average transfer time
更新时间（刷新时间）　updating time，refresh time
能量管理系统　energy management system (EMS)
自动发电控制　automatic generation control
电力系统元件　power system element
电力系统事故　power system fault
报警　alarm
网络安全防护　cyber security and prevention
网络攻击　network attack
防火墙　firewall
纵向加密　encryption
认证　authentication
横向隔离　isolation
安全区　safety zone
控制区　control zone
非控制区　non-control zone
入侵检测　intrusion detection
入侵防护　intrusion prevention

(2) 机组及辅助设备控制　Control of unit and auxiliary equipment

集中控制　centralized control

现地控制　local control
以计算机为基础的控制　computer-based automation
控制级　control level
控制方式　control mode
数据采集系统　data acquisition systems
可编程控制器　programmable logic controller (PLC)
下游水位开关　downstream level switch
电极式水位开关　electrolyte level switch
投入式水位传感器　submerged water level sensor
气泡型水位传感器　bubble type water level sensor
浮子开关　float switch
压力开关　pressure switch
压力变送器　pressure transducer
液位开关　level switch
液位变送器　level transducer
振动变送器　vibration transducer
流量变送器　flow transducer
流量开关　flow switch
限位开关　limit switch
行程开关　limit switch
限位行程开关　end-of-travel limit switch
蜗壳水压传感器　spiral case water pressure sensor
模拟量指示仪表　analog indicating instrument
数字式仪表　digital indicating instrument
指示灯　lamp indicators
机械保护系统　mechanical protection system
电操作控制元件　electrically actuated control element
手动控制　manual control
自动控制　automatic control
异常状态　abnormal condition
数据采集与监视　data acquisition and monitoring
开机过程中断　starting sequence dropout
报警状态光字指示　annunciation of alarm conditions
仪表测量　metering and instrumentation
事件记录　event recording
霍尔效应传感器　Hall effect sensor
涡流传感器　eddy current sensors
磁传感器　magnetic sensors
光传感器　optical sensors
齿盘　toothed wheel
永磁发电机　permanent magnet generators (PMG)
过速开关　overspeed switch
转速信号装置　speed indicator
蠕动检测器　creep detector
模拟量测值监视　monitoring of analog measurements
主轴密封的冷却与润滑　shaft seal cooling and lubrication
超声波流量计　ultrasonic flow meter
差压流量计　differential pressure flow meter
分布控制　distributed control
下游水位变送器　downstream level transducer
控制盘　control board
与控制盘的通信　communications with control board
与辅助设备的通信　communications with auxiliary equipment
流程连锁　sequence interlocking
压缩空气系统控制　compressed air system control
控制系统接口　control system interfaces
定子绕组温度　stator winding temperature
推力轴承温度　thrust bearing temperature
轴承油温　bearing oil temperature
空气冷却器出风温度　air cooler outlet air temperature
空气冷却器进风温度　air cooler inlet air temperature
发电机磁场温度　generator field temperature
轴承振动检测器　bearing vibration detector
大轴摆度检测器　shaft run out detector
启停命令　start-stop commands
发电机润滑油系统　generator lubrication oil system
机组内加热器　generator housing heaters
制动气源　air supply for brakes
CO_2 灭火系统　CO_2 fire extinguishing system
冷却水阀门位置　cooling water valve position
变压器压力突变释放装置　transformer main tank sudden pressure relief device
油温计　oil temperature indicator
热像式绕组温度计　thermal image temperature indicator (WTI)
指针式油温计　dial type oil temperature indicator
火焰探测器　flame detector
烟雾探测器　smoke detector
冷却水源　water supply for cooling
振动探头　vibration probe
桨叶位置传感器　blade position sensor
折向器位置传感器　deflector position sensor
压水进气阀　depression air valves
电磁阀　solenoid valve
调相运行和水泵启动的压水系统　depression system for synchronous condenser mode and pump starting
开度限制器位置开关　gate limit position switch
电源消失　power supply failure

过滤器　filter，strainer
过滤器堵塞　filter obstruction
完全停机　complete shutdown
部分停机　partial shutdown
同步设备　synchronizing device（synchronizer）
辅助设备停机延时　time delay for stopping auxiliaries
调速器油压　governor oil pressure
发电机空气制动器　generator air brake
导叶锁定信号器　wicket gate lock detector
导引阀　pilot valve
火灾监测和灭火系统　detection- extinguishing system
增减命令　raise-lower command
快速减负荷　rapid unloading
投切命令　on-off command
失电停机　deenergize-to-shutdown
制动瓦未脱离　brake shoes not cleared
导叶位置开关　wicket gate position switch
快速关闭　rapid closure
上下游水位差　level difference between headwater and tailwater
全关位置　fully closed position
机组开停机流程　start and stop sequences for the unit
进水闸门开启　intake gate open
进水闸门关闭　intake gate closed
水轮机桨叶位置　turbine blade position
水轮机喷嘴位置　turbine nozzle position
推力轴承油槽油位　thrust bearing tank oil level
导轴承温度　guide bearing temperature
转子顶起系统　rotor jacking system
导轴承油流量　guide bearing oil flow
导轴承油位　guide bearing oil level
推力轴承油压　thrust bearing oil pressure
水泵充水和带负荷流程　pump priming and loading sequence
加速时间　acceleration time
开停机未完成　start-stop sequence incomplete
开停机超时　start-stop duration exceeded
常规方式自动化　conventional automation
多极开关　multipole switch
水位低于转轮　water level below the runner
水泵转向　pumping direction
充水系统　priming system
闭路循环去离子水系统　closed-loop demineralized water system
滤水器差压高　water strainer high differential pressure
备用泵启动　back-up pump start
主用空压机　lead compressor
备用空压机　lag compressor
主备控制　lead-lag control
调相方式　condensing mode
加压泵　booster pump
水位监视设备　water level monitoring equipment
净水头计算　net head calculations
导叶空载开度　speed-no-load wicket gate position
灭火系统　fire extinguishing system
辅助设备启动　auxiliaries start
开机前检查　pre-start inspection
状态监视　monitoring of status
尾水压水控制开关　tailwater depressing control switch
水轮机排气阀　turbine vent valve
测温电阻　resistance temperature detector（RTD）
热交换器的冷却水源　cooling water supply for heat exchanger
全开位置　fully open position
转轮密封温度传感器　runner seal temperature sensor
转轮下环排水阀　runner band drain valve
制动气压　brake air pressure
换相开关　phase reversal switch
水库水位允许水泵运行　reservoir levels OK for pump
制动复归　brakes released
紧急停机　emergency shutdown
紧急停机按钮　emergency push button
快速停机　quick shutdown
正常停机　normal shutdown
剪断销断裂　shear pin failure
冷却水泵　cooling water pump
启动母线　starting bus
停机按钮　shutdown push button
转速信号发生器　speed signal generator（SSG）
技术用水　service water
运行人员干预　operator intervention
制动控制开关　brake control switch
推力轴承油泵控制开关　thrust bearing oil pump control switch
空气围带　inflatable rubber seal
主轴密封水阀　shaft seal water valve

抗磨环冷却水阀 wear ring cooling water valve
最优导叶开度 optimum wicket gate position
有功和无功功率控制 control of active and reactive power
电压-无功控制 voltage-var control
联合发电控制 joint load generation control
联合无功控制 joint load var control
准备好自动开机 readiness for automatic start
机组断路器合闸 unit circuit breaker closed

8.8 继电保护系统 Relay protection system

继电保护装置 relay protection equipment
继电保护系统 relay protection system
继电保护试验 relay protection test
保护区 protected section，protected zone
保护范围 reach of protection
保护重叠区 overlap of protection
欠范围 underreach
超范围 overreach
主保护 main protection
后备保护 backup protection
辅助保护 auxiliary protection
备用保护 standby protection
瞬时保护 instantaneous protection
延时保护 delayed protection，time-delayed protection
方向保护 directional protection
电路近后备保护 circuit local backup protection
变电站近后备保护 substation local backup protection
远后备保护 remote backup protection
断路器失灵保护 circuit-breaker failure protection
双重保护 duplicate protection
继电保护可靠性 reliability of relay protection
继电保护可信赖性 dependability of relay protection
继电保护安全性 security of relay protection
继电保护选择性 selectivity of relay protection
继电保护快速性 rapidity of relay protection
继电保护灵敏性 sensitivity of relay protection
电力系统异常 power system abnormality
电力系统故障 power system fault
简单故障 simple fault
复合故障 combination fault
短路故障 short-circuit fault (shunt fault)
接地故障 earth fault
跨线故障 cross country fault，cross circuitry fault
区外故障 external fault
区内故障 internal fault
继电器 relay
突变量继电器 sudden change relay
单稳态继电器 mono-stable relay
双稳态继电器 bi-stable relay
计数继电器 counter relay
整流式继电器 rectifying relay
极化继电器 polarized relay
辅助继电器，中间继电器 auxiliary relay
信号继电器 signal relay
突变量继电器 sudden change relay
单稳态继电器 mono-stable relay
双稳态继电器 bi-stable relay
计数继电器 counter relay
整流式继电器 rectifying relay
功率继电器 power relay
电压继电器 voltage relay
电流继电器 current relay
阻抗继电器 impedance relay
频率继电器 frequency relay
同步检查继电器 synchro check relay
相位比较继电器 phase comparison relay
方向继电器 directional relay
时间继电器 time relay
定时限继电器 specified time relay，definite time relay
反时限继电器 inverse time relay
反时限最小定时限继电器 inverse definite minimum time relay (IDMT)
接地继电器 earth fault relay
零序电流继电器 zero sequence current relay，residual current relay
零序电压继电器 zero sequence voltage relay，residual voltage relay
热继电器 thermal relay
差动继电器 differential relay
速断差动保护 instantaneous (high-set) differential protection
励磁涌流 inrush current
二次谐波制动 second harmonic restrain
电流元件 current component
阻抗元件 impedance component
功率元件 power component
启动元件 starting component

测量元件 measuring component
保护元件 protection component
时间元件 time component
方向元件 directional component
电压元件 voltage component
执行元件 execute component
闭锁元件 blocking component
采样元件 sampling component
开关量输入 digital input
开关量输出 digital output
模拟量输入 analog input
模拟量输出 analog output
端子排 terminal block
电压回路 voltage circuit
电流回路 current circuit
操作回路 operation circuit
逻辑回路 logical circuit
二次回路 secondary circuit
信号回路 signal circuit
电压形成回路 voltage forming circuit
电磁继电器 electromagnetic relay
微机继电保护装置 microprocessor based protection device
整定 setting
整定计算 setting calculation
保护配合 protection coordination
时间配合 time coordination
过电流保护的选择性 over-current discrimination
继电器启动 pick up of relay，start of relay
继电器动作 operate of relay
继电器复归 reset of relay，drop out of relay
误动作 unwanted operation
越级误动作 unwanted operation for an external fault
误跳闸 nuisance trip，false trip
拒绝动作 failure to operation ，failure to trip
延时动作 delayed operation
退出运行〔动词〕 be out of service
躲过外部故障〔动词〕 remain stable for an external fault
死区 dead zone
触点 contact
常开触点 normally open contact
常闭触点 normally closed contact
动合触点 make contact
动断触点 break contact
特性量 characteristic quantity
特性量整定值 setting value of the characteristic quantity
特性量整定范围 setting range of the characteristic quantity
特性角 characteristic angle
动作时限 operation time limit
给定误差 assigned error
启动值 starting value
整定值 setting value
切换值 switching value
返回值 disengaging value
复归值 reset value
动作值 operating value
不动作值 non-operating value
动作时间 operate time
返回系数 disengaging ratio
动断触点的闭合时间 closing time of a break contact
动合触点的闭合时间 closing time of a make contact
返回时间 reset time
采样频率 sampling frequency
差模干扰电压 differential mode disturbance voltage
共模干扰电压 common mode disturbance voltage
电磁兼容设计 design of electromagnetic compatibility
故障自动检测 automatic diagnosis
数-模变换器 digital-to-analog converter
模-数变换器 analog-to-digital converter
数字滤波器 digital filter
噪声滤波器 noise filter
电流保护 current protection
电压保护 voltage protection
阻抗保护 impedance protection
零序保护 zero sequence protection
负序保护 negative sequence protection
频率保护 frequency protection
温度保护 temperature protection
气体保护 gas protection，Buchholz protection
行波保护 traveling wave protection
单元式保护 unit protection
非单元式保护 non-unit protection
分相保护 phase segregated protection
不分相保护 non-phase segregated protection
允许式保护 permissive protection
闭锁式保护 blocking protection

远方跳闸 intertripping，transfer tripping
智能继电保护 intelligent relay protection
广域保护 wide-area protection
发电机保护系统 generator protection system
定子接地保护 stator earth fault protection
转子接地保护 rotor earth fault protection
低频率保护 underfrequency protection
高频率保护 overfrequency protection
负序电流保护 negative sequence current protection
低频过流保护 low frequency overcurrent protection
次同步过流保护 sub-synchronous overcurrent protection
复压过流保护 compound voltage started overcurrent protection
失步保护 out-of-step protection
失磁保护 loss-of-field protection
过负荷保护 overload protection
零序电流保护 zero sequence current protection
零序电压保护 zero sequence voltage protection
逆功率保护 reverse power protection
低功率保护 underpower protection
异步运行保护 asynchronous operation protection
纵联差动保护 longitudinal differential protection
不完全差动保护 combined split phase and differential protection
比例制动差动保护 percentage differential relay
裂相横差保护 transverse differential protection，split-phase differential protection
零序电流型横差保护 zero-sequence differential protection
零序差动保护 restricted earth fault protection (REF)
突然加电压保护 inadvertent energizing protection
轴电流保护 shaft current protection
外壳漏电保护系统 frame leakage protection，case earth protection
过励磁保护 overexcitation protection
发电机-变压器单元保护 generator-transformer unit protection
振荡闭锁 power swing blocking
微机保护 microprocessor based protection
高阻抗型母线差动保护 high-impedance busbar differential protection
中阻抗型母线差动保护 medium-impedance busbar differential protection
微机母线保护 microprocessor based busbar protection
变压器电流差动保护 transformer current differential protection
定时限保护 definite time protection，specified time protection
反时限保护 inverse time protection
过电流保护 overcurrent protection
低压过流保护 undervoltage started over-current protection
带电流记忆的低压过流保护 undervoltage started overcurrent protection with current seal-in
低电压保护 undervoltage protection
过电压保护 overvoltage protection
微机变压器保护 microprocessor based transformer protection
微机电动机保护 microprocessor base motor protection
失电压保护 loss-of-voltage protection
功率方向保护 directional power protection
方向比较式保护 direction comparison protection
电流差动式纵联保护 current differential protection
相位比较保护 phase comparison protection
纵联保护 pilot protection，line longitudinal protection
超范围式纵联保护 overreach pilot protection
欠范围式纵联保护 underreach pilot protection
允许式纵联保护 permissive pilot protection
闭锁式纵联保护 blocking pilot protection
多段式距离保护 multi-stage (multi-zone) distance protection
光纤纵联保护系统 optical link pilot protection
分相电流差动保护 phase current differential protection
电力线载波 power line carrier (PLC)
电力线载波纵联保护系统 power line carrier pilot protection system
方向电流保护 directional current protection
方向距离保护 directional distance protection
导引线保护 pilot wire protection
电流速断保护 instantaneous overcurrent protection
后加速保护 accelerated protection after fault
短引线保护 stub protection
死区保护 dead zone protection
自动重合闸 auto-reclosing

单相自动重合闸　single phase auto-reclosing
三相自动重合闸　three phase auto-reclosing
综合重合闸　composite auto-reclosing
延时自动重合闸　delayed automatic reclosing
同步检定　synchro check
同步并列　synchronization
快速自动重合闸　high speed automatic reclosing
顺序重合闸　sequential reclosing
闭锁重合闸　lockout reclosing
禁止重合闸　inhibit reclosing
重合闸失败　unsuccessful reclosing
一次重合闸　single shot reclosing
无电压检定　non-voltage verification
自动重合闸断开时间　auto-reclose open time
复归时间［自动重合闸的］　reclaim time
自动重合闸中断时间　auto-reclose interruption time
自动切换装置　automatic switching control equipment
操作箱　controly box
电力系统自动装置　power system automatic control device
特殊保护系统［安全稳定装置］　special protection system (SPS)
补救控制系统［安全稳定装置］　remedial action scheme (RAS)
失步解列　trip for out of step
功角测量装置　phasor measurement unit, pmu
自动同步装置　automatic synchronizing device
自动准同步装置　automatic ideal synchronizing device
手动准同步装置　manual ideal synchronizing device
故障自动记录装置　automatic fault recording device
自动低频减负荷　automatic underfrequency load shedding
自动低压减负荷　automatic undervoltage load shedding
同步检查装置　synchronism detection unit, synchrocheck unit
故障定位器　fault locator
故障记录器　fault recorder
故障录波器　fault oscillograph
反事故措施　accident countermeasures
防误操作措施　misoperation countermeasures
继电保护投入率　utilization factor of protection system
继电保护正确动作率　correct actuation ratio of protection system
保护等电位连接　equopotential bonding

8.9 控制电源

Control power sources

单体电池　cell
蓄电池　storage battery
电池　battery
碱性电池　alkaline cell
激活　activation
全密封电池　hermetically sealed cell
极板　plate
涂膏式极板　pasted plate
负极板　negative plate
正极板　positive plate
管式极板　tubular plate
隔离层　separator
阀　valve
电池外壳　cell can
电池槽　case
电池盖　cell lid
电池封口剂　lid sealing compound
端子　terminal
负极端子　negative terminal
正极端子　positive terminal
阳极　anode
阴极　cathode
电解质　electrolyte
输出电缆　output cable
连接件　connector
额定容量　rated capacity
剩余容量　residual capacity
电池放电　discharge
放电电流　discharge current
放电率　discharge rate
自放电　self-discharge
放电电压　discharge voltage
初始放电电压　initial discharge voltage
终止电压　end-of-discharge voltage
开路电压　open-circuit voltage
并联　parallel connection

串联 series connection
铅酸蓄电池 lead dioxide lead battery
镉镍蓄电池 nickel oxide cadmium battery
阀控式铅酸蓄电池 valve regulated lead acid battery (VRLA)
密封电池 sealed cell
排气帽 vent cap
电池组架 battery rack
免维护电池 maintenance-free battery
电池充电 charging of a battery
恒压充电 constant voltage charge
恒电流充电 constant current charge
浮充电 floating charge
充电效率 charge efficiency
均衡充电 equalization charge
充电因数 charge factor
完全充电 full charge
初充电 initial charge
过充电 overcharge
充电率 charge rate
终止充电率 finishing charge rate
涓流充电 trickle charge
液位指示器 electrolyte level indicator
充电终止电压 end-of-charge voltage
选择蓄电池容量 battery sizing
选择充电器容量 charger sizing
负荷清单 load details
经常负荷 continuous load
事故照明负荷 emergency lighting load
蓄电池组 battery bank
降压二极管 voltage dropping diode
端电池 end cell
蓄电池充电器的参数 rating of battery charger
相控充电器 phase controlled charger
高频开关充电器 high frequency switch mode charger
纹波系数 ripple factor
直流配电盘 DC distribution board
直流隔离开关 DC disconnect switch
直流正极接地故障 DC positive to earth fault
直流负极接地故障 DC negative to earth fault
不间断电源 uninterruptible power supply (UPS)
绝缘监视装置 insulation monitoring device
静态旁路开关 static bypass switch
手动旁路开关 manual bypass switch
脉动直流电流 pulsating direct current
直流断路器 DC circuit breaker
恒压电源 constant voltage power supply
恒流电源 constant current power supply

8.10 通信、工业电视与门禁系统 Communication, industrial TV and access control system

通信 communication
厂内通信 intraplant communication
对外通信 communication external to plant
音频电缆 audio-frequency cable
自动电话交换机 automatic telephone exchange
纵横制自动电话交换机 crossbar automatic telephone exchange
接线盒 connection box
电缆分线箱 cable branch box
共电式交换机 common battery switch
共电式电话机 common battery telephone
交接箱 cross-connecting box
电缆分线箱 cable distribution head
公用自动电话局 community dial office
调度电话分机 dispatching telephone subset
调度电话主机 dispatching telephone control main station
接地板 earth plate
分机 extension
入中继线 incoming trunk
中间配线架 intermediate distributing frame
通信线路 line of communication
扬声器 loudspeaker
有线电话 line telephone
总配线架 main distributing frame
调制解调器 modulator-demodulator, modem
主通信站 main traffic station
多路设备 multiplexing equipment
多路复用设备 multiplexing unit
操作台 operation switchboard
自动电话用户小交换机 private automatic branch exchange (PABX)
专用自动小交换机 private automatic exchange (PAX)
专用小交换机 private branch exchange (PBX)
接收机 receiver
程控交换机 stored program control exchange

程控数字交换机 stored program control digital exchange
用户话机 subsciber's set
电源设备 supply equipment
发射机 transmitter
转接交换机 through switchboard
话务台 traffic switchboard
数据传输 transmitting data
波导 waveguide
波导耦合器 waveguide coupler
分频滤波器 freqency division filter
高频电缆 high frequency cable
高频振荡器 high frequency oscillator
高频引入架 high frequency patching bay
耦合电容器 coupling capacitor
结合滤波器 coupling filter
线路阻波器 line trap
载波电话终端机 carrier telephone terminal equipment
地面站 ground station
地球站 earth station
通信卫星 communication satellite
微波收发信机 microwave transmitting and receiving equipment
无线电台 radio station
无线电收发信机 radio receiver-transmitter
全向天线 omni antenna
光纤 optical fibre, fibre optic
多模光纤 multi-mode optical fibre
单模光纤 single-mode optical fibre
光缆 optical cable , fibre optic cable
光纤连接器 optical fibre connector, joint box
出中继线 outgoing trunk
光接收机 optical receiver
光中继器 optical repeater
光端机 optical transmission terminal equipment
全介质自承式光缆 all dielectric self- supporting optical cable (ADSS)
光纤复合架空地线 composite overhead groundwire with optical fibre (OPGW)
光纤复合相线 optical phase conductor (OPPC)
金属自承光缆 metallic aerial self supporting optical cable (MASS)
附挂式光缆 optical attached cable (OPAC)
光单元 optical unit
光纤元件 optical element
缆芯 cable core
光缆护套 cable jacket
内垫层 inner jacket
松套管 buffer tuber
光纤软线 optical fibre cord
尾纤 pigtail
纤芯直径 core diameter
光纤色散 fibre dispersion
光纤偏振模色散 polarization mode dispersion (pmd)
光衰减 light attenuation
光损耗 light loss
光纤衰减系数 fibre attenuation coefficient
光纤应变 fibre strain
光学连续性 attenuation uniformity
光纤接头 optical fibre splice
光缆段 optical cable section
链路 link
光缆终端 optical cable terminal
门禁系统 access control system
入侵报警系统 intrusion alarm system
防盗报警系统 burglar alarm system
视频安防监控系统 video surveillance and control system
门禁控制 access control
活动侦测 activity detection
盲区 blind zone
磁性门开关 magnetic door contact
监控中心 surveillance and control centre
防拆报警 tamper alarm
电控锁 electric strikes
入口 entrance
红外探测器 infrared detector
视频识别装置 video identification device
巡更 night patrol
考勤系统 time attendance
视场 field of view
视角 angle of view
视频信号 video signal
信噪比 signal-to-noise ratio
闭路电视 closed-circuit television
工业电视 industrial television
数字视频 digital video
视频探测 video detection
视频监控 video monitoring

视频传输 video transmission
报警图像复核 video alarm verification
视频主机 video controller, video switcher
云台 pan tilt
摄像机 camera
自动光圈镜头 automatic iris lens
自动电子快门 automatic electronic shutter
前端设备 front-end device
图像质量 picture quality
图像分辨率 picture resolution
报警联动 action with alarm
环境照度 environmental illumination
视频压缩格式 video compressed format
门禁系统 entrance control system
智能卡 smart card
防盗探测器 anti-theft detector
读卡器 card reader
门禁控制器 entrance controller
中央控制设备 central control equipment
远程中心 remote center
报警接收中心 alarm receiving center
收集点 collector point
组合系统 combined system
报警信号传输系统 alarm transmission system
手动控制装置 manual controls
实体防护 physical protection
门禁系统的设防状态 set condition of access control system
撤防状态 unset condition
视频音频同步 synchronization of video and audio
终端设备 terminal device
入侵侦测 intrusion detection
可视对讲机 visual intercom
非可视对讲机 unvisule intercom
周界防范 perimeter precaution
监视区域 surveillance area
防拆装置 tamper device
电磁锁 electromagnetic lock
像差 aberration
自动光线补偿 ALC control

8.11 电气设备布置 Electric equipment layout

设备布置 equipment layout (equipment arrangement)
设备安装 equipment erection
主厂房 main powerhouse, machine hall
副厂房 auxiliary plant, service area
发电机层 generator floor
母线层 busbar floor
水轮机层 turbine floor
蜗壳层 spiral case floor
电缆夹层 cable spreading room
主变室 main transformer room
主变洞 main transformer cavern
安装间 erection bay
母线廊道 busbar gallery
母线洞 busbar cavern
空压机室 air compressor room
中控室 central control room
计算机室 computer room
控制盘室 control panel room
保护盘室 protection panel room
直流盘室 DC panel room
蓄电池室 battery room
通信设备室 communication device room
通信电源室 communication power supply room
资料室 archive room
电工实验室 electric test room
开关楼 switchyard building
中控楼 central control building
变电站总布置 layout of substation
变电站构架 substation gantry
高型布置 high-profile layout
半高型布置 semi-high profile layout
中型布置 medium-profile layout
启闭机室 hoist building

8.12 主要图表名称 Titles of figures and tables

主(副)厂房机电设备布置图 Electro-mechanical equipment layout in the main powerhouse (auxiliary plant)
主接线图 Single-line diagram, one-line diagram
厂用电接线图 Single-line diagram of station service system, Single-line diagram of AC auxiliary system
全厂接地系统总图 General earthing system layout
全厂电缆走向规划图 General cable routing planning
开关站设备布置图 Equipment layout of switchyard

计算机监控系统结构图　Computer supervision and control system configuration
保护和测量配置图　Single line diagram with protection and metering scheme
直流系统图　Single line diagram of DC system
通信系统配置图　Communication system configuration
工业电视系统图　Industrial television system configuration
原理接线图　Schematic diagram
发电机出力图　generator capability diagram
发电机 P/Q 曲线　generator P/Q chart
端子图　Terminal block diagram
安装图　Installation diagram，Assembly diagram
埋件图　Embedded part diagram
用户手册　User manual

第9章 土建工程施工
Construction of civil works

9.1 施工布置与交通运输
Construction layout, access and transportation

施工组织设计 construction planning, construction method statement
施工方案 construction scheme
施工准备 construction preparation
施工总布置 general construction layout
施工用地 occupied land for construction
永久用地 permanent land requisition
临时用地 temporary land occupation
施工管理及生活区 contractor camp
工程建设管理区 employer camp
安全文明施工 HSE (health, safety and environment) construction
施工分区规划 plan of construction zoning
坝址施工区 construction area at damsite
施工营地 construction camp
工地生活设施 site accommodation
砂石加工系统 artificial aggregate system, aggregate processing plant
混凝土生产系统 concrete production system
综合加工厂 comprehensive processing plant
机械修配厂 machine repair workshop
木材加工厂 timber processing plant, carpenter shop
汽车保养厂 automobile service workshop
钢筋加工厂 reinforcement fabrication plant
钢管加工厂 penstock fabrication plant (steel tube processing plant)
混凝土预制厂 precast concrete plant
金属结构拼装厂 fabrication plant of steel structure
土料场 borrow area
采石场 quarry
储料场 stockyard (stockpile area)
堆料场（堆渣场） dumping site
暂存场 temporary stack yard
弃渣场 spoil area, waste disposal area
存弃渣场规划 planning and layout for stockpile and spoil area
仓库 warehouse
炸药库 explosive magazine
施工供水系统 water supply system for construction
施工供电系统 power supply system for construction
施工供风系统 compressed air supply system for construction
集中供气 centralized compressed air supply
分散供气 decentralized compressed air supply
施工通讯系统 communication system for construction
场地平整 site leveling
三通一平 supply of water and power, road connection and site leveling
土石方平衡 earth-rock cut and fill balance, excavation-fill balancing
填挖平衡的土方工程 balanced earthworks
物料平衡 construction material supply and demand balance
料源规划 material source planning
施工运输 construction transportation
内陆运输 inland transportation
场内交通 on-site access
对外交通 site access
公路运输 road transport
铁路运输 railway transport
水路运输 waterway transport
铁路接轨站 track connection station
转运仓库 in-transit depot
转运站 transfer station
桥梁 bridge
隧道 tunnel
码头 wharf
渡口 ferry
活动便桥 bailey bridge

吊桥 suspension bridge
公路等级 road class
高速公路 expressway，freeway
等级公路 classified highway
等外公路 non-graded road
乡村道路 rural road
对外交通专用公路 dedicated road to site，external traffic highway
进场公路 access road to site
场内干线公路 onsite trunk road
过坝道路 roadway across the dam
上坝道路 access road to dam
运输坡道 haul ramp
路堤 embankment
路堑 cutting
路基 subgrade
路肩 shoulder
车道 lane
错车道 passing bay
紧急车道 emergency lane
回车道 turnaround loop
通行能力 traffic capacity
行车密度 traffic density
纵坡 longitudinal slope
转弯半径 turning radius
视距 sight distance
建筑限界 structure gauge
超高[道路/铁路] superelevation
加宽 widening
竖曲线 vertical curve
平曲线 horizontal curve
路面 pavement
混凝土路面 concrete pavement
沥青路面 bituminous pavement
碎石路面 macadam pavement
泥结碎石路面 clay-bound macadam pavement
重大件 heavy-outsized pieces，oversize and weight cargo
散装货物 bulk cargo
包装货物 packed goods
危险货物 dangerous goods
分瓣运输 transport of segmentation（transport in segments）
整体运输 transport in assembly
拖挂运输 tractor-trailer transport
甩挂运输 drop and pull transport
集装运输 containerized transport
托盘运输 palletized transport
运输强度 transport intensity
货运强度 freight traffic intensity
货运量 freight traffic tonnage，ton-volume
年运输量 yearly transport volume

9.2 建筑材料

Construction materials

石料 rock material
土料 earth material
碎石 crushed stone
砂砾料 sand and gravel material
开挖料 excavated material
填筑料 embankment material
砂 sand
人工砂 artificial sand
掺合料 admixture，blend
活性掺合料 active admixture
矿物掺合料 mineral admixture
火山灰 pozzolan
粉煤灰 fly ash
粒化高炉炉渣 granulated blast-furnace slag
磷渣粉 phosphorous slag powder
凝灰岩粉 tuff powder
黏性土 cohesive soil
膨润土 bentonite
骨料 aggregate
小石 fines
中石 medium grain
大石 bulky grain
特大石 boulder
豆砾，细砾 pea stone，pebble
天然骨料 natural aggregate
人工骨料 artificial aggregate
粗骨料 coarse aggregate
细骨料 fine aggregate
超径骨料 oversize aggregate
逊径骨料 undersize aggregate
碱活性骨料 alkali-reactive aggregate
透水性较强的材料 highly pervious material
针片状颗粒 elongated and flaky particle
渣砾料 dregs and gravel
贫胶渣砾料 lean cemented dregs and gravel

重骨料　heavyweight aggregate
轻骨料　lightweight aggregate
嵌缝骨料　key aggregate
活性骨料　reactive aggregate
级配碎石　graded broken stone；graded crushed stone
天然级配骨料　pit-run aggregate
毛石骨料　rubble aggregate
不分级骨料　ungraded aggregate
级配良好的骨料　well-graded aggregate
密级配骨料　dense-graded aggregate
间断级配骨料　gap-graded aggregate
开级配骨料　open-graded aggregate
二级配骨料　aggregate with maximum grain size of 40mm
三级配骨料　aggregate with maximum grain size of 80mm
成品骨料　finished aggregate
骨料坚固性试验　soundness test of aggregate
碱活性试验　alkali reactivity test
碱-骨料反应　alkali-aggregate reaction
砾石　gravel
轧石　crushed rock
琢石　ashlar
块石　rubble
毛石　quarry-run rock
石屑　aggregate chips
填料　filler
可压缩填料　compressible-type filler
嵌缝料　caulk
含泥量　silt content
石粉含量　rock flour content
云母含量　mica content
有效碱含量　effective alkali content
破碎比　ratio of reduction
成品率　rate of finished products
胶凝材料　cementitious material
胶结料　binder
胶状颗粒　colloidal particles
防渗料　impervious material
反滤料　filter material
垫层料　cushion material
过渡料　transition material
排水体料　drainage material
坝壳料　dam shell material
固壁土料　wall-stabilizing soil
水泥　cement
高强度水泥　high-grade (strength) cement
低强度水泥　low-grade (strength) cement
普通硅酸盐水泥　Portland cement
高炉矿渣水泥　blast-furnace cement
膨胀水泥　expanding cement
火山灰质水泥　pozzolana cement
快硬水泥　quick-hardening cement
低热微膨胀水泥　low heat expansive cement
中热水泥，改良硅酸盐水泥　modified Portland cement
抗硫酸盐水泥　sulphate-resisting cement
铝酸盐水泥　aluminate cement
明矾石膨胀水泥　alunite expanding cement
自应力水泥　self-prestressed cement
水硬性水泥　hydraulic cement
熟料水泥　clinker cement
对比胶砂　reference mortar
水工沥青　hydraulic bitumen
改性沥青　modified bitumen
沥青涂层　bitumen coat
乳化沥青　emulsified bitumen
稀释沥青　dilute bitumen
热沥青　hot bitumen
沥青混合料　bituminous mixture
沥青玛蹄脂　bituminous mastic
沥青砂浆　bituminous mortar
冷底子油　adhesive bitumen primer
防水卷材　waterproof membrane
防水涂料　waterproof coating
防腐蚀涂料　anticorrosive coating
防水密封胶　waterproof sealant
土工织物　geotextile
土工膜　geomembrane
土工格栅　geogrid
土工无纺布　nonwoven geotextile
复合土工布　composite geotextile
复合土工膜　composite geomembrane
外加剂　additive，agent
改性剂　modifier
固化剂　hardener
早强剂　early-strength admixture
稀释剂　diluting agent
增塑剂　plasticizer

速凝剂 accelerating admixture
缓凝剂 retarding admixture
高温缓凝剂 high temperature retarding admixture
减水剂 water reducing admixture
早强减水剂 accelerating and water reducing admixture
缓凝减水剂 retarding and water reducing admixture
高效减水剂 high-range water-reducing admixture
引气剂 air-entraining admixture
水中不分离剂（抗分散剂） non-dispersible agent
黏着剂 bonding agent
防冻剂 anti-freezing agent
保温剂 thermal insulating agent
养护剂 curing agent
膨胀剂 expansive agent
消泡剂 defoaming agent
泵送剂 pumping admixture
止水 waterstop
止水片 waterstop strip
铜止水 copper waterstop
塑性止水材料 plastic sealer
橡胶止水 rubber waterstop
聚氯乙烯（PVC）止水带 PVC waterstop
不锈钢止水带 stainless steel waterstop
复合密封止水材料 composite waterstop
膨胀止水 expanded waterstop
止水片鼻子 nose of waterstop
止水片平段 wing flat of waterstop
止水片立腿 erecting end of waterstop
塑料排水带 plastic drain
塑料盲管 plastic blende drain
异型接头 heterogeneous jointer
木材 timber，lumber
方木 sawn timber
原木 unsawn timber
加工木材 factory lumber
胶合板 laminated wood
汽油 gasoline
柴油 diesel oil
润滑油 lubricating oil
机油 mobile oil
碳素钢 carbon steel
铸钢 cast steel
优质钢 fine steel
工字钢 H-section steel
角钢 angle steel
型钢 shaped steel
钢筋 reinforcement bar (rebar)
光面钢筋 plain bar
带肋钢筋 ribbed bar
螺旋钢筋 spiral rebar
冷拉钢筋 cold-drawn bar
螺旋钢筋 spiral rebar
钢筋束，锚索 tendon
钢绞线 steel strand
波纹板 corrugated sheet
铅丝 galvanized wire
镀锌管 galvanized pipe
铸铁管 cast iron pipe
无缝钢管 seamless steel pipe
平头螺栓 flush bolt
沉头螺栓 sunk bolt
凸头螺栓 raised head bolt
六角头螺栓 hex bolt
有眼螺栓 eye bolt
钩头螺栓 hook bolt
双端螺栓 stud bolt
贯穿螺栓 through bolt
固定螺栓 set bolt
连接螺栓 connecting bolt
锚定螺栓 anchor bolt
锁紧螺栓 lock bolt
丝头 threaded sector
电焊条 electrode
焊丝 wire
焊剂 flux
防锈漆 antirust paint
防腐漆 anticorrosive paint
防水漆 waterproof paint
防火漆 fire retardant paint
绝缘漆 insulated paint
底漆 prime paint
合成漆 synthetic paint
聚脲 polyurea
火工材料 explosive materials
抗水炸药 waterproof explosive
铵锑炸药 ammonium nitrate explosive
铵油炸药 ammonium nitrate fuel oil explosive
胶质炸药 dynamite
雷管 blasting cap，detonator

火雷管 spark blasting cap
电雷管 electric blasting cap
瞬发雷管 instantaneous detonator
迟发雷管（延期雷管） delay detonator
毫秒雷管 ms delay blasting cap
药卷 cartridge
导火索 safety fuse
导爆索 primacord
塑料导爆管 plastic primacord tube
继爆管 relay primacord tube
密封（嵌缝）材料 sealant
沥青杉木板 asphalt Chinese fur board

9.3 施工导流 Diversion during construction

导流流量标准 diversion discharge criterion
导流建筑物级别 grade of diversion structure
洪水设计标准 flood design standard
超标准洪水 over-standard flood
水位变动期 alternate period of high and low water level
水位骤降期 rapid drawdown period of water level
N年一遇洪水 N-year flood
度汛 flood protection in flood season
春汛（融雪洪水） spring flood，snowmelt flood
夏（伏）汛 summer flood
施工截流 river closure
导流 diversion
基坑排水 foundation pit dewatering
下闸 gating
封堵 plugging
水库蓄水 reservoir impoundment (filling)
全年导流 whole year diversion
枯水期导流 dry season diversion
分期导流 stage diversion
初期导流 early-stage diversion
中期导流 midterm diversion
后期导流 later stage diversion
施工期蓄水 impounding in construction period
初期蓄水 initial impounding
分期蓄水 impounding in stages
导流方式 diversion mode
围堰一次拦断河床 non-stage river closure by cofferdam
围堰分期拦断河床 staged river closure by cofferdam
枯水期围堰挡水 retaining water by cofferdam in dry season
全年围堰挡水 retaining water by cofferdam all around the year
通过交替施工的低坝块导流 diversion through alternate construction blocks
明渠导流 channel diversion
隧洞导流 tunnel diversion
涵管导流 culvert diversion
渡槽导流 aqueduct diversion
坝体底孔导流 bottom outlet diversion
缺口导流 dam-gap diversion
机械抽排导流 pump diversion
导流孔（洞）封堵 plugging of diversion opening (tunnel)
拦洪高程 retention structure elevation
临时拦洪断面 particular section of embankment for flood retaining
临时拦蓄暴雨径流 temporary storage of spate runoff
束窄河道 narrowed river course
预进占 bank-off advancing
龙口 closure gap
抛投强度 dumping intensity
截流最大落差 maximum drop during closure
单宽流量 discharge per unit width
单宽功率 stream energy per unit width
龙口平均流速 average discharge velocity at closure-gap
截流戗堤 closure dike
立堵截流 end-dump closure
平堵截流 full width rising closure
单戗立堵 single closure dike plus end-dumping closure
双戗立堵 double closure dikes plus end-dump closure
钢筋笼 steel gabion
铅丝笼 wire box
混凝土四面体 concrete tetrahedron
合龙 final gap-closing
闭气 leakage stopping
截流围堰 river closure cofferdam
上游围堰 upstream cofferdam
下游围堰 downstream cofferdam

纵向围堰 longitudinal cofferdam
横向围堰 transversal cofferdam
过水围堰 overflow cofferdam
不过水围堰 non-overflow cofferdam
土石围堰 earth rockfill cofferdam
浆砌石围堰 cemented masonry cofferdam
混凝土围堰 concrete cofferdam
胶凝砂砾石围堰 cement-sand-gravel cofferdam
板桩围堰 sheet-pile cofferdam
钢板桩框格围堰 steel sheet-pile cellular cofferdam
预留岩坎 reserved rock sill
导流隧洞 diversion tunnel
导流明渠 diversion channel
导流底孔 diversion bottom outlet
旁通洞 bypass tunnel
施工导（截）流水力学模型试验 hydraulic model test of construction diversion (river closure) works
施工期通航水力学模型试验 hydraulic model test of navigation during construction

9.4 主体工程施工 Construction of main structures

主体工程 main structures (main civil works)
施工辅助工程（临时工程） temporary works
信息法施工 construction method from information
隐蔽工程 concealed works
结尾工程 winding-up works
土石方工程 earth-rock works
料场开采 exploitation of natural construction materials
设计需要量 design exploitative reserve
规划开采量 planned exploitative reserve
可开采储量 exploitable reserve
剥离表土 stripping top soil
杂填土 miscellaneous fill
表观密度 apparent density
体积密度（容重） bulk density
出渣线路 mucking route
爆破 blasting
控制爆破 controlled blasting
裸露装药爆破 adobe blasting
梯段爆破，台阶爆破 bench blasting
预裂爆破 presplit blasting
光面爆破 smooth blasting
挤压爆破 squeezed blasting, tight blasting
松动爆破 loosening blasting
定向爆破 directional blasting
岩塞爆破 rock plug blasting
延时爆破 delay blasting
缓冲爆破 buffer blasting
集中药包爆破 concentrated charge blasting
药壶爆破 pot hole blasting
微差爆破（毫秒爆破） millisecond blasting
齐发爆破 simultaneous blasting
瞬发爆破 short-delay blasting
拆除爆破 demolition blasting
深孔爆破 deep-hole blasting
浅孔爆破 short-hole blasting
爆力 explosive strength
爆速 detonation velocity
猛度 brisance factor
殉爆距离，诱爆性 flash-over tendency
爆破有害效应 adverse effect of blasting
压缩圈 crushing zone
破坏圈 fragmental zone
爆破漏斗 explosion crater
单段爆破药量 charge for one interval, charge amount per delay interval
单响最大段起爆药量 maximum primer charge in single shot
耦合连续装药 coupled continuous charging
不耦合连续装药 decoupled continuous charging
间隔装药 discontinued charging
分段装药 deck charging
爆破孔装药结构图 charging structure in blasthole
起爆 initiation
炮孔 blast hole
周边孔 peripheral hole
掏槽孔 cut hole
残孔率（半孔率） residual hole rate
土石方开挖 earth-rock excavation
开挖料利用率 availability of excavated material
折方系数 coefficient of measure conversion
土石方明挖 open earth-rock excavation
明挖方量 excavation in open-cut
自上而下开挖 top-down excavation
阶梯式开挖 benched excavation
超挖 over-excavation
欠挖 under-excavation

扩挖 enlargement
松方 loose measure
自然方 bank measure
压实方 compacted measure
压实系数 compaction coefficient
松散系数 bulk factor
平整度 evenness
不平整度 unevenness
声波波速 acoustic wave velocity
地下开挖 underground excavation
隧洞开挖 tunnel excavation
平洞，施工支洞 adit
钻爆法 drill-blast tunneling method
顶管法 pipe jacking method
掘进机法 tunnel boring machine method (TBM)
新奥法 new Austrian tunneling method (NATM)
盾构法 shielding tunneling method (shield method)
导洞掘进法 heading and cut method
中心导洞掘进 center drift tunneling
台阶掘进法 heading and bench method
全断面掘进法 full face driving method
部分断面（分部开挖）掘进法 partial face driving method
隧洞扩大掘进法 tunnel reaming method
上导洞法 top drift method
下导洞法 bottom heading method
分层开挖 excavation in layers
开槽施工 trench installation
不开槽施工 trenchless installation
竖井 vertical shaft
斜井 inclined shaft
埋藏式斜井 shaft with underground collar
露天斜井 shaft with open air collar
掌子面（工作面） heading face
超前钻孔法 advance borehole method
超前支撑掘进 forepoling
导井法 pilot bore method
正井法 shaft-sinking method
反井钻进法 raise boring method
贯通 break through
爬罐 raise climber (Alimak)
吊罐 raising cage
出渣 mucking
有轨运输 track haulage
无轨运输 trackless haulage
皮带机运输 belt conveyor delivery
超前支护 advance support
初期支护（一期支护） initial support
二期支护 secondary support
永久支护 permanent support
组合式支护 composite support
临时支护 temporary support
系统支护 systematic support
喷锚支护 bolt-shotcrete support
柔性支护 flexible support
管棚 pipe-shed
超前管棚 forepoling pipe-shed
超前小导管 advanced ducting
衬砌隧洞 lined tunnel
无衬砌隧洞 unlined tunnel
管道系统严密性试验 tightness test of pipeline
施工期通风 ventilation during construction
排烟 smoke removal
粉尘浓度 dust concentration
压入式通风 ventilation by forced pressure
吸出式通风 ventilation by exhaust
巷道式通风 gallery ventilation
压、吸混合式通风 combined pressure and exhaust ventilation
防尘 dust control
钢支撑 steel braces，steel support
格栅钢架 reinforcing-bar truss
衬砌 lining
拱腰 haunch
挡土板 breast board
自稳时间 stand-up time
悬挂式防渗结构 suspended seepage control structure
地基截水墙 foundation cutoff
坝基截水槽 key trench
侧翼截水槽 wing trench
明挖回填截水槽 open-cut and backfilled cutoff trench
混凝土防渗墙 concrete cutoff wall
高喷防渗墙 jet grouted cutoff wall
局部截水墙 partial cutoff
泥浆防渗墙 slurry cutoff
截渗环 cutoff collar，seepage collar
导墙 guide-wall
泥浆固壁 slurry wall stabilizing

灌浆 grouting
灌浆试验 grouting test
可灌性 groutability
灌浆廊道 grouting gallery
超前灌浆 advance grouting
回填灌浆 backfill grouting
固结灌浆 consolidation grouting
铺盖灌浆 blanket grouting
帷幕灌浆 curtain grouting
接触灌浆 contact grouting
接缝灌浆 joint grouting
化学灌浆 chemical grouting
超细水泥灌浆 superfine cement grouting
分段灌浆 stage grouting
一期灌浆孔 primary grout hole
二期灌浆孔 secondary grout hole
全孔一次灌浆法 full depth one stage method
自下而上分段灌浆 ascending stage grouting
自上而下分段灌浆 descending stage grouting
分序逐步加密灌浆 split spacing grouting
闭路循环灌浆 closed-circuit grouting
开路循环灌浆 open-circuit grouting
单液灌浆法 single liquid grouting
双液灌浆法 double liquid grouting
贴嘴灌浆法 port-adhesive grouting
纯压式灌浆 non-circulation grouting
劈裂灌浆 fracturing grouting
周边灌浆 perimeter grouting
限制灌浆 containment grouting
重复灌浆 repeated grouting
真空灌浆 vacuum grouting
循环钻灌法 circulation drilling and grouting method
预埋花管法 embedded perforated pipe method
套管灌浆法 casing grouting method
打管灌浆法 pipe driving grouting method
孔口封闭灌浆法 orifice-closed grouting method
高压喷射灌浆（高喷灌浆） high-pressure jet grouting
综合灌浆法 comprehensive grouting method
压浆试验 slurry pressure test
压水试验 water pressure test
灌浆记录仪 grouting recorder
大循环方式 major cycling way
小循环方式 minor cycling way
旋喷 rotating jet grouting
摆喷 oscillating jet grouting
定喷 directional jet grouting
板桩灌注墙 sheet pile cell wall
粉喷桩 dry jet mixing pile
聚灰比 polymer-cement ratio
砂浆材料 mortar material
砌体砂浆 masonry mortar
干硬性砂浆 dry mortar
丙乳砂浆 acrylic mortar
预缩砂浆 preshrinking mortar
锚固砂浆 anchoring mortar
干拌砂浆 dry-mixed mortar
水泥浆液 cement grout
胶体浆液 colloidal grout
粒状材料浆液 particulate grout
环氧树脂砂浆 epoxy resin mortar
稳定浆液 stable grout
固化灰浆 solidification slurry
自凝灰浆 self-hardening slurry
自应力浆液 self-stressing grout
膨胀性浆液 expanding grout
干硬性浆液 no-slump grout
聚合物改性水泥砂浆 polymer-modified cement mortar
聚合物分散剂 polymer latex
聚氨酯浆液 polyurethane grout
环氧树脂浆液 epoxy resin grout
丙烯酸盐浆液 acrylate grout
水玻璃浆液 sodium silicate grout
注入量 injection rate
单位注入量 unit injection rate
耗灰量 cement consumption
耗浆量 grout consumption
排浆量 grout discharge
浆液注入率 grout injection rate
吃浆率 acceptance of grout
吃浆量 grout take
拒浆标准 refusal criterion
灌浆强度值 grouting intensity number
浆液渗透距离 grout travel
接缝灌浆温度 closure temperature
灌浆压力 grouting pressure
屏浆 measure for keeping pressure to stage
闭浆 measure for keeping closed stage

串浆 grout interconnection (grout leaking)
冒浆 grout emitting
灌浆结石 set grout
造孔 drilling hole
先导孔 pilot hole
供浆-集浆-回浆管线 supply-header-return pipeline
孔口封闭器 packing gland
止浆塞 packer
打板桩 sheet piling
跳打桩 staggered piling
预制桩 precast pile
现浇桩 mix-in-place pile
灌注桩 filling pile
贯入桩 driving pile
沉管贯入桩 mandrel driven pile
钻孔灌注桩 drill hole grouting pile
扩孔桩 belled-out pile
挖孔桩 hand-dug pile
夯实土桩 rammed-soil pile
挤密砂桩 densification by sand pile
梅花桩 quincuncial pile
端承桩 point bearing pile
扩脚桩，支座桩 pedestal pile
抗拔桩 uplift pile
格式板桩 cellular sheet pile
垫块，替打 cushion block
送桩 follower
接桩 pile extension
开口沉箱 open caisson
强夯法 dynamic compaction method
振冲复合地基 vibro-composite foundation
振冲复合土体 vibo-treated composite soil
振冲桩 vibroflotation pile
振冲置换 vibro-replacement
面积置换率 replacement rate
振冲置换砂石桩 vibro-replacement stone and sand column
留振时间 compaction time
振冲器导管 follow-up tube
桩间土试验 post-soil test
换土垫层 cushion of replaced soil
预压加固 surcharge preloading consolidation
局部挖补处理 dental treatment
喷混凝土 shotcrete
干喷混凝土 dry shotcrete
湿喷混凝土 wet shotcrete
水泥裹砂喷射混凝土 cement paste wrapping sand shotcrete
潮料掺浆法喷射混凝土 cement paste wrapping aggregate shotcrete
挂网喷混凝土 wire mesh with shotcrete
钢纤维喷混凝土 steel-fiber shotcrete
劈钻成槽法 trenching by percussion and splitting
抓钻成槽法 trenching by drilling and grabbing
抓取成槽法 trenching by grabbing
铣削成槽法 trenching by cutting
反坡 reverse slope
倒坡（逆坡） adverse slope
不稳定坡 unstable slope
削坡 slope cutting
修坡 slope dressing，slope trimming
填方边坡 filled slope
挖方边坡 excavated slope
抛石护坡 riprap slope protection
干砌石护坡 dry rubble masonry slope
浆砌石护坡 masonry slope
混凝土护坡 concrete block protection
水泥土护坡 soil-cement slope
草皮护坡 grass (sodding) slope protection
框架护坡 framed revetment
三维植被网，土工网垫 three-dimensional geonet
后张锚 post-tensioned anchor
锚具 anchorage
锚头 anchor head
锚墩 anchor block
外锚头 outer fixed end
全长黏结型锚杆 anchor bolt bonded all length
无黏结环锚 unbonded circular anchored tendon
预应力环锚 prestress circular anchor
端头锚固型锚杆 anchor bolt anchored at head
摩擦型锚杆 friction anchor bolt
张拉型锚杆 tension anchor bolt
预应力锚杆 prepressed anchor bolt
树脂锚杆 resin anchor bolt
砂浆锚杆 grouted anchor bolt
水泥卷锚杆 cement cartridge rock bolt
胀壳式锚杆 expending shell anchor bolt
楔缝式锚杆 slot-and-wedge anchor bolt
倒楔式锚杆 inverted wedge anchor bolt

缝管锚杆　slot-tube anchor bolt
楔管锚杆　wedge-and-slot-tube anchor bolt
水胀式锚杆　water expansion anchor bolt
管式锚杆　tube anchor bolt
超前锚杆　forepoling bolt
自钻式注浆锚杆　self-drilling grouted anchor bolt
对穿锚索　hole-through tendon
锚筋束（桩）　anchor bundle
灌浆岩锚　grouted rock bolt
嵌入式岩锚　recessed rock anchor
土锚钉　soil dowel
锚杆注浆密实度　anchor bolt satiation degree
锚杆饱满度　anchor bolt satiation degree
锚杆模拟试验　anchor bolt simulation test
拉拔试验　pulling test
预应力损失率　prestressing loss rate
锁定损失率　locking loss rate
锁定后损失率　locked loss rate
张拉伸长值　tensile value
钻芯法检测　testing with drilled core
超声法　ultrasonic method
衰减　attenuation
衰减率　attenuation rate
声速　velocity of sound
主频　main frequency
声波反射法　sonic reflection method
有效长度法　effective length method
能量法　energy method
铺料　placing and spreading
填筑　filling，placing
骨料露天堆放　aggregate open stockpiled
粗骨料堆遮荫　shading of coarse aggregate stockpile
骨料级配　aggregate gradation
间断级配　gap gradation
连续级配　continuous gradation
均匀级配，窄级配　narrow gradation
宽级配　spreading gradation
骨料平均粒径　average size of aggregates
洛杉矶试验机磨耗试验　Los Angeles machine abrasion test
筛分　sieving (screening)
二次筛分　finish screening
筛分试验　screen test
筛分级配　screen size gradation
筛分曲线　screening curve
过筛百分率　percent passing
过筛累积百分率　cumulative percent passing
筛分效率　efficiency of screening
自下而上填筑施工　filled from the bottom up
进占法卸料　end-dump advance method
跳堆法　interval dumping method
斜坡推进法　advancing slope method
堆石　rock filling
砌石　stone masonry
抛石　riprap
夯实　tamping
碾压　rolling
压实　compacting
刨毛　scarifying
水力冲填　hydraulic excavation and filling
普氏压实曲线　Proctor compaction curve
最优含水量　optimum moisture content
填筑含水量　fill moisture content
吸水率　water-absorbing ratio
总空隙量　total volume of voids
架空［堆石体］　large voids left in the rockfill
骨料离析　aggregate segregation
土料翻晒　soil tedding
填筑标准　placement criterion
铺层厚度　lift thickness
压实标准　compaction criterion
压实厚度　compacted thickness
压实度　compactness
相对压实度　relative compactness
最大干容重　maximum dry unit weight
填筑干容重　fill dry unit weight
湿容重　moist unit weight
浮容重　submerged unit weight
饱和容重　saturated unit weight
最大干密度　maximum dry density
基准表观密度　basic apparent density
饱和面干表观密度　saturated-surface-dry apparent density
最优普氏密度　Proctor optimum density
压缩模量　compression modulus
压碎指标　crush index
非脆性不透水带　nonbrittle impervious zone
排水铺盖　drainage blanket

水平反滤层　horizontal filter
坝脚压重　toe support fills
反滤压坡　ballasting filter
面板垫层　bedding layer
过渡层　transition layer
防渗底层　watertight bottom layer
防渗面层　watertight surface layer
找平层（整平层）　leveling coat
胶结层　binder coat
封闭层　seal coat
黏结涂层　tack coat
翻模固坡　turning-over formwork and consolidating slope
楔板　wedge plate
初碾　initial rolling
终碾　final rolling
锚筋　anchor
拉筋　lacing
挤压边墙　extrusion concrete side wall
上行振动碾压　vibrating compaction upward
下行无振碾压　non-vibrating compaction downward
碾压效率　roller compaction efficiency
填筑间隔时间　intermittent time between placing layers
直接填筑允许时间　permissible time interval between placing layers
斜层铺筑　slopping placement
振动压实指标　vibrating compaction value (VC)
激振力　exciting vibration force
线压力　linear pressure
鼓胀　bulging
凹陷　depression
鼓泡　blistering
砂井排水　sand drain
卵石排水沟　cobble drain
块石排水沟　rubble drain
水平排水沟　contour ditch
工作面排水沟　service drain
排水干沟　main drain
排水支沟　subsidiary drain
排水井　drainage well
截流井　intercepting well
排水孔　drain hole
排水幕　drain curtain
排水沟　drainage ditch，gutter
截水沟　catch drain
集水沟　collector
集水坑　collecting sump
预制排水管　formed drain
排水不畅　impeded drainage
沥青含量　bitumen content
油石比　bitumen aggregate ratio
沥青针入度　bitumen penetration
沥青延度　bitumen ductility
沥青软化点　bitumen softening point
沥青溶解度　bitumen solubility
沥青闪点　bitumen flash point
沥青弗拉斯脆点　Fraass breaking point of bitumen
沥青混凝土水稳定系数　immersion coefficient of asphalt concrete
热稳定系数　coefficient of thermal stability
黏附性　adhesiveness
斜坡流淌值　slope flow
沥青混凝土流变　rheology of asphalt concrete
亲水系数　hydrophilic coefficient
马歇尔稳定度　Marshall stability
沥青劲度模量　stiffness of bitumen materials
沥青蜡含量　bitumen paraffin content
乳化沥青的破乳速度　demulsibility of emulsified bitumen
常态混凝土　conventional concrete
碾压混凝土　roller compacted concrete (RCC)
大体积混凝土　mass concrete
流动性混凝土　flowing concrete
自流平　self-leveling
自密实　self-condensing
现浇混凝土　cast-in-situ concrete，placed-in-situ concrete
新鲜混凝土　fresh concrete
新浇混凝土　green concrete
光面混凝土　fair face concrete
人工捣实混凝土　hand-compacted concrete
模内捣实混凝土　packing concrete in forms
气压浇注混凝土　air placed concrete
混凝土拌合物　concrete mixture
预制混凝土　precast concrete
无砂混凝土　no-fines concrete
干硬性混凝土　no-slump concrete
素（无筋）混凝土　plain concrete

贫混凝土　lean concrete
预应力混凝土　prestressed concrete
加气混凝土　aerated concrete
抗磨蚀混凝土　abrasion resistance concrete
聚合物混凝土　polymer concrete
补偿收缩混凝土　expansive concrete
模袋混凝土　bagged concrete
胶凝砂砾石混凝土　cement-sand-gravel concrete
低热微膨胀混凝土　low-heat micro- expansion concrete
变态混凝土　grout enriched vibratable concrete (GEV)
水下不分散混凝土　non-dispersible underwater concrete
钢筋混凝土　reinforcement concrete
预压骨料混凝土　prepackaged concrete
预填骨料混凝土　preplaced-aggregate concrete
高标号混凝土　high grade concrete
低标号混凝土　low grade concrete
低坍落度混凝土　low slump concrete
无收缩混凝土　non-shrinkage concrete
耐火混凝土　refractory concrete
钢纤维混凝土　steel fibre concrete
沥青混凝土　asphalt concrete
泵送混凝土　pump concrete
理论输送量　theoretic delivery
泵送能力　pumpability
水灰比　water-cement ratio
水胶比　water-cementitious material ratio
标准砂（基准砂）　reference sand
砂率　fine-to-coarse aggregate ratio
稠度　consistency
黏度　viscosity
和易性　workability
渗透性　permeability
坍落度　slump
混凝土配制强度　required average strength of concrete
混凝土龄期　concrete age
混凝土成熟度　maturity degree of concrete
混凝土等级（标号）　concrete class
抗渗等级　permeation resistance class
抗冻等级　frost resistance class
立方体强度等级　class of cube strength
混凝土立方试样　cube concrete test specimen
强度保证率　assurance factor of strength
混凝土缺陷　concrete defect
取芯　coring
干缩　drying shrinkage
抗风化　weathering resistance
抗化学侵蚀　resistance to chemical deterioration
冻融循环　freezing and thawing cycle
冻融试验　freezing and thawing test
干湿交替　alternate wetting and drying
混凝土剥蚀　concrete disintegration
配合比设计室内试验　lab test of mix design
配合比设计现场铺筑试验　trial placement of mix design
配合比设计生产性试验　productive test of mix design
布莱恩试验，水泥细度试验　Blaine test
维卡针入度试验　Vicat needle test
按体积配料　batching by volume
按重量配料　batching by weight
胶凝材料用量　cementitious material consumption
供料线　feeding system
平仓　spreading and leveling
振捣　vibrating
凿毛　surface roughening
毛面　rough surface
连续浇筑　continuous concreting (placement)
平浇法　horizontal placing method
跳仓浇筑　sequence placement
柱状浇筑　concreting with longitudinal joint
通仓浇筑　concreting without longitudinal joint
导管浇筑　tremie placing
混凝土分块浇筑　block concreting
混凝土分层浇筑　concreting in lifts
后继浇筑层　succeeding lift
浇筑间歇期　delay between placement
薄层间歇浇筑　placement in shallow lifts with delays between lifts
浇筑层高度　placement lift
浇筑层间缝　lift joint
错缝浇筑　overlapping placing
斜缝浇筑　inclined joint placing
浇筑时的漏振　skipped vibration
浇筑块间的高差　height differentials between blocks
浇筑混凝土的季节性限制　seasonal limitations on placing of concrete
混凝土冬季施工　concreting in cold weather

混凝土夏季施工 concreting in hot weather
相邻浇筑层 adjacent lift
相邻浇筑块 adjacent block
齿状混凝土（混凝土塞） dental concrete
后浇带 post-cast strip
收缩缝 contraction joint
灌浆的收缩缝 grouted contraction joint
不灌浆的收缩缝 ungrouted contraction joint
有键槽的收缩缝 keyed contraction joint
无键槽的收缩缝 unkeyed contraction joint
沉降缝 settlement joint
永久缝 permanent joint
施工缝 construction joint
临时缝 temporary joint
诱导缝 inducing joint
冷缝 cold joint
键槽 key
面膜 protective membrane
无黏性填缝料 cohesionless sealer
预冷混凝土 precooled concrete
预热混凝土 preheated concrete
温控措施 temperature control measure
混凝土出机口温度 concrete temperature at outlet of mixer
浇筑温度 concrete placing temperature
混凝土内外温差 temperature difference of inner and outside concrete
绝热温升 adiabatic temperature rise
导温系数 thermal diffusivity
混凝土自生体积增长 autogenous growth of concrete
温度骤降 sudden temperature drop
寒潮 cold wave
蓄热法 method of heat accumulation
综合蓄热法 comprehensive method of heat accumulation
冷却管 cooling pipe
通水冷却 cooling by circulating water
喷淋冷却 spray cooling
初始冷却 initial cooling
中期冷却 intermediate cooling，midterm cooling
后期冷却 final cooling
降温速率 cooling rate
加冷水拌和 chilled water mixing
加冰拌和 ice mixing
预冷骨料 precooled aggregate
风冷骨料 air-cooled aggregate
水冷骨料 water-cooled aggregate
冰屑 slush ice
片冰 flake ice
碎冰 crushed ice
表面保温 surface heat preservation
预热骨料 preheated aggregate
烘干骨料 oven-dried aggregate
混凝土含气量 concrete air content
导热系数 thermal conductivity
水化热 heat of hydration
体积安定性 soundness
速凝 quick set
假凝 false set
初凝 initial set
终凝 final set
离析 segregation
泌水，泛浆 bleeding
浮浆膜 laitance
返霜 cement cream
拆模 form removal
养护 curing
养护覆盖物 curing mat
绝热养护 adiabatic curing
自然养护 dry curing
喷水养护 wet curing
暴晒 solarization
稳定温度场 stable temperature field
温度梯度 temperature gradient
基础约束裂缝 foundation restraint crack
蜂窝 honeycomb
麻面 pitted surface
错台 offset
凹坑 hollow spot
裂缝 crack
网状裂缝 map crack
细裂缝 hairline crack
顶破强度 burst strength
抗刺穿力 puncture resistance
热空气吹干 drying by forced hot air
热黏合 heat bonding
双热楔 dual-hot-wedge
热对接 heat sealing
挤出焊接 extrusion welding

化学黏合 chemical bonding
公称直径 nominal diameter
配筋率 ratio of reinforcement
配筋有效面积 effective area of reinforcement
配筋间距 bar spacing
配筋不足 under-reinforced
配筋过多 over-reinforced
钢筋与混凝土的握裹力 grip between concrete and steel
分布钢筋 distribution rebar
受力钢筋 bearing rebar
箍筋 stirrup
弯起钢筋 bent-up rebar
架立钢筋 hanger rebar
单向配筋 one-way reinforcement
复式配筋 compound reinforcement
三向配筋 three-way reinforcement
伸出钢筋，搭接钢筋 starter bar
并筋 twin bars
钢筋接头 splice，joint
钢筋锥螺纹接头 taper threaded splice of rebar
完整丝扣 one complete screwthread
钢筋机械连接 rebar mechanical splicing
套筒连接 pressed sleeve splicing of rebars
绑扎连接 splicing joint
冷弯 cold bend；cold gagging
接头抗拉强度 tensile strength of splice
接头残余变形 residual deformation of splice
接头试件的总伸长率 total elongation of splice sample
焊接 welding joint
平焊 flat welding
搭接焊 lap welding
对接焊 butt welding
角焊 fillet welding
俯焊 down-hand welding
仰焊 overhead welding
点焊 spot welding
熔焊 fusion welding
弧焊 arc welding
氩弧焊 argon arc welding
惰性气体电弧焊 inert gas arc welding
手工电弧焊 shielded metal arc welding (SMAW)
埋弧自动焊 submerged arc welding (SAW)
气体保护焊 gas metal welding (GMAW)
撕裂强度 tearing strength
滑动模板 sliding formwork (slip form)
翻转模板 turnover form
定型组合式模板 shaped composite form
无轨滑模 flexible slip form
移置模板 shifted formwork
悬臂式模板 cantilever form
平台式模板 deck form
抽屉式模板 drawer-type formwork
爬升（顶升）模板 climbing (jacked) form
保温模板 insulated form
真空模板 vacuum mat
围圈（围图） form waling
提升架 lift yoke
收分装置 draw in or disport device
脱模剂 concrete remover
滑框倒模 shifted form with sliding frame
随升井架 headframe lifting along with working deck
缆机平台 cable crane platform
工作平台，操作平台 operation platform
移动式工作平台 mobile working platform
悬挑式平台 overhanging platform
吊车起吊荷重 crane capacity load
吊车竖向冲击荷载 crane vertical impact load
吊车水平刹车力 crane thrust
吊车轮压 crane wheel load
起重吊运指挥信号 commanding signal for lifting and moving
脚手架 scaffold
悬臂式脚手架 bracket scaffold
自承脚手架 self-supporting scaffold
重载脚手架 heavy duty scaffold
扣件式钢管脚手架 steel tubular scaffold with couplers
施工便道 construction road
人行便道 sidewalk
便桥 auxiliary bridge
浮筒栈桥 pontoon causeway
人行便桥 pedestrian bridge
跳板 gangway
二期混凝土 phase Ⅱ concrete
预留孔洞 blockout
安全防护装置 security protection unit
安全防护用品 protective appliance

个人防护装备　personal protective equipment (PPE)
安全帽　safety helmet
安全网　safety net
安全带　safety rail
护栏　guard rail
人员防坠落系统　personnel fall prevention system
触电　electric shock

9.5 施工工厂设施及施工机械 Construction plants and machinery

设备技术规程　equipment specification
施工设备进场　mobilization of construction machinery
施工设备退场　demobilization of construction machinery
日常保养（例行保养）　routine maintenance
定期保养　periodical maintenance
机械化施工　construction mechanization
附属设施　appurtenant facilities
空压机站　air compressor plant
泵站　pump station
潜水泵　subaqueous (submergible) pump
高扬程水泵　high-lift pump
机械通风系统　mechanical ventilation system
机械排风　mechanical air discharge
排风机　exhaust
自备电源　self-powered system
事故备用电源　emergency backup power supply
土石方机械　earth rock excavation machinery
臂式挖掘机　boom excavator
斗式挖掘机　bucket excavator
履带式挖掘机　caterpillar excavator
抓斗挖掘机　grab excavator
多瓣抓斗式挖掘机　orange peel excavator
正铲挖土机　face shovel
反铲　backhoe shovel
两用铲　convertible shovel
索铲　cable drag scraper
抓斗挖土机　clamshell shovel
翻斗铲　rocker shovel
铲运机　bow scraper
多臂钻车　multiple boom jumbo
竖井钻车　shaft jumbo
风钻　air drill
手风钻　hand drill
气腿钻　pneumatic drill
液压钻　hydraulic drill
风动支架凿岩机　air leg-mounted drill
履带式钻机　track-mounted drill
潜孔钻机　down-the-hole drill (DTH)
旋挖钻机　rotary rig
反循环钻机　reverse rotary rig
冲击反循环钻机　percusive reverse circulation drill
螺旋钻孔机　auger drilling machine
钢模台车　formwork jumbo
钢筋台车　steel pallet
轮胎吊　rubber-tyred gantry crane (RTG)
桅杆式起重机　derrick
双轮铣槽机　trench cutter with double wheels
掘进机　tunnel boring machine (TBM)
全断面岩石掘进机　full face rock tunnel boring machine
支撑式全断面岩石掘进机　gripper type full face rock TBM
护盾式全断面岩石掘进机　shielded full face rock TBM
扩孔式全断面岩石掘进机　full face rock TBM with reaming type
立爪式装岩机　gathering-arm rock loader
蟹爪式装岩机　crab rock loader
装载机　loader
手推车　barrow
皮带装料机　belt loader
前挖后卸式装载机　backhoe front end loader
斗式装料机　hopper loader
旋臂装料机　jib loader
翻斗式装料机　skip loader
堆料机　stacker
堆取料机　stacker reclaimer
推土机　bulldozer
重型履带式推土机　heavy crawler dozer
侧铲推土机　angling dozer
混凝土喷射机械手　shotcrete robot
混凝土喷射机　concrete sprayer
喷浆车　gunite car
破碎机　crusher，breaker
颚式破碎机　jaw crusher
回转破碎机　gyratory crusher
圆锥破碎机　cone crusher

锤式破碎机 hammer crusher
冲击式破碎机 impact crusher
粗碎机 primary breaker
初碎机 preliminary crusher
二次破碎机 secondary crusher
棒磨机 rod mill
进料筛 feeder screen
圆盘筛 disc screen
振动筛 shaking screen, vibrating screen
往复筛 reciprocating screen
双层筛 double deck screen
共振筛 resonance screen
水平固定筛 horizontal stationary screen
倾斜固定筛 sloping stationary screen
旋转筛 revolving screen
冲洗筛 washing screen
脱水筛 dewatering screen
往复式给料机 reciprocating feeder
振动给料机 shaking feeder
板式给料机 table feeder, apron feeder
混凝土制备 concrete preparation
混凝土构件场 concrete blockyard
混凝土搅拌站 concrete mixing plant
周期式（间歇式）混凝搅拌站 periodic (intermittent) concrete mixing plant
连续式混凝土搅拌站 continuous concrete mixing plant
防寒混凝土拌和厂 winterized concrete plant
干混砂浆搅拌站 dry mortar mixing plant
混凝土预拌厂 ready-mix plant
水泥罐 cement silo
配料筒仓 batching silo
混凝土搅拌机 concrete mixer
分批搅拌机 batch mixer
连续式搅拌机 continuous mixer
间歇式搅拌机 intermittent mixer
自落式搅拌机 gravity mixer
反转式搅拌机 reversing drum mixer
倾翻式搅拌机 tipping drum mixer
浆式（卧轴式）搅拌机 paddle mixer
配料器 batcher
配料斗 batch hopper
称量斗 weighing hopper
混凝土输送管 elephant trunk, concrete conveying pipe
混凝土搅拌车 concrete truck mixer, agitating lorry
混凝土搅拌输送斗 concrete transport skip
混凝土泵车 truck mounted concrete pump
带布料杆的混凝土泵 concrete pump with distributor
混凝土吊罐 concrete bucket
送料溜槽 feed chute
转运溜槽 transfer chute
负压溜槽（管） negative-pressure chute (pipe)
导管［浇混凝土用］ tremie pipe
混凝土摊铺机 concrete spreader
混凝土布料机 concrete distributor
混凝土振捣器 concrete vibrator
平板式振捣器 flat-plate vibrator
插入式振捣器 immersion vibrator, poker vibrator
附着式振捣器 form vibrator, surface-type vibrator
抹面机，修整机 finisher
水枪 hydraulic gun
喷雾器 sprayer
带式输送机 belt conveyor
螺旋输送机 screw conveyor
往复式皮带输送机 shuttle conveyor
拖带式输送机 trailing conveyor
斜坡喂料车 slope feeding vehicle
摊铺机 spreader
斜坡摊铺机 sloping spreader
振动碾 vibratory roller
振动平碾 smooth drum vibrating roller
双滚筒夯实碾 dual-drum tamping roller
振动平板 vibrating plate
平板式打夯机 flat beater
气动打夯机 air earth hammer
蛙式打夯机 frog hammer
压路机 pavement roller, road roller
羊脚碾 sheep-foot roller, taper foot roller
平碾 drum roller
凸块振动碾 pad foot vibratory roller
串联式碾压机 tandem roller
夯实机 tamper
压实机 compactor
平板夯 plate compactor
混凝土预冷系统 concrete precooling system
制冷厂 refrigeration plant

隔热保冷措施 heat-insulation and cold-preservation measure
混凝土预热系统 concrete preheating system
骨料加热 aggregate heating
钢筋切割机 bar cutter
钢筋弯曲机 bar bender
钢筋调直机 bar straightener
电焊机 welder
侧卸车 side dump truck
底卸车 hopper wagon
自卸汽车，翻斗车 dump truck
罐车 tanker；tank truck
散装水泥运输车 bulk cement truck
拖车 trailer
平板拖车 flatbed trailer
半挂车 semitrailer
岩壁吊车梁 crane rock beam，crane girders anchored to rockmass
桥式起重机（大车） bridge crane，overhead crane
缆索起重机 cable crane
平移式缆机 parallel-traveling cable crane
辐射式缆机 radial-traveling cable crane
往复移动式缆索道 shuttle cableway
龙门吊 frame crane
悬臂式起重机 cantilever crane，overhang crane
门式起重机 gantry crane
轨道起重机 track crane
塔式起重机 tower crane
履带式起重机 crawler crane
汽车式起重机 truck crane
叉车 forklift
移动式高架起重机 overhead traveling crane
高空作业车 hydraulic aerial cage
液压千斤顶 hydraulic jack
卷扬机 cable hoist，winch
移动式卷扬机 traveling hoist
焊接设备 welder
自动焊机 automatic welder
焊枪 welding gun

9.6 施工进度
Construction schedule

施工工期 construction duration
计划工期 planned duration
实际工期 as-built schedule
工程筹建期 pre-preparatory period
工程准备期 preparatory period
主体工程施工期 construction period of main structures
工程完建期，竣工期 completion period
施工高峰期 peak construction period
试运行期 commissioning period
施工总进度 general construction schedule，construction master schedule
首台机组发电工期 duration of first unit putting into operation
形象进度 graphic progress
控制性进度 critical schedule
关键工序 critical sequence
关键路径 critical path
进度表 progress chart
横道图 bar chart
网络图 network chart
箭线图 arrow diagram
工期延长 extension of time
流水作业法 flow operation method
平行作业法 parallel operation method
施工程序 construction sequence
生产率 productivity
施工作业 work activity
机动时间 float time
施工有效工日 available working day
施工强度 construction intensity
施工高峰强度 peak intensity
施工平均强度 average intenisty
坝体平均月上升速度 monthly average rise of dam placement

9.7 主要报告及图表名称
Titles of reports，figures and tables

施工对外交通图 Access roads to project site
施工总布置图 General construction layout plan
施工总进度图 Chart of general construction schedule
用关键路径法（横道图法）编制的施工进度图 Construction schedule using critical path method (bar chart)
施工控制性进度表 Chart of construction critical path
带时标的施工进度图 Construction schedule

diagram with time scales

施工用地范围图 Boundary of land use for construction

施工导流建筑物布置图 Arrangement of construction diversion structures

截流工程布置图 Layout of closure works

施工期通航建筑物布置图 Layout of navigation structures in construction period

导流工程施工布置图 Construction layout of diversion works

料场布置及料场开采布置图 Layout of borrow area (quarry) and exploitation

土石方平衡及流向图 Cut and fill balance and delivery zones

存弃渣场规划图 Layout of stockpile and spoil areas

主要建筑物混凝土施工程序、施工方法及施工布置示意图 Sketch of concrete placement sequence, method and layout for main structures

大坝施工形象面貌图 Graphic progress chart of dam placement

主体工程施工大型设施布置图 Arrangement of large facilities for construction of main structures

主要建筑物施工道路及施工支洞布置图 Arrangement of access roads and adits for construction of main structures

砂石加工系统生产工艺及布置图 Production process and layout for aggregate processing system

混凝土制冷（热）系统工艺及布置图 Process and layout for concrete cooling (heating) system

供水系统工艺及布置图 Process and layout for water supply system

场内供电线路规划布置图 Layout of power supply facilities at site

灌浆系统细部图 Grouting system details

收缩缝中止水片的连接 Seals and connections at contraction joints

施工总布置专题报告 Special report on general construction layout

施工规划报告 Report on construction planning

施工分标报告 Report on sub-bid scheme

施工技术要求 Construction specifications

施工操作规程 construction operation instruction

防洪度汛技术要求 Specifications of flood control during construction

对外交通运输专题报告 Special report on access roads to project site

混凝土原材料、配合比及性能试验报告 Test report on materials, mix proportion and performance of concrete

混凝土坝温度控制专题报告 Report on temperature control of concrete dam placement

第10章　工业与民用建筑
Industrial and civil architecture

10.1　建筑与结构
Architectonics and structure

建筑物　architecture，building
建筑设计　architectural design
建筑结构设计　structural design
建筑设备设计　building service design
场地设计　site design
园林设计　landscape design
建筑构造设计　construction design
建筑室内设计　interior design
概念设计　concept design
工业建筑　industrial architecture
民用建筑　civil architecture
居住建筑，住宅　residential building
古典建筑　classic architecture
仿古建筑　pseudo-classic architecture
多层建筑物　multistory building
公共建筑物　public building
海上建筑物　offshore structure
纪念性建筑　monumental architecture
历史建筑　historical building
保护建筑　listed building for conservation
辅助建筑物　accessory building
附属建筑物　auxiliary building
建筑总体效果　architectural ensemble
建筑节能设计　energy-saving design
建筑环保　architectural environmentally friendly
无障碍设计　barrier-free design
人防设计　civil defense design
建筑外观　architectural appearance
建筑模型　building model
建筑结构　building structure
组合结构　composite structure
砖混结构　masonry-concrete structure
砖木结构　masonry-timber structure
石砌结构　stone masonry structure
砖砌结构　brick masonry structure
砌块砌体结构　block masonry structure
墙板结构　wall-slab structure
板柱结构　slab-column structure
框架结构　frame structure
剪力墙结构　shearwall structure
框架支撑结构　braced frame structure
拱结构　arch structure
折板结构　folded-plate structure
薄板结构　thin-slab structure
壳体结构　shell structure
桁架结构　truss structure
索结构　cable structure
膜结构　membrane structure
筒体结构　tube structure
单框筒结构　framed tube structure
悬挂结构　suspended structure
连体结构　connected structure
高耸结构　high-rise structure
塔式结构　tower structure
附建公共用房　accessory assembly occupancy building
设备层　mechanical floor
地下室　basement
建筑面积　floor area
使用面积　usable area
使用面积系数　usable area coefficient
基本层楼板面积　basic floor area
用地红线　property line
建筑控制线　building line
建筑密度　building density
容积率　plot ratio
绿地率　green space rate
层高　storey height
室内净高　interior storey height
开间　bay width
进深　depth
室内外高差　indoor-outdoor elevation difference
文化娱乐建筑　cultural and recreation building
文化宫　cultural palace

剧院　theatre
音乐厅　concert hall
电影院　cinema
博物馆　museum
展览馆　exhibition room
美术馆　art museum
科技馆　science museum
图书馆　library
会展中心　meeting and exhibition center
宿舍　dormitory
旅馆　hotel
汽车旅馆　motel
体育场　stadium
体育馆　gymnasium
游泳馆　natatorium hall
别墅　villa
跃层住宅　duplex apartment
联排式住宅　row house，terrace house
单元式住宅　apartment building
低层住宅　low-rise apartment
多层住宅　multi-stories apartment
高层住宅　high-rise apartment
套型　dwelling unit type
居住空间　habitable space
多功能厅　multi-functional hall
客厅　parlor
餐厅　dining room
厨房　kitchen
食堂　canteen
备餐间　pantry
盥洗室　washroom
卧室　bedroom
起居室　living room
卫生间　bathroom，toilet
更衣室　dressing-room，locker
浴室　bathroom
车库　garage
停车场　parking lot
门厅　lobby
门廊　porch
走廊　corridor
联系廊　inter-unit gallery
前室　anteroom
中庭　atrium
回廊　cloister
门斗　air lock
天井　patio
庭院　courtyard
屋顶花园　roof garden
裙房　podium
过街楼　overhead building
挑檐　overhanging eaves
女儿墙　parapet
天沟　gutter
窗井　window well
屋面防水　roof waterproofing
地下室防水　basement waterproofing
屋面排水　roof drainage system
雨水口　water outlet
雨水管　down pipe
泛水　flashing
勒脚　plinth
踢脚　baseboard
墙裙　dado
散水　apron
电梯厅　elevator hall
电梯机房　elevator machine room
自动扶梯　escalator
标准层　typical floor
架空层　elevated storey
避难层　refuge storey
细部结构　detailed structure
承重墙　structural wall，bearing wall
非承重墙　partition wall
活动隔断　movable partition
无筋砌体构件　masonry member
配筋砌体构件　reinforced masonry structure
蒸压加气混凝土砌块　autoclaved aerated concrete block
屋面板　roof board
屋面檩条　roof purlin
屋架　roof truss
天窗　skylight
屋盖支撑　roof bracing
空间桁架　space truss
网状壳体　reticulated shell
立体桁架　spatial truss
楼板　floor slab
主梁　main beam

次梁 secondary beam
横梁 cross beam
大梁 girder
过梁 lintel
圈梁 ring beam
系梁 collar beam
栋梁 ridge beam
交叉梁 beam grid
上承梁 deck beam
下承梁 through beam
等截面梁 uniform cross-section beam
变截面梁 non-uniform cross-section beam
两端固定梁 beam fixed at both ends
悬臂梁 cantilever
连续梁 continuous beam
叠合梁 superposed beam
预制梁 precast beam
山墙 gable
等截面柱 constant cross-section column
变截面柱 non-uniform cross-section column
环饰柱 annulated column
菱形柱 cant column
单立柱 detached column
阶形柱 stepped column
柱廊 colonnade
楼梯 stair
转梯 helical stairs，spiral stairs
太平梯 escape stair
室外楼梯 external stair
多梯段楼梯 staircase with several flights
单跑式楼梯 staircase with straight flight
楼梯踏步 stair step
楼梯踢板 stair riser
楼梯平台 stair landing
楼梯间 stairway
台阶 step
组合构件 built-up member
组合屋架 composite roof truss
下撑式组合梁 down-stayed composite beam
组合楼盖 composite floor system
转换层 transfer story
加强层 strengthened storey
结构缝 structural joint
变形缝 deformation joint
假缝 suppositious seam
通缝 continuous seam
阳台 balcony
雨篷 canopy
檐口 eaves
水塔 water tower
烟囱 chimney
胸墙 front wall
坡道 ramp
栏杆 railing
吊顶 suspended ceiling
管道井 pipe shaft
烟道 smoke flue
通风道 ventilating duct
装修 decoration
装饰物 ornament
采光 daylight
采光系数 daylight factor
百叶窗 blind window
双扇窗 double casement window
直拉窗 sash window
平开窗 casement window
防盗门 safety door
窗台 window sill
窗下墙 window spandrel
隔声材料 sound insulation material
净空高度 headway
楼层平面图 floor plan
门洞 door opening
门框 door frame
内墙 interior wall
外墙 exterior wall
轻质隔墙 light-weight partition
交叉支撑 cross brace（X-brace）
交叉拱顶 cross vault
室内地平标高 elevation of leveled ground
室外装修 exterior finishing
砂浆找平 mortar leveling
砂浆涂层 dash-bond coat
水泥抹面 finishing cement，cement plaster
饰面层 finish coat
素土夯实 packed soil
保护层 protective coating
扩展基础 spread footing
力矩面积法 area-moment method
吊物孔 hatch

轻型钢结构　light weight steel construction
轻质混凝土结构　light concrete structure
外倾结构面　out-dip structural plane
纵向垂直支撑　longitudinal vertical bracing
不对称荷载　anti-symmetrical load
均布动荷载　moving uniform load
均布活荷载　uniformly distributed live loads
有效宽度系数　effective width factor
配筋砖砌体　reinforcement brickwork
砌体强度　masonry strength
建筑物理学　architectural physics
建筑力学　architectural mechanics
建筑光学　architectural optics
绿色照明　green lighting
景观照明　landscape lighting
眩光　glare
直接眩光　direct glare
反射眩光　glare by reflection
光幕反射　veiling reflection
建筑热工学　building thermotics
围护结构　building envelope
围护结构传热系数　overall heat transfer coefficient
外墙平均传热系数　average overall heat transfer coefficient
围护结构热惰性指标　thermal inertia index of building envelope
材料蓄热系数　material heat store coefficient
遮阳系数　shading coefficient (SC)
建筑物体形系数　shape coefficient of building
窗墙面积比　area ratio of window to wall
窗地面积比　area ratio of glazing to floor
换气次数　air change time per hour
建筑声学　architectural acoustics
混响声　reverberation sound
计权隔声量　weighted sound reduction index
建筑隔声　sound insulation
建筑吸声　sound absorption

10.2 供暖、通风与空调 Heating, ventilation and air conditioning (HVAC)

(1) 供暖　Heating

干球温度　dry-bulb temperature
湿球温度　wet-bulb temperature
露点温度　dew-point temperature
累年最冷月　normal coldest month
累年最热月　normal hottest month
日较差　daily range
水蒸气分压力　partial pressure of water vapor
室内设计温度　indoor design temperature
冬季通风室外计算温度　outdoor design temperature for winter ventilation
冬季空气调节室外计算温度　outdoor design temperature for winter air conditioning
冬季空气调节室外计算相对湿度　outdoor design relative humidity for winter air conditioning
冬季围护结构室外计算温度　outdoor design temperature for calculated envelope in winter
采暖室外计算温度　outdoor design temper-ature for heating
采暖室外临界温度　outdoor critical air temperature for heating
采暖度日数　heating degree day based on 18℃ (HDD18)
夏季通风室外计算温度　outdoor design temperature for summer ventilation
夏季通风室外计算相对湿度　outdoor design relative humidity for summer ventilation
夏季空气调节室外计算日平均温度　outdoor design mean daily temperature for summer air conditioning
夏季空气调节室外计算逐时温度　outdoor design hourly temperature for summer air conditioning
空调度日数　cooling degree day based on 26℃ (CDD26)
新风系统　fresh air system
总冷量　total cooling capacity
冷却负荷　cooling load
热负荷　heat load
传热阻　resistance of heat transfer
供暖　heating
热媒　heating medium
热惰性指标　index of thermal inertia
水暖　hot-water heating
蒸汽供暖　steam heating
热风供暖　warm-air heating
辐射供暖　radiant heating
对流采暖　convection heating

采暖管线　heating pipe line
采暖供水管　heating supply water pipe
采暖回水管　heating return water pipe
采暖立管　heating riser
同程式系统　reversed return system
异程式系统　direct return system
采暖期天数　days of heating period
传热　heat transfer
总体传热系数　total heat transfer coefficient
传热量　capacity of heat transmission
传热阻　resistance of heat transfer
导热性，导热系数　thermal conductivity
稳态传热　steady-state heat transfer
非稳态传热　unsteady-state heat transfer
附加耗热量　additional heat loss
高压热水供暖系统　high-pressure hot water heating system
高压蒸汽供暖系统　high-pressure steam heating system
机械热水供暖系统　mechanical hot water heating system
隔热层　thermal insulation layer
建筑遮阳　building sun shading
建筑保温　building heat preservation
内保温　internal thermal insulation
外保温　external thermal insulation
建筑防热　building thermal shading
防冰设施　anti-icing device
防潮层　damp-proof course
给水温度　temperature of water supply
供回水温差　temperature difference between supply and return water
供暖负荷　heating load
供暖锅炉　heating boiler
供暖机组　heating unit
供暖面积　heating area
供暖能力　heating capacity
供暖期　heating duration
供热成本　heating cost
供热管网　heating piping network
管道保温　insulation of piping
过热　superheat
过热蒸汽　superheated steam
恒温恒湿　constant temperature and humidity
换气系统　air renewal system
回风系统　air return system
回流阀　reflux valve
回水阀　return valve
排气阀　exhaust valve
三通阀　three-way valve
止回阀　non-return valve (NRV)
丝堵　screwed plug
回水管　return pipe
回水总管　return main pipe
排气管　exhaust pipe
回风方式　air return method
局部采暖　local heating
区域采暖　district heating
连续供暖　continuous heating
热泵式采暖　heat pump heating
热损失　hot loss
热效率　thermal efficiency
热稳定性　thermal stability
热阻系数　thermal resistance coefficient
比摩阻　specific frictional resistance
局部阻力系数　coefficient of local resistance
散热器供热支管　heating branch to radiator
散热器回水支管　return branch of radiator
双管采暖系统　two-pipe heating system
水平单管采暖系统　one (single) -pipe loop circuit heating system
沿踢脚板铺设的散热器　baseboard radiator
围护结构热损失　heat loss of building envelope
围护结构温差修正系数　temperature difference correction factor of building structure
泄水管　drain pipe
循环管　circulation pipe
活接头　union
管接头　coupling
异径管接头　reducing coupling
泄压装置　pressure relief device
中分式系统　midfeed system
轴流式暖风机　unit heater with axial fan
自然循环采暖　natural circulation heating

(2) 通风　Ventilation

通风　ventilation
自然通风　natural ventilation
高窗自然通风　high window natural ventilation
机械通风　mechanical ventilation

事故通风　emergency ventilation
气流组织　air distribution
射流　jet
受限射流　jet in a confined space
回流区　return flow zone
防火　fire prevention
排烟　smoke extraction
防烟　smoke prevention
防爆　explosion proofing
置换通风　displacement ventilation
防火风管　fire-proof duct
风机出口　fan outlet
风机对流加热器　fan convection heater
风机排（送）风量　fan delivery
机械排风系统　mechanical exhaust system
机械排烟系统　mechanical smoke exhaust system
机械送风　mechanical air supply
进风道　inlet duct
进风口面积　air inlet area
局部排风　local exhaust ventilation
全面排风　general exhaust ventilation (GEV)
局部通风　local ventilation
连续通风　continuous ventilation
全面通风　general ventilation
联合通风　natural and mechanical combined ventilation
排风机室　exhaust fan room
排风竖井　exhaust shaft
排风温度　exhaust temperature
墙内风道　wall duct
通风支管　branch duct
通风止回阀　check damper
轴流风机　axial fan
离心风机　centrifugal fan
过滤效率　filter efficiency
烟羽　plume
筒形风帽　cylindrical ventilator
伞形风帽　cowl
锥形风帽　conical cowl
对开式多叶阀　opposed-blade damper
平行式多叶阀　parallel-blade damper
插板阀　slide damper
条缝型风口　slot outlet
旋流风口　twist outlet; swirl diffuser

(3) 空调　Air conditioning

舒适性空气调节　comfort air conditioning
局部区域空气调节　local air conditioning
全室性空气调节　general air conditioning
分层空气调节　stratificated air conditioning
全空气系统　all-air condition
制冷　refrigeration
冷冻水　chilled water
冷却水　cooling water
多联机空调系统　multi-connected split air conditioning system
空气分布特性指数　air diffusion performance index (ADPI)
空气源热泵　air-source heat pump
水源热泵　water-source heat pump
水环热泵空气调节系统　water-loop heat pump air conditioning system
低温送风空气调节系统　cold air distribution system
地源热泵系统　ground-source heat pump system
地热能交换系统　geothermal exchange system
传热介质　heat-transfer fluid
地埋管换热系统　ground heat exchanger system
分区两管制水系统　zoning two-pipe water system
复合通风系统　integrated ventilation system
蓄冷-制冷周期　period of charge and discharge
区域供冷　district cooling
风管配件　duct fittings
风管部件　duct accessory
漏风率　air system leakage ratio
漏风量　air leakage rate
系统风管允许漏风量　air system permissible leakage rate
净化空调系统　air cleaning system
漏光检测　air leak check with lighting
整体式制冷设备　packaged refrigerating unit
组装式制冷设备　assembling refrigerating unit
空气洁净度等级　air cleanliness class
保温　thermal insulation
保温材料　thermal insulating material
保温层　heat insulating layer
保温系数　coefficient of heat preservation
变风量系统　variable air volume system
变水量系统　variable water flow system
表面换热系数　coefficient of surface heat transfer

补水泵 make-up water pump
风道安装 air duct installation
风道加热器 duct heater
风道摩擦损失 duct friction loss
风道压力损失 duct pressure loss
风道阻力 duct resistance
空气冷却风道 air cooling duct
空气冷却系统 air cooling system
空气流量 air flow rate
冷凝水流量 condensate flow rate
冷凝温度 condensation temperature
冷却塔 cooling tower
两管制水系统 two-pipe water system
冷凝水管路 drip pipe，condensate line
盘管冷凝器 coil condenser
旁通风道 by-pass air duct
喷口送风 air supply through nozzle
喷淋冷却塔 spray cooling tower
强制换气 forced air change
强制通风冷却塔 forced draft cooling tower
双风道系统 double duct system
水冷冷却塔 water-cooling tower
水冷式空调器 water-cooled air conditioner
水冷式冷凝器 water-cooled condenser
送风管道 supply air duct
送风机室 supply fan room
送风温差 effective temperature difference
卧式壳管式冷凝器 closed shell and tube condenser
卧式壳管式蒸发器 closed shell and tube evaporator
新风风道 outside air intake duct
新风机组 fresh air handling unit
新风负荷 fresh air load
新风量 fresh air rate
蓄冷 cold accumulation
蓄热 heat accumulation
压缩式制冷 compression-type refrigeration
压缩冷凝机组 compression-type condensing unit
压缩式冷水机组 compression type water chiller
一次风 primary air
一次风机 fan for primary air
一次回风 primary return air
蒸发冷却 evaporation cooling
逐时冷负荷 hourly cooling load
逐时综合温度 hourly sol-air temperature
自然通风冷却塔 natural draft cooling tower
焓湿图 psychrometric chart
比焓 specific enthalpy
干工况 dry cooling condition
湿工况 wet cooling condition
热湿交换 heat and moisture transfer
喷水段 spray chamber
变风量末端装置 variable air volume (VAV) terminal box
转轮除湿机 rotary dehumidifier

10.3 给排水 Water and wastewater system

给水系统 water supply system
用水量 water consumption
用水定额 water consumption norm
居民生活用水 water for residential domestic use
综合生活用水 water for domestic and public use
工业企业用水 water for industrial enterprise use
浇洒道路用水 water for road washing
绿地用水 water for green belt
消防用水 water for fire fighting
自用水量 water consumption in water-works
供水量 supplying water
最小服务水头 minimum service head
岸边式取水构筑物 riverside intake structure
河床式取水构筑物 riverbed intake structure
直流水系统 once through system
复用水系统 water reuse system
井群 battery of wells
管井 deep well
大口井 open well
渗渠 infiltration gallery
进水流道 inflow runner
自灌充水 self-priming
输水管渠 delivery pipe
配水管网 distribution system
环状管网 loop pipe network
枝状管网 branch system
给水管 water supply pipe
给水立管 water riser pipe
给水总管 water main
管道防腐 corrosion prevention of pipes
水处理 water treatment
原水 raw water
小时用水量 hourly water consumption

倒流防止器 backflow prevent
真空破坏器 vacuum breakers
叠压供水 additive pressure water supply
中水 reclaimed water
集中热水供应系统 central hot water supply system
局部热水供应系统 local hot water supply system
全日热水供应系统 all day hot water supply system
定时热水供应系统 fixed time hot water supply system
热泵热水供应系统 heat pump hot water system
太阳能保证率 solar fraction
太阳辐照量 solar irradiation
设计小时耗热量 design heat consumption of maximum time
设计小时供热量 design heat supply of maximum time
同程热水供应系统 reversed return hot water system
第二循环系统 hot water circulation system
管道直饮水系统 purified drinking water system
水质阻垢缓蚀处理 water quality treatment of scale-prevent and corrosion-delay
中水设施 installation of reclaimed water
建筑物中水 reclaimed water system for building
水量平衡 water balance
集中供热水系统 collective hot water supply system
分散供热水系统 individual hot water supply system
太阳能直接系统 solar direct system
太阳能间接系统 solar indirect system
自然循环系统 natural circulation system
强制循环系统 forced circulation system
直流式系统 series-connected system
给水配件 water supply fittings
地表积水 surface ponding
积水面积 accumulated water area
满管压力流雨水排水系统 full pressure storm system
虹吸式屋面雨水排水系统 roof siphonic drainage system
排水体制 sewerage system
排水定额 wastewater flow norm
合流制管道溢流 combined sewer overflow
自循环通气 self-circulation venting
真空排水 vacuum drain
同层排水 same-floor drainage
杂排水 gray water
排水系统 wastewater system
自然沉淀 plain sedimentation
凝聚沉淀 coagulation sedimentation
卫生器具 sanitary fixtures
阻火圈 firestop collar
压力管道 pressure pipeline
无压管道 non-pressure pipeline
刚性管道 rigid pipeline
柔性管道 flexible pipeline
管渠 ditch
压力管道水压试验 water pressure test for pressure pipeline
无压管道闭水试验 water obturation test for non-pressure pipeline
无压管道闭气试验 pneumatic pressure test for non-pressure pipeline

10.4 消防 Fire protection

安全防范系统 security alarm system (SAS)
火灾报警系统 fire alarm system (FAS)
监控中心 monitoring and controlling center
建筑防火 building fire protection
建筑火灾 building fire
建筑物的耐火等级 fire-resistant grade of buildings
耐火性 fire resistance
总体结构耐火性分析 global structural analysis of fire resistance
间接火灾作用 indirect fire actions
绝热性 insulation
隔离功能 separating function
隔离构件 separating element
标准耐火性 standard fire resistance
温度分析 temperature analysis
热作用 thermal actions
高级火灾模型 advanced fire model
简单火灾模型 simple fire model
防火墙 fire wall
燃烧系数 combustion factor
外部火灾曲线 external fire curve
火灾激化风险 fire activation risk
火灾荷载密度 fire load density

火灾场景 fire scenario
局部火灾 localized fire
开口系数 opening factor
发热率 rate of heat release
升温曲线 temperature-time curves
形状系数 configuration factor
对流传热系数 convective heat transfer coefficient
辐射率 emissivity
净热通量 net heat flux
钢筋临界温度 critical temperature of reinforcement
明火 naked flame
耐火面 refractory surface
耐火极限 duration of fire resistance
耐火材料 refractory material
耐火隔板 refractory shield
耐火隔热砖 refractory and insulating fire brick
耐火黏土砌块 refractory clay block
耐火稳定性 refractory stability
耐火完整性 refractory integrity
耐火隔热性 refractory insulation
防护的破坏时间 failure time of protection
常温设计 normal temperature design
耐火等级 fire resistance classification
耐火电缆 fire resistant cable
耐火分隔 fire resistant barrier
防火窗 fire resisting window
防火卷帘 fire resisting shutter
防火门 fire door
薄钢板防火门 sheet-metal fire door
防火幕 fire curtain
防火间距 fire separation distance
防火分区 fire compartment
防火绝缘 fire resistance insulation
防火密封 fire seal
安全区 place of safety
疏散楼梯 emergency stair case
额定耐火时间 rated fire-resistance period
防烟分区 smoke bay
防烟楼梯间 smoke proof staircase
封闭的火灾 contained fire
钢结构防火涂料 fire-resistant coating for steel structure
固定吸水设施 fixed suction installation
管道阻火器 flame arrester for pipe
保温涂层 insulating coating
保温性能 insulating property
安全出口 exit door
安全标志 safety sign
安全疏散 safe evacuation
安全竖井 escape shaft
安全通道 fire escape
安全疏散距离 exit distance，safety evacuation distance
阻隔区 baffle area
阻燃材料 flame retardant material
阻燃分隔 fire retarding division
防火间隔 fire break
防火堤 fire dike
室外消防栓 outdoor fire hydrant
室外消防通道 outside fire escape
疏散路线 evacuation route
出口方向标志 exit direction sign
最终安全出口 final exit
消防泵 fire pump
消防泵房 fire pump room
消防泵接水口 fire-service connection
备用泵 emergency pump
消防分区 fire prevention zone
消防给水 fire water supply
消防供水持续时间 fire water duration
消防控制室 fire control room
消防联动控制装置 integrated fire control device
消防设施 fire fighting facilities
消防救援器材 fire rescue equipment
消防水池 fire pool
消防水压 fire pressure
消防水枪 fire water branch
充实水柱 full water spout
消防水带 fire hose
消防吸水管 fire suction hose
消防破拆工具 fire forcible entry tool
消防梯 fire ladder
消防通道 fire fighting access
消防用水量 fire consumption
感温火灾探测器 heat fire detector
感烟火灾探测器 smoke fire detector
感光火灾探测器 optical flame fire detector
气体火灾探测器 gas fire detector
复合式火灾探测器 combination type fire

detector
自动灭火系统　automatic fire extinguishing system
自动喷水灭火系统　automatic sprinkler system
闭式系统　close-type sprinkler system
湿式系统　wet pipe system
干式系统　dry pipe system
预作用系统　pre-action system
水幕系统　drencher system
细水雾灭火系统　water mist fire suppressing system
全淹没灭火系统　total flooded extinguishing system
防火分隔水幕　water curtain for fire compartment
防护冷却水幕　drencher for cooling protection
自动喷水-泡沫联用系统　combined sprinkler-foam system
雨淋系统　deluge system
蒸汽灭火系统　steam smothering system
局部应用灭火系统　local application extinguishing system
作用面积　area of sprinklers operation
响应时间指数　response time index (RTI)
标准喷头　standard sprinkler
闭式喷头　automatic sprinkler
闭式洒水喷头　sealed sprinkler head
配水干管　feed main
配水管　cross main
配水支管　branch line
配水管道　system pipe
短立管　sprig-up
信号阀　signal valve
准工作状态　condition of standing by
稳压泵　pressure maintenance pump
末端试水装置　end water-test equipment
城市给水干管　city main
城市消防　urban fire protection
城镇公共消防设施　public city fire facility
出口阀　landing valve
出口楼梯　exit stairway
出口通道　exit access
初起火　incipient fire
初期灭火　initial attack
初始燃烧　initial burning
垂直蔓延　vertical extension
大卷盘水带　double donut
挡热板　heat barrier
挡烟板　smoke barrier
低水位报警　low water alarm
消防栓　fire hydrant
墙式消防栓　wall hydrant
地上式消防栓　stand post hydrant
地下式消防栓　sunk hydrant
室内消防栓　indoor fire hydrant
室内灭火设施　indoor fire installation
报警器　alarm apparatus
报警开关　alarm switch
报警阀　alarm valve
报警标志　alarm mark
报警信号　alarm signal
报警水位　alarm water level
报警延迟　alarm delay
报警监测装置　alarm detector
备用水源　alternate source
机械控烟　mechanical smoke control
建筑物外部排水系统　guttering
紧急电源　emergency power source
排烟系统　smoke evacuation system
排烟竖井　smoke shaft
喷水密度　water discharge density
强制对流　forced convection
强制通风　forced ventilation
全部燃烧　fully involved
全面疏散　complete evacuation
燃烧持续时间　burning duration
生活消防水泵房　domestic water and fire pump house
室外火灾　outdoor fire
逃生人孔　escape hatch
外部火灾　exposure fire
严重过火　severe bum
主动火灾预防　active fire prevention
主动灭火　active fire control
消防环形主管　fire-fighting ring main
火口，燃烧层穿孔　fire hole
防火完整性　fire integrity
燃烧层　fire layer
着火点　fire point
火灾传播　fire-propagation
向火面［烟气侧表面］　fireside surface

火烟监测 fire-smoke detection
防火屏障 fire stop
表层火 skin fire
表面火 surface fire
表面火焰蔓延 surface flame spread
表面燃烧 surface burn
表面闪燃 surface flash
不易燃材料 nonflammable material
超高温耐火材料 extreme temperature refractory
耐火 4 小时的防火墙 firewall with 4-hour fire rating
灭火器配置场所 distribution place of fire extinguisher
保护距离 protective range
灭火级别 fire rating
便携式灭火器 portable fire extinguisher
备用配电盘 emergency switchgear
备用系统 redundant system
备用照明 stand-by lighting
火警疏散信号 fire alarm evacuation signal
火警显示装置 fire alarm indicating device
火警远距离显示设备 fire alarm remote indicating equipment
火警探测系统实验开关 fire detection system sets switch

10.5 主要图表名称
Titles of figures and tables

建筑设计效果图 Architectural rendering
建筑平面图 Building plan
竖向布置图 Vertical planning
建筑详图 Architectural details
建筑大样图 Architectural detail drawing
楼梯详图 Stair details
给水排水系统平面图 Plan of water and wastewater system
消防平面图 Fire protection plan
疏散图 Security map
用地红线图 Map of property line
土方图 Earth work drawing
管线综合图 Integral pipeline longitudinal and vertical drawing
建筑施工图 Construction drawing
暖通专业施工图 HVAC working drawing
设计任务书 Design assignment statement
暖通空调设备明细表 Detail list of HVAC system

第11章 工 程 投 资
Project investment

11.1 基础价格
Base price

人工单价　unit price of labor
普工　labor
计日工　daywork labor
当地工人　local labor
熟练工　skilled labor
半熟练工　semi-skilled labor
高级熟练工　senior skilled labor
特种作业人员　special operation personnel
基本工资　basic wages
辅助工资　auxiliary wage
福利费　benefits
劳动保护费　labor protection expense
地区津贴　district allowance
施工津贴　field subsidy
加班津贴　overtime allowance
交通补贴　commuting allowance
差旅费　travel expense
护照费用　passport fees
签证费用　visa fees
个人所得税　personal income tax
预扣所得税　pay as you earn (PAYE)
材料预算价格　material estimate price
当地材料　local materials
当地采购　local purchase
国外进口　foreign imports
价格水平　price level
市场价格　market price
材料原价　material original price
包装费　packing charge
运输保险费　transportation premium
运杂费　freight and miscellaneous charges
运输费　freight costs
装货费　loading charge
卸货费　unloading charge
毛重系数　coefficient of gross weight
采购及保管费　purchase and storage expenses
装置性材料　necessary accessories
周转性材料　revolving material
进出口税费　import and export tax
包装品回收　packaging recycling
摊销费　amortized cost
电、水、风预算价格　estimated price of electricity, water and compressed air supply
电网电价　on-grid tariff
施工用电价格　construction electricity price
施工用水价格　construction water price
施工用风价格　construction compressed air price
施工机械台时费　machinery daily working unit price
施工机械使用费　machinery operation expense
基本折旧费　basic depreciation cost
设备修理费　machinery repair cost
安装拆卸费　assembly and disassembly cost
大修费　cost of overhaul
日常修理费　cost of routine maintenance
机上人工费　operator wage
动力燃料费　cost of power fuel
残值率　rate of residual value
租赁　lease
砂石料单价　unit price of sand and gravel
混凝土材料单价　unit price of concrete materials
工程造价信息　guidance of cost information
工程造价指数　cost index
工程造价鉴定　construction cost verification
工程造价咨询费　consultancy service charge
全过程工程造价管理咨询　cost management consultancy of whole construction process
造价工程师　cost engineer
计量工程师　quantity surveyer
造价员　cost engineering technician

11.2 建筑及安装工程单价
Price of civil works and installation works

工程单价　unit price

综合单价 all-in unit price
计日工费用 daywork rate
工程计价定额 pricing code and index
工期定额 norm and standard of construction period
工程消耗量定额 quantity of consumption norm and standard
劳动定额 labor norm and standard
施工定额 construction norm and standard
人工消耗定额 labor consumption norm
材料消耗定额 material consumption norm
机械消耗定额 machinery consumption norm
直接费 direct cost
基本直接费 basic direct cost
其他项目费 sundry cost
间接费 indirect cost
措施项目费 expense of preliminaries
冬雨季施工增加费 additional cost of work in winter and rainy season
特殊地区施工增加费 additional cost for construction in special areas
夜间施工增加费 additional cost for night work
二次搬运费 double-handling freight
小型临时设施摊销费 amortization charge of small temporary facilities
安全文明施工措施费 measure fee for HSE construction
施工工具用具使用费 fee for utilizing tools and apparatus
检验试验费 cost of inspection and test
工程定位复测费 resurvey fee of project location
竣工场地清理费 clean-up costs of completion site
工程项目移交前的维护费 maintenance cost of the project before handover
标高价 mark up
办公费 office expense
固定资产使用费 fee for utilizing fixed assets
企业管理费 enterprise overhead
总承包服务费 main contractor's attendance
进场费 mobilization cost
退场费 demobilization cost
业务招待费 business entertaining expense
技术转让费 technology transfer fee
技术开发费 technology innovation expense
技术引进费 incidental cost of technology introduction
专利及专有技术使用费 royalty for patents and proprietary technology
社会保障费 social security cost
企业计提费 welfare withdrawal
人身意外伤害保险费 personal accident insurance
社会保险 social insurance
基本养老保险费 basic endowment insurance
医疗保险费 medical insurance
工伤保险费 work injury insurance
失业保险费 unemployment insurance
职业病防治费 occupational disease prevention fee
财务费 financial charges
手续费 charge for trouble
规费 statutory fees
工程排污费 sewage disposal charge
利润 profit
税金 tax
地方税 local tax
企业所得税 enterprise income tax
营业税 business tax
增值税 value-added tax
利润税 profit tax
代扣所得税 withholding tax

11.3 设备费 Equipment cost

设备购置费 cost of equipment procurement
设备原价 original price of equipment
离岸价 free on board (FOB)
到岸价 cost, insurance and freight (CIF)
成本加运费价 cost and freight (CFR)
国际结算 international settlements
国际市场价格 international market price
含税金价格 price including tax
含佣金价格 price including commission
含增值税价格 price including value- added tax
价格波动 price fluctuation
价格差异 price variance
关税 customs duty
关税税率 customs duty rate
关税优惠 preferential duty
出口关税 export duty
进口关税 import duty
进口附加税 import surtax

银行手续费 bank charges
进出口公司手续费 commission-charge for import and export company
商检费 commodity inspection fee
港口费 port charge
清关费 customs clearing fee
海运费 sea freight
海运保险费 marine premium
出口退税 value added tax refund for exported goods
避税 tax avoidance
减税 tax reduction
免税 duty free, tax exclusion
全额免税 full exemption
特大（重）件运输增加费 additional expense of heavy cargo transport
特大（重）件运输措施费 measure expense of heavy cargo transport
现场拼装加工费用 cost of site assembly and processing
目的地交货价格 free on board destination
包装绑扎费 packing and lashing expense
变压器充氮费 filling nitrogen cost of tansformer

11.4 投资估算 Investment estimation

(1) 枢纽工程费用 Project cost

工程量清单 bill of quantities
单位造价指标 unit cost index
发电工程每千瓦造价 cost of power generation project per kW
建筑工程费 civil works cost
安装工程费 installation woks cost
建设项目投资 project construction investment
建设预算 construction budget
工程费用 construction cost
独立核算 independent accounting
施工辅助工程 temporary works
环境保护和水土保持工程 environmental protection works, soil and water conservation works
机电设备及安装工程 E&M equipment and installation works
金属结构、设备及安装工程 hydraulic steel structure and mechanical equipment, and installation works
专项投资 specific investment
年度投资费 annualized capital cost
年度预算 annual budget
年度维修费用 annual maintenance cost
年度折旧费 annual amortization cost

(2) 建设征地和移民安置补偿 Compensation for land requisition and resettlement

重置成本（重建费） replacement cost
迁建费 relocation cost
补偿补助费 compensation subsidy
安置补助费 settlement allowances
征用土地补偿 compensation for land requisition
水库淹没补偿 compensation for reservoir inundation
货币补偿 monetary compensation

(3) 独立费用 Independent expense

其他费用 miscellaneous expense
工程前期费 preliminary engineering fees
开办费 preliminary expense
开发权转让费 development rights transfer fee
代理费 agent fee
佣金 factorage
担保费用 guarantee cost
工程建设监理费 construction supervision cost
设备监造费用 equipment supervision cost
咨询服务费 consulting service cost
项目技术经济评审费 evaluation fee for project technology and economy
项目验收费 project acceptance cost
工程保险费 project insurance premium
保险费率 insurance rate
建筑工程一切险 contractor's all risk insurance
安装工程一切险 erection all risk insurance
第三方责任险 third party insurance
人身伤害和财产损失险 insurance against injury to person and damage to property
保险期限 insurance period
专项审查费 special works review fee
招标代理费 tender agency charge
标底编制费 expense of pretender preparation
竣工决算 final account for completed project, as-built settlement

项目后评价 post-evaluation of completed project
场地准备费 cost of site preparation
生产准备费 operation start-up cost
生产人员提前进厂费 expense of production personnel advancing into the plant
培训费 training expense
管理用具购置费 expense for purchasing management appliances
备品备件购置费 expense for purchasing spare parts
工器具及生产家具购置费 expense for purchasing tools, apparatus and furniture
临时设施费 expense for temporary facility
联合试运转费 expense of joint commissioning
科研勘察设计费 cost for investigation, design and testing
施工科研试验费 cost for construction scientific test
勘察设计费 cost for project investigation and design
市政公用设施费 fee of municipal public facilities
税费 tax and fee
东道国税费 tax and fee of host country
政治保险 political insurance
战争险附加费 additional expenses of war risk
海外投资保险费 foreign investment insurance cost
海外承包工程险 overseas contract engineering risks insurance (OCER)
建设质量监督管理费 expense for construction quality supervision and management
安保措施费 expense for security measures

(4) 总投资 Total investment

当地货币 local currency
外币 foreign currency
静态投资 static investment
投资估算 investment estimate
投资指标 investment index
单位投资 unit investment cost
送出线路工程 transmission works
单位千瓦静态投资 static investment per kilowatt
单位千瓦投资 investment per kilowatt
基本预备费 physical contingency
价差预备费 price contingency
汇兑风险 exchange risk
汇兑损益 exchange loss or gain
汇率风险 exchange rate risk
资金流量 fund flow
现金流量 cash flow
预付款 advance payment
保留金 retention
资本金 capital fund
银行贷款 bank loan
承诺费 commitment fee
建设期利息 interest incurred during construction
年利率 annual interest rate
年税额 annual tax
年息 interest per annum
年终股息 year-end dividend
优惠利率 bank prime rate
本利和 compound amount
分割系数 division coefficient
物价指数 price index

11.5 主要图表名称 Titles of figures and tables

主要技术经济指标表 Main technical and economic indexes
总投资估算表 Table of total cost estimate
建筑工程估算表 Table of civil works estimate
机电设备及安装工程估算表 Table of E&M equipment and installation works estimate
建设征地处理主要实物量、工程量汇总表及补偿投资费用总表 Physical indices, summary bill of quantities, compensation cost of land acquisition
环境保护和水土保持工程估算表 Table of environmental protection, soil and water conservation works estimate
独立费用估算表 Table of independent works estimate
施工辅助工程估算表 Table of temporary works estimate
分年度投资概算表 Table of annual cost estimate
工程单价计算表 Calculation of project unit price
主要材料预算价格计算表 Calculation of main materials estimated price
主要施工机械台时费价格计算表 Calculation of machine-hour expense of main construction machinery

主要进口设备原价计算表 Calculation of prime cost of main imported equipment

主要建筑工程量汇总表 Summary of quantities of main civil works

永久及施工临时用地汇总表 Summary of land requisition for permanent use and construction temporary use

单价分析表 Table of unit price analysis

第12章　项　目　管　理
Project management

12.1　招投标
Tender and bid

国际竞争性投标　international competitive bidding
非竞争性投标　noncompetitive bid
项目法人　project legal person，project entity
项目发包人　project employer
项目承包人　project contractor
项目承包　project contracting
项目分包　subcontract
项目管理体系　project management system
项目成本管理　project cost management
项目信息管理　project information management
项目风险管理　project risk management
项目采购管理　project procurement management
招标　invitation for tender (bidding)
投标　bidding
招标控制价　tender sum limit
招标文件　bid document
招标公告　announcement of tender
招标函　letter of tender
投标书的提交　submission
招标文件澄清　clarification
询标　bid inquiry
投标人合格性　eligibility
投标担保　bid guarantee
投标保证金　bid bond
投标保函　tender security
母公司保函　parent company guarantee
履约保函，履约担保　performance guarantee
履约保证金　performance bond，performance security
银行保函　bank guarantee
无条件保函，见索即付保函　first demand guarantee
有条件保函　accessory guarantee
雇主支付保函　payment guarantee by employer
即付保函　demand guarantee
预付款保函　advance payment guarantee
保留金保函　retention money guarantee
当事人　party concerned
委托人　principal
受益人　beneficiary
担保人　guarantor
继承人　successor
受让人　assignor
转让　assignment
独立第三方　independent third party
独立的外部同行评审　independent external peer review
独立的政府评估　independent government estimate
投标资格预审　prequalification of bidder
核准制　approval system
备案制　record system
审查　review
开标　bid opening
标底　bid price
评标　bid evaluation
中标　winning bid
合同协议书　contract agreement
知识产权和工业产权　intellectual and industrial property rights
无形资产　immaterial assets

12.2　合同管理
Contract management

建设工程项目　construction project
合同　contract
一揽子合同　all-in contract，package deal contract
总价合同　lump-sum contract
固定单价合同　fixed unit-price contract
单一货币合同　single currency contract
成本加酬金合同　cost plus award-fee contract，cost reimbursement contract
成本加固定费合同　cost plus fixed-fee contract
议标合同，协商合同　negotiation contract
无附加条件的合同　absolute contract

专门技术合同 know-how contract
服务合同 contract for service
承包商带资承包合同 contractor-financed contract
分项（分段）发包合同 separate contract
分包合同 sub-contract
现场施工分包合同 field subcontract
建设-经营-转让，特许权协议 build-operate-transfer (BOT)
设计-采购-施工合同 Engineering- Procurement-Construction contract (EPC)
交钥匙工程合同 turnkey contract
建设-转让 build-transfer (BT)
设计-施工总承包 Design and building (D-B)
设计-采购总承包 Engineering and Procurement (E-P)
采购-施工总承包 Engineering and Construction (P-C)
电力购买协议 power purchasing agreement (PPA)
专利（权） patent
特许（权） franchise
过程控制模式 process control mode
事后监督模式 afterwards supervision mode
简明合同格式 Short Form of contract
施工合同条件 Conditions of Contract for Construction
合同协议书 contract agreement
合同通用条件 general conditions
合同专用条件 particular conditions
雇主要求 employer's requirements
裁决规则 rules for adjudication
裁决员协议书 adjudicator's agreement
解约条款 cancelling clause
雇主责任 employer's liabilities
施工合同条件 Conditions of Contract for Construction
合同管理 contract management
计划管理 program management
合同变更 contract variation
合同工期 contract time limit
撤销合同日期 cancelling date
终止 termination
到期日 expiry date
保修期 maintenance period
工程变更 variation of works
业主修改通知 change order
设计修改通知 amendment notice
设计错误 design error
合同价格的充分性 sufficiency of the contract price
雇主免费供应的材料 free-issue materials
雇主的资金安排 employer's financial arrangement
拨款 appropriation
付款 disbursement
付款计划表 schedule of payment
工程预付款 advance payment，prepayment
工程进度款 progress payment
到期支付款 payment due
期中付款 interim payment
最终付款 final payment
延误的付款 delayed payment
月报表 monthly statement
竣工报表 statement on completion
最终报表 final statement
履约证明 performance certificate
结清证明 written discharge
质保金保函 guarantee for retention bond
商业发票 commercial invoice
可报销费用 reimbursable expenditures
薪酬 salary
报酬 remuneration rate
工资 wage
保险证书 insurance certificate
不可抗力 force majeure
索赔 claim
工期索赔 claim for extension of time
费用索赔 claim for loss and expense
分歧（歧义） discrepancy
争端 dispute
友好解决 amicable settlement
仲裁 arbitration
争端裁决委员会 Dispute Adjudication Board (DAB)
争端裁决协议书 Dispute Adjudication Agreement

12.3 现场管理 Management of construction

水利水电工程 water resources and hydropower project
单项工程 sectional works
单位工程 unit works
分部工程 divisional works

分项工程 work element
单元工程 separated item works
扩大单元工程 combined item works
永久工程 permanent works
临时工程 temporary works
施工现场（工地） site
进场 mobilization
退场 demobilization
现场清理 general clearance（clearance of site）
施工区封闭管理 closed management of construction site
现场管理费 site overhead，construction overhead
建设管理费 overhead of client，construction administration cost
总部管理费 head office overhead
业主 owner，client
雇主 employer
雇主代表 employer's representative
雇主人员 employer's personnel
雇主委托人 delegated person
驻工地设计代表 design representative at site
监理 supervision
工程师代表（总监） engineer's representative
监理机构 supervision organization
施工监理 construction supervision
监理人员 supervisory staff
监理大纲 supervision outline
监理规划 supervision plan
监理实施细则 supervision executive detailed rules
承包商 contractor
承包商代表 contractor's representative，site agent
承包商人员 contractor's personnel
分包商 subcontractor
指定分包商 nominated subcontractor
设计联络会 design liaison meeting
项目启动会 kick-off meeting
工地例会 site regular meeting
施工图设计交底 technical interpretation of design intention
进度报告 progress report
期中报告 interim report
说明报告 supporting report
现场作业 operation on site
现场保安 security of the site
承包商监督 contractor's superintendence
例行维护与检查 routine maintenance and inspection
定期检查 periodic inspections
修补工作 remedial work
拒收 rejection
施工质量 construction quality
质量检验 quality inspection
质量评定 quality assessment
质量管理 quality control
全面质量管理 total quality control
综合质量管理 comprehensive quality control
质量管理体系 quality control system
质量保证 quality assurance
产品质量标准 product quality standard
施工质量控制等级 category of construction quality control
外观质量 quality of appearance
见证取样 evidential sampling
工程计量 measurement
工程质量优良品率 final percentage of superior works
合格性评定 evaluation of conformity
质量事故 quality accident
缺陷工程 defective work
质量缺陷 quality defect
一般缺陷 common defect
严重缺陷 serious defect
收尾工作 outstanding work
修补缺陷 remedying defect
返修 repair
返工 rework
工程照管 care of works
进场验收 site acceptance
见证取样检测 evidential testing
交接检验 handing over inspection
主控项目 dominant item
一般项目 general item
抽样检验 sampling inspection
计数检验 counting inspection
计量检验 quantitative inspection
结构性能检验 inspection of structural performance
阶段验收 stage acceptance，interim acceptance
投运前试验 pre-commissioning test
试运行 commissioning

移交　hand-over
接收　taking-over
接收通知　taking-over notice
接收证书　taking-over certificate
竣工试验　test on completion
竣工　completion
竣工后试验　test after completion
竣工验收　final acceptance
竣工文件　as-built document
项目进度管理　project schedule management
项目进度控制　project schedule control
施工生产计划　overall construction plan
年度计划　annual plan
作业计划　operation schedule
基准日期　base date
开工令　order of commencement
开工通知　notice of commencement
开工日期　commencement date
竣工日期　completion date
计划指标　plan target
工程进度　rate of progress
暂时停工　suspension of work
季节性停工　seasonal shutdown
复工　resuming of work
赶工　expediting of work
加速施工　acceleration
赶计划　catch up with the schedule
窝工区域　slack area
施工调度　operation dispatching
进度偏差　program variance of project
施工延误　construction delay
总生产工日　total working days
高峰人数　peak labor force
货物　goods
承包商设备　contractor's equipment
雇主设备　employer's equipment
物流　logistics
物流管理　logistics management
配送　distribution
应急物流　emergency logistics
托运　consignment
承运　carriage
直达运输　through transportation
中转运输　transfer transportation
仓单　warehouse receipt
货垛　goods stack
物品分类　sorting
装卸　loading and unloading
搬运　handling carrying
货损率　cargo damages rate
理货　tally
国际多式联运　international multimodal transport
陆空联运　land-air through transportation
陆海联运　land-sea through transportation
水陆联运　water-land through transportation
集装箱　container
国际航空货物运输　international airline transport
海关　custom house
报关　customs declaration
退关　shut out
放行　release of imports and exports
结关　customs clearance
保税货物　bonded goods
原产地证明　certificate of origin
装运单据　shipping documents
进出口商品检验　commodity inspection
检疫　quarantine
健康、安全与环境管理体系　HSE management system
项目安全管理　project safety management
职业健康管理　occupational health management
施工安全管理　safety management during construction
安全程序　safety procedure
应急预案　emergency preparedness plan（EPP）
应急救援演练　emergency rescue rehearsals
危险源辨识　hazard identification
重大危险源　major hazard
重大危险源评价　major hazard assessment
危险物质　hazardous substance
危险环境　hazardous environment
临界量　threshold quantity
危险因素　hazardous factor
隐患　potential hazard
事故隐患　accident potential
伤亡事故（工伤事故）　casualty accident
合同价格　contract price
议定价格　price negotiated
可调整的报价　adjustable price quotation
报价明细表　bid schedule of prices
暂列金额　provisional sum

定期付款 due payment
分期付款 installment payment
罚款 penalty
违约金 liquidated damages
不可退还的款项 nonrefundable payment
不符合合同要求引起的索赔 claim based on lack of conformity of the goods
故障造成的损失费 cost of failure
停产损失费 production suspension loss
窝工费用 expenditure of idleness
弥补亏损 recovery of loss
已完工程保护费 protection expense of completed work
缺陷责任期 defects liability period
缺陷通知期 defects notification period

12.4 档案管理 Archives management

档案 archives
档案价值 archival value
档案工作 archives work
档案管理 archives management
文书档案 administrative archives
科学技术档案 scientific and technical archives
工程档案 project archive
城建档案 urban construction archive
专业档案 specialized archives
音像档案 audio-visual archives
照片档案 photographic archives
地质档案 geological archives
文件 document, record
电子文件 electronic record
原件 original document
副本，复印件 copy, duplicated document
文稿 draft
文本 text, version
正文 official text
手稿 manuscripts
附件 appendix, attachment
条形码 bar code
档案人员 archivist
档案行政管理部门 archival administrative department
档案室 record office
档案馆 archives
综合档案馆 comprehensive archives
专业档案馆 specialized archives
企业档案馆 business archives
企业档案 business archives, business records
工程档案著录 description of project archives
工程文件 records of project
工程前期文件 prephase records of project
工程竣工文件 records on completion of project
工程施工文件 construction records of project
工程监理文件 records of supervision of project
竣工验收文件 handing over document, completion acceptance documents
原始地质资料 original geological data
蓝图 blueprint
工程档案移交 transfer of project archives
工程档案验收 acceptance of project archives
工程档案接收 accession of project archives
档案收集 acquisition, collection
归档 filing, archiving
馆藏 holdings
鉴定 appraisal
保管期限 retention period
保管期限表 records retention schedule
销毁 destruction
销毁清册 destruction list
档案整理 archival arrangement
档案实体分类 physical archives classification
立卷单位 fonds constituting unit
全宗 fonds
案卷 file
立卷 filing
卷内备考表 file note
档号 archival code
分类号 classification code
案卷题名，文件名 file name
案卷夹，文件夹 folder
题名 title
检索 retrieval
编目 cataloguing, description
档案信息分类 archival information classification
条目 entry
著录 description
标引 indexing
分类标引 classified indexing
主题标引 subject indexing

关键词　keyword
主题词　descriptor
档案主题词表　archives thesaurus
检索工具　finding aid
目录　catalogue，inventory
案卷目录　folder list，file list
卷内文件目录　innerfile item list
分类目录　classified catalogue
主题目录，专题目录　subject catalogue
索引　index
全宗指南　guide to an archival fonds
专题指南　guide to subject record
著录格式　description form and format
著录项目　item of description
阅览室　reading room
密级　security classification
降密级　downgrade
解密　declassification
公开级　public class
限制级　limiter stage
国家秘密　national security
秘密级　confidential secret
机密级　classified level
绝密级　most confidential level
保管　custody
保护　conservation
档案库房　archival repository
密集架　compact shelving
加湿　humidification
去湿　dehumidification
去污　cleaning
档案缩微品　archival microform
档案虫霉预防　pests and mould prevention in archives
档案虫霉除治　pests and mould control in archives
短期文献　short-term document
长期文献　long-term document
永久文献　permanent document
文献统计　statistics
文献登记　registration
档案信息化　archival informationization
纸质档案数字化　digitization of paper-based records
数字图像　digital image
图像压缩　image compression
数字化加工　digitize processing
电子档案　electronic archives
数字签名　digital signature
数据审计　data audit
逻辑卷　logic volume
真实性　authenticity
完整性　integrity
元数据　metadata
在线式归档　on-line filing
离线式归档　off-line filing
全文数据库　context database
目录数据库　catalogue database
档案信息资源　archival information resources

附录 1

技术标准编制常用词汇

通用

标准化　standardization
最新技术水平　state of the art
公认的技术规则　acknowledged rule of technology
协商一致　consensus
适用性　fitness for purpose，serviceability
兼容性　compatibility
互换性　interchangeability
反之亦然原则　vice versa principle
规范性文件　normative document
标准，规范　standard，code
规程　specification
导则　guide
指南　guideline
规定　rule
准则　criterion
法律　law
法规　regulation
技术要求　technical requirement，specifications
国际单位制（SI 制）　international system of units (SI)
强制性标准　mandatory standard
推荐性标准　voluntary standard
国际标准　international standard
国家标准　national standard
行业标准　industrial standard，professional standard
地方标准　provincial standard
企业标准　enterprise standard，company standard
基础标准　basic standard
术语标准　terminology standard
试验标准　testing standard
产品标准　product standard
过程标准　process standard
服务标准　service standard
接口标准　interface standard
采用　adoption
等同采用　identical (IDT)
修改采用　modified (MOD)
非等效采用　not equivalent (NEQ)
编辑性修改　editorial change
技术性差异　technical deviation
强制性条文　compulsory provision
规范性要素　normative element
必备要素　required element
可选要素　optional element

标准文件的结构及编制用语

前言　Foreword
目录（次）　Content
总则　General provision
范围　Scope
术语　Term
定义　Definition
符号　Symbol
基本规定　Basic requirement
特殊要求　Particular requirement
引用标准名录　Normative references
本规范用词说明　Explanation of Wording in This Code
附录　Appendix
条文说明　Additional explanation of the provisions
发布日期　Issue Date
施行日期　Implementation date
主编单位　Chief development organization
参编单位　Participating development organization
批准部门　Approval department
主编　editor-in-chief
主要起草人　Chief drafting staff
归口　under the jurisdiction of
备案　put on record

对执行标准严格程度的要求（摘自《ISO/IEC 标准编写规定》）

（1）要求　requirement

应　shall

不应　shall not

（2）推荐　recommendation

宜　should

不宜　should not

（3）允许　permission

可　may

不可　need not

（4）可能　possibility，capability

能够　can

不能　cannot

部分国家工程技术标准代号

ISO　国际标准化组织标准

IEC　国际电工技术委员会标准

AASHTO　美国州公路及运输协会标准

ACI　美国混凝土学会标准

ASTM　美国试验及材料学会标准

AWS　美国焊接学会标准

IEEE　美国电子电机工程师协会标准

EN　欧盟协调标准

Eurocodes　欧洲建筑和土木工程技术标准

BS　英国标准

IEE　英国电气工程师学会标准

DIN　德国标准

VDE　德国电工标准

NF　法国标准

AS　澳大利亚标准

CAN　加拿大国家标准

JIS　日本工业标准

JEC　日本电气学会标准

JEM　日本电机工业协会标准

IS　印度标准

KS　韩国标准

СНиП　俄罗斯联邦建筑法规

ГОСТ　独联体跨国标准

ГОСТР　俄罗斯建筑领域的国家标准

AENT　巴西技术标准

附录 2

部分国外机构、团体名称

联合国开发计划署　United Nations Development Programme (UNDP)
联合国环境规划署　United Nations Environment Programme (UNEP)
世界贸易组织　World Trade Organization (WTO)
世界气象组织　World Meteorological Organization (WMO)
世界卫生组织　World Health Organization (WHO)
世界银行　World Bank
国际货币基金组织　International Monetary Fund (IMF)
亚洲开发银行　Asia Development Bank (ADB)
亚洲基础设施投资银行　Asian Infrastructure Investment Bank (AIIB)
欧洲复兴开发银行　European Bank of Reconstruction and Development
美洲开发银行　Inter-America Development Bank
国际商会　International Chamber of Commerce (ICC)
经济合作与发展组织（经合组织）　Organization for Economic Co-operation and Development (OECD)
欧洲经济合作组织　Organization for European Economic Cooperation (OEEC)
欧洲经济共同体委员会 (EEC)
亚太经合组织　Asia-Pacific Economic Cooperation (APEC)
太平洋经济合作理事会　Pacific Economic Cooperation Council (PECC)
东非共同体　East African Community (EAC)
中非国家经济共同体　Economic Community of Central African States (CEEAC)
西非国际经济共同体　Economic Community of West African States (ECOWAS)
欧洲联盟（欧盟）　European Union (EU)
非洲联盟（非盟）　African Union (AU)
阿拉伯国家联盟（阿盟）　League of Arab States (LAS)
东南亚国家联盟（东盟）　Association of Southeast Asian Nations (ASEAN)
上海合作组织（上合组织）　The Shanghai Cooperation Organization (SCO)
世界自然保护联盟　International Union for Conservation of Nature and Natural Resources (IUCN)
世界自然基金会　World Wide Fund For Nature (WWF)
国际自然及自然资源保护联盟　International Union for Conservation of Nature and Natural Resources (IUCN)
国际水资源协会　International Water Resources Association (IWRA)
大自然保护协会　The Nature Conservancy (TNC)
世界动物保护协会　World Animal Protection
国际野生生物保护学会　Wildlife Conservation Society (WCS)
世界野生动物基金会　World Wildlife Fund (WWF)
国际鸟盟　Birdlife International
防止空气污染协会国际联合会　International Union of Air Pollution Prevention Association (IUAPPA)
国际能源机构　International Energy Agency (IEA)
国际标准化组织　International Organization for Standardization (ISO)

国际电工技术委员会 International electrotechnical commission (IEC)
国际电信联盟 International Telecommunication Union (ITU)
国际岩石力学学会 International society for Rock Mechanics
国际土力学与基础工程学会 International society for Soil Mechanics and Foundation Engineering
国际大坝委员会 International Commission on Large Dams (ICOLD)
国际咨询工程师联合会（菲迪克） International Federation of Consulting Engineers (FIDIC)
国际认证联盟 Association - The International Certification Network (IQNet)
美国土木工程师学会 American Society of Civil Engineers (ASCE)
美国混凝土学会 American Concrete institution (ACI)
美国试验及材料学会 American Society for Testing and Materials (ASTM)
美国州公路及运输协会 American Association of State Highway and Transportation Officials (AASHTO)
美国电子电机工程师协会 Institute of Electrical and Electronics Engineers (IEEE)
美国机械工程师学会 American Society of Mechanical Engineers (ASME)
美国焊接工程学会 American Welding Society (AWS)
美国联邦能源委员会 Federal Energy Regulatory Commission (FERC)
美国垦务局 United States Bureau of Reclamation (USBR)
美国陆军工程兵团 United States Army Corps of Engineers (USACE)
欧洲标准化委员会 Comité Européen de Normalisation (CEN)
英国标准学会 British Standards Institution (BSI)
法国标准化协会 Association Francaise de Normalisation (AFNOR)
德国标准化学会 Deutsches Institut für Normung e. V (DIN)
日本工业标准协会 Japanese Industrial Standards Committee (JISC)
印度标准局 Bureau of indian standards (BIS)
非洲区域标准化组织 African Regional organization for standardization (ARSO)

汉语排序部分

A

阿基米德原理　Archimendes' principle
阿列维常数　Allievi constant
阿太堡试验　Atterberg test
阿太堡限度　Atterberg limit
矮式机墩　short pier of turbine
安保措施费　expense for security measures
安-秒特性　Ampere-time characteristics
安培　Ampere
安全标志　safety sign
安全程序　safety procedure
安全出口　exit door
安全带　safety belt
安全岛　safety island
安全电流　safety current
安全度　degree of safety
安全阀　safety valve
安全防范系统　security alarm system（SAS）
安全防护用品　protective appliance
安全防护装置　security protection unit
安全隔离通信　security isolation communication
安全供水　safe water supply
安全关闭　safety shutdown
安全机构　safety mechanism
安全监测　safety monitoring
安全鉴定　safety appraisal
安全帽　safety helmet
安全区　safety zone
安全疏散　safe evacuation
安全疏散距离　exit distance，safety evacuation distance
安全竖井　escape shaft
安全通道　fire escape
安全网　safety net
安全文明施工　HSE（health，safety and environment）construction
安全文明施工措施费　measure fee for HSE construction
安全性　safety
安全裕度　safety margin
安全照明　safely lighting
安全制动器　safety brake
安山岩　andesite
安息角（休止角）　angle of repose
安匝　Ampere-turn
安置补助费　settlement allowances
安置区　host area
安装测量　erection survey，installation survey
安装拆卸费　assembly and disassembly cost
安装场　erection bay
安装高程　setting elevation
安装工程费　installation woks cost
安装工程一切险　erection all risk insurance
安装工艺　installation technique
安装公差　erection tolerance
安装间　erection bay
安装精度　installation accuracy
安装误差　installation error
鞍形支座　saddle support
铵锑炸药　ammonium nitrate explosive
铵油炸药　ammonium nitrate fuel oil explosive
岸边侵蚀　bank erosion
岸边式厂房　powerhouse on river bank
岸边式取水构筑物　riverside intake structure
岸边溢洪道　river-bank spillway
岸边淤积　bank deposit，inwash
岸坡式进水口　intake with inclined gate slots in the bank
岸坡表层坍落　bank sloughing
岸墙　side wall
岸塔式进水口　intake tower against the bank
按钮操纵起重机　pendant-operated crane
按体积配料　batching by volume
按重量配料　batching by weight
案卷　file
案卷夹（文件夹）　folder
案卷目录　folder list，file list
案卷题名　file name
暗杆闸阀　inside screw non-rising stem type gate valve
暗沟排水　blind drainage，buried drain
暗管排水　subsurface pipe drain
暗河　underground river
暗褐黄色　dark brownish yellow
暗梁　concealed beam
暗色矿物　dark mineral
凹岸　concave bank
凹槽　trough
凹坑　hollow spot

凹坡　concave slope
凹凸形　concavo-convex
凹陷　depression，hollow
坳沟　shallow flat ravine
拗陷盆地　depression basin
奥斯特　Oersted
奥陶纪（系）　Ordovician period（system）

B

巴氏合金瓦　babbitt bearing
坝底　dam base
坝地　farmland formed in silt storage dam
坝顶　dam crest
坝顶超高　freeboard
坝顶超填　camber
坝顶溢洪道　crest spillway
坝顶溢流　crest overflow
坝段　dam section
坝后背管　penstock laid on downstream face of dam
坝后厂房顶溢流　discharging through a spillwayon the powerhouse roof at dam toe
坝后式厂房　powerhouse at dam toe
坝后式水电站　hydropower station at dam-toe
坝基　dam foundation
坝基编录　dam foundation mapping
坝基截水槽　key trench
坝基开挖　excavation of dam foundation
坝基廊道　foundation gallery
坝基面渗透压力　seepage pressure on foundation surface
坝基排水孔　drainage holes in dam foundation
坝基渗流　seepage of dam foundation
坝基渗流量　seepage flow in foundation
坝基渗漏　leakage through dam foundation
坝肩　abutment
坝脚压重　toe support fills
坝壳　dam shell
坝壳料　dam shell material
坝块　monolith，dam block
坝面附加质量　added mass at dam face
坝面泄流　discharging along the downstream dam face
坝内埋管　embedded penstock within dam
坝内式厂房　powerhouse within dam
坝内式水电站　hydropower station in dam
坝坡　dam slope
坝坡出逸段　embankment escape area
坝区现地控制单元　dam LCU
坝身孔口泄流　flow discharge through dam outlet
坝身泄水孔　outlet through dam
坝式进水口　intake integrated with the dam
坝式开发　dam type development
坝式水电站　dam type hydropower station
坝体底孔导流　bottom outlet diversion
坝体结构　dam structure
坝体排水　drainage of embankment
坝体溢流段　overflow dam section
坝外基准点　off-dam reference
坝下淘刷　scour beneath dam
坝址洪水　flood at dam site
坝址施工区　construction area at damsite
坝趾　dam toe
坝趾排水　toe drain
坝踵　dam heel
白垩纪（系）　Cretaceous period（system）
白榴石　leucite
白云母　muscovite
白云石（岩）　dolomite
白云质灰岩　dolomitic limestone
百年一遇洪水　one percent chance flood，a 100-year recurrence flood
百叶窗　blind window
摆喷　oscillating jet grouting
斑晶结构　phenocryst texture
斑脱土（岩）　bentonite
斑岩　porphyry
斑状结构　porphyritic texture
搬迁人口　relocation population
搬运　handling carrying
板块　plate
板块缝合带　plate suture zone
板块构造环境　plate geotectonic setting
板块碰撞　plate collision
板劈理　slaty cleavage
板式给料机　table feeder，apron feeder
板式基础　slab footing
板岩　slate
板柱结构　slab-column structure
板桩灌注墙　sheet pile cell wall

板桩式挡土墙　sheet-pile retaining wall
板桩围堰　sheet-pile cofferdam
板状构造　tabular structure，platy structure
板状灰岩　platy limestone
办公费　office expense
半导体变流器　semiconductor converter
半导体阀器件　semiconductor valve device
半地下式厂房　semi-underground powerhouse
半地下式水电站　semi-underground hydropower station
半高型布置　semi-high profile layout
半固定式压力钢管　semi-fixed penstock
半固态　semi-solid state
半挂车　semitrailer
半剪刀式隔离开关　semi-pantograph disconnector
半经验公式　semi-empirical formula
半露天式厂房　semi-outdoor powerhouse
半伞式水轮发电机　semi-umbrella hydrogenerator
半熟练工　semi-skilled labor
半挖半填的边坡　cut and fill slope
半致死浓度　lethal concentration 50%（lc_{50}）
半中毒浓度　effect concentration 50%（ec_{50}）
半重力式挡土墙　semi-gravity retaining wall
伴生矿物　associated mineral
帮条焊　article for the welding
绑扎连接　splicing joint
棒磨机　rod mill
棒式电流互感器　bar primary type current transformer
棒式绝缘子　rod insulator
棒图　bar graph
傍山深路堑　deep cut in hillside
傍山隧洞　mountainside tunnel
包角焊　seal welds
包气带　aeration zone
包体结构　inclusion texture
包装绑扎费　packing and lashing expense
包装费　packing charge
包装货物　packed goods
包装品回收　packaging recycling
薄板结构　thin-slab structure
薄板堰　thin-plate weir
薄层间歇浇筑　placement in shallow lifts with delays between lifts
薄层状结构　thin layer structure
薄腹梁　thin-webbed beam
薄钢板防火门　sheet-metal fire door
薄拱坝　thin arch dam
薄膜加热试验　thin-film oven test
薄膜式减压阀　diaphragm reducing valve
饱和度　degree of saturation
饱和抗压强度　saturated compressive strength
饱和密度　saturated density
饱和面干表观密度　saturated-surface-dry apparent density
饱和容重　saturated unit weight
饱和湿胀应力　saturation swelling stress
饱和输沙（平衡输沙）　saturated sediment transport
饱和特性　saturation characteristic
饱和土　saturated soil
保安电源　emergency power supply
保持命令　maintained command
保管　custody
保管期限　retention period
保护　conservation
保护层　protective layer（coating）
保护导体　protective conductor（PE）
保护等电位连接　equopotential bonding
保护电阻器　protective resistor
保护范围　reach of protection
保护火花间隙　protective spark gap
保护建筑　listed building for conservation
保护角　angle of shade
保护接地　protective earthing，protective earthing
保护金具　protective fitting
保护距离　protective range
保护盘室　protection panel room
保护配合　protection coordination
保护区　protected section，protected zone
保护用互感器　protective instrument transformer
保护元件　protection component
保护重叠区　overlap of protection
保护装置的保护水平　protection level of protective device
保留金　retention
保留金保函　retention money guarantee
保留样品　retention of sample
保税货物　bonded goods
保梯电抗　Potier reactance

保温　thermal insulation
保温材料　thermal insulating material
保温层　heat insulating layer
保温剂　thermal insulating agent
保温模板　insulated form
保温式阀　steam jacket type valve
保温涂层　insulating coating
保温系数　coefficient of heat preservation
保温性能　insulating property
保险费率　insurance rate
保险期限　insurance period
保险证书　insurance certificate
保修期　maintenance period
保证出力　firm output
保证供水线　upper critical guide curve
保证水位　highest safety stage
保证效率　guaranteed efficiency
保证转矩　guaranteed torque
报酬　remuneration rate
报关　customs declaration
报警　alarm
报警标志　alarm mark
报警阀　alarm valve
报警监测装置　alarm detector
报警接收中心　alarm receiving center
报警开关　alarm switch
报警联动　action with alarm
报警器　alarm apparatus
报警水位　alarm water level
报警图像复核　video alarm verification
报警信号　alarm signal
报警信号传输系统　alarm transmission system
报警延迟　alarm delay
报警状态光字指示　annunciation of alarm conditions
暴露剂量　exposure dose
暴露时间　exposure time
暴晒　solarization
暴雨　rainstorm
暴雨历时　rainstorm duration
暴雨模式　rainstorm model
暴雨强度　rainstorm intensity
暴雨强度历时曲线　storm intensity-duration curve
暴雨侵蚀　storm erosion
暴雨时、面、深关系　depth-area-duration (DAD) relationship of storm
暴雨移置法　storm transposition method
暴雨中心　rainstorm center
爆力　explosive strength
爆破　blasting
爆破漏斗　explosion crater
爆破片装置　bursting disk device
爆破有害效应　adverse effect of blasting
爆速　detonation velocity
爆炸作用　explosion action
贝壳状断口　conchoidal fracture
贝叶斯概率法　Bayes probability method
备案制　record system
备餐间　pantry
备品备件　spare parts
备品备件购置费　expense for purchasing spare parts
备选方案　alternative
备选方案评估　alternative appraisal
备用保护　standby protection
备用泵　emergency pump
备用泵启动　back-up pump start
备用变压器　standby transformer
备用电源　stand-by power supply
备用电源自动投入　automatic bus transfer
备用供电系统　standby supply system
备用空压机　lag compressor
备用母线　reserve busbar
备用配电盘　emergency switchgear
备用容量　reserve capacity
备用容量曲线　reserve capacity curve
备用时间　stand-by duration
备用水源　alternate source
备用系统　redundant system
备用照明　stand-by lighting
背靠背启动　back-to-back starting (BTB starting)
背面排水　back drain
背面台阶式挡土墙　retaining wall with stepped back
背水面坡脚排水　counter drain
背斜　anticline
被动土压力　passive earth pressure
被控参数　controlled variable
被控制变量范围　controlled variable range
被控制系统　controlled system
被控制系统自调节系数　controlled system

selfregulation coefficient
本底地震　background earthquake
本底噪声　background noise
本底值　background value
本地居民　indigenous inhabitant
本地人口　indigenous population
本地文化　indigenous culture
本地终端　local terminal
本构关系　constitutive relation
本利和　compound amount
崩解（崩解性）　disintegration
崩解试验　slake test，disintegration test
崩塌　collapse，avalanche
泵盖　casing cover
泵基准面　reference plane
泵输出功率（有效功率）　pump effective power
泵送混凝土　pump concrete
泵送剂　pumping admixture
泵送能力　pumpability
泵效率　pump efficiency
泵站　pump station
泵轴　pump shaft
泵轴功率（输入功率）　pump shaft power
鼻端固定导叶　nose vane
比表面积　specific surface area
比表面积试验　specific surface area test
比尺效应　scale effect
比贯入阻力　specific penetration resistance
比焓　specific enthalpy
比例阀　proportional valve
比例放大系数　proportional amplification
比例-积分调速器　proportional-integral governor (PI governor)
比例-积分-微分调速器　proportional-integral-derivative governor (PID governor)
比例控制阀　proportional control valve
比例误差　scale error
比例增益　proportional gain
比例制动差动保护　percentage differential relay
比摩阻　specific frictional resistance
比能　specific energy
比选方案　alternative
比重　specific gravity
比重瓶法　pycnometer method
比重瓶（密度瓶）　pycnometer
比重试验　specific gravity test
比转速　specific speed
必需气蚀余量　net positive suction head required (NPSHR)
毕托管　Pitot tube
毕肖普法　Bishop method
闭合　closed
闭合导线　closed traverse
闭合褶皱　closed fold
闭环控制　closed-loop control
闭环控制同步纠偏回路　synchro-control circuit with closed-cycle
闭浆　measure for keeping closed stage
闭路电视　closed-circuit television
闭路循环灌浆　closed-circuit grouting
闭路循环去离子水系统　closed-loop demineralized water system
闭门力　closing force
闭气　leakage stopping
闭式喷头　automatic sprinkler
闭式洒水喷头　sealed sprinkler head
闭式系统　close-type sprinkler system
闭式叶轮　closed impeller
闭锁式保护　blocking protection
闭锁式纵联保护　blocking pilot protection
闭锁元件　blocking component
闭锁重合闸　lockout reclosing
壁面　wall surface
避雷器　arrester
避雷器　surge arrester，lightning arrester
避雷器标称放电电流　nominal discharge current of arrester
避雷器标准雷电冲击放电电压　standard lightning impulse sparkover voltage of arrester
避雷器残压　residual voltage of arrester
避雷器持续运行电压　continuous operating voltage of arrester
避雷器的保护特性　protective characteristics of arrester
避雷器电导电流　conducting current of arrester
避雷器额定电压　rated voltage of arrester
避雷器阀片　valve disc of arrester
避雷器工频参考电压　power frequency reference voltage of arrester
避雷器工频放电电压　power frequency sparkover

voltage of arrester
避雷器均压环 grading ring of arrester
避雷器泄漏电流 leakage current of arrester
避雷器压力释放装置 pressure-relief device of arrester
避雷器直流参考电压 direct current reference voltage of arrester
避雷线 overhead earthing wire
避雷针 lightning rod
避难层 refuge storey
避税 tax avoidance
臂式挖掘机 boom excavator
臂柱 arm column, strut
臂柱加强杆（臂柱连接系） strut bracing
边侧支撑 side shoring
边墩 abutment pier
边界条件 boundary condition
边界温度 boundary temperature
边梁 side beam
边坡 slope
边坡地质模型 geological model of slope
边坡结构模型 slope structure model
边坡开裂 slope cracking
边坡坍塌，坍坡 slope failure, slope collapse
边墙 side wall
边墙塌落 side wall collapse
边柱 side column
编码器 encoder
编目 cataloguing, description
编织层 braid
扁豆体 lenticle
扁平状 flat
扁千斤顶试验 flat jack test
变半径拱坝 variable radius arch dam
变比 transformation ratio
变电 transformation of electricity, power transformation
变电站 substation
变电站构架 substation gantry
变电站近后备保护 substation local backup protection
变风量末端装置 variable air volume terminal box
变风量系统 variable air volume system
变厚度拱 variable-thickness arches
变极调速 pole changing speed control
变极式发电电动机 pole changing motor-generator
变截面梁 non-uniform cross-section beam
变截面柱 non-uniform cross-section column
变晶结构 crystalloblastic texture
变量泵 variable displacement pump
变流量泵 variable displacement pump
变流设备 converter equipment（converter assembly）
变流因数 conversion factor
变频调速 variable frequency speed control
变频启动 variable frequency starting
变频器 frequency converter
变水量系统 variable water flow system
变水头渗透试验 variable head permeability test
变态混凝土 grout enriched vibratable concrete（GEV）
变温层 epilirnnion
变形 deformation
变形边坡 deforming slope
变形缝 deformation joint
变形监测 deformation observation
变形模量 deformation modulus
变形验算 deformation analysis
变压器充氮费 filling nitrogen cost of tansformer
变压器电流差动保护 transformer current differential protection
变压器调压装置 voltage regulator of transformer
变压器-线路组接线 transformer-line configuration
变压器相位移 phase difference for transformer
变压器压力突变 transformer sudden pressure rising
变压器压力突变释放装置 transformer main tank sudden pressure relief device
变异系数 coefficient of variation
变余结构 palimpsest texture
变质砾岩 metaconglomerate
变质泥岩 metapelite
变质砂岩 metasandstone
变质岩 metamorphic rock
变中心拱坝 variable centre arch dam
便桥 auxiliary bridge
便携式灭火器 portable fire extinguisher
标底 bid price
标底编制费 expense of pretender preparation
标杆电价 benchmark tariff
标高价 mark up
标积 scalar product

标记（标志） sign，mark
标石 markstone，monument
标示牌 signboard
标题栏 title block
标幺值 per unit value
标幺制 per unit system
标引 indexing
标志层 marker bed
标准操作冲击波 standard switching impulse
标准操作冲击耐受电压 standard switc-hing impulse withstand voltage
标准层［建筑］ typical floor
标准层［地质］ index bed，guide bed
标准冲击电流 standard impulse current
标准大气条件 standard reference atmos-pheric condition
标准地面运动分析 standard ground motion analysis
标准短时工频耐受电压 standard short duration power-frequency withstand voltage
标准贯入试验 standard penetration test（SPT）
标准贯入试验击数 SPT blow count
标准化石 guide fossil，index fossil
标准换算系数 standard conversion factor（SCF）
标准击实试验 standard compaction test（SCT）
标准件 standard component
标准径流小区 standard runoff plot
标准绝缘水平 standard insulation level
标准雷电冲击截波 standard chopped lightning impulse
标准雷电冲击耐受电压 standard lightning impulse withstand voltage
标准雷电冲击全波 standard full lightning impulse
标准耐火性 standard fire resistance
标准喷头 standard sprinkler
标准偏差 standard deviation
标准球隙 standard sphere gap
标准砂 reference sand
标准正态分布的反函数 inverse function of the standard normal distribution
标准正态分布概率密度函数 probability mass function of the standard normal distribution
标准值 characteristic value
标准组合 characteristic combination，nominal combination
表层火 skin fire
表观密度 apparent density
表计压力 gauge pressure
表面保温 surface heat preservation
表面换热系数 coefficient of surface heat transfer
表面火 surface fire
表面火焰蔓延 surface flame spread
表面监测 surface monitoring
表面清洁度 surface cleanness
表面燃烧 surface burn
表面闪燃 surface flash
表面预处理 surface preparation
表面张力 surface tension
表土 regolith，top soil
别墅 villa
滨海相 littoral facies
濒危物种 endangered species
冰坝 ice dam
冰崩 ice avalanche
冰川沉积 glacial deposit
冰川地形 glacial landform
冰川谷 glacier valley
冰川湖 glaciers lake
冰川泥石流 glacier debris flow
冰川时期 Glacial epoch（period）
冰冻隆起 frost upheaval
冰斗 glacial cirque
冰盖 ice cover
冰荷载 ice load
冰湖溃决洪水 glacial lake outburst flood
冰积层 glacial deposit
冰期 ice age
冰碛 glacial drift
冰碛阶地 glacial drift terrace
冰碛土 moraine soil
冰情 ice regime
冰情预报 ice regime forecast
冰水沉积 glaciofluvial deposit
冰水堆积体 outwash deposit
冰屑 slush ice
冰压力 ice pressure
丙乳砂浆 acrylic mortar
丙烯酸盐浆液 acrylate grout
饼状岩芯 disk-shaped rock core
并筋 twin bars

并励静止整流装置　potential source static exciter
并励绕组　shunt winding
并联　parallel connection
并联补偿　shunt compensation
并联电抗器　shunt reactor
并联电容补偿装置　shunt capacitive compensator
并联电容器　shunt capacitor
并联 PID 调速器　parallel PID governor
并联结构　parallel structure
并联控制　parallel control
并联谐振　parallel resonance
并联运行　parallel operation
并网运行　paralleling operation，grid-connected operation
拨款　appropriation
波长　wave length
波传递速度　wave travel speed
波导　waveguide
波导耦合器　waveguide coupler
波动或掺气后水深　water depth in fluctuation or after aeration
波动量　fluctuation of quantity
波动稳定断面　cross-section area of oscillating stability
波动状态　oscillation condition
波浪度　waviness
波浪反压力强度　wave counter pressure intensity
波浪爬高　wave runup
波浪侵蚀　wave erosion
波绕组　wave winding
波头截断冲击波　impulse chopped on the front
波尾截断冲击波　impulse chopped on the tail
波纹板　corrugated sheet
波纹管平衡式安全阀　bellows seal balance safety valve
波纹管式阀　bellows seal type valve
波纹管式阀杆密封　bellows stem sealing
波纹管式减压阀　bellows seal reducing valve
波纹伸缩节　bellow expansion joint
波形畸变　waveform distortion
波状层理　wave bedding
波状结构面　undulating structural plane
玻璃光泽　vitreous luster
玻璃绝缘子　glass insulator
玻璃质结构　vitreous texture
剥采比　stripping ratio
剥离表土　stripping top soil
剥离层　stripping layer
剥离量　stripped volume
剥落　spalling，exfoliation
剥蚀　abrasion，denudation
剥蚀面　denudation plane
剥蚀平原　denudation plain
剥夷面，夷平面　planation surface
播种面积　sown area
伯努利方程式　Bernoulli equation
博物馆　museum
补偿补助费　compensation subsidy
补偿调节　compensative regulation
补偿费用概算　compensation cost estimate
补偿收缩混凝土　expansive concrete
补偿张拉　compensatory tension
补救控制系统［安全稳定装置］　remedial action scheme（RAS）
补气试验　air admission test
补气系统　air admission system
补水泵　make-up water pump
补贴　subsidy
补贴政策　subsidy policy
捕虏体　xenolith
哺乳动物物种　mammal species
不饱和土　unsaturated soil
不变价格　constant price
不冲流速　non-scouring velocity
不动作值　non-operating value
不堵式叶轮　non-clogging impeller
不对称短路　asymmetrical short-cicuit
不对称负荷　asymmetrical load
不发育　undeveloped
不分级骨料　ungraded aggregate
不分相保护　non-phase segregated protection
不固结不排水剪试验　unconsolidated undrained shear test（UU）
不灌浆的收缩缝　ungrouted contraction joint
不规则进水口结构　irregular intake structure
不过水围堰　non-overflow cofferdam
不间断电源　uninterruptible power supply（UPS）
不均匀变形　inhomogeneous deformation，nonuniform deformation
不均匀沉降　differential settlement，unequal

settlement
不均匀系数　coefficient of uniformity
不开槽施工　trenchless installation
不可避免的不利环境影响 unavoidable adverse environmental impact
不可换算的损失　relative non-scalable loss
不可恢复变形（永久变形）　irreversible deformation
不可抗力　force majeure
不可控阀器件　non-controllable valve device
不可逆正常使用极限状态　irreversible serviceability limit states
不可逆转环境影响　irreversible environmental impact
不可退还的款项　nonrefundable payment
不可用时间　down duration，outage（duration）
不可再生资源　non-renewable resources
不利环境影响　adverse environmental impact
不利环境影响最小化　minimization of adverse environmental effects
不连续级配土　gap-graded soil
不灵敏度　insensitivity
不耦合连续装药　decoupled continuous charging
不平衡电流　unbalanced current
不平衡河槽［冲淤］　non-regime channel
不平衡推力法　method of non-equilibrium thrust
不平衡运行　unbalanced operation
不平整度　unevenness
不燃性　non-combustibility
不溶解固形物　indissolved solid matter
不完全差动保护　combined split phase and differential protection
不稳定　unstable
不稳定河床（动床）　mobile channel
不稳定流　unsteadiness
不稳定流和动态行进法　unsteady flow and dynamic routing method
不稳定坡　unstable slope
不锈钢　stainless steel
不锈钢复合钢板　composite stainless steel plate
不锈钢活塞杆　stainless rod
不锈钢止水带　stainless steel waterstop
不易燃材料　nonflammable material
不淤流速　non-silting velocity
不整合　unconformity
布莱恩试验，水泥细度试验　Blaine test
布氏（硬度）试验　Brinell test
部分断面（分部开挖）掘进法　partial face driving method
部分回转驱动装置　part-turn actuator
部分平衡式　partial balanced type
部分停机　partial shutdown

C

擦痕面　slickenside
材料表　bill of materials
材料消耗定额　material consumption norm
材料性能　material performance
材料性能分项系数　partial factor of mate-rial properties
材料蓄热系数　material heat store coefficient
材料预算价格　material estimate price
材料原价　material original price
财务补贴　financial subsidy
财务费　financial charges
财务分析　financial analysis
财务价格　financial price
财务净现值　financial net present value（FNPV）
财务可持续性　financial sustainability
财务内部收益率　financial internal rate of return（FIRR）
财务效益　financial benefit
裁决规则　rules for adjudication
裁决员协议书　adjudicator's agreement
裁弯工程　cut off works
采购及保管费　purchase and storage expenses
采购-施工总承包　engineering and construction（P-C）
采光　daylight
采光系数　daylight factor
采暖度日数　heating degree day based on 18℃（HDD18）
采暖供水管　heating supply water pipe
采暖管线　heating pipe line
采暖回水管　heating return water pipe
采暖立管　heating riser
采暖期天数　days of heating period
采暖室外计算温度　outdoor design temperature for heating
采暖室外临界温度　outdoor critical air temperature

for heating
采石场　quarry
采样频率　sampling frequency
采样速率　sampling rate
采样元件　sampling component
采样周期　scan period
菜地　vegetable land
参考节点　reference node
参照点　reference point
餐厅　dining room
残积　eluvium
残积土　residual soil
残孔率（半孔率）　residual hole rate
残丘　inselberg
残压起励建压　voltage build up from residual levels
残余强度　residual strength
残余扬压力　residual uplift pressure
残余扬压力强度系数　coefficient of residual uplift pressure intensity
残余应力　residual stress
残值　residual value
残值率　rate of residual value
仓单　warehouse receipt
仓库　warehouse
操动机构　operating mechanism
操作冲击半峰值时间　time to half value of switching impulse
操作冲击波保护比　protection ratio against switching impulse
操作冲击波前时间　time to peak of switching impulse
操作冲击截断时间　time to chopping of switching impulse
操作冲击绝缘水平　basic switching impulse insulation level
操作冲击耐压试验　switching impulse voltage withstand test
操作过电压　switching overvoltage
操作回路　operation circuit
操作架　cross head
操作票　operation order
操作器　position operator，gate operator
操作顺序　operating sequence
操作台　operation switchboard
操作条件　operating condition
操作箱　controly box
操作循环　operating cycle
操作员工作站　operator workstation
糙率　coefficient of roughness
槽型母线　channel busbar
草场退化　grassland degradation
草皮护坡　grass (sodding) slope protection
草田轮作　grass and crop rotation
侧槽式溢洪道　side channel spillway
侧铲　side shovel
侧铲推土机　angling dozer
侧盖式泵　side cover type pump
侧轨　side guide
侧式进/出水口　side intake/outlet
侧向荷载　lateral load
侧向挤压　lateral compression
侧向挠度　lateral deflection
侧向排水　lateral drainage
侧向收缩　lateral contraction
侧向土压力　lateral earth pressure
侧向稳定　lateral stability
侧向压力　lateral pressure
侧向应变　lateral strain
侧卸车　side dump truck
侧压力　lateral pressure
侧移刚度　lateral displacement stiffness
侧翼截水槽　wing trench
测缝计　joint gauge
测回　observation set
测回差　discrepancy between observation sets
测角网　angular network
测力扳手　torque wrench
测量标志　survey mark
测量觇标　observation target，observation tower
测量方法　survey method
测量通道极限频率　limit frequency of measuring channel
测量误差　true error
测量用互感器　measuring instrument transformer
测量元件　measuring component
测量资料（数据）　survey data
测频单元　frequency module
测速发电机　tachogenerator
测速信号源　speed signal source
测速装置　speed sensing device

测速装置放大系数 speed sensor amplification
测温电阻 resistance temperature detector (RTD)
测压管 piezometer
测压头 pressure tap
测站归心 reduction to station center
层高 storey height
层间(层内)节理 intraformational joint
层间错动带 interlayer shear zone
层间挤压面(带) crushed bedding plane (zone)
层节理 bedded joint
层理 bedding, stratification
层流 laminar flow
层面(层理面) bedding plane
层面裂隙 bedding joint
层面劈理 bedding cleavage
层析成像 computer tomography (CT)
层移质 laminated load
层状反向结构 reverse bedding structure
层状横向结构 transverse bedding structure
层状结构 stratified structure, bedding structure
层状平叠结构 horizontal bedding structure
层状同向结构 consequent bedding structure
层状斜向结构 oblique bedding structure
层状岩体 stratified rock
叉车 forklift
叉管 branch pipe
叉积 cross product
插板阀 slide damper
插入式振捣器 immersion vibrator, poker vibrator
插值法 interpolation method
插装阀 cartridge valve
查明 ascertain
差动电阻式传感器 differential resistive transducer
差动继电器 differential relay
差动式调压室 differential surge chamber
差动式鼻坎 slotted spillway bucket
差积曲线 residual mass curve
差旅费 travel expense
差模干扰电压 differential mode disturbance voltage
差围岩 poor surrounding rock
差压流量计 differential pressure flow meter
差异性隆升 differential uplift
差值曲线的斜率 slope of the droop graph
拆除爆破 demolition blasting
拆模 form removal
拆卸 disassembly
拆卸法兰 dismantling flange
柴油 diesel oil
柴油发电机 diesel generator
掺合料 admixture, blend
掺气 aeration
掺气槽 aeration slot
掺气减蚀模型试验 model test for aeration cavitation resistance
掺气设施 aerating facility
掺气水流 aerated flow
掺气水舌 aerated nappe
产流 runoff yield
产卵场 spawning ground
产卵鱼 spawner
产品质量标准 product quality standard
产沙模量 modulus of sediment yield
产业结构 industrial structure
产状 orientation, attitude
铲运机 bow scraper
颤振 flutter
长 length
长顶堰 long crested weir
长期观测 long-term observation
长期强度 long term strength
长期使用库容 long-term storage capacity of reservoir
长期水文预报 long-term hydrological forecast
长期水下环境 permanently under water environment
长期文献 long-term document
长期稳定性 long-term stability
长期组合系数 coefficient of long-term combination
长石砂岩 arkose, feldspathic sandstone
长条状 elongated
长细比 slenderness ratio
长远经济发展规划 long-term economic development plan
常闭触头(点) normally closed contact
常闭式阀 normally closed type valve
常规处理 conventional treatment
常规方式自动化 conventional automation
常规观测 routine observation

常规检查　routine inspection
常开触头（点）　normally open contact
常开式阀　normally open valve
常年河　perennial river
常年基流　perennial base flow
常态混凝土　conventional concrete
常态侵蚀　normal erosion
常温阀门　normal temperature valve
常温设计　normal temperature design
常住居民　permanent resident
常住人口　resident population
偿债备付率　debt service coverage ratio (DSCR)
厂顶溢流式水电站　overflow hydropower station
厂房地基应力计算　foundation stress calculation of powerhouse
厂房抗浮稳定计算　stability against floating calculation of powerhouse
厂级监控信息系统　supervisory information system for plant level
厂级信息系统　information system for plant level
厂内通信　intraplant communication
厂内装焊转轮　shop fabricated runner
厂前挑流　spillway with the flip bucket in front of the powerhouse at dam toe
厂用变压器　station service transformer
厂用电　station service (service power)
厂用电动机　station electric motor
厂用电接线　station service single line diagram
厂用电开关设备　station service switchgear
厂用电率　internal consumption rate, service power rate
厂用配电系统　station service distribution system
场地复杂程度　complexity of site
场地类别　site category
场地平整　site leveling
场地设计　site design
场地准备费　cost of site preparation
场内干线公路　onsite trunk road
场内交通　on-site access
敞开式安全阀　openly sealed safety valve
敞开式变电站　open-type substation
超标准洪水　over-standard flood
超导体　superconductor
超低温阀门　cryogenic valve
超范围　overreach
超范围式纵联保护　overreach pilot protection
超高［道路/铁路］　superelevation
超高库容　surcharge storage
超高温耐火材料　extreme temperature refractory
超高压　extra-high voltage (EHV)
超固结比　over-consolidation ratio
超过压力　over pressure
超基性岩　ultra-basic rock
超径骨料　oversize aggregate
超静定梁　statically indeterminate beam
超孔隙水压力　excess pore pressure
超临界流　supercritical flow, rapid flow
超前导洞　pilot heading
超前管棚　forepoling pipe-shed
超前灌浆　advance grouting
超前孔　guide hole
超前锚杆　forepoling bolt
超前小导管　forepoling ducting
超前支撑掘进　forepoling
超前支护　advance support
超前钻孔法　advance borehole method
超渗产流　runoff yield under excess infiltration
超声波流量计　ultrasonic flow meter
超声波探伤　ultrasonic testing (UT)
超声法　ultrasonic method
超瞬态短路电流　subtransient short-circuit current
超挖　over-excavation
超细水泥灌浆　superfine cement grouting
超行程　contacting travel, overtravel
超岩石圈断裂　trans-lithospheric fault
超越概率　exceeding probability
超张拉力　extra design tension
潮料掺浆法喷射混凝土　cement paste wrapping aggregate shotcrete
潮流　load flow
潮流计算　load flow calculation
潮汐电站　tidal power station
潮汐发电　tidal power
潮汐发电开发方式　tidal power generation development mode
潮汐水电站　tidal hydropower station
炒干法　fried dry method
车（船）载无线电话机　mobile radiophone
车道　lane

车库　garage
车轮　wheel
车轮上的载荷　load on wheel
车轮轴距　wheel pitch
车轮组平衡架　balancing stand of wheels
撤防状态　unset condition
撤销合同日期　cancelling date
沉淀池　sedimentation basin
沉管贯入桩　mandrel driven pile
沉积层理构造　sedimentary bedding structure
沉积建造　sedimentary formation
沉积岩　sedimentary rock
沉降　settlement
沉降缝　settlement joint
沉降粒径　settling diameter
沉井基础　open caisson foundation
沉排　mattress
沉沙池　sand trap
沉速　settling velocity
沉头螺栓　sunk bolt
衬砌　lining
衬砌隧洞　lined tunnel
称量斗　weighing hopper
成本回收　Cost Recovery
成本加酬金合同　cost plus award-fee contract，cost reimbursement contract
成本加固定费合同　cost plus fixed-fee contract
成本加运费价　cost and freight（CFR）
成本效果比　cost effectiveness ratio
成本效果分析　cost effectiveness analysis（CEA）
成本效益分析　cost Benefit analysis（CBA）
成品骨料　finished aggregate
成品率　rate of finished products
成土母岩　soil-forming rock
成岩作用　diagenesis
成因分类　genetic classification
成因分析　genetic analysis
成因类型　genetic type
承包商　contractor
承包商代表　contractor's representative，site agent
承包商带资承包合同　contractor-financed contract
承包商监督　contractor's superintendence
承包商人员　contractor's personnel
承包商设备　contractor's equipment
承船车（承船架）　ship carriage，platform
承船厢　ship chamber
承船厢调平装置　chamber leveling device
承船厢干舷高　chamber freeboard
承船厢结构制造对称性偏差　symmetry error of ship chamber
承船厢有效尺度　valid size of ship chamber
承船厢整体挠度　deflection of ship chamber
承诺费　commitment fee
承压板试验　bearing plate test
承压地层　confining stratum
承压水　artesian water，confined water
承压水流　confined flow
承压水头　confined water head，artesian head
承运　carriage
承载能力极限状态　ultimate limit states
承重墙　structural wall，bearing wall
承轴巢　pintle socket
承轴台　pintle shoe
城建档案　urban construction archive
城市给水干管　city main
城市集镇迁建　urban and town relocation
城市集镇新址　new town site
城市扩张　urban expansion
城市消防　urban fire protection
城市总体规划　urban master plan
城镇公共消防设施　public city fire facility
程控交换机　stored program control exchange
程控数字交换机　stored program control digital exchange
吃浆量　grout take
吃浆率　acceptance of grout
迟发雷管（延期雷管）　delay detonator
迟相运行　lagging operation，lagging power factor operation
迟滞　hysteresis
持久状况　persistent situation
持力层　bearing stratum
持续弹性模量　sustained modulus of elasticity
持续命令　persistent command
持住力　holding force
尺寸　dimension，size
尺寸比　length scale ratio
尺寸效应　scale effect
齿槽式挑流鼻坎　slotted spillway bucket
齿轮泵　gear pump

齿轮齿条机构 rack and pinion mechanism
齿轮齿条式 rack and pinion
齿轮传动式 gear driven type
齿轮副侧隙 side clearance of both gears
齿轮增速箱 gear box (speed increaser)
齿盘 toothed disk
齿状混凝土（混凝土塞） dental concrete
赤平极射投影 stereographic projection
赤铁矿 hematite
充电法 Mise-a-la-masse method
充电率 charge rate
充电效率 charge efficiency
充电因数 charge factor
充电终止电压 end-of-charge voltage
充实水柱 full water spout
充水阀 filling valve
充水阀总图 General drawing of filling valve assembly
充水系统 priming system
充填物 infilling，filling
充泄水系统 water filling and sluicing system
充压式水封 pressure-actuated seal
冲沟 gully
冲击波波前放电电压 impulse front discharge voltage
冲击电流 impulse current
冲击电流半峰值时间 time to half value of impulse current
冲击电流波前时间 front time of impulse current
冲击电流发生器 impulse current generator
冲击电流分流器 impulse current shunts
冲击电流试验 impulse current test
冲击电压发生器 impulse voltage generator
冲击电压分压器 impulse voltage divider
冲击反循环钻机 percusive reverse circulation drill
冲击放电电压 impulse sparkover voltage
冲击负荷 impact load
冲击接地电阻 impulse earthing resistance
冲击耐压试验 impulse voltage withstand test
冲击式机组排出高度 static discharge head of impulse turbine
冲击式破碎机 impact crusher
冲击式水轮机 impulse turbine
冲击式消能工 impact-type energy dissipator
冲击钻进 percussion drilling
冲积 alluvium
冲积河槽 alluvial channel
冲积平原 alluvial plain
冲积扇 alluvial fan
冲剪 punching shear
冲角 attack angle
冲坑水垫厚度 water cushion depth of plunge pool
冲坑水垫深度及范围 depth and extent of water cushion at the scour pit
冲坑淘刷 pool scouring
冲沙槽 flushing channel
冲沙流量 sediment flushing flow
冲沙隧洞 sluice tunnel，flushing tunnel
冲沙闸 desilting sluice
冲沙闸门 flush gate
冲刷 scour
冲刷漏斗 scouring funnel
冲洗筛 washing screen
冲泻质（非造床质） wash load
冲抓锥钻进 churn and grabbing drilling
重复利用库容 common storage，shared storage
虫媒病 vector-borne disease
抽出式泵 pull-out type pump
抽检试验 random test
抽水孔 pumping well
抽水试验 pumping test
抽水蓄能电站 pumped storage power station
抽水蓄能机组 pump storage unit
抽水蓄能开发 pumped storage development
抽屉式断路器 withdrawable circuit breaker
抽屉式模板 drawer-type formwork
抽咸换淡 pump out the saline water and recharge the fresh water
抽样调查 sampling investigation
抽样检验 sampling inspection
稠度 consistency
稠度试验 consistency test
稠度状态 consistency state
臭氧层 ozone layer
出厂试验 shop test
出厂验收 factory acceptance
出口标志灯 exit sign luminaire
出口阀 landing valve
出口方向标志 exit direction sign

出口关税　export duty
出口楼梯　exit stairway
出口通道　exit access
出口退税　value added tax refund for exported goods
出口消力戽　outlet bucket
出口总水头（排出扬程）　total discharge head
出库流量　reservoir outflow
出库泥沙　outflowing sediment
出力　output
出力保证率曲线　Output dependability curve
出力系数　coefficient of output
出流堰　effluent weir
出水口闸门　outlet gate
出水渠　outlet channel
出逸坡降　escape gradient
出渣　mucking
出渣线路　mucking route
出中继线　outgoing trunk
初步查明　preliminarily identify
初查　preliminary investigation
初充电　initial charge
初次蓄水　initial impounding，initial filling
初碾　initial rolling
初拧　early screw
初凝　initial set
初期导流　early-stage diversion
初期灭火　initial attack
初期排水　initial dewatering
初期蓄水　initial impounding
初期支护（一期支护）　initial support
初起火　incipient fire
初生空化数　inception number of cavitation
初生空化系数　incipient cavitation coefficient
初生托马数　incipient Thoma number
初始读数　initial reading
初始放电电压　initial discharge voltage
初始冷却　initial cooling
初始燃烧　initial burning
初始（瞬时）沉降　initial settlement，immediate settlement
初始条件　initial condition
初始温度　initial temperature
初始压力　initial pressure
初始压缩曲线　initial compression curve，virgin compression curve
初始应力　initial stress
初始转速　initial speed
初碎机　preliminary crusher
初损　initial loss
初蓄期　initial impound period
除湿剂　dehumidizer
厨房　kitchen
储料场　stockyard（stockpile area）
储能操动机构　stored energy operating mechanism
储能机构　energy storage mechanism
储油柜　oil conservator
触点　contact
触电　electric shock
触发　triggering
触发超前角　trigger advance angle
触发器触发设备　triggering device
触发器件　trigger device
触发延迟角　trigger delay angle
触发因素　triggering factor
触发装置　trigger set
触头　contact
触头的行程　travel of contacts
触头开距　clearance between open contacts
穿杠　tie bolt
穿过钻杆提取　retrieved through the drill rod
穿通　break-through
穿越阻抗　through impedance
传递函数　transfer function
传递系数　transmission coefficient
传动端　drive end
传染性疾病　infectious disease
传热　heat transfer
传热介质　heat-transfer fluid
传热量　capacity of heat transmission
传热阻　resistance of heat transfer
传输安全性　communication security
船舶挤靠力　ship breasting force
船舶系缆力　mooring force
船舶撞击力　ship impact force
船队　ship fleet
船队吨位　shipping tonnage
船厢水平度　chamber levelness
船厢总重　gross weight of ship chamber
船闸　navigation lock，ship lock

船闸充水和泄水　filling and emptying
船闸输水系统　conveyance system of lock
船闸闸门　lock gate
喘振　surging
串灌串排　irrigation and drainage from one field to another
串级工频试验变压器　cascade power frequency testing transformer
串级直流高压发生器　cascade high-voltage DC generator
串浆　grout interconnection (grout leaking)
串励绕组　series winding
串联　series connection
串联补偿　series compensation
串联电抗启动　reactor starting
串联电抗器　series reactor
串联电容补偿装置　series capacitive compensator
串联 PID 调速器　series PID governor
串联结构　series structure
串联式碾压机　tandem roller
串联谐振　series resonance
串联谐振试验装置　series resonant testing equipment
窗地面积比　area ratio of glazing to floor
窗井　window well
窗墙面积比　area ratio of window to wall
窗台　window sill
窗下墙　window spandrel
床面形态　bed form
床沙质（造床质）　bed material load
垂线观测仪　coordinatorgraph for plummet observation
垂直板式蝶阀　vertical disc type butterfly valve
垂直度　perpendicularity
垂直防渗措施　vertical impervious measure
垂直缝　vertical joint
垂直荷载　vertical load
垂直加速度　vertical acceleration
垂直均布压力　vertical uniform pressure
垂直蔓延　vertical extension
垂直排列　vertical arrangement
垂直排水　chimney drain
垂直升船机　vertical ship lift
垂直土压力　vertical earth pressure
垂直钻进　vertical boring
锤式破碎机　hammer crusher
春汛（融雪洪水）　spring flood, snowmelt flood
纯抽水蓄能电站　pure pumped storage power station
纯压式灌浆　non-circulation grouting
瓷绝缘子　porcelain insulator
瓷套管　porcelain bushing
磁饱和　magnetic saturation
磁场　magnetic field
磁场电流变送器　field current transducer
磁场电流设定　field current set point
磁场电压变送器　field voltage transducer
磁场断路器　field breaker
磁场放电电阻　field discharge resistor
磁场强度　magnetic field strength
磁场绕组　field winding
磁传感器　magnetic sensors
磁吹断路器　magnetic blow-out circuit-breaker
磁吹阀式避雷器　magnetic blow-out valve type arrester
磁导率　permeability
磁轭　magnetic yoke
磁粉探伤　magnetic particle testing (MT)
磁感应　magnetic induction
磁化　magnetization
磁化电流　magnetizing current
磁化率　magnetic susceptibility
磁化强度　magnetization intensity
磁化曲线　magnetization curve
磁极　magnetic pole
磁力启动器　magnetic starter
磁路　magnetic circuit
磁屏蔽　magnetic screen
磁体　magnet
磁铁矿　magnetite
磁通量　magnetic flux
磁通密度　magnetic flux density
磁性门开关　magnetic door contact
磁性吸铁装置　magnetic particle unit
磁异常　magnetic anomaly
磁油位计　magnetic oil level gauge
磁致伸缩式传感器　magnetostrictive transducer
磁滞　magnetic hysteresis
磁滞回线　magnetic hysteresis loop
磁滞损耗　magnetic hysteresis loss

磁阻　reluctance
次固结　secondary consolidation
次固结沉降　secondary consolidation settlement
次级方法　secondary method
次块状结构　sub-massive structure，sub-blocky structure
次棱角状　subangular
次梁　secondary beam
次生节理　secondary joint
次生矿物　secondary mineral
次同步过流保护　sub-synchronous overcurrent protection
次要建筑物　secondary structure
次应力　secondary stress
次圆状　subrounded
从属运行　slave operation
凑合节　compensating joint
粗糙的　rough
粗糙度　roughness，asperity
粗骨料　coarse aggregate
粗骨料堆遮荫　shading of coarse aggregate stockpile
粗晶结构　coarsely-crystalline texture
粗砾　coarse gravel
粗砾质土　cobble soil
粗粒结构　coarse grained texture
粗粒径含沙量　coarse sediment concentration
粗面岩　trachyte
粗砂　coarse sand
粗碎机　primary breaker
粗玄岩　dolerite
脆性　brittleness
脆性断裂　brittle fracture
脆性破坏　brittle failure
村民小组　village group
存弃渣场规划　planning and layout for stockpile and spoil area
措施项目费　expense of preliminaries
错车道　passing bay
错动层面　faulted bedding plane
错动劈理，应变滑劈理　strain-slip cleavage
错缝　staggered joint
错缝浇筑　overlapping placing
错台　offset
错位　dislocation，malposition

D

搭接焊　lap welding
搭接接头　lapped splices
达西定律　Darcy's law
达西-韦斯巴赫公式　Darcy-Weisbach equation
打板桩　sheet piling
打管灌浆法　pipe driving grouting method
打印机　printer
打桩　piling
大坝失事后果　consequence of dam failure
大暴雨　heavy rainstorm
大波动　large perturbation
大车　railroad car
大地构造　geotectonics
大地构造单元　geotectonic element
大地构造体系　geotectonic system
大地基准点　geodetic datum
大地控制点　geodetic control point
大地水准面　geoid
大规模农业生产　large-scale agricultural production
大洪水　major flood
大卷盘水带　double donut
大口井　open well
大口径孔［孔径＞0.6m］　large-diameter hole，calyx hole
大口径钻进　large well drilling
大理岩　marble
大梁　girder
大陆　continent
大气过电压　lightning overvoltage
大气污染控制　air pollution control
大气污染物综合排放标准　integrated emission standard of air pollutants
大气压力　atmospheric pressure（ambient pressure）
大扰动　large disturbance
大石　bulky grain
大体积混凝土　mass concrete
大体积混凝土坝　mass concrete dam
大头坝　solid-head buttress dam
大修费　cost of overhaul
大循环方式　major cycling way
大值平均值　average of the higher half values

大轴摆度检测器 shaft run out detector
大轴找正 alignment of shaft
代表性露点 representative dewpoint
代（界） Era（Erathem）
代扣所得税 withholding tax
代理费 agent fee
带补充载荷的安全阀 supplementary loaded safety valve
带不平衡负荷运行 unbalanced loading operation
带布料杆的混凝土泵 concrete pump with distributor
带电显示装置 voltage presence indicating device
带负荷试验 load test
带负荷运行 load operation
带集电环感应电动机 slip-ring induction motor
带肋钢筋 ribbed bar
带励磁失步运行 out-of-step operation with excitation
带滤清器的呼吸器 breather cap with air filter
带式输送机 belt conveyor
带通滤波器 band pass filter
带胸墙的实用堰 practical weir with breast wall
带旋转整流器的交流励磁机 AC exciter with rotating diodes
带状构造 banded structure
带状间作 strip intercropping
带状结构 banded texture
带状淤积 belt deposit
带阻滤波器 band stop filter
袋装扰动土样 disturbed bag sample
担保费用 guarantee cost
担保人 guarantor
单臂伸缩式隔离开关 semi-pantograph disconnector
单导线 single conductor
单电源供电 single supply
单独导叶接力器 individual guide vane servomotor
单独接力器控制 individual servomotor control
单段爆破药量 charge for one interval（charge amount per delay interval）
单费率电价 flat rate tariff
单腹板梁 single girder
单个构件 individual member
单个作用 single action
单荷载系数法 single load factor method
单回路 single circuit line
单回路塔 single circuit steel tower
单级泵 single-stage pump
单级可逆式水轮机 single stage pump-turbine
单极隔离开关 single pole disconnector
单价分析表 Table of unit price analysis
单孔抽水试验 single well pumping test
单库单向开发 single-lagoon one-way tidal power development
单库双向开发 single-lagoon two-way tidal power development
单宽功率 stream energy per unit width
单宽流量 discharge per unit width
单框筒结构 framed tube structure
单立柱 detached column
单梁电动葫芦 monorail hoist，monorail crane
单梁桥架 single-girder bridge
单面焊接 one side welding
单模光纤 single-mode optical fibre
单母线带旁路接线 single-bus with transfer bus configuration（main and transfer bus configuration）
单母线分段接线 sectionalized single-bus configuration
单母线接线 single-bus configuration
单跑式楼梯 staircase with straight flight
单戗立堵 single closure dike plus end-dumping closure
单曲拱坝 single-curvature arch dam
单体电池 cell
单体调校 adjustment and calibration of individual component
单位飞逸转速 unit runaway speed
单位工程 unit works
单位面积产量 yield per unit area
单位千瓦静态投资 static investment per kilowatt
单位千瓦投资 investment per kilowatt
单位水力矩 unit hydraulic torque
单位水推力 unit hydraulic thrust
单位投资 unit investment cost
单位吸水量 specific water absorption
单位造价指标 unit cost index
单位注入量 unit injection rate
单稳态继电器 mono-stable relay
单吸泵 single-suction pump
单相变压器 single phase transformer

单相感应电动机　single phase induction motor
单相接地短路　phase-to-earth fault，single line-to-earth short-cicuit
单相自动重合闸　single phase auto-reclosing
单响最大段起爆药量　maximum primer charge in single shot
单向阀　check valve
单向阀门　unidirectional valve
单向连续板　one-way continuous slab
单向门机　gantry crane with fixed hoist
单向配筋　one-way reinforcement
单项工程　sectional works
单斜构造　uniclinal structure
单心拱　single-centered arch
单芯电缆　single core cable
单液灌浆法　single liquid grouting
单一安全系数法　single safety factor method
单一货币合同　single currency contract
单一目的项目　single-purpose project
单元工程　separated item works
单元式保护　unit protection
单元式住宅　apartment building
单闸板　single gate
单轴抗拉强度　uniaxial tensile strength
单轴抗压强度　uniaxial compressive strength
单柱式隔离开关　single-column disconnector
蛋白石　opal
氮气　nitrogen
氮氧化物　nitrogen oxide（NO_X）
当地材料　local materials
当地采购　local purchase
当地工人　local labor
当地货币　local currency
当地居民参与　engagement of local residents
当量计算排量　equivalent calculated capacity
当前生计　current livelihood
当事人　party concerned
挡热板　heat barrier
挡水坝段　non-overflow section，water retaining section
挡水建筑物　water retaining structure
挡套　interstage sleeve
挡土板　breast board
挡烟板　smoke barrier
档案　archives
档案虫霉除治　pests and mould control in archives
档案虫霉预防　pests and mould prevention in archives
档案工作　archives work
档案馆　archives
档案管理　archives management
档案价值　archival value
档案库房　archival repository
档案人员　archivist
档案实体分类　physical archives classification
档案室　record office
档案收集　acquisition，collection
档案缩微品　archival microform
档案信息分类　archival information classification
档案信息化　archival informationization
档案信息资源　archival information resources
档案行政管理部门　archival administrative department
档案整理　archival arrangement
档案主题词表　archives thesaurus
档号　archival code
档距　span length
导爆索　primacord
导洞　pilot drift tunnel
导洞掘进法　heading and cut method
导管［浇混凝土用］　tremie pipe
导管浇筑　tremie placing
导航建筑物　guide structure
导火索　safety fuse
导井法　pilot bore method
导卡　miter guide
导流　diversion
导流底孔　diversion bottom outlet
导流方式　diversion mode
导流建筑物级别　grade of diversion structure
导流壳体　diffuser casing
导流孔（洞）封堵　plugging of diversion opening (tunnel)
导流流量标准　diversion discharge criterion
导流明渠　diversion channel
导流隧洞　diversion tunnel
导流闸门　diversion gate
导纳　admittance
导墙　guide-wall
导热系数　thermal conductivity

导热性，导热系数　thermal conductivity
导沙丁坎　groin sill
导沙顺坎　silt training sill
导水墙　guide wall
导水率　transmissibility
导水系数　transmissivity
导体　conductor
导体屏蔽　conductor screen
导通方向　conducting direction
导通状态　on state，conducting state
导温系数，散热系数　thermal diffusivity
导线边　traverse leg
导线测量　traverse survey
导线的夜间警告灯　night warning light for conductor
导线点　traverse point
导线对塔净空距离　clearance between conductor and structure
导线排列　conductor configuration
导线振动　conductor vibration
导线最大弧垂　maximum sag of conductor
导向轮　guide roller
导向套　guide sleeve (bush)
导向装置　guiding device
导叶　guide vane
导叶泵　diffuser pump
导叶臂　guide vane lever
导叶端面密封　guide vane end seal
导叶过载保护　guide vane overload protection
导叶角度　guide vane angle
导叶接力器　guide vane servomotor
导叶开度　guide vane opening
导叶空载开度　speed-no-load wicket gate position
导叶力矩　guide vane torque
导叶力特性　guide vane force character
导叶连杆　guide vane link
导叶扭矩脉动　guide vane pulsation
导叶锁定信号器　wicket gate lock detector
导叶位置开关　wicket gate position switch
导叶限位块　guide vane end stop
导叶止推轴承　guide vane thrust bearing
导叶轴　guide vane stem
导叶轴密封　guide vane stem seal
导叶轴套　guide vane bearing
导引阀　pilot valve
导引线保护　pilot wire protection
导轴承　guide bearing
导轴承分块瓦　guide bearing pad
导轴承温度　guide bearing temperature
导轴承油流量　guide bearing oil flow
导轴承油位　guide bearing oil level
导轴瓦　guide bearing shoe (guide bearing segment)
岛弧带　island arc belt
倒虹吸管　inverted siphon
倒流防止器　backflow prevent
倒坡（逆坡）　adverse slope
倒铅垂线　inverse plummet
倒三角排列　delta configuration
倒楔式锚杆　inverted wedge anchor bolt
倒悬　overhang
倒悬度　overhanging degree
倒转背斜　overturned anticline
倒转褶皱　overturned fold
到岸价　cost，insurance and freight (CIF)
到期日　expiry date
到期支付款　payment due
道路照明灯具　luminaire for road lighting
灯具　luminaire
灯具效率　luminarie efficiency
灯泡贯流式机组　bulb tubular unit
灯泡体　bulb
灯泡体支柱　bulb support
等电位连接　equipotential bonding
等高耕作　contour tillage
等高距　contour interval
等高线　contour
等高植物篱　contour living hedgerow
等级公路　classified highway
等价年度费用　equivalent annual cost
等角投影　equal-angle projection
等截面梁　uniform cross-section beam
等截面柱　constant cross-section column
等粒结构　granulitic texture
等面积投影　equal-area projection
等容粒径　volume equivalent diameter
等时水位线　isochrone
等水压线　isopiestic line
等速加荷固结试验　consolidation test under constant loading rate
等外公路　non-graded road
等位机架　equipotential frame

等位面 equipotential surface
等位体 equipotential volume
等位线 equipotential line
等效放热系数 equivalent coefficient of heat evolution
等效均布活荷载 equivalent uniform live load
等效率曲线 iso-efficiency curve
等效网络 equivalent network
等效线性温差 equivalent linear temperature difference
等压面 equi-pressure surface
等应变率固结试验 consolidation test under constant rate of strain
等震线 isoseismals
低标号混凝土 low grade concrete
低层住宅 low-rise apartment
低电压保护 undervoltage protection
低电压穿越 low voltage ride through (LVRT)
低负荷短时运行范围 low-load temporary operation range
低功率保护 underpower protection
低谷负荷 valley load
低合金结构钢 low-alloy structural steel
低励及失磁 underexcitation and loss of excitation
低频反复作用 low-frequency cyclic action
低频过流保护 low frequency overcurrent protection
低频率保护 underfrequency protection
低频振荡 low frequency oscillation
低强度水泥 low-grade (strength) cement
低热微膨胀水泥 low heat expansive cement
低水力比能短时运行范围 low specific hy-draulic energy temporary operating range
低水位报警 low water alarm
低水位泄洪进水口 low-level drawdown intake
低坍落度混凝土 low slump concrete
低碳 low carbon
低碳生活方式 low-carbon lifestyle
低洼地 swale
低洼盐碱地 saline-alkali swale
低温阀门 sub-zero valve
低温送风空气调节系统 cold air distribution system
低压 low voltage (LV)
低压电缆线路 low voltage cable line
低压电器 low voltage apparatus
低压过流保护 undervoltage started over-current protection
低压绕组 low voltage winding
低压缩性 low compressibility
低压脱扣器 undervoltage release
滴水 drip
抵抗力矩 resisting moment
底板隆起 bottom heaving
底阀 foot valve/bottom valve
底环 bottom ring
底槛 bottom sill
底孔闸门 ground gate
底砾岩 basal conglomerate
底流消能 bottom flow energy dissipation
底栖生物 benthonic organism
底涂层 primer coat
底漆 prime paint
底切岸坡 undercut slope
底枢 pintle assembly
底卸车 hopper wagon
地表冻胀量 amount of frost-heaving of ground surface
地表积水 surface ponding
地表加速度峰值 peak ground acceleration (PGA)
地表径流 surface runoff
地表水 surface water
地槽 geosyncline
地电位 earth potential
地盾 shield
地方病 endemic disease
地方税 local tax
地缝合带 suture belt
地基 foundation, subgrade
地基变形容许值 allowable foundation deformation
地基沉降 foundation settlement
地基承载力特征值 characteristic value of foundation bearing capacity
地基反力系数 coefficient of subgrade reaction
地基刚度 stiffness of foundation
地基截水墙 foundation cutoff
地基隆起 foundation upthrow, foundation upheaval
地基土设计冻深 design freezing depth of foundation
地坑筒式泵 pit barrel type pump
地垒 horst
地理信息系统 geographical information system (GIS)

地裂缝　ground crack
地埋管换热系统　ground heat exchanger system
地貌　geomorphy, landform
地貌类型　geomorphic type
地面操纵起重机　floor-operated crane
地面沉降　ground subsidence
地面粗糙度　terrain roughness
地面勘探　ground surface exploration
地面露点　dewpoint at earth surface
地面落雷密度　earth flash density (GFD)
地面排水　surface drainage
地面物探　surface geophysical prospecting
地面下沉　land subsidence
地面站　ground station
地堑　graben
地球站　earth station
地区津贴　district allowance
地热能交换系统　geothermal exchange system
地热异常　geothermal anomaly
地上式消防栓　stand post hydrant
地台　continental platform
地台型建造　platform type formation
地图分幅　sheet line system
地物　surface feature
地下洞室　underground cavern
地下径流　groundwater runoff
地下开挖　underground excavation
地下连续墙　underground diaphragm wall
地下排水　subsurface drainage
地下热水　hot groundwater
地下式厂房　underground powerhouse
地下式水电站　underground hydropower station
地下式消防栓　sunk hydrant
地下室　basement
地下室防水　basement waterproofing
地下水　ground water
地下水侧向补给　recharge by ground water
地下水超量开采　excessive exhaustion of ground water
地下水动态　ground water regime
地下水降深　drawdown of ground water
地下水开采　ground water exhaustion
地下水矿化度　mineralization of ground water
地下水排泄区　discharge area of ground water
地下水人工补给　artificial recharge of groundwater
地下水条件评分　rating of ground water conditions
地下水位坡降　gradient of water table
地下水下降漏斗　exhaustion cone of groundwater
地下水位　ground water table
地下水越层补给　recharge through weak permeable layer
地下水资源　ground water resources
地下自然资源　subterranean natural resource
地线支架　earth wire peak
地形　topography
地形测量　topographic survey
地形坡折　terrain slope break
地形制约　topographical constrain
地应力　geostatic stress
地应力测试　geostress measurement
地应力场　ground stress field, geostatic stress field
地源热泵系统　ground-source heat pump system
地震安全性评价　seismic safety evaluation
地震波　seismic wave
地震带　seismic belt
地震的方向分量　earthquake directional components
地震地质灾害　earthquake induced geological disaster
地震动　ground motion
地震动参数　ground motion parameter
地震动反应谱特征周期　characteristic period of the seismic response spectrum
地震动峰值加速度　ground motion peak acceleration
地震动峰值加速度　seismic peak ground acceleration
地震动水压力　seismic water pressure
地震动土压力　seismic earth pressure
地震断层　earthquake fault
地震反应谱　seismic response spectrum
地震构造　seismic structure
地震构造区　seismic structure zone
地震观测　seismological observation
地震惯性力　seismic inertia force
地震活动　seismic activity
地震活动断层　seismo-active fault
地震加速度　seismic acceleration
地震烈度　seismic intensity, earthquake intensity
地震前兆　premonitory symptom

地震区　seismic area，earthquake area
地震台　seismostation
地震台网　seismic network
地震危险性分析　seismic hazard analysis
地震液化　earthquake-induced liquefaction
地震仪　seismograph，seismometer
地震震级　earthquake magnitude
地震状况　seismic situation
地震作用　seismic action
地震作用的效应折减系数　reduction coefficient of seismic action effect
地震作用效应　seismic action effect
地质档案　geological archives
地质点测量　survey of geological observation point
地质公园　geologic park
地质构造　geological structure
地质雷达　ground penetrating radar（GPR ）
地质力学模型　geo-mechanical model
地质年代　geological time scale，geological age
地质填图　geological mapping
地质灾害　geological disaster
地质灾害调查　geologic hazards inquiry
递进式集中润滑　progressive centralization lubrication
第二循环系统　hot water circulation system
第三方责任险　third party insurance
第三纪（系）　Tertiary period（system）
第四纪（系）　Quaternary period（system）
第四系沉积物　Quaternary sediment（deposit）
第一临界转速　first critical speed
典型暴雨　typical rainstorm
点焊　spot welding
点荷载强度　point load strength
点荷载试验　point-loading test
点积　dot product
点群中心法　point group center method
点污染源　point source pollution
点型火灾探测器　spot-type fire detector
点雨量　point precipitation
电　electricity
电操作控制元件　electrically actuated control element
电测深法　electrical sounding
电厂控制级　plant control level
电场　electric field
电场测量探头　electric-field probe
电场强度　electric field intensity（strength）
电池　battery
电池槽　case
电池充电　charging of a battery
电池放电　discharge
电池封口剂　lid sealing compound
电池盖　cell lid
电池外壳　cell can
电池组架　battery rack
电磁波　electromagnetic wave
电磁波测距　electromagnetic distance measurement
电磁场　electromagnetic field
电磁发射　electromagnetic emission
电磁阀　solenoid valve
电磁辐射　electromagnetic radiation
电磁辐射污染控制　electromagnetic radiation pollution control
电磁负荷　electromagnetic load
电磁干扰　electromagnetic interference（EMI）
电磁感应　electromagnetic induction
电磁功率　electromagnetic power
电磁换向阀　solenoid operated directional valve
电磁继电器　electromagnetic relay
电磁兼容（电磁兼容性）　electromagnetic compatibility（EMC）
电磁力　electromagnetic force
电磁流量计　electromagnetic flow meter
电磁能　electromagnetic energy
电磁屏蔽　electromagnetic screen
电磁驱动装置　electromagnetic actuator
电磁式电压互感器　inductive voltage transformer
电磁锁　electromagnetic lock
电磁体　electromagnet
电磁脱扣器　magnetic release
电磁制动　electromagnetic braking
电磁制动转矩　electromagnetic braking torque
电导　conductance
电导率　conductivity
电动阀门　electrical operated valve
电动葫芦　electric hoist
电动机　electric motor
电动机操动机构　motor operating mechanism
电动机调速器　governor with motor driven gate operator
电动机工况　motor mode

电动机控制中心　motor control centers (MCC)
电动起重机　electric crane
电动势　electromotive force (e. m. f)
电动装置　electric actuator
电法勘探　electrical prospecting
电腐蚀　electro-erosion
电感　inductance
电感器　inductor
电感式传感器　inductive transducer
电感镇流器　inductive ballast，magnetic ballast
电工实验室　electric test room
电-光效应　electro-optic effect
电焊机　welder
电焊条　electrode
电荷　electric charge
电弧电流零区　arc current zero period
电弧电压　arc voltage
电弧放电　arc discharge
电化学腐蚀　electrochemical corrosion
电话谐波因数　telephone harmonic factor (THF)
电击　electric shock
电击电流　shock current
电机同步运行　synchronous operation of a machine
电机转换器　electro-mechanical converter
电极　electrode
电极式水位开关　electrolyte level switch
电接触　electric contact
电解质　electrolyte
电解质式测斜仪　electrolysis inclinometer
电介质　dielectric
电抗　reactance
电抗器　reactor
电控锁　electric strikes
电缆　cable
电缆导管　cable ducts
电缆分线箱　cable branch box
电缆敷设　cabling
电缆沟　cable trough，cable ditch
电缆故障定位　cable fault locating
电缆管道　cable trunking，cable duct
电缆夹层　cable spreading room
电缆架　cable rack
电缆架层间垂直净距　vertical clearance between cable racks
电缆交接试验　acceptance test of cable
电缆接头　cable joint
电缆卷筒　cable drum
电缆廊道　cable gallery
电缆路由　cable routing
电缆排管　cable duct bank
电缆桥架　cable tray
电缆式电流互感器　cable type current transformer
电缆试验　testing of power cable
电缆竖井　cable shaft
电缆隧道　cable tunnel
电缆托架　cable tray
电缆线路路径选择　cable route selection
电缆线路巡视检查　cable line inspection
电缆选择　cable selection，cable sizing
电缆支架　cable bearer
电缆终端头　cable terminal
电缆终端制作　preparation of cable terminal
电雷管　electric blasting cap
电力变压器　power transformer
电力电缆间水平净距　horizontal clearance between power cables
电力电量平衡　balance of power and energy
电力电容器　power capacitor
电力电子电容器　power electronic capacitor
电力电子技术　power electronics
电力电子开关　electronic power switch
电力电子设备　electronic power equipment
电力调度　power dispatching
电力负荷曲线　power load curve
电力购买协议　power purchasing agreement (PPA)
电力平衡　power balance
电力系统备用容量　reserve capacity of power system
电力系统调差系数　droop of power system
电力系统调度管理　dispatching management of power system
电力系统调度信息　information for power system dispatching
电力系统调度自动化　automation of power system dispatching
电力系统调峰　peak load regulating of power system
电力系统动态备用容量　spinning reserve of power system
电力系统动态稳定性　dynamic stability of power system

电力系统分层控制　hierarchical control of power system
电力系统负荷曲线　load curve of power system
电力系统负荷预测　load forecast of power system
电力系统功率调节特性　power characteristics of power system
电力系统故障　power system fault
电力系统经济调度　economic dispatching of power system
电力系统静态稳定性　steady state stability of power system
电力系统容量　installed capacity of power system
电力系统事故　power system accident
电力系统瓦解　power system collapse
电力系统稳定器　power system stabilizer (PSS)
电力系统无功功率平衡　reactive power balance of power system
电力系统异常　power system abnormality
电力系统异常运行　abnormal operation of power system
电力系统有功功率平衡　active power balance of power system
电力系统元件　power system element
电力系统远动技术　telecontrol technique for power system
电力系统运行　power system operation
电力系统暂态稳定性　transient stability of power system
电力系统振荡　power system oscillation
电力系统自动装置　power system automatic control device
电力线载波　power line carrier (PLC)
电力线载波纵联保护系统　power line carrier pilot protection system
电量　electrical energy
电量累积曲线　Cumulative power curve
电量平衡　energy balance
电铃　electric bell
电流　electric current
电流保护　current protection
电流变送器　current transducer
电流表　ammeter
电流差动式纵联保护　current differential protection
电流互感器　current transformer
电流回路　current circuit
电流继电器　current relay
电流零点　current zero
电流速断保护　instantaneous overcurrent protection
电流误差　current error
电流谐振　current resonance
电流型交流/直流变流器　current stiff AC/DC converter
电流引入回路　current injection circuit
电流元件　current component
电流源逆变器　current source inverter (current fed inverter)
电路　electric circuit
电路角　circuit angle
电路近后备保护　circuit local backup protection
电路图　circuit diagram
电路元件　circuit element
电纳　susceptance
电能表　energy meters
电能损耗　energy loss
电能质量　power quality
电剖面法　electrical profiling
电气安全　electrical safety
电气安全措施　electrical safety measure
电气保护屏蔽体　electrically protective screen
电气保护外壳　electrically protective enclosure
电气保护遮栏　electrically protective barrier
电气防火　electrical fire prevention
电气防火间距　electrical interval of fire prevention
电气干扰源　electrical interference source
电气故障　electrical fault (failure)
电气缓冲单元　electrical damper module
电气间隙　clearance
电气距离　electrical distance
电气开度限制单元　electrical opening limiting module
电气盘柜安装　erection of electric panel and cabinet
电气设备干燥　drying of electric equipment
电气石　tourmaline
电气事故　electrical accident
电气事故报警　electrical fault alarming
电气消防通道　fire fighting access to electrical equipment
电容　capacitance

电容分压器　capacitive voltage divider
电容器　capacitor，condenser
电容器成套装置　capacitor installation
电容式传感器　capacitive transducer
电容式电压互感器　capacitive voltage transformer
电容型套管　condenser bushing
电势　electric potential
电势差（电位差）　electric potential difference
电寿命试验　electrical endurance test
电枢短路时间常数　armature short-circuit time constant
电枢反应　armature reaction
电枢绕组　armature winding
电刷　brush
电刷磨损　wear of brush
电、水、风预算价格　estimated price of electricity, water and compressed air supply
电梯机房　elevator machine room
电梯厅　elevator hall
电网　electric power grid
电网电价　on-grid tariff
电网负荷特性　network load characteristic
电网负载特性系数　network load characteristic coefficient
电网结构　network structure，network configuration
电网解列　islanding，network splitting
电位　electric potential
电位器式传感器　potentiometric transducer
电压　voltage
电压保护　voltage protection
电压变送器　voltage transducer
电压表　voltmeter
电压波动　voltage fluctuation
电压不稳定　voltage instability
电压等级　voltage level
电压调节器　voltage regulator
电压调整　voltage regulation
电压和频率的偏离值　voltage and frequency variations
电压互感器　voltage transformer
电压恢复　voltage recovery
电压回路　voltage circuit
电压继电器　voltage relay
电压降　voltage drop，potential drop
电压控制节点　voltage controlled bus
电压偏差　deviation of voltage
电压稳定性　voltage stability
电压-无功控制　voltage-var control
电压误差　voltage error
电压消失　loss of voltage
电压谐振　voltage resonance
电压形成回路　voltage forming circuit
电压型交流/直流变流器　voltage stiff AC/DC converter
电压引入回路　voltage injection circuit
电压元件　voltage component
电压源逆变器　voltage source inverter (voltage fed inverter)
电压暂降　voltage dip，voltage sag
电压质量　voltage quality
电液比例阀　electro-hydraulic proportional valve
电液比例换向阀　electro-hydraulic proportional directional valve
电液调速器　electric-hydraulic governor
电液换向阀　electro-hydraulic directional control valve
电液伺服阀　electro-hydraulic servo-valve
电液转换器　electro-hydraulic converter
电影院　cinema
电涌保护器　surge protective device (SPD)
电源电压　source voltage
电源模件　power supply module
电源设备　supply equipment
电源消失　power supply failure
电源组成　constitution of power sources
电晕防护　corona shielding
电晕放电　corona discharge
电晕干扰　corona interference
电晕损失　corona loss
电晕效应　corona effect
电站空化系数　plant cavitation coefficient
电站输水道　plant waterway
电站托马数　plant Thoma number
电站吸出高度　static suction head
电照明　electrical lighting
电制动　electric braking
电制动　electrical braking
电制动转矩　electrical braking torque
电子档案　electronic archives

电子阀器件　electronic valve device
电子阀器件的门槛电压　threshold voltage of an electronic valve device
电子负荷调节器　electronic load controller
电子器件　electronic device
电子式互感器　electronic instrument transformer
电子文件　electronic record
电子镇流器　electronic ballast
电阻　resistance
电阻率　resistivity
电阻器　resistor
电阻式分压器　resistive divider
电阻式温度计　resistance thermometer
垫层料　cushion material
垫层区　cushion zone
垫块，替打　cushion block
垫圈　washer
垫座　concrete socket
吊车轮压　crane wheel load
吊车起吊荷重　crane capacity load
吊车竖向冲击荷载　crane vertical impact load
吊车水平刹车力　crane thrust
吊点距　distance between lifting eyes
吊顶　suspended ceiling
吊耳　lifting eye (lifting hook)
吊钩　hook
吊钩滑轮组　hook assembly
吊钩极限位置　hook approach
吊钩起重机　hook crane
吊罐　raising cage
吊桥　suspension bridge
吊物孔　hatch
调度电话分机　dispatching telephone subset
调度电话主机　dispatching telephone control main station
调度规程　dispatching regulation
调度命令　dispatching command
掉块　fallouts
掉钻　rod drop
跌落式熔断器　drop-out fuse
跌水　hydraulic drop
跌水池　plunge pool
跌水建筑物　drop structure
跌水消能工　drop energy dissipator
叠合梁　superposed beam
叠合式受弯构件　composite flexural member
叠加定理　superposition theorem
叠梁闸门　stoplog gate
叠绕组　lap winding
叠瓦断层　imbricate fault
叠瓦构造　decken structure
叠压供水　additive pressure water supply
蝶阀　butterfly valve
蝶式止回阀　butterfly swing check valve
丁坝　spur dike，groin
丁腈橡胶　acrylo nitrile butadiene rubber (NBR)
顶部滤油阀　top filter valve
顶部溢流式溢洪道　overflow spillway
顶盖　head cover
顶管法　pipe jacking method
顶紧装置　push device
顶破强度　burst strength
顶枢　top anchorage assembly
定点定面关系［暴雨］　precipitation relationship between fixed point and fixed area
定滑轮　upper pulley，sheave
定居区　settled zone
定量泵　fixed displacement pump
定轮　fixed wheel
定轮闸门　fixed wheel gate
定喷　directional jet grouting
定期保养　periodical maintenance
定期冲洗式沉沙池　intermittent scouring sand basin
定期付款　due payment
定期深入检查　periodic inspections
定时打印　periodic logging
定时热水供应系统　fixed time hot water supply system
定时限保护　definite time protection，specified time protection
定时限继电器　specified time relay，definite time relay
定时限脱扣器　definite time-delay overcurrent release
定位桩（导桩）　guide pile
定线　alignment
定向爆破　directional blasting
定向岩芯　oriented core
定向钻进　oriented boring
定型组合式模板　shaped composite form
定圆心拱坝　constant-center arch dam

定值表　setting list
定值设计法　deterministic method
定值校验　setting verify
定中心角拱坝　constant-angle arch dam
定子端部绕组绝缘　stator end winding insulation
定子机座　stator frame
定子机座底板　soleplate of the stator frame
定子接地保护　stator earth fault protection
定子绕组　stator winding
定子绕组电阻　stator winding resistance
定子绕组温度　stator winding temperature
定子绕组匝间短路　stator winding inter-turn short circuit
定子绕组直流耐压试验　stator winding DC voltage withstand test
定子铁芯　stator core
定子铁芯的损耗发热试验　stator core loss and temperature rise test
定子线棒　stator coil bar
定子线棒接头开焊　welded joint breaks of stator bar
定子线圈　stator coil
东道国税费　tax and fee of host country
冬季雨季施工增加费　additional cost of work in winter and rainy season
动冰压力　dynamic ice pressure
动触头　moving contact
动点动面关系［暴雨］　precipitation relationship between center point and variable area
动断触点的闭合时间　closing time of a break contact
动断触头（点）　break contact
动合触点的闭合时间　closing time of a make contact
动合触头（点）　make contact
动滑轮组　movable block
动库容　dynamical storage
动力触探试验　dynamic cone penetration test
动力法　dynamic method
动力黏度　dynamic viscosity
动力黏性系数　coefficient of dynamic viscosity
动力燃料费　cost of power fuel
动力系数　dynamic coefficient
动量矩　moment of momentum
动能经济指标　energy economic indicator
动平床　moving flat bed
动三轴试验　dynamic triaxial test
动水关闭试验　closing test in dynamic water
动水启闭　closing and opening in flow
动水启动试验　starting test in dynamic water
动水位　dynamic water level
动水小开度提门充水　water filling by gate opening in gap
动水压强　hydrodynamic pressure
动态调平　leveling in dynamic state
动态负荷特性　dynamic characteristics of load
动态人口　dynamic population
动态设计法　method of information design
动态系统品质　dynamic system behavior
动态作用　dynamic action
动物迁徙路线　animal migration route
动物群　fauna
动物诱捕　animal entrapment
动载试验　dynamic test
动作时间　operate time
动作时限　operation time limit
动作性能及排量试验　operational characteristics and flow capacity testing
动作值　operating value
冻害　frost damage
冻结指数　freezing index
冻融侵蚀　freeze-thaw erosion
冻融试验　freezing and thawing test
冻融循环　freezing and thawing cycle
冻土　frozen soil
冻胀力　frost heaving force（pressure）
冻胀量　amount of frost-heaving
栋梁　ridge beam
洞顶　crown，roof
洞脸塌方　portal collapse
洞内消能　inside-tunnel energy dissipation
洞室群　cavern complex，underground complex
洞周地质测绘　peripheral geologic mapping
洞周应力状态　stress conditions around openings
洞室编录　tunnel mapping（log）
洞探　adit exploration
斗式挖掘机　bucket excavator
斗式装料机　hopper loader
陡槽消力墩　chute block
陡坡　steep slope，abrupt slope
豆砾，细砾　pea stone，pebble

毒害　toxic hazard
毒害危险性　toxic risk
毒物　toxicant
毒物浓度　toxicant concentration
毒效　toxic potency
独立的外部同行评审　independent external peer review
独立的政府评估　independent government estimate
独立第三方　independent third party
独立核算　independent accounting
独立式进水口结构　free-standing intake structure
独立坐标系　independent coordinate
读卡器　card reader
度汛　flood protection in flood season
度汛方案　flood protection scheme
渡槽　aqueduct，flume
渡槽导流　aqueduct diversion
渡口　ferry
镀铬活塞杆　chromeplate steel rod
镀镍处理　nickel plating
镀锌钝化　galvanized and passivation
镀锌钢丝绳　drawn galvanized rope wire
镀锌管　galvanized pipe
端承桩　point bearing pile
端电池　end cell
端盖　end cover
端口　port
端梁　end carriage，end truck
端面跳动　face runout
端头锚固型锚杆　anchor bolt anchored at head
端子　terminal
端子排　terminal block
端子箱　terminal box，marshalling kiosk
短立管　sprig-up
短路比　short-circuit ratio
短路点电流　current at the short-circuit point
短路电流　short-circuit current
短路电流的热效应　heat effect of short-circuit current
短路电流非周期分量　aperiodic component of short-circuit current
短路电流峰值　peak short-circuit current
短路电流交流分量　AC component of short-circuit current
短路电流允许值　short-circuit current capability
短路电流直流分量　DC component of short-circuit current
短路电流周期分量　periodic component of shortcircuit current
短路故障　short-circuit fault（shunt fault）
短路关合能力　short-circuit making capacity
短路计算　short-circuit calculation
短路开断能力　short-circuit breaking capacity
短路容量　short-circuit capacity
短路特性　short-circuit characteristics
短路装置　short-circuiting device
短路阻抗　short-circuit impedance
短期调节水电站　pondage power station
短期水文预报　short-term hydrological forecast
短期文献　short-term document
短期稳定性　short-term stability
短时电压升高试验　short time voltage rising test
短时工作制　short-time duty
短时过电流试验　short time overcurrent test
短时耐受电流　short-time withstand current
短引线保护　stub protection
短暂状况　transient situation
短轴褶皱　brachy-axis fold
断层　fault
断层擦沟　fault striae
断层擦痕　slickenside，fault striation
断层活动段　active fault segment
断层交会带　fault intersection zone
断层泥　fault gouge，fault clay
断层破碎带　fault fractured zone
断层破碎带处理　treatment of fault and fracture zone
断层三角面　fault triangular facet
断层崖　fault scarp
断层影响带　fault influenced zone
断口　fracture
断块山地　fault-block mountain
断裂槽谷　fault trough valley
断裂活动强度　intensity of faulting activity
SF_6 断路器　SF_6 circuit-breaker（sulfur hexafluoride circuit-breaker）
断路器　circuit-breaker
断路器合闸电阻　closing resistor of circuit-breaker
断路器恢复电压　recovery voltage of circuit-breaker

断路器失灵保护 circuit-breaker failure protection
断路器首开极因数 first-pole-to-clear factor of circuit-breaker
断路器瞬态恢复电压 transient recovery voltage of circuit-breaker（TRV）
断面测量 section survey
断面水工模型试验 sectional hydraulic model test
断态 off state，blocking state
断陷盆地 faulted basin
断续周期工作制 intermittent periodic duty
锻件 steel forging
堆焊 overlay welding
堆积 accumulation
堆积阶地 constructional terrace
堆积密度 bulk density
堆积平原 accumulation plain
堆积体 accumulation body
堆料场（堆渣场） dumping site
堆料机 stacker
堆取料机 stacker reclaimer
堆石 rock filling
堆石坝 rockfill dam
堆石坝体 embankment
堆石护坡 rockfill slope protection
堆石料的孔隙率 porosity of rockfill material
堆石区 rockfill zone
对比胶砂 reference mortar
对称度 symmetry
对称短路电流 symmetrical short-circuit current
对称短路视在功率 symmetrical short-circuit apparent power
对称分量法 method of symmetrical component
对称配光型灯具 symmetrical luminaire
对称三相电路 symmetric three phase circuit
对穿锚索 hole-through tendon
对地间隙 clearance to earth
对地净距 earth clearance
对地泄漏电流 earth leakage current
对夹式蝶阀 wafer type butterfly valve
对接焊 butt welding
对接接头 butt splices
对接锁定装置 locking device for connection
对接装置 connection device
对开挡环 split ring
对开法兰 split flange
对开式多叶阀 opposed-blade damper
对流采暖 convection heating
对流传热系数 convective heat transfer coefficient
对数螺线型拱坝 logarithmic spiral arch dam
对数正态分布 logarithmic normal distribution
对外交通 site access
对外交通专用公路 dedicated road to site，external traffic highway
对外通信 communication external to plant
对障碍物的净距 clearance to obstacles
盾构法 shielding tunneling method（shield method）
多瓣抓斗式挖掘机 orange peel excavator
多臂钻车 multiple boom jumbo
多边形连接 polygon connection
多层缠绕 spooling of a wire rope in multilayer
多层建筑物 multistory building
多层进水口（分层取水口） multi-level intake
多层住宅 multi-stories apartment
多叉河道 braided channel
多次重复作用 repeated action/cyclic action
多点位移计 multipoint displacement meter
多段式距离保护 multi-stage（multi-zone）distance protection
多扉闸门 multi-leaf gate
多费率电度表 multi-rate energy meter
多功能厅 multi-functional hall
多回路 multiple circuit line
多回转驱动装置 multi-turn actuator
多级泵 multi-stage pump
多级泵导叶 conveyor vane
多级泵中段 conveyor case
多级船闸 flight lock，multi-stage lock
多级可逆式水轮机 multi-stage pump-turbine
多极开关 multipole switch
多孔抽水试验 multiple well pumping test
多孔结构 hiatal texture
多跨单向板 one-way slab with multispans
多联机空调系统 multi-connected split air conditioning system
多路复用设备 multiplexing unit
多路设备 multiplexing equipment
多模光纤 multi-mode optical fibre
多年调节 overyear regulation，carryover regulation
多年平均含沙量 average annual solid content

多年平均流量　long-period average flow
多年平均年发电量　average annual energy output
多年平均年径流量　mean annual runoff
多梯段楼梯　staircase with several flights
多线船闸　multi-line (multiple) lock
多相网络的不平衡状态　unbalanced state of a polyphase network
多相制　multiphase system，polyphase system
多芯电缆　multicore cable
多旋回性　multiple cyclicities
多遇地震　frequently occurred earthquake，low-level earthquake
多元件变送器　multi-element transducer
多转速电动机　multispeed motor
躲过外部故障　remain stable for an external fault
惰化　inerting
惰性气体　inert gas
惰性气体电弧焊　inert gas arc welding

E

额定电压比　rated voltage ratio
额定电压因数　rated voltage factor
额定分接　principal tapping
额定荷载　normal load，primary load
额定开启高度　rated lift
额定励磁电流　rated field current
额定励磁电压　rated field voltage
额定耐火时间　rated fire-resistance period
额定排量　certified discharge capacity
额定起重量　rated capacity
额定起重量限制器　rated capacity limiter
额定起重量指示器　rated capacity indicator
额定容量　rated capacity
额定水头　rated head
额定压力　rated pressure
额定转速　rated speed
额定准确限值一次电流　rated accuracy limit primary current
颚式破碎机　jaw crusher
鲕状灰岩　oolitic limestone
二长岩　monzonite
二次搬运费　double-handling freight
二次调频　secondary control of active power in a system
二次回路　secondary circuit
二次破碎机　secondary crusher
二次绕组　secondary winding
二次筛分　finish screening
二次系统　secondary system
二次谐波　second harmonic component
二次谐波制动　second harmonic restrain
二道坝　auxiliary weir
二叠纪（系）　Permian period (system)
二级配骨料　aggregate with maximum grain size of 40mm
二期灌浆孔　secondary grout hole
二期混凝土　phase Ⅱ concrete
二期面板　second-stage facing
二期支护　secondary support
二维模型（平面模型）　two-dimensional model，plane model
二氧化硫　sulfur dioxide (SO_2)
二氧化碳　carbon dioxide (CO_2)

F

发电　power generation
发电厂　power plant (power station)
发电厂出力　output of power plant
发电厂接入系统　generation interconnection
发电厂送出工程　generation interconnection project
发电电动机　motor-generator
发电机保护系统　generator protection system
发电机-变压器单元保护　generator-transformer unit protection
发电机层　generator floor
发电机磁场温度　generator field temperature
电气预防性试验　electrical preventive test for generator
发电机电压设定　generator voltage set point
发电机风罩　generatorpit
发电机负载调节系数　generator load self-regulation coefficient
发电机工况　generator mode (generating mode)
发电机故障　generator fault
发电机进人孔　generator access hatch
发电机空气间隙测量　measuring of air gap between generator stator and rotor
发电机空气制动器　generator air brake

发电机冷却　generator cooling
发电机冷却系统试验　generator cooling system test
发电机润滑油系统　generator lubrication oil system
发电机特殊运行　special operation of generator
发电机效率　generator efficiency
发电机异常运行　abnormal operation of generator
发电机轴承油冷却器　generator bearing oil coolers
发电机组二次功率调节　secondary power control operation of a generating set
发电计划　generation schedule
发电量　energy output
发电隧洞　power tunnel
发电系统　power generation system
发电效益　power generation benefit
发光二极管灯　light emitting diode lamp，LED lamp
发光强度　luminous intensity
发热率　rate of heat release
发散性振荡　divergent oscillation
发射机　transmitter
发育　developed
发展目标　development objective
发展性故障　developing fault
发震断层　seismogenic fault
发震构造　seismogenic structure，seismogenic tectonics
发震机制　seismogenic mechanism
乏　var
罚款　penalty
阀瓣　valve disc
阀臂　valve arm
阀操作机构　valve-actuating mechanism
阀电压降　valve voltage drop
阀盖　valve bonnet
阀杆　stem，spindle
阀杆螺母　yoke nut
阀控式铅酸蓄电池　valve regulated lead acid battery (VRLA)
阀器件闭锁　valve device blocking
阀器件熄断　valve device quenching
阀式避雷器　valve type arrester
阀体　valve body
阀芯行程　valve spool stroke
阀座　valve seat
筏基础　raft footing
法拉　Farad
法拉第定律　Faraday law
法拉第效应　Faraday effect
翻板坝　shutter dam
翻板闸门　hinged crest gate，wicket gate
翻斗铲　rocker shovel
翻斗式装料机　skip loader
翻浆　frost boiling
翻模固坡　turning-over formwork and consolidating slope
翻转模板　turnover form
繁殖场　breeding farm
反铲　backhoe shovel
反冲盘　disc holder
反导叶　return ring vane
反电动势　counter-electromotive force，back electromotive force
反调节　re-regulation
反分析　back analysis
反复荷载　repetitive load
反轨　converse guide
反弧段水流离心力　flow centrifugal force on reverse curve section
反弧面溢洪道　ogee spillway
反击率　risk of flashback
反击式水轮机　reaction turbine
反井钻进法　raise boring method
反馈控制　feedback control
反馈设计　feedback design
反馈通道　feedback path
反馈信号　feedback signal
反馈信息　feedback information
反滤层　filter
反滤料　filter material
反滤料区　filter zone
反滤排水沟　filter drainage ditch
反滤压坡　ballasting filter
反滤准则　filter criteria
反坡　reverse slope
反坡梯田　back-slope terrace
反射比　reflectance
反射波　reflected wave
反射率　reflectivity
反射系数　reflection coefficient

反射眩光 glare by reflection
反时限保护 inverse time protection
反时限继电器 inverse time relay
反时限脱扣器 inverse time-delay overcurrent release
反时限最小定时限继电器 inverse definite minimum time relay (IDMT)
反事故措施 accident countermeasures
反相 in opposition
反向并联晶闸管 antiparallel connected thyristors
反向弧形闸门 reverse tainter gate
反向击穿 reverse breakdown
反向坡 adverse slope，reversal slope
反向推力轴承 counter thrust bearing
反向阻断阀器件 reverse blocking valve device
反向阻断状态 reverse blocking state
反循环钻机 reverse rotary rig
反循环钻进 reverse circulation drilling
反应谱特征周期 eigenperiod of response spectrum
反转式搅拌机 reversing drum mixer
返工 rework
返回时间 reset time
返回系数 disengaging ratio
返回信息 feedback information
返回值 disengaging value
返霜 cement cream
返水颜色 color of the return water
返修 repair
泛光灯 flood lamp
泛浆 bleeding
泛水 flashing
方案比选 comparison and selection of alternatives
方波电压发生器 square wave voltage generator
方差 variance
方解石 calcite
方解石充填 calcite infilled
方解石脉 calcite vein
方均根值（有效值） root-mean-square value (rms value)
方木 sawn timber
方位角 azimuth angle
方向保护 directional protection
方向比较式保护 direction comparison protection
方向电流保护 directional current protection
方向继电器 directional relay
方向角 direction angle
方向距离保护 directional distance protection
方向元件 directional component
防爆 explosion proofing
防爆灯具 luminaire for explosive atmosphere
防冰措施 preventive measure against icing
防冰设施 anti-icing device
防拆报警 tamper alarm
防拆装置 tamper device
防潮层 damp-proof course
防尘 dust-tight，dust control
防尘灯具 dust-proof luminaire
防尘圈 wiper seal
防冲槽 scour-resistant slot
防冲措施 erosion control measure
防盗报警系统 burglar alarm system
防盗门 safety door
防盗探测器 anti-theft detector
防电晕层 anti-corona coating
防冻剂 anti-freezing agent
防风固沙 windbreak and sand-shifting control
防风固沙林 sand-shifting control forest
防腐 corrosion prevention
防腐漆 anticorrosive paint
防腐蚀措施 corrosion proof measure
防腐蚀涂料 anticorrosive coating
防寒混凝土拌和厂 winterized concrete plant
防洪 flood control
防洪标准 flood control standard
防洪措施 Flood control measures
防洪堤 flood dyke
防洪对象 protected objects against flood
防洪高水位 top level of flood control
防洪规划 flood control planning
防洪库容 flood control storage
防洪墙 flood wall
防洪水库 flood control reservoir
防洪限制水位 low limit level during flood season
防洪限制线 guide curve for flood control
防洪效益 flood control benefit
防洪闸门 flood gate
防洪专用库容上限 top of exclusive flood control capacity
防护的破坏时间 failure time of protection
防护等级 degree of protection
防护工程 protection works

防护冷却水幕　drencher for cooling protection
防护林　shelter forest
防火　fire prevention
防火窗　fire resisting window
防火堤　fire dike，fire bund
防火阀　fire resisting damper
防火分隔水幕　water curtain for fire compartment
防火分区　fire compartment
防火风管　fire-proof duct
防火间隔　fire break
防火间距　fire separation distance
防火卷帘　fire resisting shutter
防火绝缘　fire resistance insulation
防火门　fire door
防火密封　fire seal
防火幕　fire curtain
防火屏障　fire stop
防火漆　fire retardant paint
防火墙　fire wall
防火完整性　fire integrity
防静电接地　static electricity protection earthing
防空蚀措施　cavitation prevention measure
防空蚀设计　cavitation prevention design
防浪墙　wave wall
防凌　ice control
防弃水线　guide curve for reducing abandoned water
防沙措施　sediment control measure
防渗层　impervious layer
防渗底层　watertight bottom layer
防渗和止水系统　seepage control and water stop system
防渗料　impervious material
防渗面层　watertight surface layer
防渗铺盖　impervious blanket
防渗墙　cutoff wall，watertight diaphragm
防渗，渗流控制　seepage control，seepage prevention
防渗帷幕　impervious curtain
防渗心墙　impervious core
防水层　sealing layer，waterproof layer
防水灯　water proof lamp，under water lamp
防水卷材　waterproof membrane
防水密封胶　waterproof sealant
防水（耐水）　waterproof
防水漆　waterproof paint
防水涂料　waterproof coating
防淘齿墙　scour prevention key wall
防跳跃装置　anti-pumping device
防污措施　trash control measure
防误操作措施　misoperation countermeasures
防锈　rust protection
防锈漆　antirust paint
防烟　smoke prevention
防烟阀　smoke damper
防烟分区　smoke bay
防烟楼梯间　smoke proof staircase
防振锤　vibration damper
防震　quakeproof
防震缝　aseismatic joint
防止管涌　protection against piping
防撞装置　anticollision device
房屋　housing
仿古建筑　pseudo-classic architecture
仿真模拟　analogue simulation
放大器不准确度　amplifier inaccuracy
放大器死区　amplifier dead band
放电电流　discharge current
放电电压　discharge voltage
放电率　discharge rate
放热率　heat release rate
放射法测井　radioactivity log
放射性测量　radioactivity survey
放射性测年　radioactive dating，radioactive age determination
放射性示踪剂　radioactive tracer，radioindicator
放水隧洞　relief tunnel
放行　release of imports and exports
放样　setting out，laying out
飞摆　pendulum
飞轮力矩　flywheel moment
飞逸工况　runaway speed operating condition
飞逸试验　runaway speed test
飞逸特性曲线　runaway speed curve
非饱和输沙（不平衡输沙）　non-saturated sediment transport
非常不稳定　very unstable
非常溢洪道　emergency spillway
非承重墙　partition wall
非传动端　non-drive end (NDE)

非脆性不透水带　nonbrittle impervious zone
非单元式保护　non-unit protection
非对称配光型灯具　asymmetrical luminaire
非峰时电价　off-peak tariff
非杆件体系　non-member system
非杆件体系结构裂缝控制　crack control of non-member system
非工程措施　non-structural measures
非晶质　amorphous substance
非竞争性投标　noncompetitive bid
非均匀结构　heterogeneous texture
非均匀沙　non-uniform sediment
非均质土坝　non-homogeneous earth dam，zoned earth dam
非可视对讲机　unvisule intercom
非控制区　non-control zone
非农业人口　non-agricultural population
非农业生产　non-agricultural production
非全相运行　open phase operation
非湿陷性黄土　non collapsible loess
非土质材料防渗体分区坝　zoned earth ro-ckfill dam with non-soil impervious core
非外贸产出和投入　non-traded output and input
非完整孔　partially penetrating well
非稳定流抽水试验　unsteady-flow pumping test
非稳态传热　unsteady-state heat transfer
非线性电路　nonlinear circuit
非线性放电电阻　non-linear discharge resistor
非线性分析　non-linear analysis
非溢流坝段　non-overflow dam section
非有效库容上限　top of inactive capacity
非正弦周期量　non-sinusoidal periodic quantity
非自动调节渠道　non-self-regulating canal
非自恢复绝缘　non-self restoring insulation
非自愿迁移　involuntary relocation
非自愿移民　involuntary resettler
非自重湿陷性黄土　non-weight collapsible loess
废河道　abandoned channel
废水处理　wastewater treatment
废物处理　waste disposal
费用索赔　claim for loss and expense
分瓣运输　transport of segmentation（transport in segments）
分包合同　subcontract
分包商　subcontractor
分辨率　resolution
分布参数电路　distributed circuit
分布钢筋　distribution rebar
分布控制　distributed control
分布式 I/O　distributed I/O
分布式光纤温度传感器　optical fiber distributed temperature transducer
分部工程　divisional works
分层分布的监控系统　hierarchically distributed supervision and control system
分层开挖　excavation in layers
分层空气调节　stratificated air conditioning
分层取水　water-taking at different levels
分岔管　manifold penstock
分岔角　fork angle
分段断路器　section circuit-breaker
分段关闭装置　step closing device
分段灌浆　stage grouting
分段装药　deck charging
分割系数　division coefficient
分-合时间　open-close time
分洪道　flood way
分洪工程　flood diversion works
分洪区　flood diversion area
分洪水位　flood diversion stage
分机　extension
分级加荷　stage loading
分接　tapping
分解结晶复合型腐蚀　decomposing-crystalline compound corrosion
分解型腐蚀　decomposing corrosion
分块极限平衡法　method of block limit equilibrium
分块瓦导轴承轴瓦支撑装置　pad supporting device
分类标引　classified indexing
分类号　classification code
分类目录　classified catalogue
分离式溢洪道　detached spillway
分励脱扣器　shunt release
分裂变压器　split-phase transformer
分裂导线　conductor bundle
分流墩　chute block
分流管　manifold
分流器［水机］　cut-in deflector
分流器［电气］　shunt

分配电屏　sub-distribution board
分配阀　dividing valve
分配分析　Distribution Analysis
分配箱　distribution box
分批搅拌机　batch mixer
分频滤波器　freqency division filter
分期导流　stage diversion
分期付款　installment payment
分期开发　staged development
分期设计洪水　staged design flood
分期蓄水　impounding in stages
分歧（歧义）　discrepancy
分区两管制水系统　zoning two-pipe water system
分散供气　decentralized compressed air supply
分散供热水系统　individual hot water supply system
分散接警　scattered receipt of fire alarms
分散性黏土　dispersive clay
分散性土　dispersive soil
分时电价　time-of-day tariff
分水堤　divide dike
分水岭　water divide，watershed
分水刃型线　cut-out profile
分水认脊角　angle of face at the back of cut-out
分网运行　separate network operation
分位数　quantile，fractile
分位值　tantile
分析仪的恒等百分比带宽　constant relative bandwidth of an analyzer
分系统调试　subsystem debagging
分相保护　phase segregated protection
分相电流差动保护　phase current differential protection
分项（分段）发包合同　separate contract
分项工程　work element
分项投资　breakdown cost
分项系数　partial safety factor
分项系数极限状态设计　limit states design with partial coefficients
分项系数设计　limit states design with partial coef-ficient
分谐波分量（次谐波分量）subharmonic component
分序逐步加密灌浆　split spacing grouting
分闸　opening
分闸过电压　breaking overvoltage
分闸时间　opening time
分闸速度　opening speed
分闸位置　opening position
分支接头　branch joint
分座式　separate baseplate type
玢岩　porphyrite
粉尘浓度　dust concentration
粉煤灰　fly ash
粉喷桩　dry jet mixing pile
粉砂（粉土）　silt
粉砂岩　siltstone
粉砂质泥岩　silt mudstone
丰水流量　high flow
丰水期　high-flow period
风成波　wind wave
风成地貌　eolian landform
风道安装　air duct installation
风道加热器　duct heater
风道摩擦损失　duct friction loss
风道压力损失　duct pressure loss
风道阻力　duct resistance
风动支架凿岩机　air leg-mounted drill
风干试样　air-dried sample
风管部件　duct accessory
风管配件　duct fittings
风荷载　wind load
风荷载体形系数　wind load shape coefficient
风化　weathering
风化程度　weathering degree
风化带　zone of weathering，weathered zone
风化分解　decomposition weathering
风化壳　weathering crust
风化裂隙　weathering fissure
风化破碎　disintegration weathering
风化土料　weathered soil
风化与蚀变　weathering and alteration
风机出口　fan outlet
风机对流加热器　fan convection heater
风机排（送）风量　fan delivery
风积物　eolian deposit
风景林　landscape forest
风景名胜区　scenic resort
风冷骨料　air-cooled aggregate
风力　wind force
风力侵蚀　wind erosion

风偏 wind deflection
风偏角 angle of wind deflection
风沙流 sand air current
风扇 ventilation fan
风蚀壁龛 tafone
风蚀残丘 eolian monadnock
风速 wind speed
风速仪 anemograph
风损 windage loss
风险分析 risk analysis
风险控制 risk control
风险评估 risk assessment（evaluation）
风向 wind direction
风压高度变化系数 variation coefficient of wind pressure with height
风壅水 wind setup
风壅水面高度 wave setup
风振 wind-induced vibration
风振系数 wind vibration coefficient
风钻 air drill
封闭层 seal coat
封闭的火灾 contained fire
封闭涂层 barrier coat
封冻 complete freezing
封冻预报 freezing forecast
封堵 plugging
封堵体 plug
封端球盖 corona bar，cap
封拱 closure of arch
封拱温度 closure temperature
封禁治理 closing hillside for erosion control
封孔 borehole backfilling
峰-峰值 peak-to-peak value
峰-谷值 peak-to-valley value
峰荷 peak load
峰荷电价 peak-load tariff
峰荷电站 peak-load power station
峰荷机组 peak load set
峰值 peak value
峰值电弧电压 peak arc voltage
峰值关合电流 peak making current
峰值加速度 peak acceleration
峰值耐受电流 peak withstand current
峰值强度 peak strength
峰值纹波因数 peak-ripple factor，peak distortion factor
蜂鸣器 buzzer
蜂窝 honeycomb
蜂窝状风化 honeycomb weathering
缝管锚杆 slot-tube anchor bolt
孵卵 hatching
弗劳得数相似 Froude similitude
弗劳德数 Froude number
伏赫比限制器 Volts per Hertz limiter（V/F limiter）
伏-秒特性曲线 volt-time characteristics
伏特 Volt
扶壁式挡土墙 counterfort retaining wall
服务合同 contract for service
服务业 service business
氟橡胶 fluoroelastomer（FPM）
浮称法 buoyancy method
浮充电 floating charge
浮动环密封 floating ring seal
浮动式球阀 floating ball valve
浮浆膜 laitance
浮密度 submerged density
浮容重 submerged unit weight
浮式拦漂排 floating boom
浮筒式 flotation tank type
浮筒栈桥 pontoon causeway
浮托力 buoyancy force
浮箱闸门（浮动闸门） floating gate
浮游动物 zooplankton
浮游植物 phytoplankton
浮子开关 float switch
符号 symbol
辐射供暖 radiant heating
辐射馈线 radial feeder
辐射率 emissivity
辐射式缆机 radial-traveling cable crane
福利费 benefits
福利费用 welfare Cost
福利效益 welfare Benefit
俯焊 down-hand welding
俯视图 top view，plan view
辅助保护 auxiliary protection
辅助电路 auxiliary circuit
辅助工资 auxiliary wage
辅助继电器，中间继电器 auxiliary relay
辅助建筑物 accessory building

辅助接力器　auxiliary servomotor
辅助设备启动　auxiliaries start
辅助设备停机延时　time delay for stopping auxiliaries
辅助消能工　auxiliary energy dissipator
辅助溢洪道　auxiliary spillway
腐蚀类型　corrosion types
腐蚀性水　corrosive water
腐蚀余量　corrosion allowance
付款　disbursement
付款计划表　schedule of payment
负搭叠　underlapped
负荷备用　load reserve
负荷备用容量　standby capacity
负荷的功率调节系数　power regulation coefficient of load
负荷电流补偿器　load current compensator
负荷电压特性　voltage characteristics of load
负荷（负载）　load
负荷隔离开关　load-disconnector switch
负荷惯性　load inertial
负荷节点　load bus
负荷开关　switch，load break switch
负荷频率特性　frequency characteristics of load
负荷清单　load details
负荷曲线　load curve
负荷试验　load test
负荷数学模型　mathematical model of load
负荷特性　load characteristics
负荷同时率　load coincidence factor
负荷统计　load estimation
负荷稳定性　load stability
负荷中心　load center
负荷转移　load transfer
负极板　negative plate
负极端子　negative terminal
负序保护　negative sequence protection
负序电流保护　negative sequence current protection
负序电流承载能力　negative sequence current withstand capability
负序短路电流　negative sequence short circuit current
负序分量　negative sequence component
负序网络　negative sequence network
负序阻抗　negative sequence impedance
负压　negative pressure
负压溜槽（管）　negative-pressure chute (pipe)
负载持续率　cyclic duration factor
负载惯性时间常数　load inertial time constant
负载换相　load commutation
负载换相逆变器　load commuted inverter (LCI)
负载加速时间常数　load acceleration constant
负载试验　load test
负载损耗　load loss
负载特性　load characteristics
负载系数　load factor
附挂式光缆　optical attached cable (OPAC)
附合导线　connecting traverse
附环闸门　ring-follower gate
附加背压力　superimposed back pressure
附加耗热量　additional heat loss
附加荷载　additional load
附加扰力　additional disturbing force
附加绕组　auxiliary winding
附加损耗　supplementary load loss
附加应力（叠加应力）　superimposed stress
附件　appendix，attachment
附建公共用房　accessory assembly occupancy building
附属建筑物　auxiliary building
附属设施　appurtenant facilities
附着式振捣器　form vibrator，surface-type vibrator
复背斜　anticlinorium
复测　repetition survey
复查　review，check
复工　resuming of work
复功率　complex power
复归时间［自动重合闸的］　reclaim time
复归值　reset value
复合故障　combination fault
复合绝缘子　composite insulator
复合密封止水材料　composite waterstop
复合农林业　agro-forestry
复合式火灾探测器　combination type fire detector
复合通风系统　integrated ventilation system
复合土工布　composite geotextile
复合土工膜　composite geomembrane
复合振动或脉动　compound vibration or pulsation

复核　reexamine
复理石建造　flysch formation
复燃　reignition
复式断面　composite section
复式配筋　compound reinforcement
复向斜　synclinorium
复压过流保护　compound voltage started overcurrent protection
复用水系统　water reuse system
复杂构件　complex assembly
复种　multiple cropping
复种面积　multiple cropping area
复种指数　multiple crop index
副坝　secondary dam (saddle dam)
副本，复印件　copy，duplicated document
副变质岩　parametamorphite
副厂房　auxiliary plant，service building
副航道　sub-waterway
副业　sideline activity
副业收入　sideline income
傅里叶变换　Fourier transformation
傅里叶级数　Fourier series
傅里叶逆变换　inverse Fourier transform
傅立叶级数基波分量　fundamental component of a Fourier series
富营养化　eutrophication
腹板　web
覆冰厚度　radial thickness of ice
覆冰区　ice coverage area
覆盖性岩溶　covered karst
覆盖种植　covering cultivation

G

改良硅酸盐水泥　modified Portland cement
改性剂　modifier
改性沥青　modified bitumen
改造与增容　rehabilitation and upgrading
钙华　calc-sinter
钙质白云石大理岩　calc-dolomite marble
钙质薄膜　calcium coated
钙质胶结　calcareous cement
钙质结核　caliche nodule，calcareous concretion
盖重区　weighted cover zone
概化洪水过程线　simplified flood hydrograph
概率　probability
概率分布　probability distribution
概率分布函数　probability distribution function
概率分布模型　probability distribution model
概率密度函数　probability density function
概率曲线　probability curve
概率设计法　probabilistic method
概念设计　concept design
干拌砂浆　dry-mixed mortar
干粉灭火剂　powder extinguishing agent
干工况　dry cooling condition
干混砂浆搅拌站　dry mortar mixing plant
干抗压强度　dry compressive strength
干流　main stream
干密度　dry density
干喷混凝土　dry shotcrete
干漆膜厚　dry paint film thickness
干砌石坝　loose-rock dam
干砌石挡土墙　dry-laid rubble masonry wall
干砌石护坡　dry rubble masonry slope
干球温度　dry-bulb temperature
干容重　dry unit weight
干湿交替　alternate wetting and drying
干式变压器　dry type transformer
干式电抗器　dry type reactor
干式系统　dry pipe system
干试验　dry test
干缩　drying shrinkage
干硬性混凝土　no-slump concrete
干硬性浆液　no-slump grout
干硬性砂浆　dry mortar
干运式升船机　dry shiplift
干燥到渗水　dry to seep
杆件体系　member system
杆塔　support (structure of an overhead line)
坩埚　crucible
赶工　expediting of work
赶计划　catch up with the schedule
感光火灾探测器　optical flame fire detector
感抗　inductive reactance
感纳　inductive susceptance
感生应力　induced stress
感温火灾探测器　heat fire detector
感烟火灾探测器　smoke fire detector
感应电动机　induction motor
感应电动势　induced electromotive force，induced

voltage
感应电压　induced voltage
感应过电压　induced overvoltage
感应雷击　indirect lightning strike
橄榄岩　peridotite，olivinite
刚度　stiffness，rigidity
刚性管道　rigid pipeline
刚性闸板　rigid gate
刚玉　corundum
钢板桩框格围堰　steel sheet-pile cellular cofferdam
钢衬钢筋混凝土管　steel lined reinforced concrete penstock
钢带铠装电缆　steel-tape armoured cable
钢管凑合节　adjustor of steel pipe
钢管加工厂　penstock fabrication plant（steel tube processing plant）
钢管圆度偏差　penstock roundness tolerance
钢绞线　steel strand
钢绞线　steel strand wire
钢结构防火涂料　fire-resistant coating for steel structure
钢筋　reinforcement bar（rebar）
钢筋调直机　bar straightener
钢筋混凝土　reinforcement concrete
钢筋混凝土杆　steel reinforced concrete pole
钢筋混凝土结构　reinforced concrete structure
钢筋混凝土框架　reinforced concrete frame
钢筋混凝土尾水管　reinforced concrete draft tube
钢筋混凝土蜗壳　reinforced concrete spiral case
钢筋机械连接　rebar mechanical splicing
钢筋计　reinforcement stress meter
钢筋加工厂　reinforcement fabrication plant
钢筋间距　spacing of bars
钢筋接头　splice，joint
钢筋连接　splice reinforcement
钢筋临界温度　critical temperature of reinforcement
钢筋笼　steel gabion
钢筋强度标准值　characteristic value of rebar strength
钢筋强度设计值　design value of rebar strength
钢筋切割机　bar cutter
钢筋束，锚索　tendon
钢筋台车　steel pallet
钢筋弯曲机　bar bender
钢筋网　reinforcing mesh
钢筋与混凝土的握裹力　grip between concrete and steel
钢筋锥螺纹接头　taper threaded splice of rebar
钢模台车　formwork jumbo
钢丝铠装电缆　steel-wire armoured cable
钢丝绳　wire rope
钢丝绳弹性模量偏差　elasticity modulus error of wire rope
钢丝绳卷扬式　winch，wire rope hoist
钢丝绳倾角　angle of inclination of rope
钢丝绳涂油器　rope lubrication device
钢丝绳张力检测设备　inspection device of wire rope tension
钢丝绳张力均衡装置　balancing device of wire rope tension
钢丝绳直径制造误差　manufacture error of wire rope diameter
钢纤维混凝土　steel fibre concrete
钢纤维喷混凝土　steel-fiber shotcrete
钢弦式传感器　string transducer
钢芯铝合金绞线　aluminium alloy conductor steel reinforced（AACSR）
钢芯铝绞线　aluminium conductor steel reinforced（ACSR）
钢支撑　steel braces，steel support
港口费　port charge
杠杆式安全阀　lever and weight loaded safety valve
杠杆式减压阀　lever reducing valve
高　height
高标号混凝土　high grade concrete
高层住宅　high-rise apartment
高差　altitude difference
高差曲线　departure curve
高程控制点　vertical control point
高程控制网　vertical control network
高窗自然通风　high window natural ventilation
高次谐波　harmonic，harmonic component
高低轨　high-low tracks
高低轮　high-low wheels
高地　highland
高电压技术　high voltage technology
高电压试验测量系统　measuring system for high voltage testing
高电压试验设备　high voltage testing equipment

高峰人数　peak labor force
高负荷短时运行范围　high-load temporary operation range
高含沙水流　flow with hyperconcentrated sediment
高级火灾模型　advanced fire model
高级熟练工　senior skilled labor
高空作业车　hydraulic aerial cage
高岭石　kaolinitic
高炉矿渣水泥　blast-furnace cement
高密度电法　resistivity imaging
高喷防渗墙　jet grouted cutoff wall
高频电缆　high frequency cable
高频开关充电器　high frequency switch mode charger
高频率保护　overfrequency protection
高频引入架　high frequency patching bay
高频振荡器　high frequency oscillator
高强度水泥　high-grade (strength) cement
高强螺栓　high-strength bolts
高水力比能短时运行范围　high specific hy-draulic energy temporary operating range
高斯　Gauss
高耸结构　high-rise structure
高速公路　expressway，freeway
高速滑坡　high speed landslide
高速水流　high-velocity flow
高速水流区　high-velocity flow area
高温阀门　high temperature valve
高温缓凝剂　high temperature retarding admixture
高效减水剂　high-range water-reducing admixture
高型布置　high-profile layout
高压　high voltage (HV)
高压标准电容器　high voltage standard capacitor
高压电桥　high voltage bridge
高压开关设备和控制设备　high voltage switchgear and controlgear
高压开关设备联锁装置　high voltage switchgear interlocking device
高压开关装置（高压配电装置）　high voltage switchgear
高压钠（蒸汽）灯　high pressure sodium (vapour) lamp
高压耦合电容器　high voltage coupling capacitor
高压喷射灌浆（高喷灌浆）　high-pressure jet grouting
高压绕组　high voltage winding
高压热水供暖系统　high-pressure hot water heating system
高压示波器　high voltage oscilloscope
高压隧洞　high pressure tunnel
高压压水试验　high pressure water test
高压蒸汽供暖系统　high-pressure steam heating system
高压整流器　high voltage rectifier
高压直流　high voltage direct current (HVDC)
高扬程水泵　high-lift pump
高原　plateau
高阻抗型母线差动保护　high-impedance busbar differential protection
戈壁　gobi，stone desert
格构锚固　anchored framework
格式板桩 cellular sheet pile
格栅灯　grille lamp
格栅钢架　reinforcing-bar truss
格栅基础　grill footing
隔板　diaphragm
隔板　interstage diaghragm
隔点设站法　setting station between two points
隔离变压器　isolating transformer
隔离层　separator
隔离功能　separating function
隔离构件　separating element
隔离剂　release agent
隔离距离　isolating distance
隔离开关　disconnector，isolating switch
隔离开关熔断器　switch-fuse-disconnector
隔膜阀　diaphragm valve
隔坡梯田　interval terrace
隔热保冷措施　heat-insulation and cold-preservation measure
隔热层　thermal insulation layer
隔舌　cut-water
隔声材料　sound insulation material
隔室　compartment
隔水层　aquifuge
隔水底层　confining underlying bed
隔水顶层　confining overlying bed
隔震　seismic isolation
隔震装置　isolation device
镉镍蓄电池　nickel oxide cadmium battery
个人防护装备　personal protective equipment (PPE)

个人所得税　personal income tax
各向同性　isotropy
各向异性　anisotropy
各向异性岩体　anisotropic rock mass
给定误差　assigned error
给定信号相对偏差　relative deviation of command signal
给定（指令）信号　command signal
给排水　water supply and sewerage system
给水度　specific yield
给水管　water supply pipe
给水立管　water riser pipe
给水配件　water supply fittings
给水温度　temperature of water supply
给水系统　water supply system
给水总管　water main
根系层　root zone
更新时间（刷新时间）　updating time，refresh time
更新世（统）　Pleistocene epoch（series）
更衣室　dressing-room，locker
耕地　cultivated land
耕种土　agricultural soil，cultivated soil
耕作层　plough horizon
耕作密度　cultivated density
工厂预组装　shop pre-assembled
工程保险费　project insurance premium
工程边坡　cut slope，engineered slope
工程变更　variation of works
工程筹建期　pre-preparatory period
工程措施　structural measure
工程单价　unit price
工程单价计算表　Calculation of project unit price
工程档案　project archive
工程档案接收　accession of project archives
工程档案验收　acceptance of project archives
工程档案移交　transfer of project archives
工程档案著录　description of project archives
工程定位复测费　resurvey fee of project location
工程费用　construction cost
工程规模　project scale（project size）
工程计价定额　pricing code and index
工程计量　measurement
工程监理文件［档案］　records of supervision of project
工程建设管理区　employer camp
工程建设监理费　construction supervision cost
工程进度　rate of progress
工程进度款　progress payment
工程竣工文件［档案］　records on completion of project
工程开发任务　project development purposes
工程勘测　engineering investigation
工程类比、经验类比　project analogue
工程量汇总表　Summary bill of quantities
工程量清单　bill of quantities
工程排污费　sewage disposal charge
工程前期费　preliminary engineering fees
工程前期文件［档案］　prophase records of project
工程师代表（总监）　engineer's representative
工程师工作站　engineer workstation
工程陶瓷涂层　ceramic coating
工程特性表　project features
工程投资　project investment
工程完建期，竣工期　completion period
工程文件［档案］　records of project
工程消耗量定额　quantity of consumption norm and standard
工程效益　project benefits
工程影响的居住区　project-affected community
工程影响人口　project affected persons（PAPs）
工程预付款　advance payment，prepayment
工程造价　construction cost
工程造价鉴定　construction cost verification
工程造价信息　guidance of cost information
工程造价指数　cost index
工程造价咨询费　consultancy service charge
工程照管　care of works
工程质量优良品率　final percentage of superior works
工程准备期　preparatory period
工地例会　site regular meeting
工地生活设施　site accommodation
工地装焊转轮　site fabricated runner
工具软件　tool software
工控机　Industrial Personal Computer（IPC）
工况点　operating point
工况观测　conditional observation
工矿企业　industrial and mining enterprises

工频　power frequency
工频过电压　power frequency overvoltage
工频接地电阻　power frequency earthing resistance
工频耐受电压　power frequency withstand voltage
工频耐压试验　power frequency voltage withstand test
工频试验变压器　power frequency testing transformer
工频谐振试验变压器　power frequency resonant testing transformer
工期定额　norm and standard of construction period
工期索赔　claim for extension of time
工期延长　extension of time
工伤保险费　work injury insurance
工业部门　industrial sector
工业电视　industrial television
工业建筑　industrial architecture
工业企业厂界噪声　industrial enterprises noise at boundary
工业企业用水　water for industrial enterprise use
工业迁建　Industrial relocation
工业用水　industrial water
工业与民用建筑　industrial and civil architecture
工艺流程　workmanship procedure
工资　wage
工字钢　H-section steel
工作参数指示器　operating parameter indicator
工作荷载　working load
工作级别　classification group
工作密封　service seal
工作面排水沟　service drain
工作母线　main busbar
工作平台（操作平台）　operation platform
工作容量　working capacity
工作寿命　service life
工作台上定压试验　bench testing
工作行程　operating stroke
工作压力　operating pressure
工作应力法　working stress method
工作油压　operating oil pressure
工作闸门　service gate
工作制　duty
工作制类型　duty type
工作周期　duty cycle
公安消防队　fire brigade of public security
公差　tolerance
公称尺寸　nominal size
公称直径　nominal diameter
公共产品　Public Goods
公共建筑物　public building
公共健康　public health
公共连接点　point of common coupling
公共绕组　common winding
公开级　public class
公路等级　road class
公路运输　road transport
公认　public acceptance
公私合作模式　public-private-partnership (PPP)
公用设备现地控制单元　common device LCU
公用自动电话局　community dial office
公众参与　public participation
功　work
功角测量装置　phasor measurement unit，pmu
功角特性　load-angle characteristic
功率　power
功率调节　power control
功率调节范围　controlling power range
功率方向保护　directional power protection
功率给定单元　power setting module
功率继电器　power relay
功率角　load angle
功率脉动　power pulsation
功率谱密度　power spectral density
功率输出调节器　power output governor
功率损耗　power loss
功率稳定性指数　power stability index
功率因数　power factor
功率因数表　power factor meter
功率因数控制器　power factor controller
功率元件　power component
功率整流器　power converter
功能　function
功能限值　limit value of function
供电　power supply
供电可靠性　service reliability
供电连续性　continuity of power supply
供电质量　quality of supply
供给价　supply price
供回水温差　temperature difference between supply

and return water
供浆-集浆-回浆管线 supply-header-return pipeline
供料线 feeding system
供暖 heating
供暖负荷 heating load
供暖锅炉 heating boiler
供暖机组 heating unit
供暖面积 heating area
供暖能力 heating capacity
供暖期 heating duration
供暖通风与空调 heating, ventilation and air conditioning (HVAC)
供热成本 heating cost
供热管网 heating piping network
供水 water supply
供水保证率 water supply dependability
供水计划 water supply scheme
供水量 supplying water
供水效益 water supply benefit
供油系统 oil supply system
拱坝 arch dam
拱坝弧高比 ratio of arch crest length to height (crest length-height ratio)
拱坝竖向曲率 vertical curvature of the arch dam
拱坝体型 arch dam configuration
拱坝整体安全性 integral safety of the arch dam
拱坝轴线 axis of arch dam
拱顶塌落 crown collapse
拱端局部加厚度拱 locally-thickened arches at both abutments
拱冠 crown
拱冠梁 crown cantilever
拱肩 spandrel
拱脚 springer
拱结构 arch structure
拱梁分载法 trial-load method
拱梁网格体系 grid system consisting of a series of arches and cantilever units
拱圈分块（楔块） voussoir
拱圈线型 arch shape
拱圈中心角 central angle of an arch
拱形闸门 arch gate
拱腰 haunch
拱座 abutment pad
共电式电话机 common battery telephone
共电式交换机 common battery switch
共轭断层 conjugated fault
共轭梁 conjugate beam
共轭裂隙（节理） conjugate joints
共轭水深 conjugate depth
共模干扰电压 common mode disturbance voltage
共生型泥石流 debris flow of integration pattern
共箱母线 non-segregated phase busbar
共用库容 common storage, shared storage
共振 resonance
共振筛 resonance screen
共振柱试验 resonant column test
共轴式 close coupled type
共座式 common baseplate type
沟边埂 ridge along gully
沟槽式排水 trench drain
沟道密度 gully density
沟道蓄水工程 water storage works in gully
沟道治理工程 gully erosion control works
沟垄耕作 furrow-ridge tillage
沟蚀 gully erosion
沟头防护工程 protective works of gully head
钩头螺栓 hook bolt
构件承载能力 bearing capacity of member
构件刚度 stiffness of structural member
构造格局 tectonic framework
构造挤压带 crushed zone of structure
构造角砾岩 tectonic breccia
构造类比 structure analog
构造裂隙（节理） tectonic joint
构造盆地 structural basin
构造穹窿 structural dome
构造体系 tectonic system
构造稳定区 tectonically stable zone
构造线 tectonic line
构造岩 tectonite
构造应力 tectonic stress
孤立电厂 isolated power plant
孤立系统 isolated power system, island in a power system
孤立运行 isolated operation
孤石 lonestone
箍筋 stirrup

箍筋肢距 spacing of stirrup legs
古代冰川作用 ancient glaciation
古地理学 paleogeography
古地下水 fossil groundwater，paleo-groundwater
古地震 paleo-earthquake
古典建筑 classic architecture
古河槽 old channel，buried channel
古河道 ancient river course，paleochannel
古滑坡 ancient landslide，old landslide
古生代（界） Palaeozoic era（erathem）
古氏坩埚 Gooch crucible
古文化遗址 historic culture site，archeological area
古新世（统） Paleocene epoch（series）
古岩溶 ancient karst
谷值 valley value
股线 wire，strand
骨料 aggregate
骨料级配 aggregate gradation
骨料加热 aggregate heating
骨料坚固性试验 soundness test of aggregate
骨料离析 aggregate segregation
骨料露天堆放 aggregate open stockpiled
骨料平均粒径 average size of aggregates
鼓泡 blistering
鼓式制动器 shoe brake
鼓形闸门 drum gate
鼓胀 bulging
固壁土料 wall-stabilizing soil
固定觇标 stationary target
固定单价合同 fixed unit-price contract
固定导叶 stay vane
固定电接触 stationary electric contact
固定铰 trunnion yoke
固定螺栓 set bolt
固定式活塞取样器 fixed-piston sampler
固定式拦污栅 fixed trash rack
固定式灭火器 fixed fire extinguisher
固定式启闭机 fixed hoist
固定式球阀 fixed ball valve
固定吸水设施 fixed suction installation
固定资产使用费 fee for utilizing fixed assets
固定作用 fixed action
固端梁 built-in beam，fixed end beam
固化灰浆 solidification slurry
固化剂 hardener
固结比 consolidation ratio
固结不排水剪试验 consolidated undrained shear test（CU）
固结不排水三轴试验 consolidated undrained triaxial test
固结沉降 consolidation settlement
固结度 consolidation degree
固结灌浆 consolidation grouting
固结排水剪试验 consolidated drained shear test（CD）
固结排水三轴试验 consolidated drained triaxial test
固结曲线 consolidation curve
固结试验 consolidation test
固结系数 consolidation coefficient
固结压力 consolidation pressure
固坡工程 slope stabilization project
固态 solid state
固碳功能 carbon sequestration
固体废物污染 solid waste pollution control
固体径流，输沙量 sediment runoff
固有流量特性 inherent flow characteristic
固有频率 inherent frequency，natural frequency
固有制动转矩 inherent braking torque
故障穿越 fault ride through（FRT）
故障导向安全 fail-safe
故障点 fault point
故障点电流 current at the fault point
故障电流 fault current
故障定位器 fault locator
故障记录器 fault recorder
故障录波器 fault oscillograph
故障清除 fault clearance
故障区段 fault section
故障停电平均持续时间 average interruption duration
故障相 fault phase
故障信号 fault signal
故障造成的损失费 cost of failure
故障状态信息 fault state information
故障自动记录装置 automatic fault recording device
故障自动检测 automatic diagnosis
故障阻抗 fault impedance
雇主 employer

雇主代表　employer's representative
雇主的资金安排　employer's financial arrangement
雇主免费供应的材料　free-issue materials
雇主设备　employer's equipment
雇主委托人　delegated person
雇主要求　employer's requirement
雇主责任　employer's liability
雇主支付保函　payment guarantee by employer
刮污环　scraper ring
挂板　clevis，tongue
挂钩　hook
挂环　link，eye
挂网喷草　spraying grass seeds with net
挂网喷混凝土　wire mesh with shotcrete
拐点电压　knee point voltage
关断臂　turn-off arm
关断期　hold-off interval
关合　making
关合-开断时间　make-break time
关合时间　make time
关键词　keyword
关键工序　critical sequence
关键路径　critical path
关税　customs duty
关税税率　customs duty rate
关税优惠　preferential duty
关死扬程　shut off head
关系数据库　relational database
观测墩　instrument pier
观测孔　observation well
观测网　observation network
观测误差　observation error
观测系列　observation series
观测站　reading station
观赏植物　decorative plant
官方汇率　official exchange rate（OER）
馆藏　holdings
管壁弹性　pipe wall flexibility
管壁等效翼缘宽度　equivalent flange width of pipe shell
管道保温　insulation of piping
管道防腐　corrosion prevention of pipes
管道井　pipe shaft
管道排水　pipe drain
管道式　inline type
管道系统严密性试验　tightness test of pipeline
管道直饮水系统　purified drinking water system
管道阻火器　flame arrester for pipe
管段　pipe segment（section）
管夹材料表　bill of material of pipe support
管接头　pipe fittings，coupling
管井　deep well
管理用具购置费　expense for purchasing management appliances
管路压力损失　pressure loss of tube
管棚　pipe-shed
管渠　ditch
管式避雷器　tube type arrester
管式极板　tubular plate
管式锚杆　tube anchor bolt
管形阀　tubular valve
管型母线　busduct
管涌　piping
管子对口　aligning of pipes
管子坡口加工　polish of tube groove
贯穿螺栓　through bolt
贯穿性裂缝　through crack
贯流式水轮机　tubular turbine
贯流式座环　stay cone
贯入桩　driving pile
贯通　break through
贯通测量　holing through survey
贯通结构面　penetrating（through） structural plane
贯通式漏斗漩涡　through funnel vortex
惯用最大操作过电压　conventional maximum switching overvoltage
惯用最大雷电过电压　conventional maximum lightning overvoltage
盥洗室　washroom
灌溉　irrigation
灌溉保证率　dependability of irrigation
灌溉典型年　typical design year for irrigation
灌溉定额　irrigation duty
灌溉回归水补给　recharge from return flow of irrigation
灌溉渠道　irrigation canal
灌溉水源　water source for irrigation
灌溉系统　irrigation system
灌溉效益　irrigation benefit

灌溉用水量 irrigation water consumption
灌溉制度 irrigation program
灌浆 grouting
灌浆的收缩缝 grouted contraction joint
灌浆记录仪 grouting recorder
灌浆结石 set grout
灌浆廊道 grouting gallery
灌浆强度值 grouting intensity number
灌浆试验 grouting test
灌浆压力 grouting pressure
灌浆岩锚 grouted rock bolt
灌木林地 shrub forest land
灌区 irrigation area
灌注桩 filling pile
罐车 tanker; tank truck
光传感器 optical sensors
光单元 optical unit
光电发射 photoelectric emission
光电式互感器 optical instrument transformer
光电效应 photoelectric effect
光度计 photometer
光端机 optical transmission terminal equipment
光滑的 smooth
光接收机 optical receiver
光缆 optical cable , fibre optic cable
光缆段 optical cable section
光缆护套 cable jacket
光缆终端 optical cable terminal
光面爆破 smooth blasting
光面钢筋 plain bar
光面混凝土 fair face concrete
光幕反射 veiling reflection
光强分布 distribution of luminous intensity
光衰减 light attenuation
光损耗 light loss
光通量 luminous flux
光纤 optical fibre, fibre optic
光纤复合架空地线 composite overhead groundwire with optical fibre (OPGW)
光纤复合相线 optical phase conductor (OPPC)
光纤光栅传感器 optical fiber grating transducer
光纤接头 optical fibre splice
光纤连接器 optical fibre connector, joint box
光纤偏振模色散 polarization mode dispersion
光纤软线 optical fibre cord
光纤色散 fibre dispersion
光纤衰减系数 fibre attenuation coefficient
光纤应变 fibre strain
光纤元件 optical element
光纤纵联保护系统 optical link pilot protection
光学测量系统 optical measuring system
光学连续性 attenuation uniformity
光中继器 optical repeater
光字信号器 annunciator
广义韦氏附加质量法 Generalized Westergaard Added-Mass
广域保护 wide-area protection
广照型灯具 wide angle luminaire
归档 filing, archiving
规定点 specified point
规费 statutory fees
规格 specification
规划阶段 planning stage
规划开采量 planned exploitative reserve
规模经济 Economies of scale
硅胶呼吸器 silicagel breather
硅质胶结 siliceous cement
轨道 crane rail, track
轨道起重机 track crane
轨道总成 rail track
轨距 track gauge
辊轮 roller
滚动支撑 roller support
滚动支座 rolling ring girder support
滚动轴承 antifriction bearing, rolling bearing
滚轮闸门 free roller gate
锅炉给水泵 boiler feed pump
国际多式联运 international multimodal transport
国际航空货物运输 international airline transport
国际结算 international settlements
国际竞争性投标 international competitive bidding
国际市场价格 international market price
国家大地测量网 national geodetic net
国家地震台网 national network of seismograph
国家公园 national park
国家环境政策 national environmental policy
国家控制测量网 national control survey net
国家秘密 national security

国家坐标系统　state coordinate system
国民收入平减指数　GDP deflator
过坝道路　roadway across the dam
过程控制级　process control level
过程控制模式　process control mode
过程模拟　process simulation
过程线上升段　hydrograph ascending limb
过程线下降段　hydrograph recession limb
过充电　overcharge
过电流保护　overcurrent protection
过电流保护的选择性　over-current discrimination
过电压　overvoltage
过电压保护　overvoltage protection
过调节　overshoot
过度放牧　overstocking
过渡层　transition layer
过渡电流　transition current
过渡过程　transient
过渡接头　transition joint
过渡料　transition material
过渡区　transition zone
过负荷保护　overload protection
过火面积　burned area
过机含沙量　solid content passing through hydroturbine
过街楼　overhead building
过励磁保护　overexcitation protection
过励磁运行　overexcitation operation
过励限制器　overexcitation limiter
过梁　lintel
过流能力　discharge capacity
过流脱扣器　over-current release
过滤精度　filter fineness
过滤能力　filter capacity
过滤器　filter，strainer
过滤器堵塞　filter obstruction
过滤效率　filter efficiency
过木　log-passing
过木建筑物　log pass structure
过热　superheat
过热蒸汽　superheated steam
过筛百分率　percent passing
过筛累积百分率　cumulative percent passing
过水断面　wetted section
过水隧洞　waterway tunnel
过水围堰　overflow cofferdam
过速保护　over speed protection
过速开关　overspeed switch
过压保护　overvoltage protection
过鱼建筑物　fish pass structure
过鱼闸　fish lock
过载能力　overload ability
过载脱扣器　overload release

H

海拔校正系数　altitude correction factor
海关　custom house
海进　marine ingression
海上建筑物　offshore structure
海水浪溅区　seawater spraying zone
海退　marine regression
海外承包工程险　overseas contract engineering risks insurance（OCER）
海外投资保险费　foreign investment insurance cost
海相　marine facies
海相碎屑岩　clastic rocks of marine facies
海运保险费　marine premium
海运费　sea freight
含泥量　silt content
含沙量　sediment concentration
含沙量沿程变化　longitudinal variation of sediment concentration
含水层　aquifer
含水量，含水率　water content，moisture content
含水率试验　water content test
含税金价格　price including tax
含佣金价格　price including commission
含增值税价格　price including value- added tax
函数发生器　function generator
焓湿图　psychrometric chart
涵管导流　culvert diversion
寒潮　cold wave
寒冷地区　cold region
寒武纪（系）　Cambrian period（system）
罕遇地震　rare earthquake，high-level earthquake
夯实　tamping
夯实机　tamper
夯实土桩　rammed-soil pile
旱地　arid land

旱季　dry season
焊缝　weld
焊缝代号　welding symbols
焊缝分类　clarification of welds
焊缝检验　weld inspection，weld testing
焊工资质　welder's qualification
焊后热处理　heat treatment after welding
焊剂　flux
焊接　welding joint
焊接变形　weld deformation
焊接工艺　welding process
焊接件　weldment
焊接设备　welder
焊接式管接头　weld fittings
焊接质量　weld quality
焊前预热　pre-heating of welding piece
焊枪　welding gun
焊丝　wire
航道等级　grade of waterway
航道断面系数　cross section factor of waterway
航道规划　waterway planning
航道及运河　waterway and canal
航道设计水位　design stage of waterway
航道通过能力　navigation capacity
航道弯曲半径　curvature radius of waterway
航道整治线　regulation line of waterway
航空警告标志　aircraft warning marker
航摄照片　aerial photograph
航运　navigation
航运效益　navigation benefit
毫秒雷管　ms delay blasting cap
好围岩　good surrounding rock
耗灰量　cement consumption
耗减补偿　depletion premium
耗浆量　grout consumption
耗水强度　intensity of water consumption
合成漆　synthetic paint
合成有机聚合物　synthetic organic polymer
合-分操作　close-open operation
合-分时间　close-open time
合格性评定　evaluation of conformity
合环　ring closing
合金结构钢　alloy structural steel
合流制管道溢流　combined sewer overflow
合龙　final gap-closing
合同　contract
合同变更　contract variation
合同工期　contract time limit
合同管理　contract management
合同价格　contract price
合同价格的充分性　sufficiency of the contract price
合同通用条件　general conditions
合同协议书　contract agreement
合同专用条件　particular conditions
合闸　closing
合闸过电压　closing overvoltage
合闸时间　closing time
合闸速度　closing speed
合闸位置　closed position
和易性　workability
河岸林地　riparian woodland
河岸内坡　inside bank slope，landside slope
河岸外坡　outside bank slope，river side slope
河槽　river channel，stream channel
河槽集流（河网集流）　concentration of channel flow
河槽阻力　resistance of river channel
河长　river length
河床　riverbed
河床抗冲能力　riverbed erosion-resistant capacity
河床式厂房　water retaining powerhouse
河床式取水构筑物　riverbed intake structure
河床式水电站　water retaining type hydropower station，hydropower station in river channel
河道安全泄量　safety discharge in river
河道比降，坡降　river gradient（slope）
河道的冲淤　scour-and-deposition in the river channel
河道治理　river training（improvement）
河底平洞　adit under river
河段　reach
河谷　river valley
河谷平原　valley flat
河湖淤泥　lake and river silt
河间地块渗漏　leakage through interfluve
河口　estuary，river mouth
河流　river
河流阶地　river terrace
河流泥沙　river sediment

河流泥沙运动力学　dynamics of river sediment movement
河流生态修复　river eco-restoration
河流水电规划报告　River hydropower planning
河流梯级开发　cascade development
河流污染　stream pollution
河漫滩　alluvial flat，flood plain
河滩地　flood plain
河湾　river bend
河网密度　drainage density
河相　fluvial facies
河源　headwaters，river source
荷载　load
荷载分布　load distribution
荷载工况　load case
荷载限制器　load limiter
荷载指示器　load indicator
荷载组合　load combination
荷重悬挂点　point of suspension of load
核定　verify，examined by
核准制　approval system
赫兹　Hertz
褐铁矿　limonite
黑启动　black start
黑云母　biotite
很差围岩　very poor surrounding rock
亨利　Henry
恒电流充电　constant current charge
恒定中心　constant center
恒荷载　dead load
恒流电源　constant current power supply
恒温恒湿　constant temperature and humidity
恒温水槽　thermostatic water bath
恒压充电　constant voltage charge
恒压电源　constant voltage power supply
桁架拱　truss arch
桁架结构　truss structure
横波　transverse wave，shear wave
横吹灭弧室　cross blast interrupter
横道图　bar chart
横缝　transverse joint
横谷　transverse valley
横节理　cross joint
横拉闸门　lateral movement gate
横梁　cross beam
横剖面图　transversal section，cross-section
横向隔离　isolation
横向荷载　transverse load
横向围堰　transversal cofferdam
横移　traversing
衡重式挡土墙　shelf retaining wall
轰燃　flash over
烘干法　drying method
烘干骨料　oven-dried aggregate
红黏土　laterite
红外探测器　infrared detector
红外线操纵起重机　infrared rays operated crane
红柱石　andalusite
虹吸式屋面雨水排水系统 roof siphonic drainage system
虹吸筒法　siphon cylinder method
洪泛平原　floodplain
洪泛区　flood plain
洪峰　flood peak
洪峰流量历时　duration of flood flow
洪积　diluvium
洪积扇　diluvial fan，proluvial fan
洪水　flood
洪水保险　flood insurance
洪水波行进时间　flood wave travel times
洪水超高　flood surcharge
洪水调节　flood regulation
洪水过程线　flood hydrograph
洪水记号　floodmark
洪水量　flood volume
洪水频率　flood frequency
洪水频率曲线　flood frequency curve
洪水设计标准　flood design standard
洪水行进　flood routing
洪水预报　flood forecast
洪水之前降低库水位　pre-flood drawdown
喉部衬套　throat bush
喉管　throat ring
喉径　throat diameter
后备保护　backup protection
后盖板　back shroud
后继浇筑层　succeeding lift
后加速保护　accelerated protection after fault
后浇带　post-cast strip
后期导流　later stage diversion

后期冷却　final cooling
后损　latter loss
后翼缘　downstream flange
后张法预应力混凝土结构　post—tensioned prestressed concrete structure
后张锚　post-tensioned anchor
厚壁冲击管　heavy wall drive barrel
厚层状结构　thick layer structure
厚高比　ratio of thickness to height
厚拱坝　thick arch dam
弧触头　arcing contact
弧垂　sag
弧垂观测　visual of sag
弧电压　arcing voltage
弧焊　arc welding
弧后电流　post arc current
弧门半径　gate radius，skin plate radius
弧形闸门　radial gate，tainter gate
弧形闸门支座　support of radial gate
胡克定律　Hooke's law
湖积平原　lacustrine plain
湖积物　lacustrine deposit
湖相　lacustrine facies，lake facies
湖鱼　lacustrine fish
蝴蝶阀（蝶阀）　butterfly valve
互层　alternating layers，interbedded layers
互层状结构　alternately bedded structure
互感　mutual-inductance
互感电动势　mutual induced e. m. f
互感器　instrument transformer
互感器额定负荷　rated burden of instrument transformer
互感器负荷　burden of instrument transformer
互感应　mutual induction
互联　interconnection
互联运行　interconnected operation
户籍人口　registered population
户内变电站　indoor substation
户内变压器　indoor type transformer
户内外绝缘　indoor external insulation
户外变电站　outdoor substation
户外变压器　outdoor type transformer
户外外绝缘　outdoor external insulation
户主　household head
护岸工程　bank protection works
护盾式全断面岩石掘进机　shielded full face rock TBM
护角　armor angle
护栏　guard rail
护坡　slope protection
护坡工程　slop protection works
护坦（海漫）　apron
护套　sheath，jacket
护舷　fender
护照费用　passport fees
戽斗式消能工　bucket-type energy dissipator
花岗斑岩　granite-porphyry
花岗片麻岩　granitie-gneiss
花岗闪长岩　granodiorite
花岗伟晶岩　granite-pegmatite
花岗岩　granite
花篮螺丝　turn buckle
华力西运动　Variscian movement
华氏温度　Fahrenheit temperature
滑差　slip
滑触线　trolley conductor
滑动测微计　sliding micrometer
滑动模板　sliding formwork（slip form）
滑动平均曲线　moving average curve
滑动闸门　sliding gate
滑动支撑　sliding support
滑阀　spool valve
滑环　collector ring，slipring
滑环盖　slipring cover
滑块（支承滑块）　slide block，bearing block
滑框倒模　shifted form with sliding frame
滑轮　sheave，pulley
滑劈理　slip cleavage
滑坡　landslide
滑坡坝　landslide dam
滑坡壁　slip cliff，landslide main scarp
滑坡侧缘　landslide flank
滑坡堆积体　landslide deposit
滑坡鼓丘　landslide bulge
滑坡后缘　landslide crown
滑坡基座　slip foundation
滑坡监测　landslide monitoring
滑坡剪出口　toe of sliding surface
滑坡裂缝　landslide fracture
滑坡泥石流监测预警　monitoring and forecasting

of landslide and debris flow
滑坡前缘　landslide deposit toe
滑坡舌　slip tongue
滑坡台阶　landslide terrance
滑坡体　slip mass
滑坡洼地　landslide graben
滑坡易发区　hazardous area of landslide
滑坡涌浪模拟　landslide-generated waves simulation
滑石　talc
滑石片岩　talc schist
滑雪道式消能工　ski-jump energy dissipator
滑雪道式溢洪道　ski-jump spillway
滑移-弯曲　sliding-buckling
滑移-压致拉裂　sliding-compressive generated tensile fracture
滑移验算　slip resistance analysis
化学成分　chemical composition
化学风化　chemical weathering
化学灌浆　chemical grouting
化学黏合　chemical bonding
化学蚀变　chemical alteration
化学需氧量　chemical oxygen demand (COD)
划格法［涂层附着力试验］　knife and tape test
话务台　traffic switchboard
环保型方式　environmentally sound manner
环带构造，带状构造　zonal structure
环刀　cutting ring
环刀取样器　ring sampler
环境保护　environmental protection
环境保护和水土保持工程　environmental protection works，soil and water conservation works
环境本底值　present environmental baseline
环境调查　environmental investigation
环境恶化　environmental deterioration
环境工程学　environment engineering
环境公约　environmental covenant
环境关注　environmental concerns
环境回顾评价　retrospective assessment of environment
环境监测　environmental monitoring
环境监测计划　environment monitoring program
环境考虑　environmental consideration
环境流量　environmental flow
环境敏感对象　environmental sensitive object
环境目标　environmental objective
环境评估系统　environmental evaluation system
环境容量　environmental carrying capacity
环境生态学　environment ecology
环境退化　environmental degradation
环境完整性　environmental integrity
环境温度　ambient temperature
环境污染综合防治　integrated control of environment pollution
环境现状评价　present situation assessment of environment
环境研究　environmental study
环境意识　environmental awareness
环境因素　environmental factor
环境影响后评价　follow-up environmental impact assessment
环境影响评价　environmental impact assessment (EIA)
环境影响评价区　environmental impact assessment area
环境影响识别　identification of environmental impact
环境友好的　environment-friendly
环境预测评价　prospective assessment of environment
环境约束　environmental constraints
环境照度　environmental illumination
环境质量　environmental quality
环境质量报告　Environmental quality report
环境质量改善　environmental quality improvement
环境综合评价　comprehensive assessment of environment
环境作用　environmental action
环球法　ring-and-ball apparatus
环式渗透仪　ring infiltrometer
环饰柱　annulated column
环形　ring
环形闸门　ring gate
环氧绝缘母线　epoxy insulated busbar
环氧树脂浆液　epoxy resin grout
环氧树脂砂浆　epoxy resin mortar
环氧填料　epoxy filler
环状管网　loop pipe network
缓闭止回阀　low speed closed check valve

缓冲爆破 buffer blasting
缓冲电路 snubber circuit
缓冲器 buffer
缓冲时间 damping time
缓冲型调速器 damping type governor
缓冲装置 buffer device
缓解方案 mitigation program
缓凝剂 retarding admixture
缓凝减水剂 retarding and water reducing admixture
缓坡 gentle slope，flat slope
缓倾角 low-angle dip
换流，变流 power conversion
换气次数 air change time per hour
换气系统 air renewal system
换算系数 conversion factor
换土垫层 cushion of replaced soil
换位 transposition
换位杆塔 transposition support
换相 commutation
换相电抗器 commutation reactor
换相电路 commutation circuit
换相电容器 commutation capacitor
换相电压 commutating voltage
换相开关 phase reversal switch
换相失败 commutation failure
换相数 commutation number
换向阀 directional control valve
荒地 barren land
荒漠化 desertification
黄铁矿 pyrite
黄铜矿 chalcopyrite
黄土 loess
黄土高原 loess plateau
黄土梁 loess ridge
黄土峁 loess hill
黄土坪 loess terrace
黄土塬 loess tableland
黄玉 topaz
煌斑岩 lamprophyre
煌斑岩脉 lamprophyre vein
灰烬 ash
灰岩，石灰岩 limestone
恢复电压 recovery voltage
恢复计划 restoration plan
辉长岩 gabbro
辉绿岩 diabase，dolerite
辉石 augite
辉岩 pyroxenite
回车道 turnaround loop
回弹仪测试 rebond hammar test
回风方式 air return method
回风系统 air return system
回归水的重复使用 reuse of return flows
回火 tempering
回廊 cloister
回流阀 reflux valve
回流区 backflow area，return flow zone
回馏 reflux distillation
回路 loop
回路的预期瞬态恢复电压 prospective transient recovery voltage of circuit
回水变动区 fluctuating backwater zone
回水阀 return valve
回水管 return pipe
回水末端 upstream end of backwater
回水总管 return main pipe
回缩量 retraction range
回填 backfill
回填灌浆 backfill grouting
回油管 oil return pipe
回油箱 sump tank
回转半径 radius of gyration
回转动力式泵 rotodynamic pump
回转环 return ring
回转环导叶 return ring vane
回转机构 slewing mechanism
回转破碎机 gyratory crusher
回转型气动装置 rotary pneumatic actuator
回座压力 reseating pressure
汇兑风险 exchange risk
汇兑损益 exchange loss or gain
汇流 concentration of flow
汇率风险 exchange rate risk
汇率溢价 Foreign Exchange Premium
会展中心 meeting and exhibition center
惠斯登电桥测试装置 Wheatstone bridge test set
混合阀 mixing valve
混合接警 combined receipt of fire alarms
混合侵蚀 mixed erosion
混合式抽水蓄能电站 mixed pumped storage

power station
混合式开发　dam and conduit type development
混合岩　migmatite
混合岩化　migmatization
混流泵　mixed flow pump
混流式水轮机　Francis turbine, radial-axial flow turbine
混凝土拌合物　concrete mixture
混凝土保护层　concrete cover
混凝土泵车　truck mounted concrete pump
混凝土剥蚀　concrete disintegration
混凝土泊松比　Poisson's ratio of concrete
混凝土布料机　concrete distributor
混凝土材料单价　unit price of concrete materials
混凝土成熟度　maturity degree of concrete
混凝土出机口温度　concrete temperature at outlet of mixer
混凝土的比热　specific heat of concrete
混凝土等级（标号）　concrete class
混凝土吊罐　concrete bucket
混凝土冬季施工　concreting in cold weather
混凝土防渗墙　concrete cutoff wall
混凝土分层浇筑　concreting in lifts
混凝土分块浇筑　block concreting
混凝土干缩影响　influence of concrete drying shrinkage
混凝土构件场　concrete blockyard
混凝土含气量　concrete air content
混凝土护坡　concrete block protection
混凝土搅拌车　concrete truck mixer, agitating lorry
混凝土搅拌机　concrete mixer
混凝土搅拌输送斗　concrete transport skip
混凝土搅拌站　concrete mixing plant
混凝土结构　concrete structure
混凝土抗冻耐久性指数　durability factor of concrete (DF)
混凝土空腹重力构　concrete hollow gravity dam
混凝土宽缝重力构　concrete slotted gravi-ty dam
混凝土拉应力限制系数　coefficient for limiting concrete tensile stress
混凝土立方试样　cube concrete test specimen
混凝土龄期　concrete age
混凝土路面　concrete pavement
混凝土面板　concrete face slab
混凝土面板堆石坝　concrete face rockfill dam
混凝土内外温差　temperature difference of inner and outside concrete
混凝土配制强度　required average strength of concrete
混凝土喷射机　concrete sprayer
混凝土喷射机械手　shotcrete robot
混凝土强度等级　strength class (grade) of concrete
混凝土缺陷　concrete defect
混凝土生产系统　concrete production system
混凝土实体重力构　concrete solid gravity dam
混凝土收缩　shrinkage of concrete
混凝土受弯构件　concrete member in bending
混凝土受压构件　concrete member in compression
混凝土输送管　elephant trunk, concrete conveying pipe
混凝土四面体　concrete tetrahedron
混凝土摊铺机　concrete spreader
混凝土碳化　carbonation of concrete
混凝土围堰　concrete cofferdam
混凝土夏季施工　concreting in hot weather
混凝土徐变　creep of concrete
混凝土预拌厂　ready-mix plant
混凝土预冷系统　concrete precooling system
混凝土预热系统　concrete preheating system
混凝土预制厂　precast concrete plant
混凝土振捣器　concrete vibrator
混凝土制备　concrete preparation
混凝土轴心抗拉强度标准值　characteristic value of concrete axial tensile strength
混凝土轴心抗拉强度设计值　design value of concrete axial tensile strength
混凝土轴心抗压强度标准值　characteristic value of concrete axial compressive strength
混凝土轴心抗压强度设计值　design value of concrete axis compressive strength
混凝土自生体积增长　autogenous growth of concrete
混响声　reverberation sound
活动便桥　bailey bridge
活动觇标　movable target
活动带　active belt
活动断层探测　surveying and prospecting of active fault

活动隔断　movable partition
活动构造　active structure
活动铰　trunnion hub
活动侦测　activity detection
活断层　active fault
活荷载　live load
活化测井　activation log
活接头　union
活塞　piston
活塞杆　piston rod
活塞式储压器　piston accumulator
活塞式减压阀　piston reducing valve
活性掺合料　active admixture
活性骨料　reactive aggregate
火场指挥部　fire commanding post
火成岩　igneous rock
火工材料　explosive materials
火花放电　sparkover
火花检测器　spark tester
火警电话　fire telephone
火警调度台　fire alarm dispatching console
火警调度专线　fire alarm dispatching proprietary wires
火警瞭望　fire lookout
火警疏散信号　fire alarm evacuation signal
火警探测系统实验开关　fire detection system sets switch
火警显示装置　fire alarm indicating device
火警远距离显示设备　fire alarm remote indicating equipment
火口　fire hole
火雷管　spark blasting cap
火山灰　pozzolan
火山灰质水泥　pozzolana cement
火山角砾岩　volcanic breccia
火山碎屑岩　pyroclastic rock
火山岩　volcanic rock
火烟监测　fire-smoke detection
火焰传播　flame spread
火焰传播速率　flame spread rate
火焰峰　flame front
火焰探测器　flame detector
火灾报警　fire alarming
火灾报警系统　fire alarm system（FAS）
火灾场景　fire scenario
火灾传播　fire-propagation
火灾分类　fire classification
火灾荷载密度　fire load density
火灾激化风险　fire activation risk
火灾监测和灭火系统　detection- extinguishing system
火灾气流　fire effluent
火灾探测和报警　fire detection and alarm
火灾自动报警系统　automatic fire alarm system
或有估价法　contingent valuation method (CVM)
货币补偿　monetary compenstion
货币补偿　monetary indemnity（compensation）
货币的时间价值　time value of money
货垛　goods stack
货损率　cargo damages rate
货物　goods
货运量　freight traffic tonnage，ton-volume
货运强度　freight traffic intensity
霍尔发生器　Hall generator
霍尔效应　Hall effect
霍尔效应传感器　Hall effect sensor

J

击穿　breakdown
击穿保护器　sparkover protective device
击实试验　compaction test
机电设备　electromechanical equipment
机动时间　float time
机墩　turbine pier
机房　machine room
机构利用等级 T　classes of utilization of mechanisms
机构组别 M　mechanism groups
机会成本　Opportunity Cost
机加工　machining，machine work
机架　base frame
机壳　housing
机坑　pit
机坑里衬　pit liner
机密级　classified level
机桥　machine bridge
机上人工费　operator wage
机械保护系统　mechanical protection system
机械测功器　mechanical brake

机械抽排导流　pump diversion
机械对接　mechanical butt splicing
机械功率　mechanical power
机械功率损失　mechanical power losses
机械化程度　mechanization level
机械化施工　construction mechanization
机械加工余量公差　machining allowance tolerance
机械开度限制机构　mechanical opening limiter
机械控烟　mechanical smoke control
机械连杆　mechanical linkage
机械密封　mechanical seal
机械排风　mechanical air discharge
机械排风系统　mechanical exhaust system
机械排烟系统　mechanical smoke exhaust system
机械热水供暖系统　mechanical hot water heating system
机械式自动抓梁　lifting beam with latching and unlatching mechanisms
机械寿命试验　mechanical endurance test
机械送风　mechanical air supply
机械损失　mechanical loss
机械通风　mechanical ventilation
机械通风系统　mechanical ventilation system
机械同步　mechanical synchronization
机械消耗定额　machinery consumption norm
机械性能　mechanical property
机械修配厂　machine repair workshop
机械液压调速器　mechanical hydraulic governor
机械制动　mechanical braking
机械制动转矩　mechanical braking torque
机油　mobile oil
机组调差系数　droop of a set
机组动平衡试验　dynamic balancing test of unit
机组段　unit bay
机组断路器合闸　unit circuit breaker closed
机组发电工况黑启动　black start of units in generating mode
机组惯性时间常数　unit inertial time constant
机组过水能力　turbine discharge
机组技术数据　unit technical data
机组加速时间常数　unit acceleration constant
机组开机时间　start-up time of unit
机组开停机流程　start and stop sequences for the unit
机组内加热器　generator housing heaters
机组启停控制　automatic control for unit start-up and shutdown
机组事故停机后备装置　backup unit emergency shutdown device
机组停机时间　shutdown time of unit
机组现地控制单元　unit LCU
机组效率　overall efficiency
机组自用变压器　auxiliary transformer of unit
积分电路　integrating circuit
积分时间常数　integral time constant
积分增益　integral gain
积分作用时间　integral action time
积水面积　accumulated water area
基本变量　basic variables
基本测站　basic survey station
基本层楼板面积　basic floor area
基本冲击绝缘水平　basic impulse insulation level (BIL)
基本导线　primary traverse
基本地震烈度　basic seismic intensity
基本风压　basic wind pressure, reference wind pressure
基本工资　basic wages
基本荷载　basic load, usual load
基本绝缘　basic insulation
基本烈度　basic seismic intensity
基本农田　capital farmland
基本频率　elementary frequency
基本气温　reference air temperature
基本设计压力　basic design pressure
基本稳定　basically stable
基本雪压　basic snow pressure /reference snow pressure
基本养老保险费　basic endowment insurance
基本预备费　physical contingency
基本折旧费　basic depreciation cost
基本振型　fundamental mode of vibration
基本直接费　basic direct cost
基本周期　fundamental period, elementary period
基本组合　basic (fundamental) combination
基波　fundamental, fundamental component
基波分量　fundamental component
基波功率　fundamental power
基波频率（基频）　fundamental frequency
基波因数　fundamental factor
基础环　foundation ring

基础设施　infrastructure
基础约束裂缝　foundation restraint crack
基础约束系数　restraint coefficient of foundation
基尔霍夫第二定律　Kirchhoff's second law
基尔霍夫第一定律　Kirchhoff's first law
基尔霍夫电流定律　Kirchhoff's current law (KCL)
基尔霍夫电压定律　Kirchhoff voltage law (KVL)
基荷　base load
基荷电厂　base load power plant
基荷电站　base-load power station
基荷机组　base load set
基坑排水　foundation pit dewatering
基坑涌水 water gushing in foundation pit
基流　base flow
基线　base line
基线测量　base measurement
基性岩　basic rock
基岩变形计　bedrock deformeter
基岩与覆盖层分界线　boundary of bedrock and overburden
基于能量的抗震设计　energy-based seismic design
基于位移的抗震设计　displacement-based seismic design
基于性能的抗震设计　performance-based seismic design
基质　matrix
基准表观密度　basic apparent density
基准觇标　reference sighting target
基准点　datum point，control point
基准面　base level，datum plane
基准日期　base date
基准收益率　hurdle cut-off rate
基准运行时间　reference duration of operation
基准值　base value
基座阶地　bedrock seated terrace
激发极化法　induced polarization
激光准直法　laser alignment method
激光准直系统　laser alignment system
激活　activation
激励　excitation，stimulus
激振力　exciting vibration force
级配反滤料铺盖　weighted graded filter
级配良好的骨料　well-graded aggregate
级配碎石　graded crushed stone
极板　plate
极点　pole
极端丰水期　extreme wet period
极端枯水期　extreme dry period
极端最低温度　extreme minimum temperature
极端最高温度　extreme maximum temperature
极好围岩　excellent surrounding rock
极化电流　polarization current
极化继电器　polarized relay
极间电气间隙　clearance between poles
极强透水　very strongly pervious
极微透水　very slightly pervious
极限安全地震动　ultimate safety ground motion
极限变形　ultimate deformation
极限承载力　ultimate bearing capacity
极限偏差　limit deviation
极限平衡法　limit equilibrium method
极限强度　ultimate strength
极限位置装置　limit switch；limiter
极限应变　ultimate strain
极限状态　limit states
极限状态方程　limit state equation
极震区　meizoseismal area
极值暴雨样本　sample of extreme storms
极值Ⅰ型分布　extreme value type Ⅰ distribution
即付保函　demand guarantee
急流　supercritical flow，rapid flow
急性毒性　acute toxicity
疾病携带动物　disease-carrying animal
集成式行程检测装置　integrated travel detector
集电环　collector ring，slipring
集流坑　catchpit
集水沟　collector
集水坑　collecting sump
集体所有制企业　collective ownership enterprise
集中参数电路　lumped circuit
集中供气　centralized compressed air supply
集中供热水系统　collective hot water supply system
集中荷载　centralized load
集中接警　centralized receipt of fire alarms
集中控制　centralized control
集中热水供应系统　central hot water supply system
集中润滑　centralization lubrication
集中药包爆破　concentrated charge blasting
集装箱　container

集装运输　containerized transport
几何参数　geometrical parameter
几何高度　geometric height
几何相似　geometric similarity
挤包绝缘电缆　cable with extruded insulation
挤出焊接　extrusion welding
挤密砂桩　densification by sand pile
挤压爆破　squeezed blasting，tight blasting
挤压边墙　extrusion concrete side wall
挤压法　extruded
挤压片理和劈理带　compression schistosity and cleavage zone
挤压片状岩　sliced rock
计划工期　planned duration
计划管理　program management
计划停运时间　planned-outage duration，scheduled-outage duration
计划指标　plan target
计价单位　Numeraire
计量工程师　quantity surveyer
计量检验　quantitative inspection
计权隔声量　weighted sound reduction index
计日工　daywork labor
计日工费用　daywork rate
计数继电器　counter relay
计数检验　counting inspection
计算高度　effective height
计算机辅助绘图　computer aided drawing
计算机辅助设计　computer aided design (CAD)
计算机监控系统　computer supervision and control system (CSCS)
计算机室　computer room
计算机系统安全性　computer system security
计算基准值　fiducial value
计算拉断力　rated tensile strength (RTS)
计算模式不定性　uncertainties of calculation model
纪念性建筑　monumental architecture
技术供水系统故障　service water system fault
技术开发费　technology innovation expense
技术引进费　incidental cost of technology introduction
技术用水　service water
技术转让费　technology transfer fee
技术最小出力　minimum output
季风雨　monsoon rain
季节冻结深度　depth of seasonal freezing
季节冻土　seasonally frozen ground
季节性电能　seasonal energy
季节性洪水　seasonal flood
季节性水道　seasonal watercourse
季节性停工　seasonal shutdown
既有结构　existing structure
继爆管　relay primacord tube
继承人　successor
继电保护安全性　security of relay protection
继电保护可靠性　rcliability of relay protection
继电保护可信赖性　dependability of relay protection
继电保护快速性　rapidity of relay protection
继电保护灵敏性　sensitivity of relay protection
继电保护试验　relay protection test
继电保护投入率　utilization factor of protection system
继电保护系统　relay protection system
继电保护选择性　selectivity of relay protection
继电保护正确动作率　correct actuation ratio of protection system
继电保护装置　relay protection equipment
继电器　relay
继电器动作　operate of relay
继电器复归　reset of relay，drop out of relay
继电器启动　pick up of relay，start of relay
加班津贴　overtime allowance
加冰拌和　ice mixing
加工精度　machining precision
加工木材　factory lumber
加工图（车间图）　fabrication drawing，shop drawing
加固　reinforcement
加筋土挡土墙　reinforced earth wall
加劲板、加筋板　stiffener
加劲环　stiffener ring
加宽　widening
加肋板　ribbed slab
加冷水拌和　chilled water mixing
加里东运动　Caledonian movement
加气混凝土　aerated concrete
加强层　strengthened storey
加强型绞线的芯　core of reinforced conductor
加权平均粒径　weighted mean diameter
加权平均水头　weighted average head

加权平均值　weighted average
加权平均资金成本　Weighted Average Cost of Capital（WACC）
加热装置　heater
加湿　humidification
加速度　acceleration
加速度-缓冲型调速器　acceleration-damping type governor
加速度计　accelerometer
加速时间　acceleration time
加速时间常数　derivative time constant
加压泵　booster pump
夹层　interlayer，intercalation
夹轨器　rail clamping device
夹式引伸计　clip typed extension meter
夹渣　slag
价差预备费　price contingency
价格波动　price fluctuation
价格差异　price variance
价格扭曲　Price Distortion
价格水平　price level
价格指数　Price Index
架间水平净距　horizontal clearance between cable racks
架空［堆石体］　large voids left in the rockfill
架空层　elevated storey
架空地线　overhead earthing wire
架空线路　overhead line
架空线路的导线　conductor of overhead line
架空线路的回路　circuit of an overhead line
架空线路的耐张段　tension section of overhead line
架立钢筋　hanger rebar
架与壁间水平净距　horizontal clearance between rack and wall
假定坐标系　assumed coordinate system
假缝　suppositious seam
假凝　false set
假设高程　assumed elevation
假整合　disconformity
尖端放电　point discharge
尖灭　wedge out
坚硬岩　hard rock
间测　intermittent gauging
间层风化　interstratified weathering
间断级配　gap gradation
间断级配骨料　gap-graded aggregate
间隔　bay
间隔式金属封闭开关设备和控制设备 compartmented switchgear and controlgear
间隔装药　discontinued charging
间接费　indirect cost
间接火灾作用　indirect fire actions
间接冷却　indirect cooling
间隙　gap
间隙密封装置　clearance seal device
间歇灌溉　intermittent irrigation
间歇河　intermittent river
间歇喷泉　geyser
间歇式搅拌机　intermittent mixer
间歇性隆升　intermittent uplift
间谐波分量　interharmonic component
间谐波频率　interharmonic frequency
监测断面　monitoring section
监测系统　monitoring system
监测仪表与自动化元件　measuring instrument and automatic element
监控级　supervision level
监控中心　monitoring and controlling center
监控中心　supervision and control centre
监理　supervision
监理大纲　supervision outline
监理规划　supervision plan
监理机构　supervision organization
监理人员　supervisory staff
监理实施细则　supervision executive detailed rules
监视区域　surveillance area
监视信息　monitored information
兼容性　compatibility
检查廊道　inspection gallery
检查扭矩　check torque
检定　verification
检索　retrieval
检索工具　finding aid
检修备用　maintenance reserve
检修间隔　maintenance interval
检修密封　standstill seal
检修闸门　bulkhead gate
检验试验费　cost of inspection and test
检疫　quarantine
剪刀式隔离开关　pantograph disconnector

剪断销断裂　shear pin failure
剪跨比　ratio of shear span to effective depth (shear span to depth ratio)
剪力　shear force
剪力墙　shear wall
剪力墙结构　shearwall structure
剪裂隙（节理）　shear joint
剪切变形　shear deformation
剪切带　shear zone
剪切断层　shear fault
剪切模量　shear modulus
剪切销装置　shear pin device
剪应变　shear strain，tangential strain
剪应力　shear stress；tangential stress
剪胀性　dilatancy
减缓措施　mitigation measure
减水剂　water reducing admixture
减税　tax reduction
减速器（减速箱）　speed reducer
减压衬套　pressure reducing bushing
减压阀　pressure reducing valve
减压井　relief well
减压套　pressure reducing sleeve
减淤　desilting
减振器　dashpot
简布法　Janbu slice method
简单故障　simple fault
简单火灾模型　simple fire model
简化毕肖普法　simplified Bishop method
简化楔体　simplified wedge method
简明合同格式　short form of contract
简式断面　simplified section
简谐振动或脉动　simple harmonic vibra-tion or pulsation
简支梁　free beam，simple beam
简支轮　simple support wheel
碱度（碱性）　alkalinity
碱-骨料反应　alkali-aggregate reaction
碱活性骨料　alkali-reactive aggregate
碱活性试验　alkali reactivity test
碱性电池　alkaline cell
碱性水　alkaline water
见索即付保函　first demand guarantee
见证取样　evidential sampling
见证取样检测　evidential testing
建弧率　arc over rate
建基面　foundation surface (interface)
建设场地清理　construction land clean-up
建设工程项目　construction project
建设管理　construction management
建设管理费　overhead of client，construction administration cost
建设-经营-转让　build-operate-transfer (BOT)
建设期利息　interest incurred during construction
建设项目投资　project construction investment
建设预算　construction budget
建设征地与移民安置　land requisition and resettlement
建设质量监督管理费　expense for construction quality supervision and management
建设-转让（BT）　build-transfer (BT)
建议值　recommended value
建筑保温　building heat preservation
建筑大样图　architectural detail drawing
建筑防火　building fire protection
建筑防热　building thermal shading
建筑隔声　sound insulation
建筑工程费　civil works cost
建筑工程估算表　Table of civil works estimate
建筑工程一切险　contractor's all risk insurance
建筑构造设计　construction design
建筑光学　architectural optics
建筑环保　architectural environmentally friendly
建筑火灾　building fire
建筑节能设计　energy-saving design
建筑结构　building structure
建筑结构设计　structural design
建筑控制线　building line
建筑力学　architectural mechanics
建筑密度　building density
建筑面积　floor area
建筑模型　building model
建筑热工学　building thermotics
建筑设备设计　building service design
建筑设计　architectural design
建筑设计效果图　architectural rendering
建筑声学　architectural acoustics
建筑室内设计　interior design
建筑外观　architectural appearance

建筑物　architecture，building
建筑物的耐火等级　fire-resistant grade of buildings
建筑物理学　architectural physics
建筑物轮廓线　outline of structure
建筑物体形系数　shape coefficient of building
建筑物外部排水系统　guttering
建筑物中水　reclaimed water system for building
建筑物自重　dead load of structure
建筑吸声　sound absorption
建筑限界　structure gauge
建筑详图　architectural details
建筑遮阳　building sun shading
建筑总体效果　architectural ensemble
健康、安全与环境管理体系　HSE management system
渐变流时均压力　time-average pressure of gradually varied flow
渐进破坏　progressive failure
渐新世（统）　Oligocene epoch（series）
鉴定　appraisal
键槽　key
箭线图　arrow diagram
浆河现象　clogging of river
浆砌石挡土墙　mortar-rubble masonry wall
浆砌石护坡　masonry slope
浆砌石围堰　cemented masonry cofferdam
浆液渗透距离　grout travel
浆液注入率　grout injection rate
桨式（卧轴式）搅拌机　paddle mixer
桨叶检修孔　blade removal opening
桨叶位置传感器　blade position sensor
降落漏斗　depression cone
降密级　downgrade
降水　precipitation，rainfall
降温速率　cooling rate
降压变电站　step-down substation
降压变压器　step-down transformer
降压二极管　voltage dropping diode
降压启动　reduced voltage starting
降雨径流的预报　rainfall runoff forecast
降雨径流关系 precipitation-runoff relation
降雨历时　duration of rainfall
降雨面积（雨区）　precipitation area（rain area）
降雨侵蚀指数　rainfall erosion index
降雨入渗补给　infiltration recharge by rainfall
降雨型泥石流　debris flow of precipitation pattern
降阻剂　resistance reducing agent
交叉补贴　Cross Subsidization
交叉拱顶　cross vault
交叉梁　beam grid
交叉支撑　cross brace（X-brace）
交错层理　cross bedding
交换机　switch
交接检验　handing over inspection
交接试验　acceptance test
交接箱　cross-connecting box
交联聚乙烯绝缘电缆　cross-linked polyethylene insulated cable（XLPE）
交流变流器　AC converter
交流变流因数　AC conversion factor
交流采样　direct acquisition from CT and VT
交流电机内角　internal angle of an alternator
交流电流　alternating current（AC）
交流电压　alternating voltage（AC Voltage）
交流励磁机　AC exciter
交流耐压试验　AC voltage withstand test
交流输电　AC power transmission
交流系统　alternating current system（AC system）
交流线路　AC line
交流线路的相　phase of AC line
交流/直流变流器　AC/DC converter
交通补贴　commuting allowance
交线　line of intersection
交易费　Transaction Costs
交钥匙工程合同　turnkey contract
交轴超瞬态电抗　quadrature-axis subtransient reactance
交轴超瞬态短路时间常数　quadrature-axis subtransient short circuit time constant
交轴超瞬态开路时间常数　quadrature-axis subtransient open-circuit time constant
交轴瞬态电抗　quadrature-axis transient reactance
交轴瞬态短路时间常数　quadrature-axis transient short-circuit time constant
交轴瞬态开路时间常数　quadrature-axis transient open-circuit time constant
交轴同步电抗　quadrature-axis synchronous reactance

浇洒道路用水　water for road washing
浇筑层高度　placement lift
浇筑层间缝　lift joint
浇筑混凝土的季节性限制　seasonal limitations on placing of concrete
浇筑间歇期　delay between placement
浇筑块间的高差　height differentials between blocks
浇筑时的漏振　skipped vibration
浇筑温度　concrete placing temperature
胶合板　laminated wood
胶结　cement
胶结层　binder coat
胶结料　binder
胶凝材料　cementitious material
胶凝材料用量　cementitious material consumption
胶凝砂砾石混凝土　cement-sand-gravel concrete
胶凝砂砾石围堰　cement-sand-gravel cofferdam
胶体浆液　colloidal grout
胶质炸药　dynamite
胶状颗粒　colloidal particles
焦耳　Joule
焦耳定律　Joule law
焦耳效应　Joule effect
焦味　smell of scorching
角撑板　gusset plate
角动量矩　angular momentum
角度不整合　angular unconformity
角钢　angle steel
角焊　fillet welding
角焊缝　fillet welds
角砾结构　brecciated texture
角砾岩　breccia
角频率　angular frequency
角闪石　hornblende
角闪石片岩　amphibole schist
角闪岩　amphibolite
角式阀　angle type valve
角形接线　ring bus configuration
角岩　hornfels
绞线　stranded conductor
铰接排架柱　hinged bent column
脚手架　scaffold
较发育　relatively developed
较破碎　relatively crushed
较完整　relatively integral
阶地　terrace
阶段验收　stage acceptance，interim acceptance
阶梯式开挖　benched excavation
阶形柱　stepped column
阶跃变化　step change
阶跃位移输入　step displacement input
阶跃响应曲线　step response curve
接触冲刷　contact scouring
接触电位差　contact potential difference
接触电压　touch voltage
接触管涌　piping on contact surface
接触灌浆　contact grouting
接触金具　contact tension fitting
接触流土　soil flow on contact surface
接触器　contactor
接地板　earth plate
接地变压器　earthing transformer
接地电刷　earthing brush
接地电压互感器　earthed voltage transformer
接地端子　earthing terminal
接地故障　earth fault
接地故障电流　earth fault current
接地和短路装置　earthing and short-circuiting equipment
接地汇流排　main earthing conductor
接地继电器　earth fault relay，earth fault relay
接地开关　earthing switch，earthing disconnector
接地体　earthing electrode
接地网　earthing grid
接地线（接地导体）　earthing conductor
接地引下线　down conductor（down lead）
接地装置　earth-termination system
接地装置对地电位　potential of earthing connection
接电持续率　duty factor
接缝灌浆　joint grouting
接缝灌浆温度　closure temperature
接警持续时间　duration of receiving alarms
接口软件　interface software
接力器　servomotor
接力器关闭时间　servomotor closing time
接力器缓冲时间　servomotor cushioning time
接力器开启/关闭规律　servomotor opening/closing law
接力器开启时间　servomotor opening time
接力器容积　servomotor volume

接力器容量　servomotor capacity
接力器响应时间　servomotor response time
接力器行程　servomotor stroke
接力器作用力　servomotor force
接收　taking-over
接收机　receiver
接收通知　taking-over notice
接收证书　taking-over certificate
接受补偿意愿　willingness to accept（WTA）
接头残余变形　residual deformation of splice
接头抗拉强度　tensile strength of splice
接头试件的总伸长率　total elongation of splice sample
接线盒　connection box
接桩　pile extension
节点　node，panel point
节点导纳矩阵　bus admittance matrix，Y bus matrix
节点阻抗矩阵　bus impedance matrix，Z bus matrix
节段式泵　sectional type pump
节间充水　water filling by gate split
节间水封　seal between sections
节理充填物　joint filling
节理连通率　joint persistence ratio
节理（裂隙）　joint
节理玫瑰图　joint rosette
节理密集带　densely jointed belt，joint-concentrated zone
节理频数　joint frequency
节理系　joint system
节理岩体　jointed rock mass
节理中等发育岩体　moderately jointed rock
节理组　joint set
节流衬套　throttle bush
节流阀　throttle valve
节流套　throat bushing
结构　texture
结构措施　structural measure
结构的权系数　weight coefficient of structure
结构的整体稳固性　structural integrity，structural robustness
结构动力特性　dynamic properties of structure
结构分析　structural analysis
结构缝　structural joint
结构构件（结构件）　structural member (component)
结构件利用等级 B　class of utilization of mechanisms
结构抗力　structural resistance
结构抗震性能　earthquake resistant behavior of structure
结构可靠度　structural reliability
结构力学法　structural mechanics method
结构面　structural plane，discontinuity
结构面发育程度　development degree of discontinuity
结构面分级　grading of structural plane
结构面强度　discontinuity strength
结构面状态评分　rating of discontinuity conditions
结构模型　structural model
结构目标可靠性指标　structure target reliability index
结构耐久性　structure durability
结构体　structural mass，structural body
结构体系　structural system
结构系数　structure partial factor
结构性能观测　structural behavior measurement
结构性能检验　inspection of structural performance
结构影响系数　influential coefficient of structure
结构振动控制　structural vibration control
结构重要性系数　coefficient of structure importance，importance factor of structure
结关　customs clearance
结合滤波器　coupling filter
结晶灰岩　crystalline limestone
结晶型腐蚀　crystalline corrosion
结晶质　crystalline substance
结清证明　written discharge
结尾工程　winding-up works
截断电流　cut-off current，let through current
截流工程布置图　Layout of closure works
截流井　intercepting well
截流戗堤　closure dike
截流围堰　river closure cofferdam
截流最大落差　maximum drop during closure
截面抵抗矩的塑性系数　ratio of plastic moment to elastic moment
截面惯性矩　second moment of area of sec-tion，moment of inertia of section

截面回转半径　radius of gyration of section
截面极惯性矩　polar second moment of area of section，polar moment of inertia of section
截面面积矩　first moment of area of section
截面模量　sectional modulus
截面内力　sectional internal force
截渗环　cutoff collar，seepage collar
截水槽，齿槽　cutoff trench
截水堤　cut-off dike
截水沟　catch drain
截止阀　globe valve，stop valve
截止式隔膜阀　globe diaphragm valve
解冻预报　ice break forecast
解环　ring opening
解理　cleavage
解密　declassification
解约条款　cancelling clause
介电常数　dielectric constant，permittivity
介质损耗试验　dielectric dissipation test，loss tangent test
介质损耗因数　dielectric loss factor
界面波　boundary wave，interface wave
界限含水率试验　side water content test
金刚石　diamond
金具　fitting
金卤灯　metal halide lamp，metal halogen lamp
金属箔电容器　metal foil capacitor
金属光泽　metallic luster
金属化电容器　metalized capacitor
金属结构拼装厂　fabrication plant of steel structure
金属塑料瓦　metal-PTFE bearing，metal-TEFLON bearing
金属套　metallic sheath
金属芯钢丝绳　independent wire rope core （IWRC） wire rope
金属性短路　dead short
金属氧化物避雷器　metal oxide arrester （MOA）
金属自承光缆　metallic aerial self supporting optical cable （MASS）
紧凑型荧光灯　compact fluorescent lamp
紧急操作　emergency operation
紧急车道　emergency lane
紧急电源　emergency power source
紧急停机　emergency shutdown
紧急停机按钮　emergency push button
紧急停机油体积　tripping oil volume
紧急停机油压　tripping oil pressure
紧密密度　compaction density
进场　mobilization
进场费　mobilization cost
进场公路　access road to site
进场验收　site acceptance
进出口端面距离　face-to-face dimension
进出口公司手续费　commission-charge for import and export company
进出口商品检验　commodity inspection
进出口税费　import and export tax
进/出水口　intake/outlet
进度报告　progress report
进度表　progress chart
进度偏差　program variance of project
进风道　inlet duct
进风口　air inlet
进风口面积　air inlet area
进口附加税　import surtax
进口关税　import duty
进口平价　Import Parity Price
进料筛　feeder screen
进深　depth
进水　water influx
进水口底板高程　intake bottom elevation
进水口平台　intake platform
进水口闸门　intake gate
进水流道　inflow runner
进水渠　approach channel
进水闸门关闭　intake gate closed
进水闸门开启　intake gate open
进线断路器　incoming breaker
进相运行　leading operation，leading power factor operation
进占法卸料　end-dump advance method
近端短路　near-to generator short-circuit
近景摄影测量　close-range photogrammetry
近期规划　short-term plan
晋宁运动　Jinniing movement
浸没区　inundation area
浸润线　phreatic line，seepage line
浸油润滑　oil-bath lubrication
禁猎期　fence time

禁猎区　sanctuary
禁渔期　fence time
禁止开垦坡度　critical cultivation slope
禁止重合闸　inhibit reclosing
经常负荷　continuous load
经常性排水　regular dewatering
经济补贴　Economic Subsidy
经济不发达地区　undeveloped region
经济电流密度　economic current density
经济发达地区　developed region
经济发展总体规划　overall economic development plan
经济费用　economic cost
经济分析　economic analysis
经济规模　economies scale
经济价格　economic price
经济净现值　economic net present value（ENPV）
经济林　non-wood forest
经济内部收益率　economic internal rate of return（EIRR）
经济评估　economic appraisal
经济生存能力　economic viability
经济寿命　economic life
经济效率　economic efficiency
经济效益　economic benefit
经济影响评价　economic impact assessment
经济资源　economic resource
经济作物　commercial crops
经纬仪　theodolite
经验参数　empirical parameter
经验公式　empirical formula
经验数据　empirical data
经验系数　empirical coefficient
晶孔　vugular pore
晶闸管桥　thyristor bridge（SCR bridge）
晶闸管桥的（n－1）冗余　（n-1）redundancy of thyristor bridges
晶闸管整流器　thyristor rectifier
精度　accuracy，precision
精密测量观测　precise surveying measurement
精确到小数点第X位　accurate to the X-th decimal place
精确到［仪器读数］　reading to
精确到±X%以内　accurate to within plus or minus X percent
精准至［计算结果］　accurate to，to the nearest
井群　battery of wells
井探　shaft exploration
景观照明　landscape lighting
警告牌　warning board
警戒水位　warning stage
径流　runoff
径流调节　runoff regulation
径流模数　runoff modulus
径流深　runoff depth
径流式水电站　run-of-river hydropower station
径流系数　runoff coefficient
径流小区观测　observation of runoff plot
径缩变形　radial contracting deformation
径向力　radial force
径向剖分泵　radially split pump
径向跳动　circular runout
径向通风　radial ventilation
径向载荷脉动　radial load pulsation
径向轴承　transverse bearing，radial bearing
径向主应力　radial principal stress
净高　net height
净化空调系统　air cleaning system
净空高度　headway
净跨度　net span
净流通面积　net flow area
净起重量　net load
净热通量　net heat flux
净水头　net head
净水头计算　net head calculations
净现值　net present value（NPV）
净油器　oil filter
净资产收益率　return on equity（ROE）
静冰压力　static ice pressure
静触头　fixed contact
静电场　electrostatic field
静电感应　electrostatic induction
静定梁　statically determinate beam
静库容　stilling storage
静力触探试验　static cone penetration test
静力水准系统　static leveling system
静平床　stationary flat bed
静平衡试验　static balance test
静水启闭　operation under balanced head
静水头　hydrostatic head

静水压强　hydrostatic pressure intensity
静态调平　leveling in static state
静态负荷特性　steady state characteristics of load
静态旁路开关　static bypass switch
静态投资　static investment
静态作用　static action
静扬程　total static head
静载试验　static test
静止变频器　static frequency convertor (SFC)
静止同步补偿装置　static synchronous compensator
静止土压力　earth pressure at rest, static earth pressure
静止无功补偿器　static var compensator (SVC)
静止移相器　static phase shifter
静止整流励磁装置　static exciter
静止止漏环　stationary seal ring
镜板　thrust runner, thrust collar
镜面的　mirror, polished
酒精燃烧法　alcohol burning method
就地保护　in-situ conservation
就近后靠　local resettlement to higher elevation
就业机会　job opportunity
就业率　employment rate
就业状况　employment status
居民迁移线　border of resident relocation
居民生活用水　water for residential domestic use
居住建筑，住宅　residential building
居住空间　habitable space
局部不稳定　locally unstable
局部采暖　local heating
局部等电位连接　local equipotential bonding
局部放电　partial discharge
局部放电检测仪　partial discharge detector
局部火灾　localized fire
局部截水墙　partial cutoff
局部排风　local exhaust ventilation
局部破坏　local damage
局部区域空气调节　local air conditioning
局部热水供应系统　local hot water supply system
局部失稳　local instability
局部受压承载力　local compression bearing capacity
局部水头损失　local head loss
局部损失　local pressure loss
局部通风　local ventilation
局部挖补处理　dental treatment
局部应力　local stress
局部应用灭火系统　local application extinguishing system
局部阻力系数　coefficient of local resistance
矩形横截面形状构件　rectangular cross-sectional shape member
矩形进水口结构　rectangular intake structure
矩形母线　rectangular busbar
矩阵方法　Matrix Approach
巨厚层状结构　giant-thick layer structure
拒浆标准　refusal criterion
拒绝动作　failure to operation , failure to trip
拒收　rejection
具有正斜率的水泵特性（驼峰区）　pump characteristic with positive slope
剧院　theatre
锯齿状连续延伸　successively zigzag spread
锯齿状褶皱　zigzag fold
聚氨酯　polyurethane (PUR)
聚氨酯浆液　polyurethane grout
聚合物分散剂　polymer latex
聚合物改性水泥砂浆　polymer-modified cement mortar
聚合物混凝土　polymer concrete
聚灰比　polymer-cement ratio
聚氯乙烯电缆　polyvinyl chloride insulated cable
聚氯乙烯（PVC）止水带　PVC waterstop
聚脲　polyurea
聚四氟乙烯　Polytetrafluoroethene (PTFE)
涓流充电　trickle charge
卷内备考表　file note
卷内文件目录　innerfile item list
卷筒　wire rope drum
卷筒直径与钢丝绳直径比　drum to wire rope diameter ratio
卷筒直径制造误差　manufacture error of drum diameter
卷筒制动器　drum brake
卷扬机　cable hoist, winch
卷扬式启闭机　wire rope hoist, cable hoist
绝对磁导率　absolute permeability
绝对高程　absolute elevation
绝对湿度　absolute humidity

绝对温度　absolute temperature
绝对型编码器　absolute rotary encoder
绝对压力　absolute pressure
绝对压强　absolute pressure
绝密级　most confidential level
绝热温升　adiabatic temperature rise
绝热性　insulation
绝热养护　adiabatic curing
SF_6 绝缘变压器　SF_6 gas insulated transformer
绝缘的耐热等级　thermal class of insulation
绝缘杆　insulating stick
绝缘隔离装置　insulated isolated device
绝缘故障　insulation fault
绝缘故障率　failure rate of insulation
绝缘击穿　insulation breakdown
绝缘监视装置　insulation monitoring device
绝缘老化　ageing of insulation
绝缘老化试验　insulation ageing test
绝缘配合　insulation coordination
绝缘配合因数　insulation coordinating factor
绝缘屏蔽　insulation shielding
绝缘漆　insulated paint
绝缘性能指标　performance criterion of insulation
绝缘子　insulator
绝缘子保护金具　insulator protective fitting
绝缘子串　insulator string
绝缘子串组　insulator set
掘进机　tunnel boring machine (TBM)
掘进机法　tunnel boring machine method (TBM)
均布动荷载　moving uniform load
均布活荷载　uniformly distributed live loads
均衡充电　equalization charge
均衡座，支承走道　roller race
均温层（冷水层）　hypolimmion
均压环　ring manifold
均压环　grading ring
均匀变形　homogeneous deformation, uniform deformation
均匀级配，窄级配　narrow gradation
均匀沙　uniform sediment
均值系数　coefficient of mean value
均质土坝　homogeneous earth dam
竣工　completion
竣工报表　statement on completion
竣工测量　finial construction survey, as-built survey
竣工场地清理费　clean-up costs of completion site
竣工后试验　test after completion
竣工决算　final account for completed project, as-built settlement
竣工日期　completion date
竣工试验　test on completion
竣工图　as-built drawing
竣工文件　as-built document
竣工验收　final acceptance
竣工验收文件　handing over document, completion acceptance documents

K

卡钻　drill rod sticking
喀斯特处理　karst treatment
开办费　preliminary expense
开标　bid opening
开槽施工　trench installation
开敞式进水口　open inlet
开敞式堰　open weir
开敞式堰面曲线　open weir surface curve
开敞式溢洪道　open spillway
开度　aperture, separation
开度限制器　opening limiter
开度限制器位置开关　gate limit position switch
开断　breaking
开断电流　breaking current
开断时间　break time
开发方式　development scheme
开发权转让费　development rights transfer fee
开工令　order of commencement
开工日期　commencement date
开工通知　notice of commencement
开关阀器件　switched valve device
开关柜　switch cubicle
开关量，二进制量　binary variable, on/off variable
开关量输出　digital output
开关量输入　digital input
开关楼　switchyard building
开关设备的极　pole of switchgear
开关站　switchyard
开关站现地控制单元　switchyard LCU
开环控制　open-loop control
开环控制同步纠偏回路　synchro-control circuit

with open-cycle
开环增益　opening loop gain
开机过程中断　starting sequence dropout
开机前检查　pre-start inspection
开级配骨料　open-graded aggregate
开级配沥青混凝土　open-graded asphalt concrete
开间　bay width
开口沉箱　open caisson
开口三角形连接　open-delta connection, broken-delta connection
开口系数　opening factor
开阔地　open terrain, open ground
开路电压　open-circuit voltage
开路循环灌浆　open-circuit grouting
开式齿轮　open gearing
开式叶轮　open impeller
开停机超时　start-stop duration exceeded
开停机未完成　start-stop sequence incomplete
开通　firing
开挖料　excavated material
开挖料利用率　availability of excavated material
开挖平面图　Excavation plan
开挖坡比　excavation gradient, gradient of cutting
开挖线　excavation line
铠装层　armour
铠装电缆　armoured cable
铠装金属封闭开关设备　metal-clad switchgear
勘测大纲　outline of investigation program
勘测阶段　investigation stage
勘察设计费　cost for project investigation and design
勘探洞　exploration adit, trial adit
勘探工作布置　arrangement of exploration work
勘探工作量　quantity of exploration work
勘探项目　exploration item
勘探钻孔　exploration borehole
坎德拉　candela
抗拔力　pull-out force, pull-out resistance
抗拔桩　uplift pile
抗变形　deformation resistance
抗冲刷性　scour resistance
抗冲性能，抗侵蚀性能　erosion-resisting performance
抗刺穿力　puncture resistance
抗冻等级　frost resistance class
抗风化　weathering resistance
抗腐蚀性　corrosion resistance
抗滑　sliding resistance
抗滑安全系数　factor of safety against sliding
抗滑稳定　stability against sliding
抗滑移系数　slip coefficient
抗滑桩　slide-resisting pile
抗化学侵蚀　resistance to chemical deterioration
抗剪洞　shear resisting plug
抗剪断强度（剪摩强度）　shear-rupture strength
抗剪强度　shear strength
抗剪强度（纯摩强度）　shear strength, shearfriction strength
抗剪试验　shear tests
抗剪阻力　shear resistance
抗拉强度　tensile strength
抗老化　ageing resistance
抗力的变异系数　coefficient of variation of resistance
抗力分项系数　partial factor for resistance
抗裂　crack resistance
抗硫酸盐水泥　sulphate-resisting cement
抗磨板　facing plates, wear plates
抗磨环冷却水阀　wear ring cooling water valve
抗磨蚀混凝土　abrasion resistance concrete
抗扭箍筋　torsion stirrup
抗倾覆稳定　stability against overturning
抗倾覆验算　overturning resistance analysis
抗渗等级　permeation resistance class
抗水炸药　waterproof explosive
抗撕裂强度　tearing strength
抗外压稳定临界压力　critical external compressive resistance of buckling
抗弯强度，抗折强度　bending strength
抗压强度　compressive strength
抗震安全性　aseismatic safety
抗震措施　aseismatic measure
抗震等级　seismic grade
抗震概念设计　seismic concept design
抗震构造措施　structural details of seismic design
抗震计算　aseismatic checking
抗震加固　seismic strengthening
抗震鉴定　seismic appraisal
抗震结构整体性　integral behavior of ase-ismatic structure

抗震设防标准 seismic fortification criterion
抗震设防烈度 seismic fortification intensity
抗震设防区 seismic precautionary zone
抗震设防要求 seismic precautionary requirement
考勤系统 time attendance
靠船建筑物 berthing structure
科技馆 science museum
科学技术档案 scientific and technical archives
科研勘察设计费 cost for investigation，design and testing
颗分曲线 gradation curve
颗粒比表面 specific grain surface
颗粒尺寸 grain（particle）size
颗粒大小分布曲线 particle size distribution curve
颗粒大小频率图 grain size frequency diagram
颗粒分析 grading analysis，particle size analysis
颗粒分析试验 grain size analysis test
颗粒级配 gradation of grain
颗粒级配良好 well-distributed gradation
颗粒间空隙 intergranular space
颗粒胶结 granular cementation
颗粒结构 grain structure
颗粒物 particulate matter（PM）
颗粒形状 particle shape
颗粒组成 grain composition
壳衬 casing liner side plate
壳式变压器 shell type transformer
壳体 shell
壳体结构 shell structure
壳体密封环 casing ring
壳体试验 shell test
可报销费用 reimbursable expenditures
可避免的不利环境影响 avoidable environmental impact
可编程控制器 programmable logic controller（PLC）
可变作用的伴随值 accompanying value of a variable action
可变作用的频遇值 frequent value of a variable action
可变作用的准永久值 quasi-permanent value of a variable action
可变作用的组合值 combination value of a variable action
可变作用（荷载） variable action（load）
可持续发展框架 sustainable development framework
可调容量 adjustable capacity
可调整的报价 adjustable price quotation
可动电接触 movable electric contact
可耕地 arable land
可灌性 groutability
可换算的损失 relative scalable loss
可开采储量 exploitable reserve
可开发蕴藏量 exploitable potential
可靠电能 firm energy
可靠度分析 reliability analysis
可靠概率 probability of survival
可靠流量 dependable flow
可靠性（可靠度） reliability
可靠性指标 reliability index
可控阀器件 controllable valve device
可控平衡重 gravity counterweight
可控源音频大地电磁测深法 controlled source audio frequency magnetotellurics（CSAMT）
可控制的可变荷载 governable variable load
可利用资料 available data（information）
可耐受水平 tolerable level
可能的不利影响 possible adverse effect
可能环境因子 possible environmental factor
可能活动断层 capable fault
可能最大洪水 probable maximum flood（PMF）
可能最大降水 probable maximum precipitation（PMP）
可能最大露点 probable maximum dewpoint
可逆式电机 reversible machine
可逆式水轮机 reversible turbine
可逆正常使用极限状态 reversible serviceability limit states
可燃性 combustibility
可溶性 solubility，dissolubility
可生物降解表面活性剂混合物 biodegradable mixture of surfactant
可视对讲机 visual intercom
可听噪声 audible noise
可维修性 maintainability
可行性研究 feasibility study
可行性研究阶段 feasibility study stage
可修复性 restorability
可压缩层 compressible stratum
可压缩填料 compressible-type filler
可压缩性 compressibility

可液化土层 potential liquefaction soil layer
可移动的集中荷载 movable concentrated load
可移动的局部荷载 movable partial load
可移式灯具 portable luminaire
可以自由处置的收入 disposable income
可饮用水 potable water
可用流量 available flow
可用气蚀余量 net positive suction head available (NPSHA)
可用时间 up duration
可用性 availability
可用油体积 usable oil volume
可再生资源 renewable resources
可治理面积 erosion area suitable to control
可追溯性 traceability
可钻性 drillability
克利夫兰开口杯 Cleveland open cup
刻槽法 narrow slot method
刻度 division
客货运量 passenger and freight volume
客厅 parlor
坑（槽）探 pit（trench） exploration
空调度日数 cooling degree day based on 26℃ （CDD26）
空腹重力拱坝 hollow gravity arch dam
空化 cavitation
空化基准面 cavitation reference level
空化试验 cavitation test
空化系数 cavitation coefficient
空化裕量 cavitation margin
空间工作性能 spatial behavior
空间桁架 space truss
空间继承性 succession in space
空-空冷却发电机 air-to-air cooled machine
空气断路器 air circuit breaker
空气分布特性指数 air diffusion performance index (ADPI)
空气间距 air clearance
空气洁净度等级 air cleanliness class
空气冷却风道 air cooling duct
空气冷却器 air cooler
空气冷却器出风温度 air cooler outlet air temperature
空气冷却器进风温度 air cooler inlet air temperature
空气冷却系统 air cooling system
空气流量 air flow rate
空气湿度 air humidity
空气围带 inflatable rubber seal
空气污染 air pollution
空气源热泵 air-source heat pump
空蚀 cavitation erosion（cavitation pitting）
空蚀保证期 cavitation pitting guarantee period
空蚀保证运行时间 cavitation pitting gu-arantee duration of operation
空蚀破坏 cavitation damage
空-水冷却发电机 air-to-water cooled machine
空闲容量 idle capacity
空箱式挡土墙 chamber retaining wall
空隙量 volume of voids
空心电抗器 air-core reactor
空心射流泄荷阀 hollow-jet valve
空心锥形泄荷阀 hollow-cone valve（Howell-Bunger valve）
空压机室 air compressor room
空压机站 air compressor plant
空载电流 no-load current
空载工况 no-load operating condition
空载励磁电流 no-load field current
空载励磁电压 no-load field voltage
空载扰动试验 no-load disturbing test
空载试验 no-load test
空载试运行 trial running without load
空载损耗 no-load loss
空载特性 no-load characteristics，open circuit characteristics
空载运行 no-load operation，idling operation
空注阀 hollow jet valve
空转 idling
孔板 orifice plate
孔底高程 borehole bottom elevation
孔口封闭灌浆法 orifice-closed grouting method
孔口封闭器 packing gland
孔口高程 collar elevation
孔口净高 opening height
孔口净宽 opening span（opening width）
孔口型式 opening type
孔内残留岩芯 amount of core left in the hole
孔内掉块 material caving into the borehole
孔深 borehole depth
孔隙比 void ratio

孔隙率　porosity
孔隙潜水　pore groundwater
孔隙水　pore water
孔隙水压力　pore pressure
孔隙压力计（盒）　pore pressure meter（cell）
孔隙压力消散试验　test of pore pressure dissipation
孔斜　borehole deviation
孔斜计　clinograph
控导工程　river control works
控制爆破　controlled blasting
控制比降固结试验　consolidation test under controlled gradient
控制测量设备　control and metering equipment
控制触头　control contact
控制点　control point
控制电缆　control cable
控制方式　control mode
控制功能　control function
控制观测　control observation
控制环　regulating ring
控制级　control level
控制开关　control switch
控制流路　main current control
控制盘　control board
控制盘室　control panel room
控制盘台　control panel and console
控制器　controller
控制区　control zone
控制网　control network
控制网原点　origin of control network
控制系统本体　control systems proper
控制系统不动时间　control system dead time
控制系统部件　control systems component
控制系统接口　control system interfaces
控制性进度　critical schedule
控制中心　control center
口岸价　Border Price
扣件式钢管脚手架　steel tubular scaffold with couplers
枯水调节　low flow regulation
枯水流量　low flow
枯水期　low-flow period
枯水期导流　dry season diversion
枯水期围堰挡水　retaining water by cofferdam in dry season
枯水预报　low flow forecast
枯枝落叶层　litter layer
库岸　reservoir bank
库朗条件　Courant condition
库仑　Coulomb
库盆防渗设计　reservoir basin seepage control design
库容分配　storage allocation
库容系数　coefficient of storage
库水位年周期变化　annual periodic fluctuation of reservoir level
库尾　reservoir head
跨步电压　step voltage
跨接电路　crowbar circuit
跨孔测试　cross-hole test
跨孔层析摄影　crosshole tomography
跨孔地震探测　crosshole seismic probe
跨孔电阻率探测　crosshole resistivity probe
跨流域调水　trans-basin water transfer
跨流域开发　transbasin development，interbasin development
跨流域引水工程　trans-basin diversion project
跨线故障　cross country fault，cross circuitry fault
跨越　crossing
块裂结构　blocky-fractured structure
块石　rubble
块石　rock block
块石排水沟　rubble drain
块式制动器　shoe type brake
块状变质岩　massive metamorphic rock
块状构造　massive structure
块状结构　massive structure，blocky structure
块状破裂岩体　blocky and seamy rock
快剪试验　quick shear test
快速傅里叶变换　fast Fourier transform（FFT）
快速关闭　rapid closure
快速减负荷　rapid unloading
快速接地开关　high speed earthing switch
快速励磁　high response excitation
快速熔断器　ultra-rapid fuses
快速瞬态过电压　very fast transient overvoltage（VFTO）
快速停机　quick shutdown
快速闸门　quick-acting shutoff gate，stop gate

快速自动重合闸　high speed automatic reclosing
快硬水泥　quick-hardening cement
宽　width
宽顶堰　broad-crested weir
宽高比　aspect ratio
宽级配　spreading gradation
宽尾墩　end-flared pier
矿产资源　mineral resources
矿化度　mineralization，salinity
矿化水　mineral water
矿物掺合料　mineral admixture
框架横梁　spanning member
框架护坡　framed revetment
框架结构　frame structure
框架梁　frame beam
框架梁中间节点　intermediate nodes of frame beam
框架梁柱节点　beam-column nodes in frame
框架式房屋　frame-type house
框架支撑　shored with cribbing
框架支撑结构　braced frame structure
框架柱　frame column
框图　block diagram
馈线　feeder
馈线断路器　feeder breaker
馈线隔离开关　feeder disconnector
溃坝分析　dam break study
溃坝洪水　dambreak flood
溃屈　buckling
扩大单元工程　combined item works
扩大电流值　extended rating current
扩脚桩，支座桩　pedestal pile
扩孔　enlarge boring
扩孔式全断面岩石掘进机　full face rock TBM with reaming type
扩孔桩　belled-out pile
扩散式消力戽　diffusion bucket
扩散与束流墙　spray walls and training wall
扩挖　enlargement
扩展基础　spread footing
阔叶林　broad-leaved forest

L

拉拔试验　pulling test
拉杆　rod，hanger
拉格朗日法　Lagrangian method
拉筋　lacing
拉开法［涂层附着力试验］　pull-off test
拉力　tensile force
拉力分散型锚索　tensioned multiple-head tendon
拉力型锚索　tensioned grout tendon
拉裂面　pulling apart plane
拉普拉斯变换　Laplace transformation
拉普拉斯逆变换　inverse Laplace transform
拉索桩　cable-stay pile
拉线　guy，stay
拉线棒　anchor rod
拉线盘　anchor
拉压复合型锚索　tension-compression combined tendon
拉张区　tension zone
喇叭形进水口　bell mouth inlet
蜡封法　wax-sealing method
蜡封试样　wax-sealed sample
拦洪高程　retention structure elevation
拦门沙　bar
拦泥库　sediment detention reservoir
拦沙坝　sediment trapping dam
拦沙坎　silt sill
拦沙堰　sediment detention weir
拦石网　protecting wire mesh
拦污漂　trash boom
拦污栅　trash rack
拦渣工程　tailing hold structure
栏杆　railing
蓝图　blueprint
缆机平台　cable crane platform
缆索起重机　cable crane
缆芯　cable core
廊道　gallery
浪压力（波浪力）　wave pressure，wave force
劳动安全与工业卫生　labor safety and industrial hygiene
劳动保护费　labor protection expense
劳动保险　labor insurance
劳动定额　labor norm and standard
劳动力　labor force
劳动力机会成本　Opportunity Cost of Labor
劳动力密集的行业　labor intensity industry
劳动力市场　labor market

劳务收入 service income
涝 surface waterlogging
勒夫波 Love wave
勒脚 plinth
勒克斯 lux
雷暴日 thunderstorm day
雷暴小时 thunderstorm hour
雷电冲击半峰值时间 time to half value of a lightning impulse
雷电冲击波保护比 protection ratio against lightning impulse
雷电冲击波前时间 front time of a lightning impulse
雷电冲击截波耐受电压 withstand voltage of chopped lightning impulse
雷电冲击截波试验 chopped lightning impulse test
雷电冲击全波 full lightning impulse
雷电电磁脉冲 lightning electromagnetic pulse (LEMP)
雷电过电压 lightning overvoltage
雷电流 lightning current
雷电流峰值 peak value of lightning current
雷电流平均陡度 average steepness of lightning current
雷电流总电荷 total charge of lightning current
雷管 blasting cap，detonator
雷击跳闸率 lightning outage rate
雷诺数 Reynolds number
雷诺数相似 Reynolds similitude
雷诺相似定律 Reynolds' similarity law
类复理石建造 flyschoid formation
累积频率 cumulative frequency
累积曲线 mass curve
累计输沙量 cumulative sediment load
累计影响 cumulative impact
累年最冷月 normal coldest month
累年最热月 normal hottest month
棱角状 angular
棱体排水 prism drainage
楞次定律 Lenz law
冷备用 cold standby reserve
冷底子油 adhesive bitumen primer
冷冻水 chilled water
冷缝 cold joint
冷拉钢筋 cold-drawn bar
冷凝水管路 drip pipe，condensate line
冷凝水流量 condensate flow rate
冷凝温度 condensation temperature
冷却负荷 cooling load
冷却管 cooling pipe
冷却媒质 cooling medium
冷却器 cooler
冷却水 cooling water
冷却水泵 cooling water pump
冷却水阀门位置 cooling water valve position
冷却水源 water supply for cooling
冷却塔 cooling tower
冷却系统 cooling system
冷态工作压力 cold working pressure (CWP)
冷态试验差压力 cold differential test pressure
冷弯 cold bend；cold gagging
离岸价 free on board (FOB)
离差系数 coefficient of skew
离散量 discrete quantity
离散元分析 discrete element method (DEM)
离析 segregation
离线式归档 off-line filing
离相封闭母线 isolated-phase bus (IPB)
离心泵 centrifugal pump
离心风机 centrifugal fan
离子毫克当量 milligram equivalent of ion
离子毫克数 milligram value of ion
里氏震级 Richter magnitude
理货 tally
理论排量 theoretical discharge capacity
理论输送量 theoretic delivery
理论扬程 theoretical pump head
理论蕴藏量 theoretical potential
理想变压器 ideal transformer
理想电流源 ideal current source
理想电压源 ideal voltage source
力 force
力矩 moment of force
力矩面积法 area-moment method
力矩平衡重 torque counterweight
力矩系数 torque coefficient
力矩因数 torque factor
力特性 force character
力特性试验 force characteristic test
历时曲线 duration curve
历时系列 duration series

历史洪水灾害　historical flood damage
历史建筑　historical building
历史数据存储　historical data memory
历史数据库服务器　historical data server
历史文化遗产　historical and cultural heritage
历史文化遗址　historical and cultural site
立堵截流　end-dump closure
立方体强度等级　class of cube strength
立卷　filing
立卷单位　fonds constituting unit
立式水轮发电机　vertical hydrogenerator
立视图　elevation view
立体桁架　spatial truss
立体图　stereogram
立轴机组　vertical shaft unit
立轴式弧形闸门　sector gate，vertical axes tainter gate
立爪式装岩机　gathering-arm rock loader
励磁变压器　excitation transformer
励磁电流调节器　excitation current regulator（ECR），field current regulator（FCR）
励磁控制　excitation control
励磁绕组　excitation winding，field winding
励磁绕组端子　field winding terminals，excitation leads
励磁系统　excitation system
励磁系统顶值电流　excitation system ceiling current
励磁系统顶值电压　ecitation system ceiling voltage
励磁系统额定电流　excitation system rated current
励磁系统额定电压　excitation system rated voltage
励磁系统额定响应　excitation system nominal response
励磁系统故障　excitation system fault
励磁系统空载顶值电流　excitation system no-load ceiling current
励磁系统空载顶值电压　excitation system on-load ceiling voltage
励磁涌流　inrush current
励磁装置　exciter
利润　profit
利润税　profit tax
利息备付率　interest coverage ratio（ICR）
利益相关者参与　stakeholder engagement
利用系数　utilization factor
沥青弗拉斯脆点　Fraass breaking point of bitumen
沥青含量　bitumen content
沥青混合料　bituminous mixture
沥青混凝土　asphalt concrete
沥青混凝土流变　rheology of asphalt concrete
沥青混凝土面板　asphalt concrete facing
沥青混凝土水稳定系数　immersion coefficient of asphalt concrete
沥青混凝土心墙　asphalt concrete core
沥青劲度模量　stiffness of bitumen materials
沥青蜡含量　bitumen paraffin content
沥青路面　bituminous pavement
沥青玛蹄脂　bituminous mastic
沥青溶解度　bitumen solubility
沥青软化点　bitumen softening point
沥青砂浆　bituminous mortar
沥青杉木板　asphalt Chinese fur board
沥青闪点　bitumen flash point
沥青涂层　bitumen coat
沥青延度　bitumen ductility
沥青针入度　bitumen penetration
例行维护与检查　routine maintenance and inspection
砾砂岩　conglomeratic sandstone
砾石　gravel
砾岩　conglomerate
砾质土　gravelly soil
粒度　granularity
粒度模数　grain modulus（GM）
粒化高炉炉渣　granulated blast-furnace slag
粒状材料浆液　particulate grout
粒状灰岩　granular limestone
粒状结构　granular texture
粒组　fraction
连拱坝　multiple arch dam
连接法兰　coupling flange
连接件　connector
连接金具　link fitting，insulator set clamp
连接螺栓　coupling bolt（connecting bolt）
连接箱　link box
连接组标号　connection symbol
连体结构　connected structure
连通率　continuity
连通试验　hydraulic connectivity test

连续冲洗式沉沙池　continuous scouring sand basin
连续倒塌　progressive collapse
连续工作制　continuous running duty
连续供暖　continuous heating
连续级配　continuous gradation
连续浇筑　continuous concreting（placement）
连续梁　continuous beam
连续梁中间支座 intermediate support of continuous beam
连续取样　continuous sampling
连续式混凝土搅拌站　continuous concrete mixing plant
连续式搅拌机　continuous mixer
连续挑坎　continuous flip bucket
连续通风　continuous ventilation
帘面积　curtain area
联板　yoke plate
联合发电控制　joint load generation control
联合试运转费　expense of joint commissioning
联合通风　natural and mechanical combined ventilation
联合无功控制　joint load var control
联合消能　combined energy dissipation
联络变压器　system interconnection transformer
联络线　interconnection line
联络线负荷　connection line load，tie-line load
联络线输送容量　transmission capacity of a link
联排式住宅　row house，terrace house
联锁　interlock
联锁控制　interlock control
联系廊　inter-unit gallery
联用消防车　universal fire truck
联轴螺栓　coupling bolts
联轴器　coupling
联轴器使用系数　coupling service factor
联轴器罩　coupling guard
链板　chain plate
链路　link
链式启闭机　chain hoist
链条　roller chain
粮食补贴　grain allowance
粮食产量　grain output
粮食作物　grain crops
两瓣式岩芯管　double split barrel
两岔管　bifurcated pipe
两电动势间相角差　angle of deviation between two e. m. f.'s
两端固定梁　beam fixed at both ends
两管制水系统　two-pipe water system
两系统同步　synchronization of two systems
两相短路　phase-to-phase short-cicuit，line-to-line short-cicuit
两相对地短路　two phase-to-earth short-cicuit，double-line-to-earth short-cicuit
两相运行　two-phase operation
两用铲　convertible shovel
亮度　luminance
亮度计　luminance meter
亮灰绿色　light grayish green
亮色矿物　light mineral
量程　range
量水堰　measuring weir
了解　learn about
料场开采　exploitation of natural construction materials
料源规划　material source planning
列维准则　Levy's criteria
列线图（图算法）　nomography
劣化　degradation
劣化模型　degradation model
裂缝　crack
裂缝充填物　fissure filling，crack filling
裂缝开度　crack aperture
裂缝控制等级　crack control class
裂缝宽度控制　crack width control
裂缝宽度限值　allowable value of maximum crack width
裂缝愈合　crack healing
裂谷　rift valley
裂纹　crack
裂隙产状修正评分　rating adjustment for joint orientation
裂隙承压水　fissure artesian groundwater
裂隙膨胀与压缩　joint dilation and contraction
裂隙潜水　fissure groundwater
裂隙水　fissure water
裂相横差保护　transverse differential protection，split-phase differential protection
邻谷渗漏　leakage to adjacent valley
邻近效应　proximity effect

林场 forest farm
林带 forest belt
林地 forest land
临界档距 critical span
临界地下水埋深 critical depth of groundwater table
临界空化系数 critical cavitation coefficient
临界量 threshold quantity
临界流 critical flow
临界破坏面 critical failure surface
临界气蚀余量 critical net positive suction head (NPSH)
临界水力坡降 critical hydraulic gradient
临界水深 critical water depth
临界阻尼 critical damping
临空面 free face
临时缝 temporary joint
临时工程 temporary works
临时护壁，临时套管 temporary casing
临时拦洪断面 particular section of embankment for flood retaining
临时拦蓄暴雨径流 temporary storage of spate runoff
临时弃物存放 temporary spoil storage
临时设施费 expense for temporary facility
临时性建筑物 temporary structure
临时性预应力锚杆 temporary prestressed anchor
临时悬吊松动岩石 false hanging loose rock
临时淹没区 infrequent flooded zone
临时用地 temporary land occupation
临时支护 temporary support
淋溶侵蚀 leaching erosion
磷化处理 phosphating
磷灰石 apatite
磷渣粉 phosphorous slag powder
灵敏度漂移 sensitivity shift
菱形柱 cant column
零搭叠 zero-lapped
零点漂移 zero shift
零件 component
零起升压 voltage build up from zero
零星林木 scattering trees
零序保护 zero sequence protection
零序差动保护 restricted earth fault protection (REF)
零序电流保护 zero sequence current protection
零序电流互感器 zero sequence current transformer, residual current transformer
零序电流继电器 zero sequence current relay, residual current relay
零序电流型横差保护 zero sequence differential protection
零序电压保护 zero sequence voltage protection
零序电压继电器 zero sequence voltage relay, residual voltage relay
零序短路电流 zero sequence short-circuit current
零序分量 zero sequence component
零序网络 zero sequence network
零序阻抗 zero sequence impedance
留振时间 compaction time
流变 rheology, flowing deformation
流变特性 rheological behavior
流场 stream field
流程联锁 sequence interlocking
流道面积 flow area
流动监测 mobile monitoring
流动人口 migratory population
流动性混凝土 flowing concrete
流量 flow rate
流量变送器 flow transducer
流量阀 flow valve
流量恒定泵 constant displacement pump
流量计 flow meter
流量开关 flow switch
流量历时曲线 flow-duration curve
流量系数 discharge coefficient
流量因数 discharge factor
流明 lumen
流劈理 flow cleavage
流水作业法 flow operation method
流速 flow velocity
流速场 velocity field
流速脉动 velocity fluctuation
流速仪 current meter
流速仪法 current meter method
流体动力径向轴承 hydrodynamic radial bearing
流体动力止推轴承 hydrodynamic thrust bearing
流体动力轴承 hydrodynamic bearing
流土 soil flow
流网 flow net

流纹片麻岩　rhyolite-gneiss
流纹岩　rhyolite，liparite
流纹状构造　rhyotaxitic structure
流线　stream line
流域　river basin，watershed
流域产沙量　watershed sediment yield
流域地理位置示意图　Sketch of geographic location of the river basin
流域分水线　river basin divide
流域湖泊率　percentage of lakes in drainage area
流域面积（集水面积）　catchment area，drainage area
流域输沙量　amount of sediment delivery
流域沼泽率　percentage of swamps in drainage area
流域植被率　percentage of vegetations in drainage area
流状结构　fluidal texture
硫化氢　hydrothion
硫磺华　sulphur-sinter
硫酸镁型腐蚀　magnesium sulfate corrosion
硫酸盐腐蚀　sulfates corrosion
六角头螺栓　hex bolt
龙口　closure gap
龙口平均流速　average discharge velocity at closure-gap
龙门吊　frame crane
龙抬头泄洪洞　inlet-raised discharge tunnel
笼型感应电动机　cage induction motor，squirrel cage induction motor
隆升　uplift
隆升速率　uplift amplitude
楼板　floor slab
楼层平面图　floor plan
楼梯　stair
楼梯间　stairway
楼梯平台　stair landing
楼梯踏步　stair step
楼梯踢板　stair riser
楼梯详图　stair detail drawing
漏磁通　leakage flux
漏风量　air leakage rate
漏风率　air system leakage ratio
漏光检测　air leak check with lighting
漏水　water loss
漏水损失　leakage loss
陆海联运　land-sea through transportation
陆空联运　land-air through transportation
陆面蒸发　evaporation from land surface
陆生物种　terrestrial species
陆相　continental facies
陆相剥蚀平原　plain of subaerial denudation
陆相红层　red bed of continental facies
路堤　embankment
路基　subgrade
路肩　shoulder
路面　pavement
路堑　cutting
路线测量　route survey
路缘石　curb stones
滤波电抗器　filter reactor，smoothing reactor
滤波器　filter
滤水器差压高　water strainer high differential pressure
滤芯强度　filter element strength
滤油车　oil filter vehicle
滤油器　oil filter
露点　dewpoint
露点温度　dew-point temperature
露顶式闸门　emersed gate
露天环境　outdoor environment
露天开关站　open switchyard
露天式厂房　outdoor powerhouse
露天斜井　shaft with open air collar
吕梁运动　Luliang movement
吕荣值　Lugeon value
旅馆　hotel
旅游　tourism
铝包钢加强铝绞线　aluminium clad steel reinforced aluminium conductor
铝合金绞线　all aluminium alloy conductor（AAAC）
铝绞线　all aluminium conductor（AAC）
铝酸盐水泥　aluminate cement
铝土页岩　bauxitic shale
履带式起重机　crawler crane
履带式挖掘机　caterpillar excavator
履带式闸门（链轮闸门）　caterpillar gate，roller chain gate
履带式钻机　track-mounted drill
履约保函，履约担保　performance guarantee
履约保证金　performance bond，performance security

履约证明　performance certificate
率定（校准）　calibration
绿地率　green space rate
绿地用水　water for green belt
绿化　afforestation
绿帘石　epidote
绿帘石脉　epidote vein
绿泥石　chlorite
绿色照明　green lighting
绿岩（绿闪石片岩）　noricite
氯离子在混凝土中的扩散系数　chloride diffusion coefficient of concrete
卵石　cobble，pebble
卵石排水沟　cobble drain
卵石土　cobble soil
轮距　wheel track
轮胎吊　rubber-tyred gantry crane（RTG）
轮压　wheel load
轮叶　impeller blade
轮叶接力器　blade servomotor
逻辑回路　logical circuit
逻辑卷　logic volume
螺杆启闭机　Screw stem hoist
螺孔　screw hole
螺母　screw nut
螺栓　screw
螺栓连接　bolted connection
螺旋钢筋　spiral rebar
螺旋输送机　screw conveyor
螺旋钻　auger drill
螺旋钻孔机　auger drilling machine
裸露型岩溶　bare karst
裸露装药爆破　adobe blasting
洛杉矶试验机磨耗试验　Los Angeles machine abrasion test
落差　drop
落地罐式断路器　dead tank circuit-breaker
落水洞　sinkhole

M

麻花钻　helical
麻粒岩　granulite
麻面　pitted surface
马道　berm
马蹄形断面　horseshoe section
马歇尔稳定度　Marshall stability
码头　wharf
埋藏阶地　buried terrace
埋藏式斜井　shaft with underground collar
埋管　buried penstock
埋弧自动焊　submerged arc welding（SAW）
埋件图　Embedded part diagram
埋入式测温计　embedded temperature detector
埋设仪器系统　embedded instrument system
埋葬型岩溶　buried karst
埋钻　drill rod burying
麦加利烈度　Mercalli intensity
麦克斯韦　Maxwell
脉波数　pulse number
脉冲　pulse
脉冲变压器　pulse transformers
脉冲传感器　impulse transducer
脉冲放大器　pulse amplifier
脉冲控制　pulse control
脉冲命令　pulse command
脉动量　pulsation of quantity
脉动压力　fluctuating pressure
脉动压力计　pulse pressure gauge
脉动因数　pulsation factor
脉动直流电流　pulsating direct current
脉宽调制控制　pulse width modulation control（PWM）
脉岩　vein rock，dyke rock
脉状水　veinwater
满发利用小时数　full output power hours
满负荷关机　full-load shut down
满负荷试运行　trial running with full load
满管流　full conduit flow
满管压力流雨水排水系统　full pressure storm system
满载　full load
满载值　full load value
曼宁公式　Manning equation
慢剪试验　slow shear test
慢性毒性　chronic toxicity
漫射型灯具　diffused luminaire
盲区　blind zone
毛面　rough surface
毛石　quarry-run rock
毛石骨料　rubble aggregate

毛石圬工坝　rubble masonry dam
毛水头　gross head
毛细管带　zone of capillarity
毛细上升高度　capillary rise
毛细水　capillary water
毛重系数　coefficient of gross weight
锚定梁　anchor beam
锚定螺栓　anchor bolt
锚定装置　anchor
锚墩　anchor block
锚杆饱满度　anchor bolt satiation degree
锚杆模拟试验　anchor bolt simulation test
锚杆式挡土墙　anchor retaining wall
锚杆应力计　rock bolt stress meter
锚杆注浆密实度　anchor bolt satiation degree
锚固　anchorage
锚固洞　retaining concrete plug
锚固段　anchored section
锚固砂浆　anchoring mortar
锚筋　anchor
锚筋束（桩）　anchor bundle
锚具　anchorage
锚索（杆）测力计　anchorage cable (rock bolt) dynamometer
锚头　anchor head
冒顶　roof fall
冒浆　grout emitting
冒水喷沙　water spraying and sand emitting
梅花桩　quincuncial pile
煤层　coal seam
每段进尺记录　record of each run
美术馆　art museum
门背联结系，背拉杆　diagonal
门槽　gate slot (gate groove)
门槽型式　gate slot type
门洞　door opening
门斗　air lock
门架　gantry
门禁控制　access control
门禁控制器　access controller
门禁系统　access control system
门禁系统的设防状态　set condition of access control system
门框　door frame
门廊　porch
门楣　lintle
门式启闭机（门机）　gantry crane
门式起重机　gantry crane
门厅　lobby
门腿　support column
门叶结构　gate leaf
门轴柱　quoin post
门轴柱支枕垫　quoin block
猛度　brisance factor
蒙脱石　montmorillonite
弥补亏损　recovery of loss
迷宫密封　labyrinth seal
迷宫式溢洪道　labyrinth spillway
糜棱结构　mylonitic texture
糜棱岩　mylonite
米赛斯准则　Von Mises criteriaon
觅食地　feeding area
泌水　bleeding
秘密级　confidential secret
密度　density
密度传感器　density sensor
密度计法　densimeter method
密度继电器　density relay
密度试验　density test
密封电池　sealed cell
密封隔离　hermetic separation
密封面　sealing face
密封面积　seat area
密封面斜角　seat angle
密封（嵌缝）材料　sealant
密封圈（密封件）　sealing ring (element)
密封试验　closure test
密级　security classification
密级配骨料　dense-graded aggregate
密级配沥青混凝土　dense-graded asphalt concrete
密集架　compact shelving
密实度　compactness
免税　duty free, tax exclusion
免维护电池　maintenance-free battery
面板垫层　bedding layer
面板接缝　face joint
面板脱空　separation between concrete slab and cushion
面波　surface wave
面分布力　force per unit area

面积置换率　replacement rate
面接触钢丝绳　facial contact lay wire rope
面流消能　surface flow energy dissipation, roller bucket type energy dissipation
面膜　protective membrane
面漆　surface coat
面蚀　surface erosion
面污染源　non-point source pollution
面雨量　areal precipitation
苗木　nursery stock
苗圃　nursery garden
灭磁回路　field suppression circuit, de-excitation circuit
灭弧管　arc-extinguishing tube
灭弧室　arc-extinguishing chamber
灭弧装置　arc-control device
灭火级别　fire rating
灭火剂　extinguishing agent
灭火器　fire extinguisher
灭火器配置场所　distribution place of fire extinguisher
灭火系统　fire extinguishing system
灭火装置　fire-extinguishing device
民用建筑　civil architecture
民族意识　ethnic awareness
敏感对象　sensitive object
敏感性分析　Sensitivity Analysis
敏感性土　sensitive soil
敏感性指标　Sensitivity indicator
名义价格　nominal price
名义值　nominal value
明矾石膨胀水泥　alunite expanding cement
明杆闸阀　outside screw rising stem type gate valve
明管　exposed penstock
明火　naked flame
明渠导流　channel diversion
明渠流　open channel flow
明渠排水　open-channel drain
明挖方量　excavation in open-cut
明挖回填截水槽　open-cut and backfilled cutoff trench
模袋混凝土　bagged concrete
模量　modulus
模内捣实混凝土　packing concrete in forms
模拟降雨试验　simulated rainfall experiment
模拟量　analog variable
模拟量测值监视　monitoring of analog measurements
模拟量输出　analog output
模拟量输出模件　analog output (AO) module
模拟量输入　analog input
模拟量输入模件　analog input (AI) module
模拟量指示仪表　analog indicating instrument
模拟屏　mimic board
模拟通信　analogue communication
模-数变换器　analog-to-digital converter
模型到原型换算　model to prototype conversion
模型对比试验　comparative model test
模型分析　model analysis
模型试验　model test
模型水轮机　model turbine
膜结构　membrane structure
膜应力　film stress
摩擦力矩　friction torque
摩擦面　friction surface
摩擦通风损耗　friction and windage loss
摩擦系数　coefficient of friction
摩擦型锚杆　friction anchor bolt
摩根斯顿-普赖斯法　Morgenstern-Price
磨拉石建造　molasse formation
磨蚀　abrasion, combined erosion by sand and cavitation
磨圆碎屑　rounded fragment
蘑菇石　mushroom stone
蘑菇轴头　pintal
抹面机，修整机　finisher
末端试水装置　end water-test equipment
莫尔-库仑准则　Mohr-Coulomb criteria
莫尔破裂包络线　Mohr's rupture envelope
母公司保函　parent company guarantee
母联断路器　bus tie breaker
母线　busbar
母线槽　busway
母线层　busbar floor
母线洞　busbar cavern
母线分段隔离开关　busbar section disconnector
母线固定金具　busbar support clamp
母线故障　busbar fault
母线接地开关　earthing switch for busbar

母线金具　busbar fitting
母线廊道　busbar gallery
母线伸缩节　busbar expansion joint
母线式电流互感器　bus type current transformer
母线转换断路器　switched busbar circuit-breaker
母岩　host rock
木材　timber，lumber
木材加工厂　timber processing plant，carpenter shop
木柴　firewood
木房　wood house
木笼坝　crib dam
木笼填石坝　rock-crib dam
目标函数　objective function
目测　eye observation
目测草图　visual sketch
目的地交货价格　free on board destination
目录　catalogue，inventory
目录数据库　catalogue database

N

纳维-斯托克斯方程式　Navier-Stokes equations
耐电弧性　arc resistance
耐电压试验　withstand voltage test
耐腐蚀泵　anti-corrosive pump
耐火材料　refractory material
耐火等级　fire resistance classification
耐火电缆　fire resistant cable
耐火分隔　fire resistant barrier
耐火隔板　refractory shield
耐火隔墙　fire resisting partition
耐火隔热性　refractory insulation
耐火隔热砖　refractory and insulating fire brick
耐火管道　fire resisting duct
耐火混凝土　refractory concrete
耐火极限　duration of fire resistance
耐火面　refractory surface
耐火黏土砌块　refractory clay block
耐火竖井　fire resisting shaft
耐火完整性　refractory integrity
耐火稳定性　refractory stability
耐火 4 小时的防火墙　firewall with 4-hour fire rating
耐火性　fire resistance
耐久性　durability
耐雷水平　lightning withstand level
耐受电压　withstand voltage
耐受概率　probability of withstand
耐污绝缘子　anti-pollution type insulator
耐压试验　pressure test
耐压性　resistance to pressure
耐油　oil resistance
耐张绝缘子串　tension insulator string
耐张绝缘子串组　tension insulator set
耐张塔　tension support，angle support
耐张线夹　tension clamp，dead-end clamp
难确定的（无形）洪水灾害　intangible flood damage
难燃的　difficult-flammable
囊状风化　scrotiform weathering
挠度　deflection
挠曲　bending fold
内保温　internal thermal insulation
内部观测　internal monitoring
内部侵蚀　internal erosion
内部收益率　Internal Rate of Return（IRR）
内插法　interpolation
内导水环　inner guide ring
内垫层　inner jacket
内过电压　internal overvoltage
内弧面，拱腹　intrados
内加强月牙肋岔管　crescent-rid reinforced bifurcation
内绝缘　internal insulation
内壳　inner casing
内窥镜　endoscope
内力调整系数　adjustment coefficient of internal force
内陆运输　inland transportation
内锚固段　inner anchored section
内摩擦角　angle of internal friction
内墙　interior wall
内水压力　internal water pressure
内缩量　drawn-in
内泄漏试验　inner-leakage test
内业　office work，indoor work
内因　internal cause
内营力　endogenetic agent
内置式行程检测装置　built-in travel detector
能　energy

能耗制动　dynamic braking
能量法　energy method
能量管理系统　energy management system (EMS)
能量系数　energy coefficient
能量效益　energy benefit
能量指标　energy indexes
尼龙　polyaminde (PA), Nylon
泥化夹层　argillized seam
泥灰岩　marl
泥浆　slurry
泥浆泵　sludge pump
泥浆防渗墙　slurry cutoff
泥浆固壁　slurry wall stabilizing
泥结碎石路面　clay-bound macadam pavement
泥流阶地　mudflow terrace
泥盆纪（系）　Devonian period (system)
泥沙冲淤平衡　sediment erosion and deposition balance
泥沙防治　sediment control
泥沙颗粒分析　grain size analysis
泥沙矿物成分　solid mineral composition
泥沙粒径　solid grain size, silt grain size
泥沙流　silt flow
泥沙磨损　sand erosion
泥沙起动　incipient motion of sediment
泥沙输移　sediment transport
泥沙输移比　sediment delivery ratio
泥沙形态系数　shape coefficient of sediment
泥沙性质　property of sediment
泥沙淤积　sediment accumulation
泥沙预报　sediment forecast
泥石流　debris flow
泥石流堆积区　debris flow accumulating region
泥石流防治工程　debris flow control works
泥石流流通区　debris flow transporting region
泥石流形成区　debris flow forming region
泥石流易发区　susceptible area of debris flow
泥岩　mudstone
泥质灰岩　argillaceous limestone
泥质胶结　argillaceous cement
泥质结构　pelitic texture
泥质物充填　argillaceous infilled
拟定移民方针　development of a resettlement concept
拟静力法　pseudo-static method
逆变　inversion
逆变灭磁　de-excitation by inversion
逆变器　inverter
逆变因数　inversion factor
逆断层，上冲断层　upthrown fault, thrust fault
逆功率保护　reverse power protection
逆掩断层　overthrust fault
年变幅　annual amplitude
年变化　annual variation
年代测定　age dating
年单向通过能力　annual through capacity in one-way
年调节　yearly regulation, annual regulation
年度计划　annual plan
年度投资费　annualized capital cost
年度维修费用　annual maintenance cost
年度预算　annual budget
年度折旧费　annual amortization cost
年金化值　annuities value
年利率　annual interest rate
年利用小时数　annual utilization hours
年漏气率　yearly gas leakage rate
年平均发生率　annual incidence rate
年平均气温　mean annual temperature
年日照时数　annual solar radiation hours
年税额　annual tax
年息　interest per annum
N 年一遇洪水　N-year flood
年运输量　yearly transport volume
年终股息　year-end dividend
黏弹性变形　viscoelastic deformation
黏弹性模量　viscoelastic modulus
黏度　viscosity
黏附性　adhesiveness
黏结强度　bond strength
黏结涂层　tack coat
黏聚力，凝聚力　cohesion
黏土　clay
黏土充填　clay infilled
黏土夹层　clay seam
黏土斜墙土石坝　earth-rock fill dam with sloping clay core
黏土心墙土石坝　earth-rock fill dam with clay core
黏土岩　clay stone (rock)

黏土页岩　clay shale
黏土质结构　argillaceous texture
黏土质砂岩　argillaceous sandstone
黏性泥石流　viscous debris flow
黏性土　cohesive soil，clayey soil
黏滞塑性变形　visco-plastic deformation
黏着剂　bonding agent
碾压　rolling
碾压混凝土　roller compacted concrete（RCC）
碾压混凝土坝　RCC dam
碾压式土石坝　rolled earth-rock fill dam
碾压效率　roller compaction efficiency
鸟瞰图　aerial view
凝灰岩　tuff
凝灰岩粉　tuff powder
凝结水泵　condensate pump
凝聚沉淀　coagulation sedimentation
牛顿　Newton
牛轭湖　ox-bow lake
扭矩　torque
扭矩传感器　torque sensor
扭矩仪　torque meter
扭曲　distortion
扭曲式挑坎　distorted type flip bucket，torsioned flip bucket
扭曲褶皱　contorted fold
扭性结构面　torsion structural plane
扭转结构　twisted structure
扭转效应　torsional effect
农村家庭　rural household
农村移民　rural resettler
农村移民安置　rural resettlement
农田防护林　shelter-belt on farmland
农田林网　forest net in farmland
农田排水　farmland drainage
农业安置　agricultural resettlement
农用地　farmland
女儿墙　parapet
暖泵　warming-up
诺模图　nomogram

O

欧拉平衡方程　Euler's equilibrium equation
欧拉数　Euler number
欧姆　Ohm
欧姆定律　Ohm law
偶然误差　accidental error
偶然状况　accidental situation
偶然组合　accidental combination
偶然作用　accidental action
耦合　coupling
耦合电容器　coupling capacitor
耦合连续装药　coupled continuous charging
耦合系数　coupling coefficient

P

爬电比距　specific creepage distance
爬罐　raise climber（Alimak）
爬距　creepage distance
爬升（顶升）模板　climbing（jacked）form
帕氏量水槽　Parshall flume
帕斯卡定律　Pascal's law of pressure
拍　beat
拍频　beat frequency
排冰排漂　ice and floating debris release
排出压力　discharge pressure
排放背压力　built-up back pressure
排放和粉尘控制　emission and dust control
排放面积　discharge area
排放压力　relieving pressure
排风机　exhaust
排风机室　exhaust fan room
排风竖井　exhaust shaft
排风温度　exhaust temperature
排洪槽　over-chute
排架　bent frame
排架桩墩　bent pile pier
排浆量　grout discharge
排量　displacement
排气阀　air release valve，exhaust valve
排气管　exhaust pipe
排气帽　vent cap
排沙孔　sediment flushing outlet
排水不畅　impeded drainage
排水定额　wastewater flow norm
排水干沟　main drain
排水沟　drainage ditch，gutter
排水井　drainage well
排水孔　drain hole
排水廊道　drainage gallery

排水幕　drain curtain
排水铺盖　drainage blanket
排水区　drainage zone
排水体料　drainage material
排水体制　sewerage system
排水系统　wastewater system
排水支沟　subsidiary drain
排污阀　blow-down valve
排烟　smoke extraction
排烟风机　smoke extractor exhaust fan
排烟口　smoke vent
排烟竖井　smoke shaft
排烟系统　smoke evacuation system
排油阀　drain valve
排针闸门　pin gate
盘车　barring（turning）machine for alignment
盘车装置　barring gear
盘管冷凝器　coil condenser
盘式制动器　disc brake
盘香式启闭机　incense coil hoist
旁路母线　transfer busbar
旁通洞　bypass tunnel
旁通风道　by-pass air duct
旁通管充水　water filling by pass pipe
旁压试验　pressure-meter test（PMT）
抛石　riprap
抛石护坡　riprap slope protection
抛石区　rip-rap zone
抛投强度　dumping intensity
刨毛　scarifying
炮孔　blast hole
泡沫灭火剂　foam extinguishing agent
泡田　steeping field
泡田定额　duty of steeping field
培训费　training expense
培训工作站　training workstation
配电装置　power distribution unit
配管　piping
配筋不足　under-reinforced
配筋过多　over-reinforced
配筋间距　bar spacing
配筋率　reinforcement ratio
配筋砌体构件　reinforced masonry structure
配筋图　reinforcement drawing
配筋有效面积　effective area of reinforcement
配筋砖砌体　reinforcement brickwork
配料斗　batch hopper
配料器　batcher
配料筒仓　batching silo
配水干管　feed main
配水管　cross main
配水管道　system pipe
配水管路　intake pipe
配水管网　distribution system
配水渠道　water distribution canal
配水支管　branch line
配送　distribution
配压阀　distributing valve
配重式（平衡重式）　counterweight type
喷出岩　extrusive rock，eruptive rock
喷管　nozzle pipe
喷混凝土　shotcrete
喷浆车　gunite car
喷口送风　air supply through nozzle
喷淋冷却　spray cooling
喷淋冷却塔　spray cooling tower
喷锚支护　bolt-shotcrete support
喷泉　fountain
喷水段　spray chamber
喷水密度　water discharge density
喷水养护　wet curing
喷雾器　sprayer
喷油润滑　spray lubrication
喷针　needle
喷针杆　needle rod
喷针接力器　needle servomotor
喷针头　needle tip
喷针行程　needle stroke
喷针折向器定位装置　needle-deflector positioner
喷针折向器连杆　needle deflector link
喷嘴　nozzle，injector
喷嘴保护罩　nozzle shield
喷嘴口环　nozzle tip ring
盆地　basin
膨润土　bentonite
膨润土泥浆，皂土液　bentonite slurry
膨胀　expansion，swelling
膨胀剂　expansive agent
膨胀率　expansion ratio，specific expansion
膨胀试验　expansion test，swelling test

膨胀水泥　expanding cement
膨胀速率　rate of swelling
膨胀土　swelling soil，expansive soil
膨胀系数　coefficient of expansion（swelling）
膨胀性浆液　expanding grout
膨胀岩体　swelling rock
膨胀止水　expanded waterstop
批准　approved by
劈理　cleavage
劈裂剥落，片帮　splitting and peeling off，scaling
劈裂灌浆　fracturing grouting
劈裂强度　splitting strength
劈钻成槽法　trenching by percussion and splitting
皮带传动式　belt driven type
皮带机运输　belt conveyor delivery
皮带装料机　belt loader
皮肤病　skin disease
疲劳强度　fatigue strength
疲劳验算　fatigue analysis
匹配用电压互感器　voltage matching transformer
片冰　flake ice
片麻岩　gneiss
片麻状构造　gneissic structure
片岩　schist
片状构造　schistose structure
片状颗粒　flake-shaped particle
偏流器　diverter
偏斜　declination
偏心距　eccentricity
偏心率　relative eccentricity
偏心受压构件　eccentric compressed member
偏转角　angle of band，deflection angle
漂凌　drift ice
漂石（漂砾）　erratic boulder
拼合地带　collage belt
贫混凝土　lean concrete
贫胶渣砾料　lean cemented dregs and gravel
贫困线　poverty line
贫穷地区　impoverished area
频带　frequency band
频率　frequency
频率保护　frequency protection
频率变送器　frequency transducer
频率表　frequency meter
频率调节　frequency control
频率给定单元　frequency setting module
频率继电器　frequency relay
频率偏差　deviation of frequency
频率特性　frequency characteristics
频率稳定性　frequency stability
频率响应　frequency response
频敏电阻启动　frequency-sensitive rheostat starting
频谱　frequency spectrum
频闪灯　stroboscopic light
频闪效应　stroboscopic effect
频跳　chatter
频域分析　frequency-domain analysis
平板坝　flat slab dam
平板夯　plate compactor
平板式打夯机　flat beater
平板式振捣器　flat-plate vibrator
平板拖车　flatbed trailer
平板仪　plane-table
平板载荷试验　plate bearing test
平波电抗器　smoothing reactor
平仓　spreading and leveling
平顶堰　flat-topped weir
平洞　adit
平堵截流　full width rising closure
平方和方根法　square root of sum square method
平焊　flat welding
平衡衬套　balancing bushing
平衡阀　counterbalance valve
平衡鼓式　balancing piston type
平衡管　balancing pipe
平衡河槽［冲淤］　regime channel
平衡滑轮　equalizer pulley（sheave）
平衡节点　balancing bus
平衡孔式　balancing hole type
平衡链　balance chain
平衡盘式　balancing disc type
平衡室盖　cover of balancing chamber
平衡套　balancing sleeve
平衡温度　equilibrium temperature
平衡挟沙能力　regime sediment charge
平衡重　counterweight
平缓河道　mild channel
平浇法　horizontal placing method
平均传送时间　average transfer time

平均厚度法　average thickness method
平均粒径　mean diameter，mean particle size
平均流量　mean flow
平均失效间隔时间　mean time between failures
平均输沙量　mean sediment load
平均输沙率　mean sediment discharge
平均水头　average head
平均无故障工作时间　mean time to failure
平均运行张力　everyday tension
平均增量财务费用　average incremental financial cost（AIFC）
平均增量成本　average incremental cost（AIC）
平均增量经济费用　average incremental economic cost（AIEC）
平均值　mean value
平开窗　casement window
平面布置图　plan，layout
平面测量　plane survey
平面度　flatness
平面滑动　planar slide
平面滑动支座　sliding ring girder support
平面角　plane angle
平面控制测量　horizontal control survey
平面控制网　horizontal control network
平面闸门　plain gate，vertical-lift gate
平碾　drum roller
平曲线　horizontal curve
平台式模板　deck form
平头螺栓　flush bolt
平稳切换　bump-free switch-over
平卧褶皱　recumbent fold
平行测定　parallel measure
平行度　parallelism
平行断面法　parallel section method
平行式多叶阀　parallel-blade damper
平行式双闸板　parallel double gate
平行式闸阀　parallel gate valve（parallel slide valve）
平行试验　comparative test，parallel test
平行作业法　parallel operation method
平压方式　method of water pressure balance
平移式缆机　parallel-traveling cable crane
平移型滑坡　horizontal-moving landslide
平移（走滑）断层　displacement fault，strike-slip fault
平原　plain
平整度　evenness
平直的　planar
平直河道　straight channel
平直结构面　planar structural plane
平趾板　flat plinth
评标　bid evaluation
评价系统　assessment system
屏蔽电泵　canned motor pump
屏蔽环　shielding ring
屏浆　measure for keeping pressure to stage
坡边线　slope margin line
坡道　ramp
坡度　gradient，slope
坡积　slope wash
坡积裙　talus fan
坡口焊（斜角焊）　bevel welding，groove welding
坡面集雨工程　rainfall harvesting works on the slope
坡面截流沟　water intercepting and drainage ditch on the slope
坡面水系工程　slope water works
坡面治理工程　slope treatment for erosion control
坡式梯田　sloping terrace
坡体结构类型　type of slope structure
坡中谷　valley in slope
破冰式隔离开关　ice-breaking disconnector
破坏包线　failure envelope
破坏荷载　failure load
破坏圈　fragmental zone
破坏性地震　destructive earthquake
破坏性放电　disruptive discharge
破坏性放电电压　disruptive discharge voltage
破坏性放电试验　disruptive discharge voltage test
破劈理　fracture cleavage
破碎　crushed
破碎比　ratio of reduction
破碎机　crusher，breaker
破碎角砾岩　cataclastic breccia
剖口　groove
剖面图　sectional drawing
铺层厚度　lift thickness
铺盖　blanket
铺盖灌浆　blanket grouting
铺盖排水（褥垫排水）　blanket drain

铺料　placing and spreading
蒲福风力　Beaufort force
普工　labor
普氏击实试验　Proctor compaction test
普氏坚固系数　M. M. Protogiakonov's coefficient
普氏压实曲线　Proctor compaction curve
普通硅酸盐水泥　Portland cement

Q

栖息地改变　habitat change
栖息地消失　habitat loss
期（阶）　Stage (formation)
期望值　expected value
期中报告　interim report
期中付款　interim payment
齐发爆破　simultaneous blasting
其他费用　miscellaneous expense
其他项目费　sundry cost
气垫机制　air-cusion mechanism
气垫式调压室　air cushion surge chamber
气动操动机构　air operating mechanism
气动打夯机　air earth hammer
气动阀门　pneumatically operated valve
气动装置　pneumatic actuator
气动装置的行程　stroke of pneumatic actuator
气候变化　climate change
气候区　climatic zone
气化　gasify
气孔　porosity
气孔状构造　vesicular structure
气流组织　air distribution
气囊式储压器　bladder accumulator
气泡型水位传感器　bubble type water level sensor
气体保护　gas protection, Buchholz protection
气体保护焊　gas metal welding (GMAW)
气体放电　gas discharge
气体含量　gas content
气体火灾探测器　gas fire detector
气体介质击穿　gas dielectric breakdown
气体绝缘管道输电线（气体绝缘管道母线）　gas insulated line (GIL)
气体绝缘介质　insulating gas
气体绝缘金属封闭变电站　gas insulated metal-enclosed substation
气体绝缘金属封闭组合电器　gas insulated metal enclosed switchgear and controlgear (GIS)
气腿钻　pneumatic drill
气温　air temperature
气象站　meteorological station
气压浇注混凝土　air placed concrete
企业档案　business archives, business records
企业档案馆　business archives
企业管理费　enterprise overhead
企业计提费　welfare withdrawal
企业所得税　enterprise income tax
弃渣场　spoil area, waste disposal area
汽车保养厂　automobile service workshop
汽车旅馆　motel
汽车式起重机　truck crane
汽化压力　vapour pressure
汽蚀比转速　suction specific speed
汽蚀核　nuclie
汽油　gasoline
启闭机室　hoist building
启闭件　valvedisc
启闭容量（启闭力）　hoisting capacity
启闭速度　hoist speed
启闭压差　blowdown
启动母线　starting bus
启动器　starter
启动元件　starting component
启动值　starting value
启动转矩　start torque
启门力　lifting force
启停命令　start-stop commands
砌块砌体结构　block masonry structure
砌石　stone masonry
砌体强度　masonry strength
砌体砂浆　masonry mortar
起爆　initiation
起动流速　incipient velocity
起动拖曳力，临界拖曳力　incipient tractive force
起伏的　undulating
起伏度　fluctuation, waviness
起火（引燃）　ignite
起居室　living room
起升范围　lifting range
起升高度　load-lifting height
起升高度指示器　gate position indicator

起升机构　lifting mechanism
起升速度　load-lifting speed
起始边　initial side
起始数据　initial data
起始水力坡降　beginning hydraulic gradient
起始瞬态恢复电压　initial transient recovery voltage (ITRV)
起算震级　lower limit earthquake
起重吊运指挥信号　commanding signal for lifting and moving
起重葫芦　hoist
起重机的额定行走速度　nominal travel speed of appliance
起重机的谱等级 Q　sectrum classes for cranes
起重机轨　crane rail
起重机轨距　track center
起重机轨面高度　crane track height
起重机跨距偏差　divergence in span of crane
起重机械利用等级 U　class of utilization of lifting appliances
起重机运行机构　crane travel mechanism
起重机组别 A　crane groups
起重力矩　load moment
起重小车　crab，trolley
起重循环　hoisting cycle
恰贝冲击试验　Charpy test
千枚岩　phyllite
千枚状构造　phyllitic structure
千糜岩　phyllonite
迁地保护　ex-situ conservation
迁建补偿标准　relocation compensation standard
迁建费　relocation cost
迁建规划　relocation plan
迁移物种　migratory species
迁移着的空化　travelling cavitation
牵引绞车　winch
牵引型滑坡　landslide of dragging type
铅垂线竖井　plumbline well
铅丝　galvanized wire
铅丝笼　wire box
铅酸蓄电池　lead dioxide lead battery
签证费用　visa fees
签字栏　authentication block
前池　forebay
前端设备　front-end device
前方交会　forward intersection method
前盖板　front shroud
前寒武纪（系）　Precambrian period (system)
前室　anteroom
前挖后卸式装载机　backhoe front end loader
前翼缘　upstream flange
前震　foreshock
前震旦纪（系）　Presinian period (system)
前趾　fore toe
钳式电流互感器　split core type current transformer
潜坝　submerged dike
潜伏毒性　delayed toxicity
潜孔锤　hammer down the hole (HDT)
潜孔式闸门　submerged gate
潜孔钻机　down-the-hole drill (DTH)
潜蚀　internal scour，underground corrosion
潜水　phreatic water
潜水泵　subaqueous (submergible) pump
潜水位　water table，groundwater table
潜水蒸发　phreatic water evaporation
潜液电泵　submergible motor pump
潜在不稳定体　potential instable rock mass
潜在的下游危害　downstream hazard potential
潜在破坏面　potential failure surface
潜在影响　potential impact
潜在震源区　potential seismic source zone
浅变质岩　epimetamorphic rock
浅层地震反射波法　shallow seismic reflection
浅层地震折射波法　shallow seismic refraction
浅层抗滑稳定　sliding stability in shallow layer
浅海相　neritic facies，shallow sea facies
浅基础　shallow footing
浅孔爆破　short-hole blasting
浅滩整治　shoal training
浅源地震　shallow-focus earthquake
欠范围　underreach
欠范围式纵联保护　underreach pilot protection
欠固结土　underconsolidated soil
欠励磁运行　underexcitation operation
欠励限制器　underexcitation limiter
欠挖　under-excavation
嵌缝骨料　key aggregate
嵌缝料　caulk
嵌入式灯具　recessed luminaire
嵌入式岩锚　recessed rock anchor

歉收年　fail year
戗台　berm
强度　strength
强度保证率　assurance factor of strength
强度设计法　ultimate strength method
强风化　highly weathered
强腐蚀　highly corrosive
强腐蚀环境　severe erosive environment
强夯法　dynamic compaction method
强迫导向油循环风冷　oil directed air forced cooling (ODAF)
强迫导向油循环水冷　oil directed water forced cooling (ODWF)
强迫风冷却　air forced cooling (AF)
强迫冷却　forced cooling
强迫停运时间　forced-outage duration
强迫油循环风冷却　oil forced air forced cooling (OFAF)
强迫油循环水冷却　oil forced water forced cooling (OFWF)
强透水　strongly pervious
强卸荷　highly relaxed
强行励磁　forced exciting
强震　strong earthquake
强震仪　strong-motion seismograph
强制电流　impressed current
强制对流　forced convection
强制换气　forced air change
强制通风　forced ventilation
强制通风冷却塔　forced draft cooling tower
强制循环系统　forced circulation system
强制振荡　forced oscillation
墙板结构　wall-slab structure
墙内风道　wall duct
墙牛腿（壁式牛腿）　corbel
墙裙　dado
墙式机墩　wall-type pier of turbine
墙式消防栓　wall hydrant
桥架　bridge
桥跨结构　bridge span structure
桥梁　bridge
桥式连接　bridge connection
桥式起重机（桥机）　bridge crane，overhead crane
桥形接线　bridge configuration
切层滑动　slide cutting bedding plane
切层滑坡　insequent landslide
切割面　cutting plane
切换开关　transfer switch
切换值　switching value
切土管刃　cutting shoe
切线冻胀力　tangential frost heaving force
切向主应力　tangential principal stress
侵入岩分期　intrusive rock period
侵蚀二氧化碳　corrosive carbon dioxide
侵蚀基准面　erosion basis，base level of erosion
侵蚀及剥蚀地貌单元　erosional and denudational geomorphic unit
侵蚀阶地　erosional terrace
侵蚀平原　erosion plain
侵蚀性水　aggressive water
侵蚀营力　erosion force
亲水系数　hydrophilic coefficient
青苗　young crops
轻骨料　lightweight aggregate
轻微发育　slightly developed
轻型钢结构　light weight steel construction
轻质隔墙　light-weight partition
轻质混凝土结构　light concrete structure
倾倒　toppling
倾翻式搅拌机　tipping drum mixer
倾伏背斜　plunging anticline
倾伏方向　trend
倾伏角　plunge
倾覆力矩　overturning moment
倾覆褶皱　plunging fold
倾角　dip
倾角计　inclination transducer
倾向　dip direction
倾斜固定筛　sloping stationary screen
倾斜式进水口结构　inclined intake structure
倾斜式进水塔　inclined intake tower
倾斜仪　clinometer，inclinometer
清根　back gouging
清关费　customs clearing fee
清洁度　cleanliness
清水　clear water
清污机　screen cleaning device
丘陵　hill
球阀　spherical valve，ball valve

球间隙　sphere gap
球面滑动轴承　spherical plain bearing
球头　ball
球形岔管　spherical branch pipe
球形闸门　ball gate
球状风化　spheroidal weathering
区内故障　internal fault
区外故障　external fault
区域采暖　district heating
区域地质　regional geology
区域供冷　district cooling
区域性主要断裂　regional major fault
曲率系数　coefficient of curvature
曲线拟合　curve fitting
曲折形连接　zigzag connection
驱动功率　drive power
驱动能量　actuating energy
驱动系统　drive system
驱动装置　actuator
屈服变形　yield deformation
屈服标准　yield criterion
屈服极限　yield limit
屈服强度　yield strength
趋肤效应　skin effect
趋势分析　trend analysis
渠床糙率　roughness of canal bed
渠道冲淤平衡　balance of scouring and silting in canal
渠道断面宽深比　ratio of bottom width to water depth in canal
渠道过水能力　carrying capacity of canal
渠道净流量　net discharge in canal
渠道毛流量　gross discharge in canal
渠道坡降　gradient of canal
渠道输水损失　water conveyance loss in canal
渠道允许不冲流速　permissible unscouring velocity in canal
渠道允许不淤流速　permissible unsilting velocity in canal
渠化航道　channelized waterway
渠系规划　planning of canal system
渠系建筑物　canal structure
取土坑　borrow pit
取土器　soil sampler
取物装置　load-handling device
取芯　coring
取岩芯　core extraction
取样　sampling
取样程序　sampling procedure
去湿　dehumidification
去污　cleaning
去油雾设备　oil mist exhaust equipment
圈梁　ring beam
权益资本　Equity Capital
全部燃烧　fully involved
全长黏结型锚杆　anchor bolt bonded all length
全电流　total current
全电压启动　full voltage starting, direct on line starting
全断面掘进法　full face driving method
全断面岩石掘进机　full face rock tunnel boring machine
全额免税　full exemption
全风化　completely weathered
全关位置　fully closed position
全贯流式机组　rim-generator tubular unit
全介质自承式光缆　all dielectric self-supporting optical cable (ADSS)
全径阀门　full-port valve
全开位置　fully open position
全空气系统　all-air condition
全孔一次灌浆法　full depth one stage method
全控连接　fully controllable connection
全密封电池　hermetically sealed cell
全面排风　general exhaust ventilation (GEV)
全面疏散　complete evacuation
全面通风　general ventilation
全面质量管理　total quality control
全年导流　whole year diversion
全年围堰挡水　retaining water by cofferdam all around the year
全平衡式　full balanced type
全启式安全阀　full lift safety valve
全球定位系统　global positioning system (GPS)
全燃火　fully developed fire
全日热水供应系统　all day hot water supply system
全生命周期　total life cycle
全室性空气调节　general air conditioning
全水头扬压力　full uplift pressure

全特性　complete characteristics
全文数据库　context database
全向天线　omni antenna
全新世（统）　Holocene epoch（series）
全淹没灭火系统　total flooded extinguishing system
全站仪　total station
全宗　fonds
全宗指南　guide to an archival fonds
泉　spring
泉华　sinter
缺口导流　dam-gap diversion
缺陷　defects
缺陷工程　defective work
缺陷焊缝　poor weld
缺陷通知期　defects notification period
缺陷责任期　defects liability period
确定性分析　deterministic analysis
确定性水文模型　deterministic hydrological model
阙值　threshold
裙房　podium
群体沉速　settling velocity of grains
群众支持　public support
群组　Group

R

燃点　fire point
燃弧时间　arcing time
燃烧层　fire layer
燃烧层穿孔　fire hole
燃烧持续时间　burning duration
燃烧系数　combustion factor
壤土　loam
扰动区　zone of disturbance
扰动土样，非原状土样　disturbed soil sample
绕坝渗漏　bypassing leakage，leakage around dam abutment
绕击率　shielding failure rate due to lightning stroke
绕线式转子　wound rotor
绕线转子感应电动机　wound-rotor induction motor
绕组的分级绝缘　non-uniform insulation of a winding
绕组的全绝缘　uniform insulation of a winding
绕组绝缘　winding insulation
绕组绝缘电阻测量　winding insulation resistance measurement
绕组温度计　winding temperature indicator
绕组直流电阻测量　DC winding resistance measurement
热备用　hot standby
热泵热水供应系统　heat pump hot water system
热泵式采暖　heat pump heating
热电偶式传感器　thermocouple transducer
热电偶温度计　thermocouples
热对接　heat sealing
热惰性指标　index of thermal inertia
热风供暖　warm-air heating
热辐射　thermal radiation
热负荷　heat load
热过载脱扣器　thermal overload release
热继电器　thermal relay
热交换器的冷却水源　cooling water supply for heat exchanger
热空气吹干　drying by forced hot air
热力学法　thermodynamic method
热力学温度　thermodynamic temperature
热沥青　hot bitumen
热媒　heating medium
热黏合　heat bonding
热喷涂　thermal spray
热膨胀系数　coefficient of thermal expansion
热湿交换　heat and moisture transfer
热损失　hot loss
热稳定　thermal equilibrium
热稳定系数　coefficient of thermal stability
热稳定性　thermal stability
热像式绕组温度计　thermal image temperature indicator（WTI）
热效率　thermal efficiency
热液蚀变　hydrothermal alteration
热影响区　heat affected zone
热转移媒质　heat transfer agent
热阻系数　thermal resistance coefficient
热作用　thermal actions
人防设计　civil defense design
人工单价　unit price of labor
人工捣实混凝土　hand-compacted concrete
人工堆积　artificial deposit
人工骨料　artificial aggregate
人工接地体　artificial earthing electrode
人工砂　artificial sand

人工湿地　artificial wetland
人工死区单元　artificial dead band module
人工填土　backfilled soil
人工污秽试验　artificial pollution test
人工消耗定额　labor consumption norm
人机接口　human-machine interface (HMI), human-computer interface
人均耕地　cultivated land per capita
人均住房面积　floor area per capita
人口稠密区　densely populated area
人口结构　population structure
人口流入　population influx
人口密度　population density
人口迁移线　relocation border
人口预测　population prediction
人口增长率　population growth rate
人类环境　human environment
人类活动　human activity
人类活动水文效应　hydrological response due to human activities
人力资本法　human capital method
人身保险　life insurance
人身伤害和财产损失险　insurance against injury to person and damage to property
人身意外伤害保险费　personal accident insurance
人为侵蚀　erosion caused by human activities
人为污染　man-made pollution
人文因素　human factor
人行便道　sidewalk
人行便桥　pedestrian bridge
人与环境和谐　harmony between human being and environment
人与野生动物冲突　human-wildlife conflict
人员防坠落系统　personnel fall prevention system
人造环境　built environment
人字形排水系统　chevron drain
人字闸门　miter gate
认证　authentication
任意堆石区　randomfill zone
韧性剪切带　ductile shear zone
韧性，延展性　ductility
日常保养（例行保养）　routine maintenance
日常修理费　cost of routine maintenance
日调节　daily regulation
日调节水库　daily regulating reservoir
日负荷　daily load
日负荷曲线　daily load curve
日较差　daily range
日平均负荷率（日负荷系数）　average daily load factor
日照辐射热　radiant heat of sunshine
日照率　percentage of possible sunshine
日照时间　sunshine duration
日最大负荷利用小时　daily maximum load utilization hours
容错　fault-tolcrance
容积法　volumetric method
容积率　plot ratio
容积损失　volumetric loss
容抗　capacitive reactance
容量效益　capacity benefit
容纳　capacitive susceptance
容许安全系数　allowable factor of safety
容许变形量　allowable deformation
容许沉降量　allowable settlement
容许承压力　allowable bearing pressure
容许工作范围　allowable operating range
容许极限　acceptable limit
容许土壤流失量　soil loss tolerance
容许应力法　permissible (allowable) stress method
容许变形　allowable deformation
容重　unit weigth
溶出型腐蚀　dissolving corrosion
溶洞　karst cave
溶斗　doline, funnel
溶沟　karren
溶解固形物　dissolved solid matter
溶解氧　dissolved oxygen (DO)
溶隙　solution crack
溶氧浓度　dissolved oxygen concentration
溶于水的气体　dissolved air
熔滴　melt drip
熔断器　fuse
熔断器监视　fuse monitoring
熔断器式隔离器　fuse-disconnector
熔焊　fusion welding
熔透焊缝　full-penetration weld
熔渣　clinker
融雪径流　snowmelt runoff
融资后分析　analysis after financing

融资前分析　analysis before financing
融资主体　Financing Entity
柔性管道　flexible pipeline
柔性填料　flexible filler
柔性支护　flexible support
揉皱　crumple
蠕变　creep
蠕变变形　creep deformation
蠕变破坏　creep rupture
蠕变曲线　creep curve
蠕变试验　creep test
蠕变岩体　creeping rock mass
蠕动检测　creep detection
蠕动检测器　creep detector
蠕滑　creeping slide
蠕滑-拉裂　creep sliding-tensile fracture
蠕行　crawling
乳化沥青　emulsified bitumen
乳化沥青的破乳速度　demulsibility of emulsified bitumen
入口　entrance
入口总水头（吸入扬程）　total suction head
入库洪峰流量　peak rate of inflow
入库洪水　inflowing flood
入库流量　reservoir inflow
入库泥沙　inflowing sediment
入库设计洪水　inflow design flood（IDF）
入流过程线的涨水段　rising limb of inflow hydrograph
入侵报警系统　intrusion alarm system
入侵防护　intrusion prevention
入侵检测　intrusion detection
入侵物种　invasive species
入侵侦测　intrusion detection
入射波　incident wave
入渗　infiltration
入中继线　incoming trunk
软磁材料　magnetically soft material，soft magnetic material
软垫层　soft cushion
软管　flexible hose
软化系数　softening coefficient
软基处理　treatment of soft foundation
软母线　flexible busbar
软启动　soft starting
软弱夹层　weak interlayer
软弱夹层抗滑稳定　sliding stability along weak intercalation
软弱结构面　soft（weak）structural plane
软弱颗粒　soft particle
软弱矿物　soft mineral
软质岩　soft rock
锐缘堰　sharp-crested weir
瑞典滑弧法　Swedish slipcircle method
瑞典条分法　Swedish slice method
瑞雷波法　rayleigh wave method
瑞利波　Rayleigh wave
润滑系统　lubrication system
润滑油　lubricating oil
弱风化　moderately weathered
弱腐蚀　weakly corrosive
弱腐蚀环境　weak corrosive environment
弱透水　weakly pervious
弱卸荷　weakly relaxed
弱震　weak earthquake

S

萨尔玛法　Sarma method
塞尺　feeler gauge
塞焊　Plug weld
三瓣式岩芯管　triple split barrel
三边测量　trilateration
三等水准测量　third-order leveling
三叠纪（系）　Triassic period（system）
三分之四断路器接线　4/3 breaker configuration
三合一减速器　gearmotor
三机式机组　tandem unit
三基色　tri-phosphor
三级配骨料　aggregate with maximum grain size of 80mm
三极隔离开关　three pole disconnector
三角测量　triangulation
三角觇标系统　triangulation target system
三角高程测量　trigonometric leveling
三角台　tribrach
三角网　triangulation network
三角网平差　triangulation net adjustment
三角形法　triangular method
三角形连接　delta connection
三角形排列　triangular configuration

三角形-星形变换　delta-wye conversion，delta-star transformation
三角洲　delta
三角洲淤积　delta deposit
三梁岔管　three-girder reinforced bifurcation
三绕组变压器　three winding transformer
三通阀　three-way valve
三通一平　supply of water and power，road connection and site leveling
三维模型（立体模型）　three-dimensional model，stereo-model
三维协同设计　computer supported collaborative 3D Design
三维植被网　three-dimensional geonet
三相变压器　three phase transformer
三相电流不平衡　three phase current unbalance
三相电压不平衡　three phase voltage unbalance
三相对称短路　symmetrical short-cicuit，three phase short-cicuit
三相四线制　three phase four wire system
三相制　three phase system
三相自动重合闸　three phase auto-reclosing
三向测缝计　three-way joint meter
三向配筋　three-way reinforcement
三向位移计　three-way displacement meter
三向应变计组　three-way extensometer group
三心拱　three-centered arch
三中心变厚度拱坝　three-centered variable thickness arch dam
三轴剪试验　triaxial shear test，triaxial compression test
三柱式隔离开关　three-column disconnector
三坐标测量仪　three dimensional coordinate measuring machine
伞式水轮发电机　umbrella hydrogenerator
伞形风帽　cowl
散落　ravelling
散热器　radiator
散热器供热支管　heating branch to radiator
散热器回水支管　return branch of radiator
散水　apron
散体结构　loose granular structure，cohesionless granular structure
散装货物　bulk cargo
散装水泥运输车　bulk cement truck
丧失能力　incapacitation
扫描速率　scan rate
埽工　fascine works
色度计　colorimeter
色温　colour temperature
森林覆盖率　forest coverage rate
森林公园　forest park
森林面积　forest area
沙波　sand wave
沙波阻力　form resistance of sand wave
沙粒阻力　grain resistance
沙垄　dunes
沙漠　desert
沙纹　sand ripple
砂　sand
砂泵　sand pump
砂沸　sand boiling
砂浆材料　mortar material
砂浆锚杆　grouted anchor bolt
砂浆涂层　dash-bond coat
砂浆找平　mortar leveling
砂井排水　sand drain
砂砾料　sand and gravel material
砂砾石　sand and gravel
砂砾石地基　sand and gravel base
砂砾岩　sandy conglomerate
砂率　fine-to-coarse aggregate ratio
砂壤土　sandy loam
砂石加工系统　artificial aggregate system，aggregate processing plant
砂石料单价　unit price of sand and gravel
砂土液化　sand liquefaction
砂性土　sandy soil
砂岩　sandstone
砂质结构　arenaceous texture
砂质黏土　sandy clay
筛分　sieving (screening)
筛分级配　screen size gradation
筛分曲线　screening curve
筛分试验　screen test
筛分效率　efficiency of screening
筛析法　test sieving method
筛余量　sample retained on sieve
山地　mountain region (area)
山谷　valley

山洪易发区 susceptible area of mountain torrent
山间盆地 intermontane basin
山口 mountain pass
山麓 mountain foot
山脉 mountain range
山坡 mountain slope, hillside
山墙 gable
山系 mountain system
山字形构造 epsilon-type structure
闪长岩 diorite
闪点 flash point
扇形闸门 sector gate
伤亡事故（工伤事故） casualty accident
商检费 commodity inspection fee
商业发票 commercial invoice
上坝道路 access road to dam
上半球投影 upper hemisphere projection
上部结构 superstructure
上部主轴 upper shaft
上层滞水 perch groundwater
上承梁 deck beam
上导洞法 top drift method
上导轴承 upper guide bearing
上盖板 upper cover
上机架 upper bracket
上基本调配线 upper critical guide curve
上密封 back seal
上密封试验 backseat test
上升泉 ascending spring
上水库 upper reservoir
上水库充水方式 impounding mode for upper reservoir
上挑鼻坎 upturned deflector
上网电价 feed-in tariff
上网电量 on-grid energy
上下游水位差 level difference between headwater and tailwater
上限解 upper bound solution
上新世（统） Pliocene epoch (series)
上行振动碾压 vibrating compaction upward
上游 upper reaches
上游堆石区 upstream rockfill zone
上游防渗铺盖 upstream impervious blanket
上游坡（迎水坡） upstream slope
上游围堰 upstream cofferdam
上轴 upper shaft
烧杯 beaker
烧毁面积 damaged area
烧焦 scorch
烧失量试验 ignition loss test
烧钻 bit burnt
勺钻 bucket auger
少数民族 ethnic minorities
少数民族居住地区 minority-inhabited areas
舌瓣闸门，拍门 flap gate
蛇绿岩带 ophiolite belt
蛇纹石 serpentine
蛇纹岩 serpenite
设备安装 equipment erection
设备布置 equipment layout (equipment arrangement)
设备层 mechanical floor
设备购置费 cost of equipment procurement
设备技术规程 equipment specification
设备监造费用 equipment supervision cost
设备修理费 machinery repair cost
设备原价 original price of equipment
设备总成套 system equipment design and supply
设定命令 set-point command
设防地震 precautionary earthquake
设计安全标准 design safety criterion, design safety standard
设计保证率 design dependability
设计暴雨 design rainstorm
设计-采购-施工合同 Engineering- Procurement-Construction contract (EPC)
设计-采购总承包 Engineering and Procurement (E-P)
设计程序 design procedure
设计错误 design error
设计地震动 design ground motion
设计地震动特征周期 design characteristic period of ground motion
设计地震加速度 design seismic acceleration
设计地震烈度 design seismic intensity
设计反应谱 design response spectra
设计灌水周期 designed interval of irrigation
设计荷载 design load
设计洪水 design flood
设计洪水标准 standard of design flood

设计洪水位　design flood level
设计基本地震加速度　design basic acceleration of ground motion
设计基准期　design reference period
设计警戒值　design threshold
设计径向负荷　design radial load
设计净雨量　design net rainfall
设计联络会　design liaison meeting
设计烈度　design seismic intensity
设计锚固力　design anchoring
设计年径流量　design annual runoff
设计年径流量年内分配　distribution of design annual runoff within a year
设计频率　design frequency
设计任务书　design assignment statement
设计-施工总承包（D-B）　design and building (D-B)
设计使用年限 design working life, design service life
设计水平年　design time horizon
设计水头　design head
设计小时供热量　design heat supply of maximum time
设计小时耗热量　design heat consumption of maximum time
设计修改通知　amendment notice
设计需要量　design exploitative reserve
设计油压　design oil pressure
设计雨型　design rainfall pattern
设计张拉力　design tension
设计状况　design situations
设计状况系数　coefficient of design situation
设站流域　gaged basin
社会保险　social insurance
社会保障费　social security cost
社会及经济方面的重建　social and economic reestablishment
社会经济基线　socio-economic baseline
社会经济问题　socio-economic issue
社会行动计划　social action plan
射流　jet
射流分布圆直径　jet circle diameter
射流相对转轮的对准度　alignment of jet to runner
射流相对转轮的偏移值　offset of jet to runner
射线探伤　radiographic testing (RT)
涉及野生动物的犯罪　wildlife-related crime
摄氏温度　Celsius temperature
摄像机　camera
申诉机制　grievance mechanism
申诉救济制度　grievance redress system
伸长计，应变计　extensometer
伸出钢筋，搭接钢筋　starter bar
伸缩缝　expansion joint
深变质岩　katametamorphic rock
深槽式转子　deep slot type rotor
深层抗滑稳定　sliding stability in deep layer
深成岩　plutonic
深大断裂　deep fault
深泓线　thalweg
深基础　deep footing
深孔爆破　deep-hole blasting
深梁　deep beam
深埋隧洞　deep buried tunnel
深受弯构件　deep flexural member
深卸荷带　deep relaxed zone
深源地震　deep-focus earthquake
深照型灯具　narrow angle luminaire
审查　review, reviewed by
渗出物　exudation
渗径　seepage path
渗流　seepage
渗流计算　seepage calculation
渗流量　seepage flow
渗流量监测　seepage flow monitoring
渗流模型　seepage model
渗流区　seepage area, vadose region
渗流速度　seepage velocity
渗流网　seepage flow net
渗流压力　seepage pressure
渗漏　leakage
渗漏量　leakage quantity
渗漏流量　leakage discharge
渗漏排水井　leakage water dewatering pit
渗滤　percolation
渗渠　infiltration gallery
渗入　influent seepage
渗水　water seeps
渗透　seepage, permeation
渗透变形　seepage deformation
渗透力　seepage force

渗透量 seepage quantity
渗透坡降 seepage gradient
渗透破坏 seepage failure
渗透剖面 seepage profile，permeability profile
渗透试验 permeability test
渗透速度 seepage velocity
渗透探伤 penetrating testing（PT）
渗透稳定计算 seepage stability calculation
渗透系数 permeability coefficient
渗透性 permeability
渗透压力监测 seepage pressure monitoring
渗透压力强度系数 coefficient of seepage pressure intensity
渗透扬压力 seepage uplift
升程（提升高度） lifting range（lift height）
升船机 shiplift（ship elevator）
升降立式止回阀 vertical lift check valve
升温曲线 temperature-time curves
升卧式闸门 lifting-lie gate
升压变电站 step-up substation
升压变压器 step-up transformer
生产安置人口 agricultural population to be job-arranged
生产安置设计 production resettlement design
生产扶持 production support
生产率 productivity
生产人员提前进厂费 expense of production personnel advancing into the plant
生产准备费 operation start-up cost
生化需氧量 biochemical oxygen demand（BOD）
生活标准 living standard
生活垃圾 domestic waste
生活区 living quarter
生活消防水泵房 domestic water and fire pump house
生活饮用水卫生标准 sanitary standard for drinking water
生活用水 domestic water
生计恢复 livelihood restoration
生境适宜性 habitat suitability
生态公益林 non-commercial forest
生态平衡 ecological balance
生态退化 ecological degradation
生态系统 ecological system
生态系统服务 ecosystem service
生物多样性管理计划 biodiversity management plan
生物结构 biogenetic texture
生物圈 biosphere
生物碎屑灰岩 bioclastic limestone
生育率 fertility rate
声波波速 acoustic wave velocity
声波测井 acoustic log
声波测试 acoustic wave survey
声波电视测井 acoustic televiewer log
声波发射器 pinger
声波反射法 sonic reflection method
声发射监测 acoustic emission monitoring
声速 velocity of sound
声学法（超声波法） acoustic method（ultrasonic method）
声学流量计 acoustic flow meter
牲畜头数 livestock population
绳索滑轮组 reeving system
圣·维南方程 Saint-Venant equation
盛行风 prevailing wind
剩磁（剩余磁化强度） remanent magnetization
剩磁通密度 magnetic remanence
剩余（不可用）油体积 residual（not usable）oil volume
剩余电流 residual current
剩余电流动作保护装置 residual current operated device（RCD）
剩余动作电流 residual operating current
剩余浪压力强度 residual wave pressure intensity
剩余容量 residual capacity
失步保护 out-of-step protection
失步解列 trip for out of step
失步运行 out-of-step operation
失磁保护 loss-of-field protection
失电停机 deenergize-to-shutdown
失电压保护 loss-of-voltage protection
失电制动器 power-off brake
失速 stall
失通 firing failure
失效概率 probability of failure
失业保险费 unemployment insurance
失业率 rate of unemployment
施工安全管理 safety management during construction

施工编录　construction mapping
施工便道　construction road
施工操作规程　construction operation instruction
施工测量　construction survey
施工程序　construction sequence
施工导线　construction traverse
施工地质工作　geological works during construction
施工地质日志　geological daily record during construction
施工地质通知　geological notification during construction
施工调度　operation dispatching
施工定额　construction norm and standard
施工方案　construction scheme
施工仿真　construction simulation analysis
施工分标报告　Report on sub-bid scheme
施工分区规划　plan of construction zoning
施工缝　construction joint
施工辅助工程　temporary works
施工辅助工程（临时工程）　temporary works
施工高峰期　peak construction period
施工高峰强度　peak intensity
施工工具用具使用费　fee for utilizing tools and apparatus
施工工期　construction duration
施工供电系统　power supply system for construction
施工供风系统　compressed air supply system for construction
施工供水系统　water supply system for construction
施工管理及生活区　contractor camp
施工合同条件　conditions of contract for construction
施工机械使用费　machinery operation expense
施工机械台时费　machinery daily working unit price
施工技术要求　Construction specifications
施工监理　construction supervision
施工截流　river closure
施工津贴　field subsidy
施工科研试验费　cost for construction scientific test
施工控制网　construction control network
施工扭矩　construction torque
施工平均强度　average intenisty
施工期通风　ventilation during construction
施工期蓄水　impounding in construction period
施工强度　construction intensity
施工区封闭管理　closed management of construction site
施工设备进场　mobilization of construction machinery
施工设备退场　demobilization of construction machinery
施工设计洪水　design flood during construction
施工生产计划　overall construction plan
施工通讯系统　communication system for construction
施工图　construction drawing
施工图设计交底　technical interpretation of design intention
施工现场（工地）　site
施工详图　detailed drawing
施工详图设计　detailed design
施工详图设计阶段　construction detailed design stage
施工延误　construction delay
施工营地　construction camp
施工用地　occupied land for construction
施工用电价格　construction electricity price
施工用风价格　construction compressed air price
施工用水价格　construction water price
施工有效工日　available working day
施工运输　construction transportation
施工支洞　adit
施工质量　construction quality
施工准备　construction preparation
施工总布置　general construction layout
施工总进度　general construction schedule, construction master schedule
施工组织设计　construction planning, construction method statement
施工作业　work activity
施工坐标系　construction coordinate system
湿地　wetland
湿度　humidity
湿工况　wet cooling condition
湿化试验　slaking test
湿喷混凝土　wet shotcrete
湿球温度　wet-bulb temperature

湿容重　moist unit weight
湿式系统　wet pipe system
湿试验　wet test
湿陷起始压力　initial collapse pressure
湿陷系数　coefficient of collapsibility
湿陷性黄土　collapsible loess
湿研磨分离制备法　wet-grind separation method
湿运式升船机　wet shiplift
十字板剪切试验　vane shear test
石粉含量　rock flour content
石膏　gypsum
石灰纪（系）　Carboniferous period（system）
石料　rock material
石笼坝　gabion dam
石墨　graphite
石砌结构　stone masonry structure
石笋　stalagmite
石屑　aggregate chips
石英　quartz
石英脉　quartz vein
石英脉充填　quartz vein infilled
石英砂岩　quartz sandstone
石英岩　quartzite
石渣坝　rock debris dam
石柱　stalacto-stalagmite，column
时变效应　time-dependent effect
时标　absolute chronology，time tagging
时程分析法　time history method
时段单位线　unit hydrograph
时价　current price
时间常数　time constant
时间-电流特性　time-current characteristic
时间继电器　time relay
时间配合　time coordination
时间偏好率　time preference rate
时间效应　time effect
时间行程特性　time travel diagram
时间元件　time component
时均压力　time-average pressure
时域分析　time-domain analysis
实测界线　surveyed boundary
实测最大降水量　maximum observed precipitation
实地调查　field investigation
实地访问　field inquiry
实际尺寸　full size
实际工期　as-built schedule
实际汇率　real exchange rate
实际价格　real price
实际价值　real value
实际热值　actual calorific valve
实际运行时间　actual duration of operation
实体防护　physical protection
实物量　volume terms
实物收入　income-in-kind
实物指标公示　soliciting public comments on inventory in-kind
实物指标确认　confirmation of inventory in-kind
食草动物　grazer
食堂　canteen
蚀变　alteration
蚀变带　altered zone
蚀变矿物　altered mineral
矢高　rise
使用价值　use value
使用面积　usable area
使用面积系数　usable area coefficient
始新世（统）　eocene epoch（series）
示范农场　demonstration farm
示坡线　slope line
示意图（略图）　sketch（schematic drawing）
示踪试验　tracing test
示踪同位素　tracer isotope
示踪元素　tracer element
世（统）　epoch（series）
市场价格　market price
市政公用设施费　fee of municipal public facilities
事故备用　emergency reserve
事故备用电源　emergency backup power supply
事故备用容量　reserve capacity for emergency
事故通风　emergency ventilation
事故隐患　accident potential
事故闸门　emergency gate
事故照明　emergency lighting
事故照明负荷　emergency lighting load
事故追忆　post disturbance review
事后监督模式　afterwards supervision mode
事件　event
事件记录　event recording
事件日志　event log

事件顺序记录　sequence of events
事件信息　event information
势能　potential energy
饰面层　finish coat
试错法　trial-and-error method，cut-and-try method
试通航　trial navigation
试验大纲　testing schedule
试验荷载　test load
试验介质　test fluid
试验介质温度　test fluid temperature
试验热值　experimental heat release
试验台　rig for model test
试验压力　test pressure
试样　sample，specimen
试运行　commissioning
试运行期　commissioning period
试载扭转分析法　trial-load twist method of analysis
视差　parallax
视场　field of view
视角　angle of view
视距　sight distance
视距测量　stadia survey，tacheometric survey
视距导线　tacheometric polygon
视频安防监控系统　video surveillance and control system
视频传输　video transmission
视频监控　video monitoring
视频识别装置　video identification device
视频探测　video detection
视频信号　video signal
视频压缩格式　video compressed format
视频音频同步　synchronization of video and audio
视频主机　video controller，video switcher
视倾角（假倾角）　apparent dip
视在功率　apparent power
视准测量　collimation measurement
视准点，照准点　collimating point，sighting point
视准线法　collimation line method
视准仪　collimator
室内潮湿环境　indoor damp environment
室内地平标高　elevation of leveled ground
室内净高　interior storey height
室内灭火设施　indoor fire installation
室内设计温度　indoor design temperature
室内外高差　indoor-outdoor elevation difference
室内消防栓　indoor fire hydrant
室内正常环境　normal indoor environment
室外火灾　outdoor fire
室外楼梯　external stair
室外消防栓　outdoor fire hydrant
室外消防通道　outside fire escape
室外装修　exterior finishing
释水系数　storage coefficient
收分装置　draw in or disport device
收集点　collector point
收敛变形　convergent deformation
收敛测量　convergence measurement
收敛检验　test for convergence
收缩缝　contraction joint
收缩试验　shrinkage test
收尾工作　outstanding work
手动变量轴向柱塞泵　manual variable displacement axial piston pump
手动操作机构　dependent manual operating mechanism
手动阀门　manual operated valve
手动火灾报警按钮　manual fire alarm call point
手动控制装置　manual controls
手动旁路开关　manual bypass switch
手动起重机　manual crane
手动运行　manual operation
手动准同步装置　manual ideal synchroni-zing device
手动-自动之间的跟踪　manual-automatic followup function
手风钻　hand drill
手稿　manuscripts
手工电弧焊　shielded metal arc welding（SMAW）
手提式灭火器　portable fire extinguisher
手推车　barrow
手续费　charge for trouble
首部闸门　head gate
首次蓄水期　first impound period
首台机组发电工期　duration of first unit putting into operation
受冲切承载力　punch shear bearing capacity
受挤压岩体　squeezing rock
受教育程度　educational background

受拉力控制截面　section controlled by tension
受力层　compression zone
受力钢筋　bearing rebar
受扭承载力　torsion bearing capacity
受扭构件　torsion member
受让人　assignor
受弯承载力　flexural bearing capacity
受弯构件的挠度限值　allowable deflection of flexural member
受限射流　jet in a confined space
受压承载力　compression bearing capacity
受压力控制截面　section controlled by compression
受益人　beneficiary
受油器　oil head
受阻容量　restricted capacity
舒适性空气调节　comfort air conditioning
疏林地　vegetable and forest land
疏散标志灯　escape sign luminaire
疏散楼梯　emergency stair case
疏散路线　evacuation route
疏散图　security map
疏散照明　escape lighting
疏水阀　steam trap
输变电工程　power transmission and distribution project
输出变压器　outgoing transformer
输出电缆　output cable
输出阻抗　output impedance
输电　transmission of electricity，power transmission
输电容量　transmission capacity
输电损耗　transmission loss
输电线路　transmission line
输电线路走廊　transmission line corridor
输电效率　transmission efficiency
输入变压器　incoming transformer
输入阻抗　input impedance
输沙率　sediment discharge
输水管渠　delivery pipe
输水渠道　water conveyance canal
输水隧洞　convey tunnel，delivery tunnel
输水系统　water conveyance system
蔬菜　vegetable
熟练工　skilled labor
熟料水泥　clinker cement
鼠笼式转子　squirrel cage rotor
束窄河道　narrowed river course
树脂锚杆　resin anchor bolt
竖井　vertical shaft
竖井编录　shaft log
竖井跌水式溢洪道　drop-inlet spillway
竖井定向测量　shaft orientation survey
竖井贯流式机组　pit tubular unit
竖井式进/出水口　shaft intake/outlet
竖井式溢洪道　shaft spillway，glory hole spillway
竖井通道　access shaft
竖井钻车　shaft jumbo
竖曲线　vertical curve
竖式排水　vertical drainage
竖向布置图　vertical planning
竖向地震作用　vertical seismic action
竖向冻胀力　vertical frost heaving force
竖向分布钢筋　vertically distributed bar
竖向排水　vertical drainage
数据安全　data security
数据采集软件　data acquisition software
数据采集系统　data acquisition systems
数据采集与监控系统　supervisory control and data acquisition（SCADA）
数据采集与监视　data acquisition and monitoring
数据处理软件　data treatment software
数据传输　transmitting data
数据传送　data transfer
数据服务器　data server
数据结构　data structure
数据库　data base
数据库管理系统　data base management system
数据库软件　database software
数据审计　data audit
数据通信　data communication
数-模变换器　digital-to-analog converter
数值分析　numerical analysis
数字大坝技术　digital technology for dam
数字地形图　digital topographic map
数字地质编录系统　digital geological logging system
数字化变电站　digital substation
数字化大坝监测管理系统　digital dam monitor management system
数字化加工　digitize processing
数字记录仪　digital recorder

数字量　digital variable
数字量输出模件　digital output (DO) module
数字量输入模件　digital input (DI) module
数字滤波器　digital filter
数字签名　digital signature
数字式仪表　digital indicating instrument
数字视频　digital video
数字图像　digital image
刷握　brush holder
衰减　attenuation, decay
衰减率　attenuation rate
甩负荷　load rejection (load shedding)
甩负荷试验　load rejection test
甩挂运输　drop and pull transport
双壁壳式泵　armoured type pump
双臂伸缩式隔离开关　pantograph disconnector
双层筛　double deck screen
双电源供电　duplicate supply
双调节机械　double regulated machine
双调整调速器　double regulating governor
双端螺栓　stud bolt
双断路器接线　double-breaker configuration
双扉闸门　double-leaf gate
双风道系统　double duct system
双腹板梁　double girder
双管采暖系统　two-pipe heating system
双滚筒夯实碾　dual-drum tamping roller
双回路　double circuit line
双回路塔　double circuit steel tower
双库单向电站　two-lagoon one-way tidal power station
双联弹簧式安全阀　duplex safety valve
双梁桥架　double-girder bridge
双轮铣槽机　trench cutter with double wheels
双面焊接　double side welding
双母线带旁路接线　double-bus with transfer bus configuration
双母线分段接线　sectionalized double-bus configuration
双母线接线　double-bus configuration
双戗立堵　double closure dikes plus end-dump closure
双曲拱坝　double-curvature arch dam
双绕组变压器　two winding transformer
双热楔　dual-hot-wedge
双扇窗　double casement window
双室式调压井　double-chamber surge shaft
双锁紧泄放阀 (DBB)　double-block-and-bleed valve
双态信息　binary state information
双通道自动电压调节器　dual automatic channel voltage regulator
双稳态继电器　bi-stable relay
双蜗形体　double volute casing
双吸泵　double-suction pump
双向测缝计　two-way joint meter
双向电能表　bidirectional energy meter
双向阀门　bidirectional valve
双向门机　trolley-mounted gantry crane
双向偏心受压　biaxial eccentric compression
双向平板　two-way flat slab
双向水封　bidirectional seal
双向应变计组　two-way extensometer group
双心拱　two-centered arch
双液灌浆法　double liquid grouting
双闸板　double gate
双重保护　duplicate protection
双重故障　double faults
双重化馈线　duplicate feeders
双重控制　dual control
双柱式墩　queen-post supporting pier
双座双向阀门　twin-seat bidirectional valve
水泵充水和带负荷流程　pump priming and loading sequence
水泵导叶　diffuser vane
水泵工况　pump mode (pumping mode)
水泵控制阀　pump control valve
水泵扩散段　pump diffuser
水泵启动　pump starting
水泵水轮机　pump-turbine
水泵水轮机全特性　complete characteristics of pump-turbine
水泵转向　pumping direction
水玻璃浆液　sodium silicate grout
水尺　staff gage
水处理　water treatment
水传疾病　waterborne disease
水锤　water hammer
水锤数　water hammer number, Allievi constant

水锤压力　water hammer pressure
水道　conduit
水的可压缩性　water compressibility
水电厂　hydropower plant
水电工程　hydropower project (hydroelectric project)
水电工程等别　rank (scale) of hydropower project
水电蕴藏量　hydropower potential
水电站　hydropower station
水电站厂房　hydroelectric powerhouse
水电站进水口　intake
水电站引水渠　headrace canal
水电站引用流量　power discharge
水垫塘　plunge pool
水斗　bucket
水斗的出水角　bucket discharge angle
水斗的倾角　bucket inclination
水斗宽度　bucket width
水斗式水轮机　Pelton turbine
水封　water sealing
水封环　water seal cage
水封压板　keeper plate, clamp bar
水封座板　base plate for seal
水工建筑物　hydraulic structure
水工建筑物级别　grade of hydraulic structure
水工建筑物结构安全级别　safety level of hydraulic structure
水工结构可靠度设计　design of reliability of hydraulic structure
水工结构可靠度水平　reliability level of hydro-structure
水工金属结构　hydraulic steel structure
水工沥青　hydraulic bitumen
水工隧洞　hydraulic tunnel
水管式沉降仪　water level settlement gauge
水化热　heat of hydration
水化学分析　hydro-chemical analysis
水环境　water environment
水环境容量　water environment capacity
水环热泵空气调节系统　water-loop heat pump air conditioning system
水灰比　water-cement ratio
水浇地　irrigated land
水胶比　water-cementitious material ratio
水库保证出水量　safe reservoir yield
水库调查　reservoir investigation (survey)
水库调度　reservoir scheduling
水库调洪　reservoir flood routing
水库调节　reservoir regulation
水库回水　reservoir backwater
水库浸没　reservoir immersion
水库库底清理　clearing of reservoir zone
水库库容面积曲线　reservoir storage capacity-area curve
水库泥沙淤积速率　rate of reservoir sedimentation
水库年限　service life of reservoir
水库起调水位　initial reservoir elevation
水库群调节　multi-reservoir regulation
水库渗漏　reservoir leakage
水库坍岸　reservoir bank-collapse, reservoir bank ruin
水库特征库容　characteristic capacity of reservoir
水库特征水位　characteristic level of reservoir
水库泄空排沙　sediment releasing by emptying reservoir
水库蓄水　reservoir impoundment (filling)
水库淹没　reservoir inundation, reservoir submergence
水库淹没补偿　compensation for reservoir inundation
水库淹没处理　reservoir inundation treatment
水库淹没区　reservoir inundation zone
水库淹没实物指标　reservoir inundation inventory
水库异重流　density current in reservoir
水库影响区　reservoir-affected area
水库诱发地震　reservoir induced earthquake
水库淤积　reservoir silting, sediment deposition in reservoir
水库淤积极限　limit state of sediment deposition in reservoir
水库淤积平衡比降　equilibrium slope of sediment deposition in reservoir
水库淤积上延（翘尾巴）　upward extension of reservoir deposition
水库综合开发利用　multipurpose development of reservoir
水冷骨料　water-cooled aggregate
水冷冷却塔　water-cooling tower
水冷式空调器　water-cooled air conditioner

水冷式冷凝器　water-cooled condenser
水力测功器　hydraulic brake
水力冲填　hydraulic excavation and filling
水力冲填坝　hydraulic fill dam
水力发电设施　hydroelectric installations
水力功率　hydraulic power
水力过渡过程　hydraulic transient
水力机械设备　hydraulic machinery
水力控制阀　hydraulic control valve
水力比降（水力坡度）　hydraulic gradient (slope)
水力侵蚀　water erosion
水力设计　hydraulic design
水力系数　hydraulic factor
水力相似　hydraulic similitude
水力性能　hydraulic performance
水力要素　hydraulic elements
水利工程（水资源工程）　water resources project
水利计算　computation of water conservancy
水利区划　zoning of water conservancy
水利水电工程　water resources and hydropower project
水量分布曲线　water distribution curve
水量平衡　water balance
水流对尾槛的冲击力　impact force of flow on end sill
水流惯性时间常数　water inertial time constant
水流空化数　flowing cavitation number
水陆联运　water-land through transportation
水路运输　waterway transport
水轮发电机　hydrogenerator
水轮发电机组试运行　trial starting of hydrogenerator set
水轮机　hydraulic turbine
水轮机比转速　specific speed of turbine
水轮机层　turbine floor
水轮机调节系统　hydraulic turbine regulating system
水轮机额定输出功率　rated output power of turbine
水轮机飞逸转速　runaway speed of turbine
水轮机工况　turbine mode
水轮机功率试验　turbine output test
水轮机机坑　turbine pit
水轮机机械效率　mechanical efficiency of turbine
水轮机桨叶位置　turbine blade position
水轮机进水流道　turbine inlet water passage
水轮机空载流量　no-load discharge of turbine
水轮机控制传递比　turbine control transmission ratio
水轮机控制系统　hydraulic turbine control system
水轮机控制系统静态特性　droop graph of turbine control system
水轮机廊道　turbine walkway
水轮机流量　turbine discharge
水轮机排气阀　turbine vent valve
水轮机喷嘴位置　turbine nozzle position
水轮机输出功率　turbine output power
水轮机输入功率　turbine input power
水轮机水力效率　hydraulic efficiency of turbine
水轮机瞬态压力　momentary pressure of turbine
水轮机自调节系数　turbine self-regulation coefficient
水轮机最大瞬态转速　maximum momenta-ry over-speed of turbine
水面线　water surface profile
水面蒸发　evaporation from water surface
水幕系统　drencher system
水能　waterpower，hydropower
水能计算　hydropower computation
水能利用规划　waterpower utilization planning
水能资源（水力资源）　waterpower resources，hydropower resources
水泥　cement
水泥罐　cement silo
水泥裹砂喷射混凝土　cement paste wrapping sand shotcrete
水泥浆液　cement grout
水泥卷锚杆　cement cartridge rock bolt
水泥抹面　finishing cement，cement plaster
水泥土护坡　soil-cement slope
水暖　hot-water heating
水平层理　horizontal bedding
水平单管采暖系统　one (single) -pipe loop circuit heating system
水平冻胀力　horizontal frost heaving force
水平度　levelness
水平度盘　horizontal circle
水平反滤层　horizontal filter
水平分布钢筋　horizontally distributed bar

水平沟　level ditch
水平固定筛　horizontal stationary screen
水平加速度　horizontal acceleration
水平阶　horizontal stage
水平均布压力　horizontal uniform pressure
水平梁单元　beam element
水平梁单元的支座　abutment of a beam element
水平年　target year
水平排列　horizontal configuration
水平排水　horizontal drainage
水平排水垫层　horizontal drainage cushion
水平排水沟　contour ditch
水平坡　horizontal slope
水平梯田　bench terrace
水平向地震作用　horizontal seismic action
水平旋转式隔离开关　center rotating disconnector
水平钻进　horizontal boring
水枪　hydraulic gun
水情遥测站　telemetry station
水情中继站　relay station
水情中心站　center station
水情自动测报系统　hydrological telemetry and forecasting system
水舌挑距　trajectory distance of nappe
水生生态　aquatic ecology
水生生态变化　aquatic ecology alteration
水生维管束植物　aquatic vascular plant
水生物　aquatic creature
水生物栖息地　aquatic habitat
水生物种　aquatic species
水生植物　aquatic plant
水声探测　sonic echo exploration
水室式调压室　water-chamber-type surge chamber
水损失　water loss
水塔　water tower
水套盖　jacket cover
水体掺气　water aeration
水田　paddy field
水听器　hydrophone
水头　water head
水头保证率曲线　head dependability curve
水头损失　head loss
水土保持　water and soil conservation
水土保持措施　soil and water conservation measures
水土保持方案　soil and water conservation program
水土保持方案报告　Soil and water conservation report
水土保持耕作措施　agricultural measure for soil and water conservation
水土保持规划　soil and water conservation planning
水土保持监督　soil and water conservation supervision
水土保持经济效益　economic benefit of soil and water conservation
水土保持林　forest for soil and water conservation
水土保持区划　soil and water conservation regionalization
水土保持设施　soil and water conservation facilities
水土保持社会效益　social benefit of soil and water conservation
水土保持生态环境建设　soil and water conservation for ecological environment rehabilitation
水土保持生态效益　ecological benefit of soil and water conservation
水土保持效益　soil and water conservation benefit
水土流失　soil erosion and water loss
水土流失观测　observation of soil erosion and water loss
水土流失监测　monitoring of soil and water conservation
水土流失类型　type of soil erosion and water loss
水土流失面积　area of soil erosion and water loss
水土流失区　region of soil erosion and water loss
水土流失预防　prevention of soil erosion and water loss
水土流失治理程度　erosion control ratio
水土流失治理面积　area of water and soil conservation
水土流失综合治理　comprehensive control of soil erosion and water loss
水推力　hydraulic thrust
水位变动期　alternate period of high and low water level
水位变动区　water level fluctuating zone
水位变幅　water level amplitude
水位变率　change rate of water level
水位低于转轮　water level below the runner

水位基点，测站基面　gauge-datum
水位计　hydrological guage
水位监视设备　water level monitoring equipment
水位降深　drawdown
水位流量关系　stage-discharge relationship
水位骤降期　rapid drawdown period of water level
水温观测　observation of water temperature
水文测验　hydrometry
水文测站　gauging station
水文代表站　representative station
水文地质　hydrogeology
水文地质试验　hydrogeological test
水文基本站　basic data station
水文基准站，参证站　bench mark station
水文计算　hydrological computation
水文记录　hydrological records
水文模型　hydrological model
水文年　water year
水文频率分析　hydrological frequency analysis
水文频率曲线　hydrological frequency curve
水文设计站　design station
水文统计　hydrological statistics
水文系列　hydrological series
水文遥测系统　hydrological telemetry system
水文站　hydrological station
水污染　water pollution
水污染控制　water pollution control
水系错动　dislocated water system
水系（河网）　hydrographic net，river system
水下不分散混凝土　non-dispersible underwater concrete
水下地形　underwater topography
水压计　hydraulic gauge
水压力　water pressure
水压试验　hydrostatic pressure test
水压致裂法　hydraulic fracturing method
水样　water sample
水样简分析　simplified analysis of water sample
水样全分析　total analysis of water sample
水硬性水泥　hydraulic cement
水源涵养林　forest for water resources conservation
水源热泵　water-source heat pump
水跃　hydraulic jump
水跃消力池　hydraulic jump basin
水跃淹没度　submergence degree of hydraulic jump
水云母　hydromica
水藻（藻类）　algae
水胀式锚杆　water expansion anchor bolt
水蒸气分压力　partial pressure of water vapor
水直接冷却发电机　direct water-cooled gnerator
水质　water quality
水质标准　water quality standard
水质参数　water quality parameter
水质分析报告　Water quality analysis report
水质监测　water quality monitoring
水质评价　water quality assessment
水质预报　water quality forecast
水质阻垢缓蚀处理　water quality treatment of scale-prevent and corrosion-delay
水中不分离剂（抗分散剂）　non-dispersible agent
水轴承　submerged bearing
水轴承套　bearing sleeve
水轴承体　bearing spider
水柱压力　water column pressure
水坠坝　sluicing-siltation earth dam
水准测量　leveling，level survey
水准测量平差　leveling adjustment
水准点　leveling point，bench mark
水准网　level net
水准仪　level
水资源　water resources
水资源保护　water resources protection
水资源分配　water resources allocation
水资源分区　water resources regionalization
水资源供需分析　demand and supply analysis of water resources
水资源管理　water resource management
水资源规划　water resources planning
水资源决策支持系统　water resources decision support system
水资源评价　water resources assessment
水资源危机　water resources crises
水资源项目（水利项目）　water resources project
税费　tax and fee
税金　tax
税收优惠政策　preferential tax policy
顺坝　training dike，parallel dike
顺层滑动　slide along bedding plane
顺层滑坡　consequent landslide

顺层节理　bedding joint
顺河断层　fault stretching along river
顺向坡　dip slope，consequent slope
顺序阀　sequence valve
顺序控制系统　sequence control system
顺序重合闸　sequential reclosing
瞬变电磁法　transient electromagnetic method (TEM)
瞬变状态　transient condition
瞬发爆破　short-delay blasting
瞬发雷管　instantaneous detonator
瞬时保护　instantaneous protection
瞬时单位线　instantaneous unit hydrograph
瞬时弹性模量　instantaneous modulus of elasticity
瞬时功率　instantaneous power
瞬时故障　transient fault
瞬时角速度　instantaneous angular speed
瞬时脱扣器　instantaneous release
瞬时值　instantaneous value
瞬态　transient
瞬态短路电流　transient short-circuit current
瞬态分量　transient component
瞬态过电压　transient overvoltage
瞬态恢复电压上升率　rate of rise of TRV，RRRV
瞬态压力变化率　momentary pressure variation ratio
瞬态转速变化率　momentary speed variation ratio
司机室　cab
司机室操纵起重机　cab-operated crane
丝堵　screwed plug
丝头　threaded sector
丝绸之路经济带　Silk Road Economic Belt
私人融资　Private Financing
斯宾塞程序　Spencer's Procedure
撕裂强度　tearing strength
死库容　inactive storage，dead storage
死区　dead band，dead zone
死区保护　dead zone protection
死水位　minimum operating level，dead water level
四边支承楼板　two-way flat slab
四分法　quartering method
伺服电动机　servomotor
伺服定位器时间常数　time constant of the servo-positioner
伺服阀　servo valve
伺服机械装置　servomechanism
伺服加速度式测斜仪　servo acceleration inclinometer
饲料作物　forage crops
饲养制度　feeding regime
松　loose
松弛节点　slack bus
松弛区　relaxed zone
松动爆破　loosening blasting
松动区　loosened zone
松动圈　relaxation zone
松动岩体　loosened rock mass
松方　loose measure
松散系数　bulk factor
松散岩体　loosed rock mass，broken rock mass
松套管　buffer tuber
送出线路工程　transmission works
送风管道　supply air duct
送风机室　supply fan room
送风温差　effective temperature difference
送料溜槽　feed chute
送桩　follower
素混凝土结构　plain concrete structure
素土夯实　packed soil
素（无筋）混凝土　plain concrete
速动时间常数　promptitude time constant
速度　velocity，speed
速度比能　velocity energy
速度计　speed gauge
速度三角形　velocity triangle
速度水头　velocity head
速断差动保护　instantaneous (high-set) differential protection
速凝　quick set
速凝剂　accelerating admixture
宿舍　dormitory
塑壳断路器　moduled case circuit breaker (MCCB)
塑料导爆管　plastic primacord tube
塑料盲管　plastic blende drain
塑料排水带　plastic drain
塑流-拉裂　plastic flow-tensile fracture
塑态　plastic state
塑限　plastic limit
塑性　plasticity
塑性变形　plastic deformation
塑性分析　plastic analysis

塑性挤出　plastic squeezing out
塑性矩　plastic moment
塑性区　plastic zone
塑性止水材料　plastic sealer
塑性指数　plasticity index
溯源侵蚀　headward erosion
溯源淘刷　retrogressive scouring，head ward scouring
溯源淤积　backward deposition
酸度　acidity
酸碱度（pH）　acidity-alkalinity，power of hydrogen（pH value）
酸洗　acid pickling
酸型腐蚀　acid corrosion
酸性水　acidic water
酸性岩　acid rock
酸雨　acid rain
随动的伺服定位器　following main servo-positioner
随动控制　follow-up control
随动系统　servo-system
随机变量　arbitrary variation，stochastic variable
随机变量当量正态分布　equivalent normal distribution of stochastic variable
随机变量的数学期望值　mathematical expectation value of stochastic variable
随机变量设计验算点　design checking point of stochastic variable
随机锚杆　spot bolting
随机特性　stochastic characteristic
随机性水文模型（非确定性水文模型）stochastic hydrological model
随机振动或脉动　random vibration or pulsation
随升井架　headframe lifting along with working deck
碎冰　crushed ice
碎部测量　detail survey
碎块状结构　clastic structure
碎裂结构　cataclastic structure，disintegrated structure
碎裂岩　cataclasite
碎石　crushed stone
碎石路面　macadam pavement
碎石土　debris soil
碎石，岩屑　debris
碎屑沉积　clastic sediment
碎屑结构　clastic texture
碎屑流机制　clastic flow mechanism
碎屑状结构　fragmental structure
隧道　tunnel
隧洞导流　tunnel diversion
隧洞定位点（桩）　tunnel station（stake）
隧洞开挖　tunnel excavation
隧洞扩大掘进法　tunnel reaming method
隧洞倾斜度　inclination of tunnel
隧洞上平段　upper horizontal section
隧洞下平段　lower horizontal section
隧洞形状　shape of tunnel
损失吨位　decreased load
损失分布系数　loss distribution coefficient
缩径阀门　reduced-port valve
缩口阀门　reduced-bore valve
缩限　shrinkage limit
索铲　cable drag scraper
索饵场　feeding ground
索结构　cable structure
索赔　claim
索引　index
锁定荷载（锁定吨位）　lock-in load
锁定后损失率　locked loss rate
锁定损失率　locking loss rate
锁定装置　dogging device（dog）
锁紧螺栓　lock bolt

T

塌坡　slope collapse
塌陷　caving
塔高　tower height
塔式结构　tower structure
塔式进水口　intake tower
塔式起重机　tower crane
踏勘　reconnaissance
踏面　rolling face
弹簧薄膜式减压阀　spring diaphragm reducing valve
弹簧垫圈　spring washer
弹簧式安全阀　direct spring loaded safety valve
弹簧座　spring plate
弹塑性变形　elastic-plastic deformation
弹性　elasticity

弹性变形　elastic deformation
弹性波测试　elasticity wave testing
弹性矩　elastic moment
弹性抗震设计　seismic elasticity design
弹性力学法　elastic mechanics method
弹性密封副　resilient seats
弹性模量　elastic modulus
弹性闸板　flexible gate disc
调峰电厂　peak load power plant
调峰容量　peak load regulating capacity
调峰填谷　peak shaving and valley filling
调洪库容　flood regulation storage
调节保证　regulating guarantee
调节池　regulating pond
调节阀（控制阀）　control valve，adjusting valve
调节负荷机组　controllable set
调节和控制柜　regulating and control cubicle
调节流量　regulated flow
调节螺母　adjusting screw
调节命令　adjusting command
调节器输出信号　governor output signal
调节圈　adjusting ring
调节系数　regulation coefficient
调节周期　period of regulation
调频方式　mode of frequency regulation
调频容量　frequency regulating capacity
调速阀　flow control valve
调速器油压　governor oil pressure
调相方式　condensing mode
调相工况　condenser mode
调相容量　synchronous condensing capacity
调相运行　condensing operation
调压方式　voltage control method
调压井　surge shaft
调压室涌浪　excursion of surge tank water level
调整板　adjusting plate
调整环　adjust ring
调制解调器　modulator-demodulator，modem
调质　quenched and tempered
台车式启闭机（台车）　trolley
台地　terrace
台阶　step
台阶结构面　stepped structural plane
台阶掘进法　heading and bench method
台阶式溢洪道　stepped spillway
太古代（界）　Archaeozoic era（erathem）
太平梯　escape stair
太沙基承载力理论　Terzaghi's bearing capacity theory
太沙基固结理论　Terzaghi's consolidation theory
太阳辐照量　solar irradiation
太阳能保证率　solar fraction
太阳能灯　solar lamp
太阳能间接系统　solar indirect system
太阳能直接系统　solar direct system
坍岸　bank sloughing
坍岸区　bank caving area
坍落度　slump
摊铺机　spreader
摊销费　amortized cost
滩涂　beach
炭　char
炭质页岩　carbonaceous shale
探槽　trial trench
探坑　trial pit
碳弧气刨　air carbon arc gouging
碳化硅阀式避雷器　silicon carbide valve type surge arrester
碳汇　carbon sink
碳素钢　carbon steel
碳素结构钢　carbon steel
碳酸型腐蚀　carbonic acid corrosion
碳酸盐岩建造　carbonatite formation
塘堰　pond
掏槽孔　cut hole
逃生滑梯　slide escape
逃生人孔　escape hatch
陶瓷活塞杆　ceramic-coated rod
淘蚀岸坡　undercut slope
淘刷　scouring
套管　bushing
套管灌浆法　casing grouting method
套管式电流互感器　bushing type current transformer
套筒连接　pressed sleeve splicing of rebars
套筒伸缩节　sleeve expansion joint
套筒式沉降仪　telescopic settlement gauge
套筒式控制阀　sleeve valve
套型　dwelling unit type
套用图　drawing use indiscriminately

特大暴雨　extraordinary rainstorm
特大石　boulder
特大（重）件运输措施费　measure expense of heavy cargo transport
特高压　ultra-high voltage (UHV)
特高压直流　ultra-high voltage direct current (UHVDC)
特雷斯卡准则　Tresca criteria
特殊保护系统［安全稳定装置］　special protection system (SPS)
特殊地区施工增加费　additional cost for construction in special areas
特殊垫层区　special cushion zone
特殊毒性　specific toxicity
特殊荷载　special load, unusual load
特斯拉　Tesla
特性参数　characteristic parameter
特性角　characteristic angle
特性量　characteristic quantity
特性量整定范围　setting range of the characteristic quantity
特性量整定值　setting value of the characteristic quantity
特性曲线　characteristic curver
特性试验　characteristic test
特性阻抗　characteristic impedance
特许（权）　franchise
特许权协议　build-operate-transfer (BOT)
特有物种　endemic species
特征方程式　characteristic equation
特征值　characteristic value
特征周期　characteristic period
特种作业人员　special operation personnel
梯段爆破，台阶爆破　bench blasting
梯级水电站　cascade hydropower station
梯田　terrace
踢脚　baseboard
提升架　lift yoke
提升力　lifting force, hoist capacity
提升式平板闸门（串轮闸门）　Stoney gate
提升速度　lifting speed
提水灌溉　pumping irrigation
题名　title
体波　body wave
体分布力　force per unit volume
体积安定性　soundness
体积密度（容重）　bulk density
体积配筋率　volumetric reinforcement ratio
体型图　shape drawing
体育场　stadium
体育馆　gymnasium
替代法　substitution approach
替代生境　alternative habitat
天窗　skylight
天沟　gutter
天井　patio
天然骨料　natural aggregate
天然含水量（含水率）　natural water content
天然级配骨料　pit-run aggregate
天然建筑材料分布图　Distribution of natural construction material sources
天然密度　natural density
天然坡面径流场　natural runoff plot
天然橡胶　natural rubber (NR)
天然有机聚合物　natural organic polymer
添加剂泥浆　slurry with additive
田间需水量　water consumption on farmland
填表调查　tabulating statistic
填补块［面板坝］　fillet
填方边坡　filled slope
填料　filler
填料垫　packing seat
填料盒　stuffing box
填料环　seal cage
填料式旋塞阀　gland packing plug valve
填料压盖　gland cover
填料轴套　packing sleeve
填挖平衡的土方工程　balanced earthworks
填筑　filling, placing
填筑坝（土石坝）　earth-rock fill dam, embankment dam
填筑标准　placement criterion
填筑分区　embankment zoning
填筑干容重　fill dry unit weight
填筑含水量　fill moisture content
填筑间隔时间　intermittent time between placing layers
填筑料　embankment material
填筑碾压标准　embankment rolling criteria
调速器　governor

挑流鼻坎　deflecting bucket，flip bucket
挑流式厂房　flyover type powerhouse
挑流水舌　trajectory nappe
挑流消能　ski-jump energy dissipation
挑檐　overhanging eaves
条分法　slice method
条缝型风口　slot outlet
条目　entry
条渠沉沙池　desilting channel
条形码　bar code
跳板　gangway
跳仓浇筑　sequence placement
跳打桩　staggered piling
跳堆法　interval dumping method
跳线线夹　jumper flag，jumper lug
贴边岔管　hem reinforced branch pipe
贴附水舌　adhering nappe
贴角［拱坝］　fillet
贴坡排水　slope face drainage
贴嘴灌浆法　port-adhesive grouting
铁磁谐振　ferro-resonance
铁磁性　ferromagnetism
铁磁性物质　ferromagnetic substance
铁轨　railroad
铁路接轨站　track connection station
铁路运输　railway transport
铁锰质浸染　iron-manganese disseminated
铁损耗　iron loss
铁塔　steel tower
铁芯故障　core fault
铁芯故障探测　stator core fault detection
铁芯及机座振动　stator core and frame vibration
铁芯松弛　core loosening
铁质浸染　iron disseminated，ferruginous imbueing
铁质砂岩　ferruginous sandstone
庭院　courtyard
停产期　period of closure
停产损失　loss in production stop
停产损失费　production suspension loss
停车场　parking lot
停电　outage，interruption
停电持续时间　interruption duration
停机按钮　shutdown push button
停运率　outage rate
通仓浇筑　concreting without longitudinal joint
通带　pass band
通断　switching
通风　ventilation
通风道　ventilating duct
通风支管　branch duct
通风止回阀　check damper
通缝　continuous seam
通过能力　through capacity
通航　navigation
通航保证率　navigation dependability
通航标准　navigation standard
通航建筑物　navigation structure
通航净空　navigation clearance
通航流量　navigation discharge
通航流速　navigation velocity
通航密度　navigation density
通航期　navigation period
通航水深　navigation depth
通航水头　head of navigation
通货膨胀率　Inflation Rate
通流式调速器　governor without pressure tank，through flow type governor
通气孔　air vent
通勤人口　commuting population
通水冷却　cooling by circulating water
通信　communication
通信电源室　communication power supply room
通信工作站　communication workstation
通信模件　communication module
通信软件　communication software
通信设备室　communication device room
通信卫星　communication satellite
通信线路　line of communication
通信协议　protocol
通行能力　traffic capacity
通用负荷开关　general purpose switch
同步　synchronization
同步并列　synchronization
同步电动机　synchronous motor
同步电机的电枢反应　armature reaction of synchronous machine
同步电机的时间常数　time constants of synchronous machine
同步电机的突然短路　sudden short-circuit of synchronous machine

同步电机的振荡　oscillation of synchronous machine
同步电机相量图　phasor diagram of synchronous machine
同步电机异步运行　asynchronous operation of synchronous machine
同步调相机　synchronous condenser
同步对时　synchronous clock
同步发电机超瞬态电势　voltage behind subtransient reactance of synchronous generator
同步发电机动态稳定　dynamic stability of synchronous generator
同步发电机功率角　power angle of synchronous generator
同步发电机静态稳定　steady state stability of synchronous generator
同步发电机瞬态电势　voltage behind transient reactance of synchronous generator
同步发电机瞬态稳定　transient stability of synchronous generator
同步观测　simultaneous observation
同步检查继电器　synchro check relay
同步检查装置　synchronism detection unit, synchrocheck unit
同步检定　synchro check
同步启动　synchronous starting
同步设备　synchronizing device (synchronizer)
同步时间　synchronous time
同步摇摆　synchronous swing
同步远动传输　synchronous telecontrol transmission
同步轴　synchronizing shaft
同层排水　same-floor drainage
同程热水供应系统　reversed return hot water system
同程式系统　reversed return system
同时系数　coincidence factor
同位素年龄　isotopic age
同位素示踪法　isotopes tracer technique
同相　in phase
同轴度　coaxiality
同组试样　companion specimen
铜铝过渡板　copper to aluminium adapter board
铜损耗　copper loss
铜止水　copper waterstop
统计参数　statistical parameter
统计操作过电压　statistical switching overvoltage
统计回归分析　statistical regression analysis
统计雷电过电压　statistical lightning overvoltage
筒式导轴承轴瓦　guide bearing shell
筒式导轴承轴瓦支撑装置　shell suppor-ting device
筒体结构　tube structure
筒形风帽　cylindrical ventilator
筒形壳体　barrel casing
投标　bidding
投标保函　tender security
投标保证金　bid bond
投标担保　bid guarantee
投标人合格性　eligibility
投标文件　bidd document
投标书的提交　submission
投标资格预审　prequalification of bidder
投光灯　spot lamp
投切命令　on-off command
投入产出法　input to output approach
投入式水位传感器　submerged water level sensor
投影网　strereonet
投运前试验　pre-commissioning test
投资估算　investment estimate
投资计划与资金筹措表　Investment plan and financing
投资指标　investment index
透长石　sinidine
透镜体　lenticle (lens)
透明度与公信力　transparency and credibility
透视图　prospective view
透水层　permeable layer
透水堤　permeable dike
透水率　permeable rate
透水性　permeability
透水性较强的材料　highly pervious material
凸岸　convex bank
凸极电机　salient pole machine
凸块振动碾　pad foot vibratory roller
凸头螺栓　raised head bolt
突变段　sharp transition
突变量继电器　sudden change relay
突扩式门槽　sudden-expand gate slot
突泥　mud eruption
突然加电压保护　inadvertent energizing protection

突水　water bursting
图根点　mapping control point
图解法　graphical solution
图例　legend
图书馆　library
图像处理　image processing
图像分辨率　picture resolution
图像识别　image recognition，pattern recognition
图像压缩　image compression
图像质量　picture quality
涂层道数　number of coats
涂层附着力　coating adhesion
涂膏式极板　pasted plate
涂料　paint；coating
涂料配套（涂层系统）　coating system
涂漆　painting
土坝　earth dam
土层剪切波速　shear-wave velocity of soil layer
土的固结　consolidation of soil
土的界限含水量试验　Atterberg test
土地　land
土地承载能力　land bearing capacity
土地的机会成本　opportunity cost of land
土地开发整理　land reclamation and improvement
土地开垦　land reclamation
土地利用规划　land use planning
土地流转　land circulation
土地扰动　land disturbance
土地沙化　land desertification
土地使用权　land use right
土地适宜性评价　land suitability assessment
土地收入　land revenue
土地所有权　land ownership，land tenure
土地占补平衡　requisition-compensation balance of land
土地征用　land requisition
土地征用标准　land requisition criterion
土地征用线　property line，border of land requisition
土钉墙　soil nail wall
土房　adobe
土工格栅　geogrid
土工膜　geomembrane
土工网垫　three-dimensional geonet
土工无纺布　nonwoven geotextile
土工织物　geotextile
土力学　soil mechanics
土料　earth（soil）material
土料场　borrow area
土料翻晒　soil tedding
土锚钉　soil dowel
土锚杆支护　soil nailing
土壤饱和含水量　saturated moisture content of soil
土壤电阻率　soil resistivity
土壤含水量（含水率）　soil moisture content
土壤抗蚀性　soil erosion resistance
土壤流失量　amount of soil loss
土壤潜育化　soil gleization
土壤侵蚀　soil erosion
土壤侵蚀程度　soil erosion degree
土壤侵蚀分区　soil erosion zoning
土壤侵蚀规律　mechanism of soil erosion
土壤侵蚀类型　type of soil erosion
土壤侵蚀量　amount of soil erosion
土壤侵蚀模数　soil erosion modulus
土壤侵蚀模型　soil erosion model
土壤侵蚀强度　soil erosion intensity
土壤侵蚀速率　soil erosion rate
土壤水　soil water
土壤通气不良　defective soil aeration
土壤污染　soil pollution
土壤蒸发量　soil evaporation
土石方工程　earth-rock works
土石方机械　earth rock excavation machinery
土石方开挖　earth-rock excavation
土石方明挖　open earth-rock excavation
土石方平衡　earth-rock cut and fill balance，excavation-fill balancing
土石混合体　soil-rock mixture
土石围堰　earth rockfill cofferdam
土体的沉降　settlement of soil mass
土压力　earth pressure
土压力计　soil pressure gauge
土样分析　soil sample analysis
土与结构相互作用　soil-structure interaction
土质防渗体　soil impervious zone
土质防渗体分区坝　zoned earth rockfill dam with impervious soil core
土质心墙堆石坝　earth core rockfill dam
土钻　soil auger

吐出壳（吐出段） discharge casing
吐出弯管 discharge elbow
湍流 turbulence
团粒状土 crumbling soil
推测界线 conjectural boundary
推车式灭火器 transportable fire extinguisher
推拉杆 connecting rod
推力墩 thrust block
推力头 thrust collar
推力瓦 thrust pad，thrust bearing shoe
推力瓦支撑 thrust pad support
推力轴承 thrust bearing
推力轴承机架 thrust-bearing bracket
推力轴承基础板 thrust bearing base plate
推力轴承油箱 thrust bearing housing
推力轴承油压 thrust bearing oil pressure
推力轴承支架 thrust bearing supporting cone
推土机 bulldozer
推悬比 ratio of bed load discharge to suspended load discharge
推移带 transporting belt
推移型滑坡 landslide of pushing type
推移质 bed load
推移质输沙率 bed load discharge
退场 demobilization
退场费 demobilization cost
退出运行 out of service
退磁 demagnetization
退耕还林还草 grain-for-green project
退关 shut out
退火 annealing
退水渠 canal for water release
托辊 roller
托轮 riding wheel
托马数 Thoma number
托盘运输 palletized transport
托运 consignment
拖车 trailer
拖带式输送机 trailing conveyor
拖挂运输 tractor-trailer transport
拖曳运动 drag movement
拖曳褶皱 drag fold
脱扣器 release
脱流 separation
脱模剂 concrete remover
脱水过滤汽化装置 water-extraction and filtration unit
脱水筛 dewatering screen
驼峰堰 hump-shaped weir
拓扑结构 topology

W

挖方边坡 excavated slope
挖孔桩 hand-dug pile
洼地 depression
蛙式打夯机 frog hammer
瓦斯继电器 Buchholz relay
瓦斯突出 gas burst
瓦特 Watt
外保温 external thermal insulation
外币 foreign currency
外部导电部分 external conductive part
外部观测 external monitoring
外部火灾 exposure fire
外部火灾曲线 external fire curve
外部效果 Externality
外部直径处的节矩 bucket pitch at outer diameter
外导水环 outer guide ring
外观检测 visual inspection
外观质量 quality of appearance
外弧面，拱背 extrados
外加变形 imposed deformation
外加剂 additive，agent
外绝缘 external insulation
外壳 outer casing
外壳带电断路器 live tank circuit-breaker
外壳防护等级 degree of protection of enclosure
外壳漏电保护系统 frame leakage protection，case earth protection
外锚头 outer fixed end
外贸货物 traded goods
外贸投入和产出 trade input and output
外摩擦角 angle of external friction
外能灭弧室 external energy interrupter
外墙 exterior wall
外墙平均传热系数 average overall heat transfer coefficient
外倾结构面 out-dip structural plane
外施电压 applied voltage
外水压力 external water pressure

外水压力作用水头　acting water head of external water pressure
外水压强　external water pressure intensity
外推法　extrapolation
外泄漏试验　outer-leakage test
外延法　extrapolation method
外业　field work
外营力　exogenic agent
弯管段损失［水头］　bend loss
弯矩　bending moment
弯起钢筋　bent-up rebar
弯曲-拉裂　bending-tensile fracture
弯曲褶皱　buckled fold
弯折销装置　buckling pin device
蜿蜒河道　meandering channel，twisting channel
完全充电　full charge
完全二次型方根法　complete quadric combination method
完全停机　complete shutdown
完整　integral
完整孔　completely penetrating well
完整试样　intact specimen
完整丝扣　one complete screwthread
完整性　integrity
完整性差　poorly integral
完整岩体　intact rock
晚第三纪（系）　Neogene period（system）
晚新生代地层　Late Cenozoic Stratum
碗头　socket
网孔　mesh
网络　network
网络安全防护　cyber security and prevention
网络变换　network transformation（conversion）
网络攻击　network attack
网络化简　network simplification
网络图　network chart
网桥　network bridge
网状壳体　reticulated shell
网状裂缝　map crack
往复筛　reciprocating screen
往复式给料机　reciprocating feeder
往复式皮带输送机　shuttle conveyor
往复移动式缆索道　shuttle cableway
危险电压　dangerous voltage
危险环境　hazardous environment
危险货物　dangerous goods
危险物质　hazardous substance
危险因素　hazardous factor
危险源辨识　hazard identification
危岩　dangerous rock，unstable rock
微波收发信机　microwave transmitting and receiving equipment
微差爆破（毫秒爆破）　millisecond blasting
微地震波测试　micro-seismic wave survey
微分电路　differential circuit
微分方程式　differential equation
微分环节时间常数　time constant of derivative module
微分增益　differential gain
微分作用时间　derivative action time
微风化　slightly weathered
微机保护　microprocessor based protection
微机变压器保护　microprocessor based transformer protection
微机电动机保护　microprocessor base motor protection
微机调速器　micro-computer based governor
微机继电保护装置　microprocessor based protection device
微机母线保护　microprocessor based busbar protection
微晶结构　microlitic texture
微矿化水　slightly mineralized water
微矿化水型腐蚀　corrosion by slightly mineralized water
微裂缝，微裂隙　microfissure
微启式安全阀　low lift safety valve
微速下降速度　precision load-lowering speed
微透水　slightly pervious
微型断路器　miniature circuit breaker（MCB）
微张　slightly open
微震　microquake
微震仪　micro-seismograph
韦伯　Weber
韦伯数　Weber number
违约金　liquidated damages
围堤　border dike
围护结构　building envelope
围护结构传热系数　overall heat transfer coefficient

围护结构热惰性指标　thermal inertia index of building envelope
围护结构热损失　heat loss of building envelope
围护结构温差修正系数　temperature difference correction factor of building structure
围圈（围囹）　form waling
围压　confining pressure
围岩　surrounding rock
围岩膨胀内鼓　swelling out of surrounding rock
围岩偏压　non-uniform rock pressure
围岩强度应力比　strength-stress ratio of surrounding rock
围岩稳定　surrounding rock stability
围岩应力　surrounding rock stress
围岩总评分　total rating of surrounding rock
围堰分期拦断河床　staged river closure by cofferdam
围堰一次拦断河床　non-stage river closure by cofferdam
桅杆式起重机　derrick
帷幕灌浆　curtain grouting
维护系数　maintenance factor
维卡针入度试验　Vicat needle test
维修时间　maintenance duration
卫生间　bathroom，toilet
卫生器具　sanitary fixtures
卫生设施　sanitation facility
卫星照片　satellite photograph
未焊透　incomplete fusion，cold weld
未开垦土地　uncultivated land
未利用地　left-over land
未溶解于水的气体　undissolved air
未熔透焊缝　skin weld
未设站流域　ungaged basin
伟晶岩　pegmatite
伟晶岩脉　pegmatite vein
位移　displacement
位移传感器　displacement sensor
位移放大系数　displacement magnification factor
位移监测　displacement observation
位移延性系数　displacement ductility ratio
位置比能　potential energy
位置水头　potential head
尾槛　end sill
尾水调压室　tailrace surge chamber
尾水管　draft tube
尾水管层　draft tube floor
尾水管扩散段　draft tube outlet part
尾水管里衬　draft tube liner
尾水管涡带（混流）　vortex rope
尾水（管）闸门　tail gate，draft-tube gate
尾水管肘管　draft tube elbow
尾水管锥管　draft tube cone
尾水平台　tailrace platform
尾水渠　tailrace channel
尾水隧洞　tailrace tunnel
尾水位　tailwater level
尾水压水控制开关　tailwater depressing control switch
尾纤　pigtail
委托人　principal
温度　temperature
温度保护　temperature protection
温度场的变化　variation of temperature field
温度分析　temperature analysis
温度监测　temperature monitoring
温度梯度　temperature gradient
温度骤降　sudden temperature drop
温度作用　temperature action（effect），thermal action
温和地区　mild region
温控措施　temperature control measure
温泉　hot spring
温升试验　temperature rise test
温室气体排放　greenhouse gas emission
文本　text，version
文稿　draft
文化宫　cultural palace
文化娱乐建筑　cultural and recreation building
文件　document，record
文丘里管　Venturi tube
文书档案　administrative archives
文物保护遗址　cultural relics site
文物古迹　culturaland historic relic
文献登记　registration
文献统计　statistics
纹波系数　ripple factor
紊流　turbulent flow
稳定电源　stabilized power supply
稳定非均匀流　steady non-uniform flow

稳定计算 stability calculation
稳定浆液 stable grout
稳定流抽水试验 steady-flow pumping test
稳定水位 steady water level
稳定未变形边坡 stable and undeformed slope
稳定温度场 stable temperature field
稳定性 stability
稳定性分析 stability analysis
稳定裕度 stability margin
稳定状态 steady state
稳流特性 stabilized current characteristic
稳态传热 steady-state heat transfer
稳态短路电流 steady state short-circuit current
稳态飞逸转速 steady-state runaway speed
稳态分量 steady state component
稳态运行 steady state operation
稳压泵 pressure maintenance pump
稳压阀 pressure sustaining valve
稳压特性 stabilized voltage characteristic
涡带 vortices
涡街流量计 vortex flow meter
涡流 eddy current
涡流测功器 eddy-current brake
涡流传感器 eddy current sensors
涡流损耗 eddy current loss
涡轮流量计 turbine flow meter
窝工费用 expenditure of idleness
窝工区域 slack area
蜗杆 worm
蜗壳 spiral case
蜗壳泵 volute pump
蜗壳层 spiral case floor
蜗壳进人孔 spiral case access
蜗壳水压传感器 spiral case water pressure sensor
蜗轮 worm gear
蜗轮传动装置 worm gear actuator
蜗形体 volute casing
卧式壳管式冷凝器 closed shell and tube condenser
卧式壳管式蒸发器 closed shell and tube evaporator
卧式水轮发电机 horizontal hydrogenerator
卧室 bedroom
卧轴机组 horizontal shaft unit
握裹力 bond stress
渥太华砂 Ottawa sand
圬工坝 masonry dam
污秽区 polluted area
污染物质 pollutant
污闪 pollution flashover
污水泵 sewage pump
污水处理厂 sewage treatment plant
污水排放标准 sewage discharge standard
污水综合排放 integrated wastewater discharge
钨金瓦 babbitt bearing
屋顶挡烟隔板 roof smoke screen
屋顶花园 roof garden
屋顶通风 roof vent
屋顶闸门 roof gate
屋盖支撑 roof bracing
屋脊式隔膜阀 weir diaphragm valve
屋架 roof truss
屋面板 roof board
屋面防水 roof waterproofing
屋面檩条 roof purlin
屋面排水 roof drainage system
无侧限抗压强度 unconfined compressive strength
无差调节 no-difference regulation, no-deviation regulation
无衬砌隧洞 unlined tunnel
无触点开关 non-contact switch
无电流时间 dead time
无电压检定 non-voltage verification
无缝钢管 seamless steel pipe
无腐蚀 non-corrosive
无附加条件的合同 absolute contract
无杆腔 rodless chamber
无功电能 reactive energy
无功电能表 var-hour meter
无功负荷 reactive load
无功功率 reactive power
无功功率变送器 reactive power transducer
无功功率表 varmeter
无功功率补偿 reactive power compensation
无功功率电压调节 reactive power voltage control
无功功率与电压控制 control of reactive power and voltage
无轨滑模 flexible slip form
无轨运输 trackless haulage
无机改性水泥浆 inorganic modified grout
无间隙金属氧化物避雷器 gapless metal oxide arrester

无键槽的收缩缝　unkeyed contraction joint
无筋砌体构件　masonry member
无铠装电缆　non-armoured cable
无励磁分接开关　off-circuit tap-changer
无梁岔管　shell type branch pipe
无黏结环锚　unbonded circular anchored tendon
无黏结预应力混凝土结构　unbonded prestressed concrete structure
无黏结预应力锚杆　unbounded - prestressed anchor
无黏性填缝料　cohesionless sealer
无黏性土　cohesionless soil
无人值班变电站　unmanned substation
无人值班（少人值守）水电厂　unmanned (with few watchers) hydropower plant (unattended (with few watchers) hydropower plant)
无人值班水电厂　unmanned hydropower plant (unattended hydropower plant)
无沙水流　sediment free flow
无砂混凝土　no-fines concrete
无收缩混凝土　non-shrinkage concrete
无刷励磁机　brushless exciter
无霜期　frost-free period
无水调试　dry test
无损检测　non-destructive testing
无条件保函　first demand guarantee
无污染的　pollution-free
无系数荷载　unfactored load
无限大母线　infinite bus
无线报警　radio alarming
无线电操纵起重机　radio-operated crane
无线电干扰测试仪器　radio interference meter
无线电收发信机　radio receiver-transmitter
无线电台　radio station
无线遥控起重机　cableless remote operated crane
无形资产　immaterial assets
无压管道　non-pressure pipeline
无压管道闭气试验　pneumatic pressure test for non-pressure pipeline
无压管道闭水试验　water obturation test for non-pressure pipeline
无压隧洞　free-flow tunnel
无应力计　non-stress meter
“无应力”应变计　“no-stress” strain meter
无用层　unavailable layer，unuseful layer
无源节点　passive bus
无源滤波器　passive filter
无源网络　passive network
无约束悬臂梁　free cantilever
无闸门控制的溢洪道　ungated spillway
无闸门堰　ungated weir
无障碍设计　barrier-free design
五极隔离开关　five-pole isolator
武开江　ice breakup due to hydraulic and climatic effect
物价增长　price inflation
物价指数　price index
物理参数　physical parameter
物理风化　physical weathering
物理化学分离制备法　physical and chemical separation method
物理力学性试验　physical and mechanical property test
物理蚀变　physical alteration
物料平衡　construction material supply and demand balance
物流　logistics
物流管理　logistics management
物品分类　sorting
物探报告　Geophysical prospecting report
物探剖面　geophysical survey profile
物探综合测井　comprehensive downhole geophysical probe
误报　false alarm
误操作　misoperation
误差　error
误动作　unwanted operation
误跳闸　nuisance trip，false trip
误通　false firing
雾闪　fog flashover

X

西门子　Siemens
吸出式通风　ventilation by exhaust
吸顶灯具　ceiling luminaire，surface-mounted luminaire
吸入盖　suction cover
吸入管　suction tube
吸入壳（吸入段）　suction casing

吸入喇叭管　suction bell
吸入弯管　suction elbow
吸入压力　suction pressure
吸收比　absorption ratio
吸水率　water-absorbing ratio
吸油管　oil suction pipe
牺牲阳极　sacrificial anode
息税前利润　earnings before interest and tax (EBIT)
稀释剂　diluting agent
稀释沥青　dilute bitumen
稀性泥石流　diluted debris flow
熄断　quenching
洗孔　borehole cleaning
铣削成槽法　trenching by cutting
喜马拉雅运动　Himalayan movement
喜马拉雅运动的第Ⅱ幕　Second episode of Himalayan movement
系船柱　bollard
系梁　collar beam
系列插补　series interpolation
系列代表性　series representativeness
系列延长　series extension
系统标称电压　nominal voltage of power system
系统短路　short-circuit
系统观测　systematic observation
系统锚杆　pattern bolting
系统锚固　systematic anchoring, pattern anchoring
系统软件　system software
系统同步运行　synchronous operation of power system
系统稳态　steady state of power system
系统误差　systematic error
系统运行电压　operating voltage of power system
系统暂态　transient state of power system
系统支护　systematic support
系统阻抗　system impedance
细部结构　detailed structure
细部详图（大样图）　detail drawing
细度模数　fineness modulus (FM)
细骨料　fine aggregate
细晶岩　aplite
细晶岩脉　aplite vein
细砾　fine gravel
细裂缝　hairline crack
细砂　fine sand
细水雾灭火系统　water mist fire suppressing system
峡谷　canyon, gorge
狭缝法　narrow slot method
下半球投影　lower hemisphere projection
下部结构　substructure
下部主轴　lower shaft
下沉式弧形闸门　submersible taiter gate
下撑式组合梁　down-stayed composite beam
下承梁　through beam
下导洞法　bottom heading method
下导轴承　lower guide bearing
下机架　lower bracket
下机架底板　soleplate of the lower bracket
下机坑　lower pit
下基本调配线　lower critical guide curve
下降泉　depression spring, gravity spring
下降深度　load-lowering height
下降速度　load-lowering speed
下料　cutting
下落断层　downthrown fault, normal fault
下切　undercutting
下渗容量曲线　infiltration capacity curve
下水库　lower reservoir
下套管　casing placement
下限解　lower bound solution
下泄流量　outflow discharge
下泄水流的流态和雾化　discharging flow pattern and atomization
下行无振碾压　non-vibrating compaction downward
下游　lower reaches
下游堆石区　downstream rockfill zone
下游混凝土防渗板　downstream concrete slab connected with the plinth.
下游坡（背水坡）　downstream slope
下游视图　downstream view
下游水位变送器　downstream level transducer
下游水位开关　downstream level switch
下游水位衔接　downstream flow connection
下游透水铺盖　downstream permeable weight blanket
下游围堰　downstream cofferdam
下游消能区　downstream energy dissipation area

下闸　gating
下轴　lower shaft
夏（伏）汛　summer flood
先导孔　pilot hole
先导式　pilot operated
先导式安全阀　pilot operated safety valve
先导式减压阀　pilot operated reducing valve
先导伺服定位器　pilot servo-positioner
先期固结压力　preconsolidation pressure
先选择后操作　select before operate
先张法预应力混凝土结构　pretensioned prestressed concrete structure
纤维芯钢丝绳　fiber core（FC）wire rope
纤芯直径　core diameter
掀斜式隆升　tilting uplift
弦式应变计　wire strain gauge
显晶质结构　phanerocrystalline texture
显色指数　colour rendering index
显示偏好　Revealed Preference
现场保安　security of the site
现场初始压缩曲线　field initial compression curve
现场调查　field investigation，site survey
现场管理费　site overhead，construction overhead
现场监测　field（in-situ）monitoring
现场检验　in-situ inspection
现场拼装　site assembly
现场拼装加工费用　cost of site assembly and processing
现场清理　general clearance（clearance of site）
现场施工分包合同　field subcontract
现场试验　field test
现场总线　fieldbus
现场作业　operation on site
现存价值　existence value
现地控制　local control
现地控制单元　local control unit（LCU）
现地控制级　local control level
现浇板　cast-in-situ concrete slab
现浇混凝土　cast-in-situ concrete，placed-in-situ concrete
现浇桩　mix-in-place pile
现金流出（量）　cash outflow（CO）
现金流量　cash flow
现金流（量）　cash flow（CF）
现金流入（量）　cash inflow（CI）
现值　present value（PV）
限负荷运行　limited load operation
限流电抗器　current-limiting reactor
限流式熔断器　current-limiting fuse
限速器　speed governor，overspeed governor
限位开关　limit switch
限位行程开关　end-of-travel limit switch
限制出力线　lower critical guide curve
限制灌浆　containment grouting
限制级　limiter stage
限制粒径　constrained diameter
限制器　limiting device，limiter
线槽　trunking
线弹性分析　linear-elastic analysis
线电流　line current
线电压　line-to-line voltage
线分布力　force per unit length
线夹　clamp
线接触钢丝绳　linear contact lay wire rope
线理　lineation
线路充电　line charging
线路故障　line fault
线路过负荷能力　overload capacity of transmission line
线路压降补偿　line drop compensation
线路转角　line angle
线路自然功率　natural load of a line
线路阻波器　line trap
线热胀系数　linear expansion coefficient
线型火灾探测器　line-type fire detector
线性电路　linear circuit
线性系统　linearized system
线压力　linear pressure
线应变　linear strain
线状流水　linear running water
乡村道路　rural road
乡镇企业　township owned enterprise
详查　detailed investigation
相变　facies transformation
相电流　phase current
相电压　line to neutral voltage，phase to neutral voltage
相对磁导率　relative permeability
相对地过电压标幺值　per unit value of

phase-to-earth overvoltage
相对地净距 phase-to-earth clearance
相对高程 relative elevation
相对隔水层 relative watertight layer
相对价格作用 Relative Price effect
相对密度 relative density
相对湿度 relative humidity
相对受压区计算高度 relative calculation depth of compression zone
相对稳定地块 relatively stable block
相对效率 relative efficiency
相对压力脉动 factor of pressure fluctuation
相对压强 relative pressure
相对压实度 relative compactness
相贯线 intersection line
相间短路 interphase short-cicuit
相间过电压标幺值 per unit value of phase-to-phase overvoltage
相间净距 phase-to-phase clearance
相间距离 phase-to-phase spacing
相角差 phase difference
相控充电器 phase controlled charger
相量 phasor
相量图 phasor diagram
相邻浇筑层 adjacent lift
相邻浇筑块 adjacent block
相邻流域 adjacent drainage area
相绕组 phase winding
相似工况 similar operating condition
相似性 similarity
相似性偏差 similarity tolerance
相位 phase
相位比较保护 phase comparison protection
相位比较继电器 phase comparison relay
相位差角 phase difference angle
相位超前 phase lead
相位角 phase angle
相位移角 displacement angle
相位滞后 phase lag
相线端子 main terminals，phase terminals
相序 phase sequence
相应烈度 equivalent intensity
箱梁 box beam
箱形基础 box footing
镶嵌结构 interlocked structure，mosaic structure
响岩 phonolite
响应 response
响应比 response ratio
响应时间 response time
响应时间指数 response time index (RTI)
向火面［烟气侧表面］ fireside surface
向量积 vector product
向平面投影 projecting to plan
向剖面投影 projecting to profile
向斜 syncline
项目安全管理 project safety management
项目备选方案 Project Option
项目采购管理 project procurement management
项目成本管理 project cost management
项目承包 project contracting
项目承包人 project contractor
项目发包人 project employer
项目法人 project legal person，project entity
项目分包 subcontract
项目风险管理 project risk management
项目管理体系 project management system
项目后评价 post-evaluation of completed project
项目技术经济评审费 evaluation fee for project technology and economy
项目建议书 Project proposal (recommendation)
项目进度管理 project schedule management
项目进度控制 project schedule control
项目框架 Project Framework
项目启动会 kick-off meeting
项目申请报告 project application report
项目信息管理 project information management
项目验收费 project acceptance cost
项目周期 project cycle
巷道式通风 gallery ventilation
像差 aberration
橡胶坝 rubber dam，inflatable dam
橡胶止水 rubber waterstop
橡塑复合水封 PTFE-covered seal
削坡 slope cutting
削坡减载 cutting slope and unloading
削头压脚 cutting head and loading at foot
消除抖动 debounce
消除应力 stress relief
消防 fire protection，fire control
消防泵 fire pump

消防泵房　fire pump room
消防泵接水口　fire-service connection
消防车　fire vehicle
消防电梯　fire lift
消防分区　fire prevention zone
消防给水　fire water supply
消防供水持续时间　fire water duration
消防环形主管　fire-fighting ring main
消防计算机通信系统　fire computer communication system
消防接口　fire coupling
消防救援器材　fire rescue equipment
消防控制室　fire control room
消防联动控制装置　integrated fire control device
消防破拆工具　fire forcible entry tool
消防设施　fire fighting facilities，fire protection equipment
消防栓　fire hydrant
消防水池　fire pool
消防水带　fire hose
消防水枪　fire water branch
消防水压　fire pressure
消防水源　water source for fire fighting
消防梯　fire ladder
消防通道　fire fighting access
消防无线通信系统　fire radio communication system
消防吸水管　fire suction hose
消防系统　fire prevention system
消防信息　fire information
消防训练塔　fire station training tower
消防用水　water for fire fighting
消防用水量　fire fighting water consumption
消防有线通信系统　fire wired communication system
消费者剩余　consumer surplus
消弧电抗器　arc suppression reactor
消弧线圈　arc suppression coil，Petersen coil
消力池　stilling basin
消力池底板　stilling-basin slab
消力池护坦抗浮稳定性　stability against floatation of stilling basin apron
消力墩　baffle block
消力槛　baffle sill
消落深度　drawdown
消能防冲设施　energy dissipation facility
消能工　dissipator
消能减震　energy dissipation and earthquake response reduction
消能器　energy dissipator
消泡剂　defoaming agent
消氢处理　de-hydrogenation treatment
消缺处理　defect elimination
消涡　vortex elimination
销毁　destruction
销毁清册　destruction list
销售电价　consumer tariff
销轴　hinge pin
小车　trolley，truck
小车轨道梁腹板　web of trolley rail girder
小车轨道中心距　rail center of crab
小流域　small watershed
小流域产流模型　runoff-yield model of small watershed
小流域综合治理　comprehensive management of small watershed
小偏心受拉构件　tensile member with a small eccentricity
小气候　microclimate
小扰动　small disturbance
小石　fines
小时用水量　hourly water consumption
小型临时设施摊销费　amortization charge of small temporary facilities
小循环方式　minor cycling way
小值平均值　average of the lower half values
校核　checked by
校核洪水位　check flood level
校准法　calibration
效果费用比　Effectiveness Cost Ratio
效率比尺效应换算　efficiency scale up
效率试验　efficiency test
效率修正　efficiency correction
效益分摊　amortization of benefits
楔板　wedge plate
楔缝式锚杆　slot-and-wedge anchor bolt
楔管锚杆　wedge-and-slot-tube anchor bolt
楔式双闸板　wedge double gate
楔式闸阀　wedge gate valve
楔体滑动　wedge slide

楔形库容　wedge storage
楔形体平衡法　wedge equilibrium method
协联工况　combined condition
协联关系　cam relationship
协联特性曲线　combination curve
协联装置　combination device
协商合同　negotiation contract
挟沙水流　sediment-laden flow
斜板式蝶阀　inclined disc butterfly valve
斜背堰　inclined weir
斜层铺筑　slopping placement
斜长片麻岩　plagio-gneiss
斜长石　plagioclase
斜（陡）槽式溢洪道　chute spillway
斜缝　inclined joint
斜缝浇筑　inclined joint placing
斜击式水轮机　Turgo turbine，inclined-jet turbine
斜接　diagonal connection，juxtaposition
斜接柱　miter end
斜接柱支枕垫　miter block
斜截面承载力　bearing capacity of oblique section
斜井　inclined shaft
斜距　oblique distance
斜孔钻进　inclined boring
斜列构造　echelon structure
斜流式水轮机　diagonal turbine
斜面升船机　inclined ship lift
斜面式鱼道　inclined plane fishway
斜坡流淌值　slope flow
斜坡摊铺机　sloping spreader
斜坡推进法　advancing slope method
斜坡喂料车　slope feeding vehicle
斜向谷　insequent valley
斜支臂　inclined radial arm
斜轴机组　inclined shaft unit
谐波次数　harmonic number，harmonic order
谐波电流源　source of harmonic current
谐波电压源　source of harmonic voltage
谐波分量　harmonic component
谐波分析　harmonic analysis
谐波含量　harmonic content
谐波频率　harmonic frequency
谐波谐振　harmonic resonance
谐波因数　harmonic factor
谐波源　harmonic source
谐振　resonance
谐振过电压　resonance overvoltage
谐振频率　resonance frequency
谐振曲线　resonance curve
泄槽　chute
泄槽的纵坡　longitudinal slope of the chute
泄荷阀　pressure relief valve
泄洪隧洞　spillway tunnel
泄洪雾化　atomization during flood discharging
泄洪闸门　flood-discharge（flood-relief）gate
泄流计　leakage tester
泄流水雾　water spray created by flood discharge
泄漏　leak
泄漏电流试验　leakage current measurement
泄漏距离［电工］　creepage distance
泄水表孔　surface outlet
泄水底孔　bottom outlet
泄水管　drain pipe
泄水建筑物　water release structure
泄水深孔　deep outlet
泄水堰　sluice weir
泄水闸板　sluice board
泄水闸门　sluice gate
泄水闸室　sluice chamber
泄水中孔　middle outlet
泄压装置　pressure relief device
泄油池　oil leakage sump
泻湖相　lagoon facies
卸荷裂隙　stress release fissure
卸荷岩体　relaxed rock mass
卸荷　stress release（relief）
卸货费　unloading charge
卸载　unloading
谢才公式　Chezy formula
蟹爪式装岩机　crab rock loader
心墙　core wall
心式变压器　core type transformer
辛普森法则　Simpson's rule
新奥法　new Austrian tunneling method（NATM）
新风风道　outside air intake duct
新风负荷　fresh air load
新风机组　fresh air handling unit
新风量　fresh air rate
新风系统　fresh air system

新构造形迹　traces of neotectonics
新构造运动　neotectonics
新构造运动分区　neotectonic zoning
新建、扩建、改建　construction, extension and renovation
新浇混凝土　green concrete
新近纪湖积台地　neogene lacustrine deposit platform
新生代（界）　Cenozoic era (erathem)
新鲜　fresh
新鲜混凝土　fresh concrete
薪酬　salary
薪炭林　firewood forest
信号阀　signal valve
信号回路　signal circuit
信号记录时间　signal recording time
信号继电器　signal relay
信息法施工　construction method from information
信噪比　signal-to-noise ratio
兴利库容　effective storage, live storage
星-三角启动　star-delta starting
星形　star
星形连接　star connection
星形-三角形变换　star-delta conversion, star-delta transformation
行波　traveling wave
行波保护　traveling wave protection
行车密度　traffic density
行程开关　limit switch
行洪区　flood flowing zone
行星齿轮减速器　planetary reduction
行走机构（运行机构）　travelling mechanism
V 形隔离开关　V-type disconnector
U 形挂板　U-clevis
U 形挂环　U-shackle
V 形绝缘子串　V string of insulator
Z 形连接　zigzag connection
V 形曲线特性　V-curve characteristic
L 形三通式阀　L-pattern three way valve
T 形三通式阀　T-pattern three way valve
形象进度　graphic progress
形状系数　configuration factor
型钢　shaped steel
H 型孔 [孔径约 99 mm/4 in.]　H size hole
B 型孔 [孔径约 60 mm/2.3 in.]　B size hole
N 型孔 [孔径约 75 mm/3 in.]　N size hole
型式试验　type test
Ω 型水封　center bulb seal
I 型水封　flat type seal
P 型橡皮水封　P-type rubber seal
杏仁状构造　amygdaloidal structure
性能曲线　performance curve
性能试验　performance test
性能特性曲线　performance hill diagram
胸墙　breast wall, parapet
休闲　recreation
休闲区　recreation zone
休闲效益　recreation benefit
修补工作　remedial work
修补缺陷　remedying defect
修复时间　repair duration
修坡　slope dressing, slope trimming
修正系数　coefficient of correction
修正值　corrected value
袖珍无线电话机　hand-held radiophone
需求价格　Demand Price
需求曲线　Demand Curve
需水模数　modulus of water demand
畜牧业　livestock husbandry
畜舍　livestock shed
续流臂　free-wheeling arm
续流二极管　freewheel diodes
续焰　after flame
絮凝　flocculation
蓄电池　storage battery
蓄电池充电器的参数　rating of battery charger
蓄电池室　battery room
蓄电池组　battery bank
蓄洪垦殖　flood storage and reclamation
蓄冷　cold accumulation
蓄冷-制冷周期　period of charge and discharge
蓄满产流　runoff yield under saturated storage
蓄能泵反向飞逸转速　reverse runaway speed of storage pump
蓄能泵机械效率　mechanical efficiency of storage pump
蓄能泵净吸上扬程（空化余量）　net positive suction head (NPSH)
蓄能泵零流量功率　no discharge input power of storage pump

蓄能泵零流量扬程　no-discharge head of storage pump
蓄能泵流量　storage pump discharge
蓄能泵输出功率　storage pump output power
蓄能泵输入功率　storage pump input power
蓄能泵水力效率　hydraulic efficiency of storage pump
蓄能泵吸入高度　static suction head of storage pump
蓄能泵吸入扬程损失　suction head loss of storage pump
蓄能泵扬程　storage pump head
蓄能泵最大瞬态反向转速　maximum momentary counter rotation speed of storage pump
蓄能器　accumulator
蓄清排浑　impounding clear water and releasing muddy flow
蓄热　heat accumulation
蓄热法　method of heat accumulation
蓄水　impounding，impoundment
蓄水保土效益　water detention and soil conservation benefit
蓄水池　water storage pool
蓄水分期　impoundment stages
蓄水灌溉　water storage irrigation
蓄水基准值　fiducial value before first impound
蓄水式水电站　reservoir power station
蓄引提结合灌溉系统　irrigation system with water storage，diversion and pumping facilities
玄武岩　basalt
悬臂构件的挠度限值　allowable deflection value of cantilever member
悬臂梁　cantilever
悬臂轮　cantilevered wheel
悬臂起重机　jib crane
悬臂式挡土墙　cantilever retaining wall
悬臂式脚手架　bracket scaffold
悬臂式模板　cantilever form
悬臂式起重机　cantilever crane，overhang crane
悬臂挑台　cantilevered extension
悬臂桩　cantilever sheet-pile (pile)
悬垂绝缘子串　suspension insulator string
悬垂绝缘子串组　suspension insulator set
悬垂线夹　suspension clamp
悬吊管　column pipe
悬浮功　suspension work
悬浮指标　suspension index
悬谷　hanging valley
悬挂结构　suspended structure
悬挂式防渗结构　suspended seepage control structure
悬河　elevated river
悬链线　catenary
悬链线常数　catenary constant
悬式水轮发电机　suspended hydrogenerator
悬挑式平台　overhanging platform
悬崖　cliff，scarp
悬移质　suspended load
悬移质输沙率　suspended load discharge
旋臂装料机　jib loader
旋流风口　twist outlet；swirl diffuser
旋扭构造　rotational shear structure
旋喷　rotating jet grouting
旋启多瓣式止回阀　multi-disc swing check valve
旋启式止回阀　swing check valve
旋塞阀　plug valve
旋挖钻机　rotary rig
旋转备用（热备用）　spinning reserve（hot reserve）
旋转磁动势　rotating magnetomotive force
旋转励磁装置　rotating exciter
旋转筛　revolving screen
旋转型传感器　rotational transducer
选择并执行命令　select and execute command
选择充电器容量　charger sizing
选择命令　selection command
选择器　selector
选择蓄电池容量　battery sizing
选址　siting
眩光　glare
雪荷载　snow load
巡测　tour gauging
巡更　night patrol
询标　bid inquiry
循环持续时间　duration of cycle
循环冲洗　circulation flush
循环管　circulation pipe
循环介质　circulating medium
循环水泵　circulating water pump
循环应力　cycle stress

循环钻灌法　circulation drilling and grouting method
汛期　flood season
汛期排沙限制水位　low limit level for sediment flushing in flood season
逊径骨料　undersize aggregate
殉爆距离　flash-over tendency

Y

压差流量计　differential pressure flow meter
压电效应　piezoelectric effect
压覆矿产　covered mineral resource
压浆试验　slurry pressure test
压脚　loading at foot
压紧式水封　clamp seal，compressed seal
压力　pressure
压力比能　pressure energy
压力变送器　pressure transducer
压力表　pressure gage
压力表开关　pressure meter switch
压力波的传播速度　propagation velocity of wave
压力波动　pressure fluctuation
压力传导系数　pressure conductivity coefficient
压力传感器　pressure sensor
压力阀　pressure valve
压力分散型锚索　compression-dispersion tendon
压力钢管　steel penstock
压力钢管反射时间　penstock reflection time
压力钢管伸缩节　penstock expansion joint
压力管道　pressure pipeline
压力管道水压试验　water pressure test for pressure pipeline
压力罐式调速器　governor with pressure tank
压力集中型锚索　compression-concentra-tion tendon
压力继电器　pressure relay
压力开关　pressure switch
压力脉动　pressure pulsation
压力脉动试验　pressure fluctuation test
压力密封　pressure seal
压力平衡管　pressure balancing pipe
压力-时间法　pressure-time method
压力-时间曲线　plot of pressure versus time
压力释放阀　pressure relief valve
压力释放装置　pressure relief device
压力水头　pressure head
压力损失　pressure loss
压力梯度　pressure gradient
压力油罐　oil pressure tank
压力增益曲线　pressure gain curve
压力重梁　pressure weighbeam
压路机　pavement roller，road roller
压模法　molded
压强水头　pressure head
压曲临界荷载　buckling load
压曲稳定性　buckling stability
压入式通风　ventilation by forced pressure
压实　compacting
压实标准　compaction criterion
压实度　compactness
压实方　compacted measure
压实厚度　compacted thickness
压实机　compactor
压实系数　compaction coefficient
压水进气阀　depression air valves
压水试验　water pressure test，Lugeon test
压碎指标　crush index
压缩变形　compressive deformation
压缩层　compression zone
压缩空气系统控制　compressed air system control
压缩冷凝机组　compression-type condensing unit
压缩模量　compression modulus
压缩圈　crushing zone
压缩式冷水机组　compression type water chiller
压缩式制冷　compression-type refrigeration
压、吸混合式通风　combined pressure and exhaust ventilation
压性结构面　compressive structural plane
压阻式传感器　piezoresistive transducer
雅丹地貌　Yardang landform
亚临界流，缓流　subcritical flow，tranquil flow
垭口　saddle back
氩弧焊　argon arc welding
烟囱　chimney
烟囱式排水　chimney drain
烟囱效应　chimney effect
烟道　smoke flue
烟气控制　smoke control
烟雾探测器　smoke detector
烟羽　plume

淹没耕地　inundated cultivated land
淹没农田　inundated farmland
淹没深度　inundation depth
淹没式拦污栅　submerged trashrack
淹没水舌　drowned nappe
淹没损失　flood damage
淹没系数　submersion coefficient
淹没影响调查　inundation inventory survey
淹没于水下的消能工　submerged energy dissipator
延度仪　ductilimeter
延缓时间　cushioning time
延伸率，伸长率　elongation
延时保护　delayed protection，time-delayed protection
延时爆破　delay blasting
延时动作　delayed operation
延时自动重合闸　delayed automatic reclosing
延误的付款　delayed payment
延性抗震设计　seismic ductility design
延性破坏　ductile failure
严寒地区　severe cold region，chilly cold region
严重过火　severe bum
严重缺陷　serious defect
岩爆　rock burst
岩崩　rockfall
岩壁吊车梁　crane rock beam，crane girders anchored to rockmass
岩层　rock formation，rock stratum
岩层产状　stratum orientation，attitude of bed
岩层间断　stratum gap
岩层交互　beds alternation
岩层倾角　stratum dip
岩层走向　stratum strike
岩床　sill
岩粉与岩屑　rock cuttings and chips
岩浆　magma
岩浆活动　magmatic activity
岩浆岩　igneous rock
岩脉　vein，dike
岩锚支护　rock bolting
岩盘　laccolith
岩溶　karst
岩溶槽谷　karst valley
岩溶充填率　rate of karst filling
岩溶地形　karst topography
岩溶洞穴网　karstic network
岩溶过程　karst process
岩溶含水层　karst aquifer
岩溶井　karst well
岩溶景观　karst landscape
岩溶率　rate of karstification
岩溶盆地　karst basin
岩溶侵蚀　karst erosion
岩溶（侵蚀）基准　karst base level
岩溶水　karst water
岩溶通道　karst channel
岩溶突水　karst declogging
岩溶洼地　karst depression
岩塞爆破　rock plug blasting
岩石崩出　blowout
岩石构造　rock structure
岩石结构　rock texture
岩石力学　rock mechanics
岩石强度　rock strength
岩石完整性指数　integrity index of rock
岩石压力　rock pressure
岩石质量指标　rock quality designation（RQD）
岩体分类Q系统　Q-system of rock mass classification
岩体评分系统　rock mass rating（RMR）
岩体强度　rock mass strength
岩体完整程度　rock mass integrity
岩体完整程度评分　rating of rock mass integrity
岩体完整性　integrity of rock mass
岩体完整性系数　intactness coefficient of rock mass
岩体原位测试　in-situ measurement of rock mass
岩土渗透性分级　classification of permeability of rocks（soils）
岩土试验报告　Geotechnical test report，Report of rock and soil tests
岩相　lithofacies
岩屑砂岩　lithite
岩芯　rock core
岩芯编录　core logging
岩芯饼化　core disking
岩芯采取率　core recovery rate
岩芯隔板卡　core record spacer
岩芯试样　core sample
岩芯箱　core box
岩芯照片　core photograph

岩芯钻进观察　observation during coring operation.
岩芯钻探　core drilling（boring）
岩性学　lithology
岩盐　halite
岩株　stock
沿程水头损失　linear head loss
沿程淤积　progressive deposition
沿面放电　discharge along dielectric surface
沿踢脚板铺设的散热器　baseboard radiator
盐碱化　salinization
盐雾作用区　salty fog acting zone
檐口　eaves
眼球状构造　augen structure
验电　live line detection
验收试验　acceptance test
验证试验　proof test
堰顶水位计　weir gauge
堰塞湖　landslide lake，dammed lake
雁列断层　en echelon fault
雁行构造　echelon structure
燕山运动　Yanshan movement
扬程系数　head coefficient
扬声器　loudspeaker
扬水管　lifting pipe
扬压力　uplift pressure
扬压力强度系数　coefficient of uplift pressure intensity
羊脚碾　sheep-foot roller，taper foot roller
阳极　anode
阳离子　cation
阳离子交换量试验　cation exchange capability test
阳起石　actinolite
阳台　balcony
仰焊　overhead welding
养护　curing
养护覆盖物　curing mat
养护剂　curing agent
氧指数　oxygen index
样板　template
腰荷　medial load
腰线　frieze
窑洞式厂房　cavern powerhouse
摇摆支座　rocking ring girder support
摇杆　rocker arm
遥测　telemetering
遥调　teleadjusting
遥感　remote sensing
遥感图像　remote sensing image
遥控　telecommand
遥控起重机　remote operated crane
遥信　teleindication，telesignalization
咬边　undercut
药壶爆破　pot hole blasting
药卷　cartridge
要求回报率　required rate of return（RRR）
野生动植物栖息地　wildlife habitat
野生生物保护　wildlife conservation
野生生物迁徙　wildlife migration
野生生物物种　wildlife species
野外查勘　field reconnaissance
业务招待费　business entertaining expense
业主　owner，client
业主修改通知　change order
叶蜡石　pyrophyllite
叶理化变质岩　foliated metamorphic rock
叶轮　impeller
叶轮轮毂　impeller hub
叶轮螺母　impeller cap
叶轮密封环（口环）　impeller ring
叶轮上冠　impeller crown
叶轮上冠腔　impeller crown chamber
叶轮上止漏环　impeller crown seal
叶轮输出功率　output power of impeller
叶轮输入功率　input power of impeller
叶轮下环　impeller skirt
叶轮下环腔　impeller chamber
叶轮叶片　impeller vane
叶轮引水锥　impeller cone
叶片泵　vane pump
叶片进口安放角　blade inlet angle
叶片进口节矩　blade inlet pitch
叶片力特性　blade force character
叶片型线　blade profile
页岩　shale
夜间施工增加费　additional cost for night work
液动阀门　hydraulically operated valve
液化指数　liquefaction index
液控单向阀　pilot operated check valve
液力耦合器传动式　hydraulic coupling driven type

液态　liquid state
液体绝缘介质　insulating liquid
液位变送器　level transducer
液位传感器　oil level sensor
液位计　oil level gage；liquidometer
液位开关　level switch
液位指示器　electrolyte level indicator
液下式泵　wet pit type pump
液限　liquid limit
液性指数　liquidity index
液压泵　pump
液压操动机构　hydraulic operating mechanism
液压顶升式　hydraulic jack
液压阀　hydraulic valve
液压放大级　hydraulic amplifier stage
液压缸　cylinder
液压管道、油管　hydraulic tubing，hydraulic pipeline
液压机械部分　hydro-mechanical part
液压马达　hydraulic motor
液压启闭机　hydraulic hoist
液压起重机　hydraulic crane
液压千斤顶　hydraulic jack
液压驱动装置　hydraulic actuator
液压系统　Hydraulic system
液压油　hydraulic oil
液压自动抓梁　lifting beam with remote hydrauliccontrol of latching and unlatching
液压钻　hydraulic drill
液柱压力计　liquid column manometer
一般缺陷　common defect
一般围岩 fair surrounding rock
一般项目　general item
一次调频　primary control of the speed of generating sets
一次二阶矩法　first - order second - moment method
一次风　primary air
一次风机　fan for primary air
一次回风　primary return air
一次绕组　primary winding
一次系统　primary system
一次性通过　pass at the first attempt
一次性完成　complete at the first attempt
一次重合闸　single shot reclosing
一带一路　One Belt One Road
一等导线　first order traverse
一等控制网　primary control network
一等三角测量　first order (primary) triangulation survey
一等三角点　primary triangulation point
一等水准　first order leveling
一个半断路器接线　one-and-a-half breaker configuration，breaker-and-a-half configuration
一揽子合同　all-in contract，package deal contract
一期灌浆孔　primary grout hole
一期面板　first-stage facing
一致性偏差　uniformity tolerance
一字闸门　single leaf swing gate
伊利石　illite glimmerton
医疗保险费　medical insurance
仪表保安系数　instrument security factor
仪表测量　metering and instrumentation
仪器观测　instrumentation
仪器最小刻度　graduated in divisions，less and accurate to within
仪用变压器　instrument transformer
移动式高架起重机　overhead traveling crane
移动式工作平台　mobile working platform
移动式卷扬机　traveling hoist
移动式拦污栅　portable trash rack
移动式灭火器　mobile fire extinguisher
移动式起重机（汽车吊）　mobile crane
移动终端　mobile terminal
移交　hand-over
移民　resettler
移民安置范围　extent of resettlement
移民安置方式　resettlement scheme
移民安置费估算　resettlement cost estimate
移民安置区　resettlement area
移民安置行动计划　resettlement action plan (RAP)
移民安置总体规划　general planning of resettlement
移民搬迁期　resettlement transition period
移民搬迁设计　relocation design
移民办公室　resettlement office
移民补贴　resettlement subsidy
移民后期扶持　post-resettlement support
移民权利　entitlement of resettlers
移民生活水平评价预测　assessment and prediction

of resettlers' living standard
移民实施组织设计　resettlement implementation and organization design
移民试点工程　pilot resettlement project
移民政策　resettlement policy
移屏法　moving screen method
移液管法　pipette method
移置暴雨　transposition storm
移置模板　shifted formwork
移轴装置　hook-release device
已完工程保护费　protection expense of completed work
以计算机为基础的控制　computer-based automation
以太网　ethernet
义务消防队　volunteer fire brigade
议标合同　negotiation contract
议定价格　price negotiated
异步电动机　asynchronous motor
异步启动　asynchronous starting
异步远动传输　asynchronous telecontrol transmission
异步运行　asynchronous operation
异步运行保护　asynchronous operation protection
异常状态　abnormal condition
异程式系统　direct return system
异径管接头　reducing coupling
异形断面　odd-shaped
异形挑坎　special-shape flip bucket
异型接头　heterogeneous jointer
异重流　density current
异重流排沙　sediment releasing by density current
易燃固体　flammable solid
易燃（可燃）液体　flammable liquid
易燃性　flammability
易熔塞装置　fusible plug device
易引燃性　ease of ignition
意愿调查评估法　Contingent Valuation
溢洪道闸墩　spillway pier
溢流坝段　overflow section
溢流坝堰面曲线　spillway surface curves
溢流阀　pressure relief valve
溢流式厂房　overflow type powerhouse
溢流式调压室　overflow surge chamber
溢流堰的流量系数　discharge coefficient of overflow weir
溢流堰反弧鼻坎　upcurved spillway bucket
翼坝　aliform dam
翼墙　wing wall
阴极　cathode
阴极保护　cathodic protection
阴离子　anion
阴燃　smouldering
音乐厅　concert hall
音频电缆　audio-frequency cable
音像档案　audio-visual archives
殷钢尺　invar tape
银行保函　bank guarantee
银行贷款　bank loan
银行手续费　bank charges
引导阀　pilot distributing valve
引航道，进水渠　approach channel
引火源　ignition source
引气剂　air-entraining admixture
引燃时间　ignition time
引燃温度　ignition temperature
引水灌溉　water diversion irrigation
引水式厂房　conduit-type powerhouse
引水式开发　conduit type development
引水式水电站（引水道式水电站）　conduit type hydropower station
引水隧洞　headrace tunnel，diversion tunnel
引张线法　tension wire alignment method
引张线仪　wire alignment transducer
饮用水水源保护区　drinking water source reserve area
隐蔽工程　concealed works
隐伏断层　buried fault
隐含价值法　Hedonic Method
隐患　potential hazard
隐晶质结构　cryptocrystalline texture
印支运动　Indosinian movement
应变　strain
应变脉动　strain pulsation
应变软化　strain softening
应变硬化　strain hardening
应急措施　emergency measure
应急灯　emergency lamp
应急机构（组织）　emergency organization
应急救援演练　emergency rescue rehearsals
应急物流　emergency logistics
应急响应　emergency response

应急预案　emergency preparedness plan (EPP)
应急照明　emergency lighting
应急准备　emergency preparation
应力　stress
应力包络线　stress envelope
应力等值线　stress contour
应力恢复法　stress restoration method
应力集中　stress concentration
应力解除法　stress relief method
应力控制指标　stress control index
应力历史　stress history
应力路径　stress path
应力脉动　stress pulsation
应力释放　stress release (relief)
应力水平　stress level
应力松弛　stress relaxation
应力应变监测　stress-strain monitoring
应力重分布　stress redistribution
应力状态　stress state
应用软件　application software
英安岩　dacite, quartz andesite
婴儿死亡率　infant mortality rate
荧光灯　fluorescent lamp
荧光示踪剂　fluorescent tracer
盈亏平衡点　break-even point
萤石　fluorite
营业税　business tax
影响半径　influence radius
影响陈述　impact statement
影响处理区　affected area to be treated
影子工资率　shadow wage rate (SWR)
影子工资系数　shadow wage rate factor (SWRF)
影子汇率　shadow exchange rate
影子汇率系数　shadow exchange rate factor (SERF)
影子价格　shadow price
硬度　hardness
硬母线　rigid busbar
硬母线加工　preparation of rigid busbar
硬砂岩　greywacke
硬石膏　anhydrite
硬性结构面　rigid structural plane
佣金　factorage
壅高水位　backwater level
壅水长度　backwater length
永磁材料　permanent magnet material
永磁发电机　permanent magnet generators (PMG)
永存荷载（永存吨位）　eternal tensile load
永久变形　permanent deformation
永久磁体　permanent magnet
永久缝　permanent joint
永久工程　permanent works
永久渗漏　permanent leakage
永久水位站　permanent gauge
永久文献　permanent document
永久性故障　permanent fault (persistent fault)
永久性建筑物　permanent structure
永久性预应力锚杆　permanent prestressed anchor
永久用地　permanent land requisition
永久支护　permanent support
永久作用（荷载）　permanent action (load)
永态差值系数　permanent droop
永态转速调节　permanent speed regulation
涌水　water gushing, inrush of water
用材林　timber forest
用地红线　property line
用电　electric power consumption
用电质量　quality of consumption
用户话机　subsciber's set
用水定额　water consumption norm
用水量　water consumption
优定斜率法　optimum slope method
优惠利率　bank prime rate
优质钢　fine steel
优质碳素结构钢　quality carbon structural steel
油封式旋塞阀　lubricated plug valve
油浸风冷却　oil natural air forced cooling (ONAF)
油浸式变压器　oil immersed transformer
油浸式电抗器　oil immersed reactor
油浸纸绝缘电缆　oil impregnated paper insulated cable
油浸纸套管　oil impregnated paper bushing
油浸自冷却　oil natural air natural cooling (ONAN)
油冷却器　oil cooler
油料作物　oiling crops
油盆　oil reservoir
油气套管　oil-SF_6 bushing
油石比　bitumen aggregate ratio

油伺服系统　oil servo system
油温计　oil temperature indicator
油系统故障　oil system fault
油箱　oil reservoir，oil tank
油页岩　kerogen shale (oil shale)
油枕　oil conservator
油枕油位计　conservator oil level indicator
油脂光泽　greasy luster
游荡河槽　interlaced channel
游离二氧化碳　free carbon dioxide
游离氧化钙　free calcium oxide
游离氧化铁试验　free iron oxide test
游移河槽　shifting channel
游泳馆　natatorium hall
友好解决　amicable settlement
有坝取水　intake with dam
有闭合节理的试样　specimen with healed joints
有擦痕的　slickensided
有侧限试样　laterally confined specimen
有差调节　difference regulation，deviation regulation
有齿轮增速箱的机组　unit with gear box
有毒气体　poisonous gas，toxic gas
有杆腔　rod chamber
有功电能　active energy
有功电能表　watt-hour meter
有功负荷　active load
有功功率　active power
有功功率变送器　active power transducer
有功功率表　wattmeter
有功功率与频率控制　control of active power and frequency
有功和无功功率控制　control of active and reactive power
有轨运输　track haulage
有害气体　harmful gas
有机肥　organic fertilizer
有机质试验　organic matter test
有机质土　organic soil
有键槽的收缩缝　keyed contraction joint
有利环境影响　favorable environmental impact
有利环境影响最大化　maximization of favorable environmental effects
有名值　actual value
有名制　ohmic system
有黏结预应力混凝土结构　bonded prestressed concrete structure
有黏结预应力锚杆　bonded prestressed anchor
有启动装置的机组　unit with starting device
有人值班水电厂　attended hydropower plant
有台坎的　stepped
有条件保函　accessory guarantee
有限差分法　finite difference method (FDM)
有限元法　finite element method (FEM)
有限元网格　finite element grid (mesh)
有线报警　wired alarming
有线电话　line telephone
有线遥控起重机　cable remote operated crane
有效长度法　effective length method
有效碱含量　effective alkali content
有效库容　effective storage，live storage
有效降雨量　effective rainfall
有效宽度系数　effective width factor
有效粒径　effective diameter
有效纹波因数　r. m. s ripple factor
有效蓄水库容上限　top of active conservation capacity
有效应力法　effective stress method
有效应力路径　effective stress path
有效应力强度　effective stress strength
有压式进水口　pressure inlet
有压隧洞　pressure tunnel
有眼螺栓　eye bolt
有焰燃烧　flaming
有用层　available layer，useful layer
有源滤波器　active power filter
有源网络　active network
有载调压变压器　on-load tap-changing transformer
有载调压开关　on load tap changer (OLTC)
有载调压开关及其驱动机构　OLTC with drive mechanism
有载分接开关　on-load tap-changer
有闸门控制的溢洪道　gated spillway
右旋走滑位错　right lateral slip offset
幼鱼　fingerling
诱爆性　flash-over tendency
诱导缝　inducing joint
诱导轮　inducer
淤地坝　check dam for farmland forming

淤堵　clogging
淤积　sedimentation，sediment deposition
淤积量　siltation volume
淤泥　mud
淤泥质黏土　muddy clay
淤塞河道　blocked channel
淤沙压力　silt pressure
余震　aftershock
鱼池（鱼塘）　fish pond
鱼道　fishway
鱼类孵化场　fish hatchery site
鱼类洄游　fish migration
鱼类增殖放流站　fish breeding and releasing station
鱼类种群　fish population
鱼鳞坑　fish-scale pit
鱼苗　fry
鱼苗场　fry raising farm
鱼饲养　fish rearing
鱼梯　fish ladder，fall and fall fishway
渔业　fishery
渔业区　fishing zone
与辅助设备的通信　communication with auxiliary equipment
与控制盘的通信　communication with control board
与稳态值的相对偏差　relative deviation from a steady-state value
羽状剪节理　pinnate shear joint
羽状节理　feather joint
羽状张节理　pinnate tension joint
雨季　rainy season，wet season
雨量计　rainagauge
雨量站　precipitation station
雨淋系统　deluge system
雨篷　canopy
雨强　rainfall intensity
雨水管　down pipe
雨水径流量　storm water runoff
雨水口　water outlet
语音报警工作站　voice alarm workstation
郁闭度　crown density
浴室　bathroom
预测长期流量　long-term prediction flow
预付款　advance payment
预付款保函　advance payment guarantee
预拱度　built-in camber
预加力矩　pre-torque
预见期　forecast lead time
预进占　bank-off advancing
预可行性研究　pre-feasibility study
预可行性研究阶段　prefeasibility study stage
预扣所得税　pay as you earn（PAYE）
预拉力　pretension force
预拉伸　pre-stretching
预冷骨料　precooled aggregate
预冷混凝土　precooled concrete
预裂爆破　presplit blasting
预留环缝　preformed circumferential welding seam
预留孔洞　blockout
预留岩坎　reserved rock sill
预埋花管法　embedded perforated pipe method
预埋件　embedded parts
预期电流　prospective current
预期峰值电流　prospective peak current
预热　preheat
预热骨料　preheated aggregate
预热混凝土　preheated concrete
预缩砂浆　preshrinking mortar
预填骨料混凝土　preplaced-aggregate concrete
预想出力　expected output
预压骨料混凝土　prepackaged concrete
预压加固　surcharge preloading consolidation
预应力　prestress
预应力钢筋束　prestressing tendon
预应力环锚　prestress circular anchor
预应力混凝土　prestressed concrete
预应力混凝土结构　prestressed concrete structure
预应力混凝土闸墩　prestressed concrete pier
预应力锚杆　prestressed anchor
预应力锚固　prestressed anchorage
预应力损失　prestress loss
预张拉　pretension
预制构件的接头形式　connection type of a precast member
预制混凝土　precast concrete
预制梁　precast beam
预制排水管　formed drain
预制桩　precast pile
预作用系统　pre-action system
元古代（界）　Proterozoic era（erathem）

元数据 metadata
园地 garden plot
园林设计 landscape design
原产地证明 certificate of origin
原地面线 natural ground line
原动机输入功率 driver input power
原级方法 primary method
原件 original document
原理接线图 Schematic diagram
原理图 schematic diagram
原木 unsawn timber
原生节理 primary joint
原始地质资料 original geological data
原始浸润面 original phreatic surface
原水 raw water
原位标定 calibrated in situ
原位试验 in-situ test
原型观测 prototype observation
原型水轮机 prototype turbine
原种场 foundation seed farm
原状土样 undisturbed soil sample
圆顶堰 round crested weir
圆拱直墙式，城门洞形 inverted U-shaped section
圆辊闸门 roller gate
圆弧滑动 circular slide
圆盘摩擦损失 disk friction loss
圆盘筛 disc screen
圆盘形晶闸管 disk type thyristors
圆筒阀（筒形阀） cylindrical valve，ring gate
圆筒式调压室 cylindrical surge chamber
圆筒式机墩 cylinder pier of turbine
圆筒闸门 cylindrical gate
圆图 circle diagram
圆形断面 circular cross section
圆形进水口结构 circular intake structure
圆周速度 peripheral velocity
圆周速度系数 speed constant
圆柱体试样 cylinder specimen
圆状 rounded
圆锥贯入试验 cone penetration test（CPT）
圆锥破碎机 cone crusher
源头坝 watershed dam
远程 I/O remote I/O
远程监视 telemonitoring
远程切换 teleswitching
远程指令 teleinstruction
远程中心 remote center
远程终端 remote terminal
远动 telecontrol
远动传送时间 telecontrol transfer time
远动配置 telecontrol configuration
远端短路 far-from generator short circuit
远方控制 remote control
远方跳闸 intertripping，transfer tripping
远后备保护 remote backup protection
远景规划 long-term plan
远迁 distant relocation
约束变形 restrained deformation
月报表 monthly statement
月平均气温 mean monthly temperature
月牙加劲板 crescent stiffener
阅览室 reading room
跃变特性 jumping characteristic
跃层住宅 duplex apartment
越冬场 wintering ground
越级误动作 unwanted operation for an external fault
越岭隧洞 mountain tunnel
云母含量 mica content
云母片岩 mica schist
云台 pan tilt
允差带 tolerance band
允许承载力 allowable bearing capacity
允许工作压差 design differential pressure
允许渗透坡降 allowable seepage gradient
允许式保护 permissive protection
允许式纵联保护 permissive pilot protection
允许水力梯度 allowable hydraulic gradient
允许误载水深 permitting water depth misloading
允许吸上真空度 allowable suction vacuum
允许载流量 permissive carrying current
运动黏度 kinematic viscosity
运动黏性系数 coefficient of kinematic viscosity
运动限制器 motion limiter
运河 canal
运输保险费 transportation premium
运输费 freight costs
运输坡道 haul ramp
运输强度 transport intensity
运算放大器 operational amplifier
运算曲线法 calculation curve method

运行安全地震动　operational safety ground motion
运行方式转换　operating mode transition
运行工况　operating condition
运行基本地震　operating basis earthquake (OBE)
运行期　operation period
运行人员干预　operator intervention
运行时间　operating duration
运行速度　travelling speed
运行与检修荷载　operation and maintenance loads
运杂费　freight and miscellaneous charges
运转特性曲线　performance curve

Z

杂排水　gray water
杂散负载损耗　stray load loss
杂砂岩　greywacke
杂填土　miscellaneous fill
杂质泵　liquid-solids handling pump
载波电话终端机　carrier telephone terminal equipment
载流能力　current carrying capacity
再生制动　regenerative braking
再同步　resynchronization
再现性试验　reproducibility test
在线式归档　on-line filing
在用试验　in-service testing
暂存场　temporary stack yard
暂列金额　provisional sum
暂时过电压　temporary overvoltage
暂时渗漏　temporary leakage
暂时停工　suspension of work
暂态差值系数　temporary droop
暂态过程　transient process
暂态响应　transient response
暂态作用阶跃　step of transient function
凿毛　surface roughening
早第三纪（系）　Eogene period (system), Paleogene period (system)
早强剂　early-strength admixture
早强减水剂　accelerating and water reducing admixture
造价工程师　cost engineer
造价员　cost engineering technician
造孔　drilling hole
造林存活率　survival rate of afforestation
造林密度　density of plantation
造林整地　land preparation for afforestation
造山带　orogenic belt
造山旋回　orogenic cycle
造山运动　main tectonic movement, orogency
噪声　noise
噪声标准　noise standard
噪声测量　noise level measurement
噪声管理　noise management
噪声级　noise level
噪声控制装置　noise control device
噪声滤波器　noise filter
噪声污染　noise pollution
噪声污染控制　noise pollution control
增减开关　raise/lower switch
增减命令　raise/lower command
增量产出　incremental output
增量命令　incremental command
增量投入　incremental input
增量危害评估　incremental hazard evaluation
增量效益　incremental benefit
增量信息　incremental information
增模区　increased-modulus zone
增塑剂　plasticizer
增稳措施　stabilizing measure
增压泵　booster pump
增值税　value-added tax
渣浆泵　slurry pump
渣砾料　dregs and gravel
轧石　crushed rock
闸板　wedge
闸墩墩头　pier nose
闸墩式厂房　pier-head powerhouse
闸阀　gate valve
闸槛　ground sill
闸门安装平台　working platform for installing gate
闸门控制的堰　gated weir
闸门廊道　gate gallery
闸门面板　skin plate
闸门竖井式进水口　intake with gate shaft
闸门型式　gate type
闸室　lock chamber
闸室式鱼道　lock chamber type fishway
闸首　lock head

栅架　rack frame
栅条　rack bar
栅叶　rack
炸药库 explosive magazine
窄缝式挑坎　slit-type flip bucket
债务资金　debt capital
展览馆　exhibition room
展视图（展开图）　outspread view
战争险附加费　additional expenses of war risk
张开　open
张开的节理　fissure
张拉　tensioning
张拉段　tensile section
张拉控制应力限值　allowable value of tension stress control
张拉伸长值　tensile value
张拉型锚杆　tension anchor bolt
张裂缝　tension crack
张裂隙（节理）　tension joint
张性断层　tension fault
张性结构面　tensile structural plane
掌子面（工作面）　heading face
胀壳式锚杆　expending shell anchor bolt
招标　invitation for tender (bidding)
招标代理费　tender agency charge
招标公告　announcement of tender
招标函　letter of tender
招标控制价　tender sum limit
招标设计　tender design
招标设计报告　Tender design report
招标设计阶段　tender design stage
招标文件　bid document
招标文件澄清　clarification
招工安置　allocation with job
找平层（整平层）　leveling coat
沼泽化　swamping
照度　illuminance
照明变压器　lighting transformers
照明功率密度　lighting power density (LPD)
照片档案　photographic archives
照片判读　photograph interpretation
照准点归心　reduction to targer center
照准仪　alidate
遮阳系数　shading coefficient (SC)
折板结构　folded-plate structure
折冲水流　deflected current
折断销装置　breaking pin device
折方系数　coefficient of measure conversion
折减系数　reduction factor, coefficient of reduction
折旧　Depreciation
折流板　baffle plate
折劈理　crenulation cleavage
折射波　refracted wave
折射法勘探　refraction survey
折射率　refractive index
折射系数　refraction coefficient
折现　Discounting
折现率　Discount Rate
折线式卷筒　lebus groove drum
折向器接力器　deflector servomotor
折向器（偏流器）　deflector
折向器位置传感器　deflector position sensor
褶断山　fault-folded mountain
褶皱　fold
褶皱带　fold zone, fold belt
褶皱基底　fold basement
褶皱脊　fold hinge
褶皱束　fold bundle
褶皱系　fold system
褶皱翼　fold limb
褶皱轴　fold axis
褶皱作用　folding
阵流　intermittent flow
针片状颗粒　elongated and flaky particle
针入度仪　penetrometer
针形阀　needle valve
针叶林　coniferous forest
针状颗粒　needle-shaped particle
诊断　diagnostics
诊断软件　diagnosis software
珍稀物种　rare species
珍珠光泽　pearly luster
真空安全阀　vacuum relief valve
真空度　vacuity
真空断路器　vacuum circuit breaker
真空灌浆　vacuum grouting
真空灭弧室　vacuum extinction chamber
真空模板　vacuum mat
真空排水　vacuum drain

真空排水预压加固软基　soft foundation reinforced by vacuum drainage preloading
真空破坏器　vacuum breakers
真空熔断器　vacuum fuse
真倾角　actual dip
真实性　authenticity
振冲　vibroflotation
振冲复合地基　vibro-composite foundation
振冲复合土体　vibo-treated composite soil
振冲器导管　follow-up tube
振冲置换　vibro-replacement
振冲置换砂石桩　vibro-replacement stone and sand column
振冲桩　vibroflotation pile
振荡闭锁　power swing blocking
振荡解列操作过电压　oscillation overvoltage due to system splitting
振荡周期　oscillation period
振捣　vibrating
振动　vibration
振动变送器　vibration transducer
振动给料机　shaking feeder
振动加速度　vibration acceleration
振动碾　vibratory roller
振动平板　vibrating plate
振动平碾　smooth drum vibrating roller
振动筛　shaking screen, vibrating screen
振动速度　vibration velocity
振动探头　vibration probe
振动位移　vibration displacement
振动污染控制　vibration pollution control
振动压实指标　vibrating compaction value (VC)
振幅　amplitude of vibration
振型　mode of vibration
振型分解法　mode-superposition method
震旦纪（系）　Sinian period (system)
震动卓越周期　predominant period of vibration
震级档　magnitude interval
震级-频度关系　magnitude-frequency relation
震陷　earthquake subsidence
震源　seismic source, seismic focus
震源机制　focal mechanism
震源深度　focus depth
震中　epicenter
镇墩　anchorage block
争端　dispute
争端裁决　dispute adjudication
征用土地补偿　compensation for land requisition
蒸发　evaporation
蒸发冷却　evaporation cooling
蒸发损失　evaporation loss
蒸发岩　evaporite
蒸汽供暖　steam heating
蒸汽灭火系统　steam smothering system
蒸腾　evapotranspiration
蒸压加气混凝土砌块　autoclaved aerated concrete block
整定　setting
整定计算　setting calculation
整定试验　set test
整定压力　set pressure
整定值　setting value
整合　conformity
整机空载试验　no-load test of the whole
整机总联调　joint debagging of the whole
整机组装　assembly of the whole
整流　rectification
整流二极管　rectifier diode
整流器　rectifier
整流式继电器　rectifying relay
整流因数　rectification factor
整体剪切破坏　general-shear failure
整体结构　integral structure
整体式制冷设备　packaged refrigerating unit
整体水工模型试验　monolithic hydraulic model test
整体稳定　overall stability
整体圆弧滑动法　mass circle sliding method
整体运输　transport in assembly
正变质岩　orthometamorphite, ortho-rock
正铲挖土机　face shovel
正长石　orthoclase
正长岩　syenite (sinaite)
正常固结土　normally consolidated soil
正常检修运行方式　normal maintenance operation mode
正常连续运行范围　normal continuous operating range
正常时差修正　normal moveout correction
正常使用极限状态　serviceability limit states

正常停机　normal shutdown
正常蓄水位　normal pool (storage) level, full supply level
正常（常用）溢洪道　service spillway
正常运行方式　normal operation mode
正常照明　normal lighting
正齿轮传动装置　cylindrical gear actuator
正搭叠　overlapped
正断层　downthrown fault, normal fault
正反向旋转　clockwise and counter clockwise rotation
正火　normalizing
正极板　positive plate
正极端子　positive terminal
正交　in quadrature
正截面承载力　bearing capacity of normal section
正截面抗裂计算　crack-resisting calcula-tion for normal section
正截面裂缝宽度控制　crack width control in normal section
正井法　shaft-sinking method
正面影响　positive impact
正排量　positive displacement
正铅垂线　right plummet
正态分布　normal distribution
正文　official text
正弦变化　sinusoidal change
正向击穿　forward breakdown
正向通道　forward path
正向阻断状态　forward blocking state
正序短路电流　positive sequence short-circuit current
正序分量　positive sequence component
正序网络　positive sequence network
正序阻抗　positive sequence impedance
正应力　normal stress
政治保险　political insurance
支臂　radial arm, end frame
支撑　supporting
支撑环　supporting ring
支撑式全断面岩石掘进机　gripper type full face rock TBM
支承　end support
支承跨度　support span
支承型式　type of end support
支持绝缘子　supporting insulator
支持软件　support software
支墩　supporting pier
支墩坝　buttress dam
支墩鼻端里衬　pier nose liner
支付能力　ability to pay (ATP)
支付意愿　willingness to pay (WTP)
支管　branch pipe
支架　yoke
支铰　trunnion
支铰大梁　trunnion girder, trunnion beam
支铰轴　trunnion pin
支流　tributary
支路　branch
支线　branch line
支柱式电流互感器　support type current transformer
枝状管网　branch system
知识产权和工业产权　intellectual and industrial property rights
执行元件　execute component
直达运输　through transportation
直动式　direct operated
直剪试验　direct shear test
直接费　direct cost
直接雷击　direct lightning strike
直接冷却　direct cooling
直接受影响住户　directly affected household
直接填筑允许时间　permissible time interval between placing layers
直接型灯具　direct luminaire
直接眩光　direct glare
直接载荷式安全阀　direct-loaded safety valve
直接作用式减压阀　direct-acting reducing valve
直拉窗　sash window
直立褶皱　upright fold, erect fold
直流变流器的转换因数　transfer factor of DC converter
直流变流因数　DC conversion factor
直流波形因数　DC form factor
直流侧纹波电压　ripple voltage on DC side
直流电机　direct current machine (DC machine)
直流电抗器　DC reactor
直流电流　direct current (DC)
直流电流互感器　DC current transformer
直流电压　direct voltage (DC voltage)
直流电压调整值　direct voltage regulation

直流断路器　DC circuit breaker
直流分量　DC component
直流负极接地故障　DC negative to earth fault
直流高压发生器　high-voltage DC generator
直流隔离开关　DC disconnect switch
直流功率　DC power
直流励磁机　DC exciter
直流耐压试验　DC voltage withstand test
直流盘室　DC panel room
直流配电盘　DC distribution board
直流起励装置　DC field flashing device
直流时间常数　DC time constant
直流式阀　Y-type valve
直流式系统　series-connected system
直流输电　DC power transmission
直流水系统　once through system
直流纹波因数　DC ripple factor
直流稳态恢复电压　DC steady-state recovery voltage
直流系统　direct current system（DC system）
直流线路　DC line
直流斩波器　DC chopper
直流正极接地故障　DC positive to earth fault
直流制动　DC injection braking
直埋电缆　direct-burial cable
直埋敷设　cable direct burial laying
直驱机组　direct-driven unit
直通　conduction through
直通式阀　through way type valve
直线度　straightness
直线杆塔　intermediate support
直线型传感器　linear transducer
直线型气动装置　linear pneumatic actuator
直心墙堆石坝　vertical core rock fill dam
直支臂　parallel radial arm
直轴超瞬态电抗　direct-axis subtransient reactance
直轴瞬态电抗　direct-axis transient reactance
直轴同步电抗　direct-axis synchronous reactance
职业病　occupational disease
职业病防治费　occupational disease prevention fee
职业分类　occupation classification
职业分析　vocational analysis
职业健康管理　occupational health management
职业健康与安全　occupational health and safety
职业教育　vocational education
职业结构　occupation structure
植被覆盖率　vegetation cover ratio
植被清除　vegetation clearance
植物措施　vegetation measure
植物群　flora
植物无固相冲洗液　vegetable glue drilling fluid without clay
止动垫圈　lock washer
止回阀　check valve，non-return valve（NRV）
止浆塞　packer
止水　waterstop
止水板　seal plate
止水间距　span of side seals
止水片　waterstop strip
止水片鼻子　nose of waterstop
止水片立腿　erecting end of waterstop
止水片平段　wing flat of waterstop
止推环　thrust collar
纸介质电容器　paper capacitor
纸质档案数字化　digitization of paper-based records
指定分包商　nominated subcontractor
指示灯　lamp indicators
指示剂法　tracer method
指示命令　instruction command
指示器　indicator
指数法　index method
指数试验　index test
指向标志灯　direction sign luminaire
指针式油温计　dial type oil temperature indicator
趾板　plinth，toe slab
趾墙　toe wall
志留纪（系）　Silurian period（system）
制动复归　brakes released
制动控制开关　brake control switch
制动喷嘴　brake nozzle
制动气压　brake air pressure
制动气源　air supply for brakes
制动器　brake
制动瓦未脱离　brake shoes not cleared
制动闸　brake，jack
制动转矩　braking torque
制冷　refrigeration
制冷厂　refrigeration plant
制图　drawn by
质保金保函　guarantee for retention bond

质量　mass
质量保证　quality assurance
质量管理　quality control
质量管理体系　quality control system
质量检验　quality inspection
质量密度　mass density
质量评定　quality assessment
质量缺陷　quality defect
质量事故　quality accident
治导线（整治线）　training alignment
治沟骨干工程　key project for gully erosion control
致密灰岩　compact limestone
蛭石　vermiculite
智能变电站　smart substation
智能电网　smart grid
智能电子设备　intelligent electronic device (IED)
智能继电保护　intelligent relay protection
智能卡　smart card
滞洪坝　detention dam
滞洪排沙　flood retarding and sediment releasing
滞洪区　flood detention area
置换　replacement
置换通风　displacement ventilation
中标　winning bid
中等发育　moderately developed
中等腐蚀　moderately corrosive
中等腐蚀环境　moderate erosive environment
中等透水　moderately pervious
中墩　intermed pier
中分式系统　midfeed system
中风化　moderately weathered
中泓线　thread of channel (stream)
中厚层状结构　moderately thick layer structure
中厚拱坝　medium-thick arch dam
中间衬套　interstage bushing
中间接力器　pilot servomotor
中间联轴器　intermediate shaft coupling
中间配线架　intermediate distributing frame
中间漆　intermediate coat
中间式电流互感器　current matching transformer
中间轴　intermediate shaft
中间主应力　intermediate principal stress
中壳（中段）　stage casing
中控楼　central control building
中控室　central control room
中立试验室　independent laboratory
中砾　medium gravel
中密　medium dense
中期导流　midterm diversion
中期冷却　intermediate cooling, midterm cooling
中期水文预报　mid-term hydrological forecast
中热水泥　modified Portland cement
中砂　medium sand
中生代（界）　Mesozoic era (erathem)
中石　medium grain
中水　reclaimed water
中水设施　installation of reclaimed water
中酸性喷发岩　intermediate acidic eruptive rock
中庭　atrium
中温阀门　moderate temperature valve
中心导洞掘进　center drift tunneling
中心支承式　centerline support type
中新世（统）　Miocene epoch (series)
中型布置　medium-profile layout
中性导体　neutral conductor
中性点　neutral, neutral point
中性点不接地系统　isolated neutral system
中性点端子　neutral terminals, neutral leads
中性点接地方式　neutral point treatment
中性点经变压器接地系统　transformer earthed neutral system
中性点位移　neutral point displacement
中性点消弧线圈接地系统　arc-suppression-coil earthed neutral system
中性点谐振接地系统　resonant earthed neutral system
中性点直接接地系统　solidly earthed neutral system
中性点阻抗接地系统　impedance earthed neutral system
中性岩　intermediate rock
中压　medium voltage (MV)
中压绕组　intermediate-voltage winding
中央控制设备　central control equipment
中央控制室　central control room
中硬岩　moderately hard rock
中游　middle reaches
中值粒径　median diameter
中转运输　transfer transportation
中阻抗型母线差动保护　medium-impedance

busbar differential protection
终端杆塔　terminal support，dead end tower
终端设备　terminal device
终端止挡器　end stop
终碾　final rolling
终拧　final screw
终凝　final set
终值　final value（FV）
终止　termination
终止充电率　finishing charge rate
终止电压　end-of-discharge voltage
钟乳石　stalactite
种植农业　plantation agriculture
种族　ethnic group
种族歧视　ethnic discrimination
仲裁　arbitration
重闭式压力释放装置　reclosing pressure relief device
重锤　counterweight，suspension set weight
重大件　heavy-outsized pieces，oversize and weight cargo
重大危险源　major hazard
重大危险源评价　major hazard assessment
重点监督区　important supervision zone
重点预防保护区　important protection zone
重点治理区　important rehabilitation zone
重复灌浆　repeated grouting
重复荷载　repeated load
重复容量　duplicate capacity
重复性试验　repeatability test
重复应力　repeated stress
重骨料　heavyweight aggregate
重合时间　reclosing time
重合闸过电压　reclosing overvoltage
重合闸失败　unsuccessful reclosing
重击穿　restrike
重晶石　barytes
重力坝　gravity dam
重力场　gravitational field
重力档距　weight span
重力墩　gravity block
重力拱坝　gravity arch dam
重力加速度　acceleration of gravity
重力勘探　gravitational prospecting
重力密度　force（weight）density
重力排水　gravity drain
重力平衡重　gravity counterweight
重力侵蚀　gravitational erosion
重力式挡土墙　gravity retaining wall
重力压力计　dead weight manometer
重力异常　gravity anomaly
重量　weight
重量法　weighing method
重现期　recurrence period（interval）
重型履带式推土机　heavy crawler dozer
重要环境问题区域　area of critical environmental concern
重载脚手架　heavy duty scaffold
重置成本，重建费　replacement cost
重置价格　replacement price
周边缝　periphery joint
周边灌浆　perimeter grouting
周边孔　peripheral hole
周边应力　circumferential stress
周调节　weekly regulation
周调节水库　weekly regulating reservoir
周负荷曲线　weekly load curve
周界防范　perimeter precaution
周期　period
周期工作制　periodic duty
周期加荷三轴试验　cyclic triaxial test
周期式（间歇式）混凝搅拌站　periodic（intermittent）concrete mixing plant
周期性振动和脉动　periodic vibration and pulsation
周转性材料　revolving material
轴承高压油顶起系统　bearing oil injection device
轴承绝缘　bearing insulation
轴承体　guide bearing housing
轴承油温　bearing oil temperature
轴承振动检测器　bearing vibration detector
轴摆度（径向跳动）　shaft runout
轴的挠度　shaft deflection
轴电流　shaft current
轴电流保护　shaft current protection
轴功率系数　shaft power coefficient
轴颈　guide bearing journal
轴距　wheel base
轴力矩　shaft torque
轴领　guide bearing collar

轴流泵　axial flow pump
轴流定桨式水轮机　axial flow fixed-blade turbine
轴流风机　axial fan
轴流式暖风机　unit heater with axial fan
轴流式水轮机　axial flow turbine
轴流式止回阀　axial flow check valve
轴流转桨式水轮机　Kaplan turbine，axial flow adjustable-blade turbine
轴面劈理　axial plane cleavage
轴面速度　meridian velocity
轴面形状　meridional contour
轴伸贯流式机组　（S型机组）shaft-exten-sion type tubular unit
轴套　bushing，sleeve
轴套螺母　sleeve nut
轴向荷载　axial load
轴向力　normal force，axial force
轴向剖分泵　axially split pump
轴向受力柱　axial loaded column
轴向通风　axial ventilation
轴向载荷脉动　axial load pulsation
轴向柱塞泵　axial piston pump
轴心受拉构件　axial tensile member
轴心受压承载力　load-carrying capacity in compression
轴心受压柱　axial compression column
帚状构造　brush structure
侏罗纪（系）　Jurassic period（system）
竹叶状灰岩　wormkalk
逐户调查　household investigation
逐时冷负荷　hourly cooling load
逐时综合温度　hourly sol-air temperature
主保护　main protection
主备控制　lead-lag control
主臂　principal arm
主变洞　main transformer cavern
主变室　main transformer room
主变压器　main transformer
主变压器场　main transformer yard
主厂房　main powerhouse，machine hall
主触头　main contact
主磁通　main flux
主电路　main circuit
主动火灾预防　active fire prevention
主动灭火　active fire control
主动土压力　active earth pressure
主堆石区　main rockfill zone
主阀　main shut-off valve
主分接　principal tapping
主干电网　main grid
主轨　track
主航道　main waterway
主横梁（横向主梁）　horizontal girder
主机间　generator hall
主计算机　main computer
主接力器　main servomotor
主接线图　Single-line diagram，one-line diagram
主绝缘　main insulation
主勘探剖面　major exploration profile（section）
主控项目　dominant item
主梁　main beam
主梁拱度　bridge girder camber
主令开关　master switch
主配电屏　main distribution board
主配压阀　main distributing valve，control valve
主配压阀中间位置　neutral position of main distributing valve
主频　dominant frequency，main frequency
主提升设备　main hoist
主题标引　subject indexing
主题词　descriptor
主题目录，专题目录　subject catalogue
主体工程　main structures（main civil works）
主通信站　main traffic station
主要大地构造运动　main tectonic movement，orogeny
主要技术经济指标表　Main technical and economic indexes
主要建筑物　main structure
主要结构面产状评分　rating of orientation of main discontinuity
主要淹没影响　major inundation impact
主引出线　main terminals，phase terminals
主应变　principal strain
主应力　principal stress
主用空压机　lead compressor
主站（控制站）　master station，controlling station
主震　main shock
主轴　main shaft
主轴密封　shaft seal

主轴密封的冷却与润滑 shaft seal cooling and lubrication
主轴密封水阀 shaft seal water valve
主轴转矩脉动 shaft torque pulsation
主纵梁（竖向主梁） vertical girder
住户收支调查 surveys on household income and expenditure
注入冲洗 injection flush
注入量 injection rate
注水 priming
注水试验 injection test
注油口 pouring orifice
贮水度 free porosity
驻波 standing wave
驻测 stationary gauging
驻工地设计代表 design representative at site
柱廊 colonnade
柱牛腿（独立牛腿） bracket
柱塞线圈 plunger coil
柱上断路器 pole-mounted circuit-breaker
柱上隔离开关 pole-mounted disconnector
柱下独立基础 independent footing under column
柱状构造 columnar structure
柱状浇筑 concreting with longitudinal joint
柱状节理 columnar joint
柱状排水［土石坝内］ pillar drain
著录 description
著录格式 description form and format
著录项目 item of description
铸钢 cast steel
铸钢件 steel casting
铸铁管 cast iron pipe
铸铁件 cast-iron
筑坝料的含水量 water content of embankment material
抓斗起重机 grabbing crane
抓斗挖掘机 grab excavator
抓斗挖土机 clamshell shovel
抓梁 lifting beam
抓取成槽法 trenching by grabbing
抓钻成槽法 trenching by drilling and grabbing
专利及专有技术使用费 royalty for patents and proprietary technology
专利（权） patent
专门技术合同 know-how contract
专题指南 guide to subject record
专项设施 special facility
专项审查费 special works review fee
专项投资 specific investment
专业档案 specialized archives
专业档案馆 specialized archives
专用负荷开关 limited purpose switch
专用工具 special tool
专用荷载 special load
专用小交换机 private branch exchange (PBX)
专用自动小交换机 private automatic exchange (PAX)
砖混房 brick-concrete house
砖混结构 masonry-concrete structure
砖木房 brick-wood house
砖木结构 masonry-timber structure
砖砌结构 brick masonry structure
转差率 slip
转动惯矩 rotary moment of inertia
转动惯量 moment of inertia，rotational inertia
转动止漏环 rotating seal ring
转换层 transfer story
转角 outer corner
转角杆塔 angle support
转角水封 corner seal
转铰水封 swing seal device
转接交换机 through switchboard
转矩波动 torque fluctuation
转矩偏差 torque deviation
转轮 runner
转轮出口开度 runner outlet width
转轮除湿机 rotary dehumidifier
转轮的机械功率 mechanical power of runner
转轮检修道 runner removal access
转轮力矩 runner torque
转轮轮盘 runner disk
转轮密封温度传感器 runner seal temperature sensor
转轮腔 runner chamber
转轮上冠 runner crown
转轮上冠腔 runner crown chamber
转轮上冠止漏环 runner crown seal
转轮室 discharge chamber
转轮室上环 discharge chamber ring
转轮输出功率 output power of runner
转轮输入功率 input power of runner

转轮体　runner hub
转轮体装配　runner hub assembly
转轮下环　runner band
转轮下环排水阀　runner band drain valve
转轮下环腔　runner band chamber
转轮下环止漏环　runner band seal
转轮泄水锥　runner cone
转轮叶片　runner blade
转轮叶片角度　runner blade angle
转轮叶片接力器　runner blade servomotor
转轮叶片力矩　runner blade torque
转轮叶片连杆　runner blade link
转轮叶片密封　runner blade seal
转轮叶片枢轴　runner blade trunnion
转轮叶片转臂　runner blade lever
转轮圆周频率　runner peripheral frequency
转轮运输门　runner transport door
转轮转运车　runner cart
转频　rotational frequency
转让　assignment
转速　rotational speed
转速调节器　speed governor
转速调节图　speed regulation graph
转速调整　speed adjustment
转速调整机械　speed adjusting mechanism
转速脉动　rotational speed pulsation
转速偏差　speed deviation
转速稳定性指数　speed stability index
转速相对偏差　relative deviation of speed
转速信号发生器　speed signal generator (SSG)
转速信号给定装置　speed signal setter
转速信号装置　speed indicator
转速因数　speed factor
转速整定值　speed set-point value
转梯　helical stairs, spiral stairs
转弯半径　turning radius
转向架　bogie
转移支付　transfer payment
转移阻抗　transfer impedance
转运仓库　in-transit depot
转运溜槽　transfer chute
转运站　transfer station
转折角　turning angle
转子　rotor
转子串联电阻启动　rotor resistance starting
转子磁轭　rotor rim, rotor yoke
转子顶起系统　rotor jacking system
转子动平衡及超速试验　dynamic balance and overspeed test for rotor
转子翻身　turning over of rotor
转子风扇　rotor fan
转子环　rim
转子环密封　rim seal
转子接地保护　rotor earth fault protection
转子绕组　rotor winding
转子绕组交流阻抗测定　AC impedance test for rotor winding
转子绕组接地　rotor winding earth fault
转子绕组匝间短路　rotor winding inter-turn short circuit
转子支架　spider
转子中心体　rotor hub, spider hub
桩号　chainage
桩基础　pile foundation
桩间土试验　post-soil test
桩孔扩底钻进　hole bottom reaming drilling
装货费　loading charge
装机容量　installed capacity
装机容量年利用小时数　annual utilization hours of installed capacity
装配场　assembly bay
装配焊接　erection welding
装配记号　assembly mark
装配件　assembly
装配螺栓　assembling bolt
装配试验　assembly test
装配图（组装图）　assembly drawing
装饰物　ornament
装卸　loading and unloading
装修　decoration
装运单据　shipping documents
装载机　loader
装置性材料　necessary accessories
状态变量　state variable
状态方程　state equation
状态监测　condition monitoring
状态监视　monitoring of status
状态矢量　state vector
状态信息　state information
撞击手轮　impact hand wheel

追忆打印　post-trip logging
锥齿轮传动装置　conical gear actuator
锥管母线　generating line of conic tube
锥体淤积　cone deposit，tapered deposit
锥形阀　fixed cone valve
锥形风帽　conical cowl
锥形烧瓶　Erlenmeyer flask
准备好自动开机　readiness for automatic start
准工作状态　condition of standing by
准股本资金　Quasi-Equity
准平原　peneplain
准确级　accuracy class
准同步并列　ideal synchronization
桌状山　table mountain
着火点　fire point
琢石　ashlar
咨询服务费　consulting service cost
资本化价值　capitalized value
资本金　capital fund
资产负债率　liability on asset ratio（LOAR）
资金成本　capital cost
资金的财务机会成本　financial opportunity cost of capital（FOCC）
资金的经济机会成本　economic opportunity cost of capital（EOCC）
资金流量　fund flow
资料室　archive room
资料收集方法　data acquisition method
资源成本　Resource Cost
子导线　sub-conductor
子地震台　seismo-substation
子午面　meridian plane
子午线　meridian
子站　slave station，controlled station
自备电源　self-powered system
自承脚手架　self-supporting scaffold
自持振荡　self-sustained oscillation
自动低频减负荷　automatic underfrequency load shedding
自动低频减负荷　automatic underfrequency load shedding
自动低压减负荷　automatic undervoltage load shedding
自动电话交换机　automatic telephone exchange
自动电话用户小交换机　private automatic branch exchange（PABX）
自动电压调节器　automatic voltage regulator（AVR）
自动电子快门　automatic electronic shutter
自动调节渠道　self-regulating canal
自动调节系统　automatic modulating control system
自动发电控制　automatic generation control
自动翻板闸门　balanced wicket，flashboard
自动扶梯　escalator
自动复位　automatic reset
自动挂脱梁　hook-release lifting beam
自动关断　automatic switching off
自动光圈镜头　automatic iris lens
自动光线补偿　ALC control
自动焊机　automatic welder
自动火灾信号　automatic fire signal
自动开通　automatic switching on
自动控制　automatic control
自动灭火系统　automatic fire extinguishing system
自动喷水灭火系统　automatic sprinkler system
自动喷水-泡沫联用系统　combined sprinkler-foam system
自动切换装置　automatic switching control equipment
自动顺序换相　auto-sequential commutation
自动同步装置　automatic synchronizing device
自动运行　automatic operation
自动增益控制　automatic gain control（AGC）
自动重合闸　auto-reclosing
自动重合闸断开时间　auto-reclose open time
自动重合闸中断时间　auto-reclose interruption time
自动准同步装置　automatic ideal synchronizing device
自放电　self-discharge
自感　self-inductance
自感电动势　self-induced e. m. f
自感应　self-induction
自灌充水　self-priming
自换相　self-commutation
自恢复绝缘　self restoring insulation
自计水位计　recording gauge
自溃堤　breaching dike
自溃式非常溢洪道　fuse-plug spillway
自留地　plot for private use
自流灌溉　gravity irrigation

自流盆地　artesian basin
自流平　self-leveling
自流泉　artesian spring
自落式搅拌机　gravity mixer
自密实　self-condensing
自凝灰浆　self-hardening slurry
自耦变压器　autotransformer
自耦变压器启动　autotransformer starting
自然保护区　nature conservation area (nature reverse)
自然边坡　natural slope
自然沉淀　plain sedimentation
自然电场法　self-potential method
自然方　bank measure
自然风冷却　air natural cooling (AN)
自然拱　natural arching
自然接地体　natural earthing substance
自然景观　natural landscape
自然控烟　natural smoke control
自然栖息地　natural habitat
自然侵蚀　natural erosion
自然通风　natural ventilation
自然通风冷却塔　natural draft cooling tower
自然循环采暖　natural circulation heating
自然循环系统　natural circulation system
自然养护　dry curing
自然遗产　natural heritage
自然灾害　natural disaster
自然增长率　natural growth rate
自燃　spontaneous ignition
自燃物　pyrophoric material
自热　self-heating
自容式充油电缆　self-contained oil- filled cable
自润滑轴承　maintenance free bearing, selflubricating bearing
自上而下分段灌浆　descending stage grouting
自上而下开挖　top-down excavation
自身平衡式　self-balancing type
自停转速　trip speed
自稳时间　stand-up time
自我润滑机制　self-lubricating mechanism
自吸式泵　self priming type pump
自吸式污水泵　self-priming sewage pump
自熄性　self extinguishbility
自下而上分段灌浆　ascending stage grouting
自下而上填筑施工　filled from the bottom up
自卸汽车（翻斗车）　dump truck
自循环通气　self-circulation venting
自应力浆液　self-stressing grout
自应力水泥　self-prestressed cement
自用水量　water consumption in water-works
自由度　degree of freedom
自由孔隙率　free porosity
自由排水　free draining
自由式压力钢管　free penstock
自由水舌　free nappe
自由溢流堰　free fall weir
自由振荡　free oscillation
自由作用　free action
自振频率，固有频率　natural frequency
自振周期　natural period of vibration
自重湿陷系数　coefficient of self weight collapsibility
自重湿陷性黄土　self weight collapsible loess
自重应力　self-weight stress
自钻式注浆锚杆　self-drilling grouted anchor bolt
渍　subsurface waterlogging
纵波　longitudinal wave, compression wave
纵吹灭弧室　axial blast interrupter
纵断面　longitudinal profile
纵缝　longitudinal joint
纵谷　longitudinal valley
纵横吹灭弧室　mixed blast interrupter
纵横制自动电话交换机　crossbar automatic telephone exchange
纵节理　longitudinal joint
纵联保护　pilot protection, line longitudinal protection
纵联差动保护　longitudinal differential protection
纵坡　longitudinal slope
纵向垂直支撑　longitudinal vertical bracing
纵向荷载　longitudinal load
纵向加密［计算机］　encryption
纵向倾覆　lengthways overturn
纵向受拉钢筋　longitudinal tensile rebar
纵向围堰　longitudinal cofferdam
总部管理费　head office overhead
总承包服务费　main contractor's attendance
总传送时间　overall transfer time
总的相对损失　relative total loss
总干渠　trunk channel

总告警　common alarm
总畸变含量　total distortion content
总畸变率　total distortion ratio
总畸变因数　total distortion factor（TDF）
总加电流互感器　summation current transformer
总价合同　lump-sum contract
总库容　gross reservoir capacity
总矿化度　total salinity
总冷量　total cooling capacity
总配线架　main distributing frame
总起重量　gross load
总生产工日　total working days
总水头　total head
总水压力　total hydrostatic load
总损耗　total losses
总体传热系数　total heat transfer coefficient
总投资收益率　return on investment（ROI）
总线　bus
总响应时间　overall response time
总谐波畸变　total harmonic distortion（THD）
总谐波率　total harmonic ratio
总谐波因数　total harmonic factor（THF）
总悬浮颗粒物　total suspended solids（TSS）
总应力法　total stress method
总应力强度　total stress strength
总硬度　total hardness
综合变形模量　comprehensive deformation modulus
综合单价　all-in unit price
综合单位线　synthetic unit hydrograph
综合档案馆　comprehensive archives
综合点　summing point
综合放大单元　summation and amplifica-tion module
综合灌浆法　comprehensive grouting method
综合加工厂　comprehensive processing plant
综合抗震能力　comprehensive aseismic capability
综合拉断力　comprehensive breaking strength（UTS）
综合利用工程　multi-purpose project
综合利用规划　multipurpose planning
综合生活用水　water for domestic and public use
综合特性曲线　combined characteristic curve
综合蓄热法　comprehensive method of heat accumulation
综合循环效率　comprehensive cycle efficiency
综合质量管理　comprehensive quality control
综合重合闸　composite auto-reclosing
走道盖板　walkway
走廊　corridor
走向　strike
租赁　lease
阻隔区　baffle area
阻隔液体　barrier liquid
阻火圈　firestop collar
阻进器、缓冲器　buffer
阻抗　impedance
阻抗保护　impedance protection
阻抗变换器　impedance converter
阻抗的模　modulus of impedance
阻抗电压　impedance voltage at rated current
阻抗继电器　impedance relay
阻抗匹配　impedance matching
阻抗式调压室　restricted orifice surge chamber
阻抗元件　impedance component
阻力曲线　system head curve
阻尼　damp
阻尼比　damping ratio
阻尼器　damper
阻尼绕组　damping winding，amortisseur winding
阻尼振荡　damped oscillation
阻尼装置　damping device
阻燃材料　flame retardant material
阻燃处理　fire retardant treatment
阻燃电缆　non-flame propagating cable，flame retardant cable
阻燃分隔　fire retarding division
阻燃性　fire retardance
阻容分压器　resistance-capacitance voltage divider
阻容吸收器　RC snubber
阻水层　aquiclude
阻水环　cut-off collar
组合暴雨　synthetic storm
组合变送器　multi-section transducer
组合告警　group alarm
组合构件　built-up member
组合结构　composite structure
组合梁　built-up beam，compound beam
组合楼盖　composite floor system
组合式互感器　combined instrument transformer
组合式支护　composite support
组合屋架　composite roof truss

组合系统　combined system
组件吊运　handling of component
组命令　group command
组装　assembly
组装式制冷设备　assembling refrigerating unit
钻爆法　drill-blast tunneling method
钻孔编录　drilling log
钻孔测斜仪　bore-hole inclinometer
钻孔地质员　inspector
钻孔电视　borehole TV
钻孔顶角　borehole vertex angle
钻孔定向　borehole orientation
钻孔灌注桩　drill hole grouting pile
钻孔孔径　borehole size/borehole diameter
钻孔倾角　borehole inclination
钻孔试样　borehole sample
钻孔位置测量　survey of borehole position
钻孔物探测试　geophysical survey in borehole
钻孔照相　borehole photography
钻孔柱状图　Borehole log/Geological log of drill hole
钻探　drilling exploration
钻芯法检测　testing with drilled core
钻压　weight on bit（WOB），bit pressure
最大动密封压力　maximum dynamic sealing pressure
最大冻土深度　maximum frozen soil depth
最大截止频率　maximum cutoff frequency
最大可信地震　maximum credible earthquake（MCE）
最大连续容量　maximum continuous capacity
最大设计地震　maximum design earthquake（MDE）
最大瞬时压力变化　maximum momentary pressure variation
最大瞬时转速变化　maximum momentary speed variation
最低动作压力试验　start-up pressure test
最低可接受收益率　minimum acceptable rate of return（MARR）
最低要求压力　minimum required pressure
最低涌浪水位　lowest surge level
最高容许连续转速　maximum allowable continuous speed
最高通航水位　maximum stage of waterway
最高涌浪水位　highest surge level
最弱边　weakest side
最小安全距离　minimum approach distance（minimum working distance）
最小二乘法　least square method
最小开启压力　minimum open pressure
最优导叶开度　optimum wicket gate position
最优工况　optimum operating condition
最优含水量（含水率）　optimum water content，optimum moisture content
最优普氏密度　Proctor optimum density
最优效率　optimum efficiency
最终安全出口　final exit
最终报表　final statement
最终沉降量　ultimate settlement
最终（成品）尺寸　finished size
最终付款　final payment
左旋扭动　left lateral wrench
左旋走滑活动　left lateral strike-slip activity
作物袭击　crop raiding
作物需水量　water demand of crop
作业计划　operation schedule
作业距离　working distance
作业周期　operation cycle time
作用的标准值　characteristic value of an action
作用的代表值　representative value of an action
作用的地震组合　seismic combination
作用的频遇组合　frequent combinations
作用的设计值　design value of an action
作用的随机特性　stochastic characteristics of actions
作用的准永久组合　quasi-permanent combinations
作用分项系数　partial factor for action
作用　action，load
作用面积　area of sprinklers operation
作用效应　action effect
坐标　coordinate
坐落　slumping
座环　stay ring

英文索引

（部分技术名词）

A

abandoned channel 32
aberration 132
abrasion 9
abrasion resistance concrete 145
abrupt slope 9
absolute chronology 123
absolute contract 168
absolute humidity 32
absolute permeability 97
absolute rotary encoder 92
absorptance 115
absorption ratio 117
abutment 22
abutment pad 60
abutment pier 62
accelerating admixture 137
accelerating and water reducing admixture 137
acceleration-damping type governor 77
accelerometer 68
access control 131
accessory building 152
accessory guarantee 168
access road to dam 135
access road to site 135
access shaft 76
accidental action 53
accidental combination 53
accident countermeasures 129
accident potential 171
accumulation plain 9
accumulator 94
acid corrosion 16
acidic water 16
acidity 16
acid pickling 94
acid rain 44
acid rock 11
acoustic log 27
acoustic wave survey 27
acrylate grout 141
actinolite 12
activation log 27
active admixture 135
active belt 7
active earth pressure 54
active energy 103
active fault 7
active fire prevention 161
active load 103
active network 99
active power 99
active power filter 107
active power transducer 122
active structure 7
actual calorific valve 86
actual dip 14
actuating energy 79
actuator 85
acute toxicity 86
additive 136
adhering nappe 63
adhesive bitumen primer 136
adiabatic curing 146
adiabatic temperature rise 146
adit 140
adit under river 27
adjacent block 146
adjacent lift 146
adjudicator's agreement 169
adjustable capacity 37
adjusting plate 112
adjusting screw 85
adjustor of steel pipe 65
adjust ring 83
admittance 98
adobe 50
adobe blasting 139
advanced ducting 140
advance grouting 141
advance payment 169
advance support 140
advancing slope method 143
adverse slope 22
aerated concrete 145
aerated flow 62
aerated nappe 63
aeration slot 62
aeration zone 15
afforestation 46
after flame 86
aftershock 7
afterwards supervision 169
age dating 6
ageing of insulation 118
ageing resistance 94
agent 136
aggregate 23
aggregate chips 136
aggregate gradation 143
aggregate open stockpiled 143
aggregate segregation 143
aggressive water 16
agricultural resettlement 49
agro-forestry 50
air admission test 74
air carbon arc gouging 91
air circuit breaker 114
air cleaning system 157
air cleanliness class 157
air clearance 117
air-cooled aggregate 146
air cooler 105
air cooling duct 158
air-core reactor 110
aircraft warning marker 112
air cushion surge chamber 64
air-cusion mechanism 22
air drill 148
air earth hammer 149
air-entraining admixture 137
air flow rate 158
air forced cooling 110
air humidity 32
air inlet 87
air leakage rate 157
air leg-mounted drill 148
air lock 153
air placed concrete 144
air pollution 44
air release valve 89
air renewal system 156

air-source heat pump 157
air supply for brakes 124
air supply through nozzle 158
air-to-air cooled machine 105
air-to-water cooled machine 105
air vent 64
alarm apparatus 161
alarm delay 161
alarm detector 161
alarm mark 161
alarm signal 161
alarm switch 161
alarm transmission system 132
alarm valve 161
alarm water level 161
algae 45
alidate 30
aliform dam 60
alignment 29
alignment of jet to runner 75
alkali-aggregate reaction 136
alkaline cell 129
alkaline water 16
alkalinity 45
all-air condition 157
all aluminium alloy conductor 112
all aluminium conductor 112
all dielectric self- supporting optical cable 131
all-in contract 168
all-in unit price 164
allocation with job 49
alluvial channel 32
alluvial fan 9
alluvial plain 9
alluvium 10
alteration 17
altered mineral 12
altered zone 17
alternately bedded structure 15
alternate source 161
alternate wetting and drying 145
alternating current 98
alternating layers 13
alternating voltage 98
alternative 4，39
alternative habitat 44
altitude difference 29
aluminate cement 136
alunite expanding cement 136
ambient temperature 32
amicable settlement 169
ammeter 122
amorphous substance 12
amortisseur winding 105
amortization of benefits 38
amortized cost 163
amphibole schist 11
amphibolite 11
amplifier dead band 79
amplifier inaccuracy 79
amplitude of vibration 58
amygdaloidal structure 13
analog indicating instrument 124
analog input 127
analog output 127
analog-to-digital converter 127
analogue communication 123
analogue simulation 28
analog variable 122
analysis after financing 40
analysis before financing 40
anchor 93
anchorage 66
anchor block 142
anchor bolt 137
anchor bundle 143
anchored framework 66
anchoring mortar 141
anchor retaining wall 69
anchor rod 112
ancient glaciation 9
ancient karst 9
ancient landslide 9
ancient river course 9
andalusite 12
andesite 11
anemograph 93
angle of repose 22
angle of shade 112
angle of view 131
angle of wind deflection 111
angle support 111
angling dozer 148
angular frequency 73
angular momentum 80
angular network 29
angular unconformity 13
anhydrite 12
animal entrapment 44
animal migration route 44
anion 16
anisotropic rock mass 19
anisotropy 19
annealing 91
annual incidence rate 8
annual solar radiation hours 33
annual utilization hours 103
annuities value 40
annulated column 154
annunciation of alarm conditions 124
annunciator 121
anode 129
anteroom 153
anticline 13
anticlinorium 13
anticollision device 95
anti-corrosive pump 82
anti-freezing agent 137
antifriction bearing 90
antiparallel connected thyristors 121
anti-pumping device 109
anti-symmetrical load 155
anti-theft detector 132
apatite 12
aplite 11
aplite vein 12
apparent density 23
apparent dip 14
apparent power 99
approach channel 64，69
appropriation 169

appurtenant facilities 148
apron 63，153
apron feeder 149
aquatic creature 44
aquatic ecology 44
aquatic habitat 44
aqueduct 69
aquiclude 15
aquifer 15
aquifuge 15
arable land 46
arc discharge 117
arc-extinguishing chamber 108
arc-extinguishing tube 108
arch dam 60
archeological area 21
architectural acoustics 155
architectural appearance 152
architectural ensemble 152
architectural mechanics 155
architectural optics 155
architectural physics 155
archival code 172
archival informationization 173
archival microform 173
archival repository 173
archival value 172
archives thesaurus 173
archiving 172
arcing contact 108
arcing time 109
arcing voltage 121
arc over rate 117
arc resistance 85
arc suppression reactor 110
areal precipitation 33
area-moment method 154
arenaceous texture 13
argillaceous cement 11
argillaceous infilled 15
argillaceous limestone 11
argillaceous sandstone 11
argillaceous texture 13
argillized seam 13
argon arc welding 147
arid land 50
arkose 11
armature reaction 105
armature winding 104
armor angle 90
armour 87
armoured cable 112
arrester 93
artesian basin 16
artesian spring 16
artesian water 15
artificial aggregate 135
artificial dead band module 77
artificial deposit 10
artificial earthing electrode 118
artificial pollution test 116
art museum 153
ascending spring 16
ascending stage grouting 141
ascertain 25
aseismatic safety 58
ash 86
ashlar 136
aspect ratio 55
asperity 14
asphalt concrete 145
assembling bolt 87
assembly 87，91
assembly bay 87
assembly mark 87
assignment 168
assignor 168
assisted safety valve 85
associated mineral 12
asynchronous motor 114
asynchronous operation 102
asynchronous starting 107
atmospheric pressure 72
atrium 153
attack angle 82
attenuation 143
attenuation uniformity 131
Atterberg limit 19
Atterberg test 24
attitude 14
audible noise 111
audio-frequency cable 130
audio-visual archives 172
augen structure 13
auger drill 26
augite 12
authentication 123
authenticity 173
automatic bus transfer 115
automatic diagnosis 127
automatic electronic shutter 132
automatic fire signal 86
automatic gain control 27
automatic iris lens 132
automatic reset 94
automatic sprinkler 161
automatic voltage regulator 80
automobile service workshop 134
auto-sequential commutation 119
autotransformer 109
auxiliary circuit 108
auxiliary energy dissipator 63
auxiliary plant 65
auxiliary relay 126
auxiliary servomotor 78
auxiliary spillway 62
auxiliary weir 63
auxiliary winding 109
availability of excavated material 139
available flow 37
available working day 150
avalanche 18
axial blast interrupter 108
axial compression column 56
axial fan 157
axial flow check valve 84
axial flow pump 81
axial flow turbine 72
axial force 57
axial load 57
axial loaded column 56
axial load pulsation 73
axially split pump 81
axial piston pump 93

axial plane cleavage 15
axial tensile member 55
axial ventilation 105
azimuth angle 14

B

babbitt bearing 105
back analysis 28
back drain 69
backfill 66
backfilled soil 10
backfill grouting 141
backflow area 64
backflow prevent 159
back gouging 91
background earthquake 8
background noise 45
background value 68
backhoe front end loader 148
backhoe shovel 148
back seal 84
backseat test 85
back shroud 83
back-slope terrace 47
back-to-back starting (BTB starting) 107
backup protection 126
back-up pump start 125
backup unit emergency shutdown device 122
backward deposition 35
backwater level 21
baffle area 160
baffle block 63
baffle plate 64
baffle sill 63
balance chain 95
balanced earthworks 134
balanced wicket 89
balancing bus 101
balcony 154
ballasting filter 144
ball gate 89
ball valve (spherical valve) 83
banded structure 13
banded texture 13
band pass filter 99
band stop filter 99
bank deposit 22
bank erosion 22
bank guarantee 168
bank loan 166
bank measure 140
bank-off advancing 138
bank prime rate 166
bank sloughing 22
bar 34
bar bender 150
bar code 172
bar cutter 150
bare karst 18
barren land 50
barrier coat 92
barrier liquid 83
barring gear 107
barrow 148
barytes 12
basal conglomerate 10
basalt 12
baseboard 153
base flow 33
base level 29
base line 29
base load power plant 104
basement 152
basic combination 53
basic depreciation cost 163
basic endowment insurance 164
basic frequency 73
basic insulation 118
basic rock 11
basic seismic intensity 7
basic survey station 29
basic variables 52
batcher 149
batch hopper 149
batching by volume 145
batching by weight 145
batching silo 149
batch mixer 149
battery 129
battery bank 130
battery of wells 158
bauxitic shale 11
Bayes probability method 70
beach 44
beaker 25
beam element 55
beam grid 154
bearing block 90
bearing capacity of member 55
bearing insulation 105
bearing rebar 147
bearing sleeve 83
bearing spider 83
bearing stratum 54
bearing wall 153
beat 100
beat frequency 100
Beaufort force 33
bedded joint 14
bedding cleavage 15
bedding joint 14
bedding layer 144
bedding plane 13
bedding structure 15
bed form 35
bed load 34
bedrock seated terrace 9
bedroom 153
beds alternation 13
belled-out pile 142
bell mouth inlet 64
bellow expansion joint 65
bellows seal reducing valve 85
bellows seal type valve 84
bellows stem sealing 85
belt conveyor 149
belt deposit 35
bench blasting 139
benched excavation 139
bench mark 29
bench mark station 35
benchmark tariff 41
bench terrace 47

bending fold 14
bending strength 20
bending-tensile fracture 22
bend loss 65
beneficiary 168
bent frame 57
benthonic organism 44
bentonite 11，135
bent pile pier 56
bent-up rebar 147
berm 60
berthing structure 69
biaxial eccentric compression 54
bid bond 168
bid document 168
bid guarantee 168
bid inquiry 168
bidirectional seal 90
bidirectional valve 84
bid opening 168
bid price 168
bifurcated pipe 65
bill of quantities 165
binary state information 123
binary variable 122
binder 136
binder coat 144
biochemical oxygen demand 45
bioclastic limestone 11
biogenetic texture 13
biosphere 44
biotite 12
Bishop method 70
bi-stable relay 126
bit burnt 27
bitumen coat 136
bitumen ductility 144
bitumen flash point 144
bitumen paraffin content 144
bitumen penetration 144
bitumen softening point 144
bitumen solubility 144
bituminous mastic 136
black start 102
bladder accumulator 79
blade inlet angle 75
blade inlet pitch 75
blade position sensor 124
blade removal opening 87
blade servomotor 78
blanket 61
blast-furnace cement 136
blasting 139
blasting cap 137
bleeding 146
blind drainage 69
blind window 154
blind zone 131
blistering 144
block concreting 145
blocked channel 32
blocking component 127
blocking pilot protection 128
blocking state 119
block masonry structure 152
blockout 147
blocky and seamy rock 15
blocky-fractured structure 15
blocky structure 15
blowdown 85
blow-down valve 84
blowout 23
body wave 7
bogie 81
boiler feed pump 82
bollard 95
bolted connection 91
bolt-shotcrete support 140
bonded goods 171
bonding agent 137
bond strength 56
bond stress 56
boom excavator 148
booster pump 82
border dike 69
border price 40
borehole deviation 26
borehole inclination 26
borehole orientation 26
borehole TV 27
borehole vertex angle 26
borrow area 23
bottom flow energy dissipation 63
bottom heaving 23
bottom outlet 62
bottom ring 76
bottom sill 90
bottom valve 84
boulder 135
boundary temperature 54
boundary wave 27
bow scraper 148
box beam 56
box footing 55
braced frame structure 152
brachy-axis fold 13
bracket 56
bracket scaffold 147
braid 113
braided channel 9
brake 81
brake air pressure 125
brake nozzle 77
brakes released 125
braking torque 114
branch 98
branch duct 157
breaching dike 69
break contact 108，127
breakdown 119
breakdown cost 51
breaker 148
break-even point 40
breaking 109
breaking current 109
breaking overvoltage 118
breast board 140
breast wall 59
breather cap with air filter 93
breccia 10
brecciated texture 13
bridge crane 80
bridge girder camber 81
Brinell test 24

brisance factor 139
brittle failure 14
brittle fracture 14
broad-crested weir 62
broad-leaved forest 44
brush 114
brush holder 105
brushless exciter 120
Buchholz relay 110
bucket 77
bucket auger 26
bucket excavator 148
bucket inclination 75
bucket-type energy dissipator 63
buckled fold 13
buckling 18
buckling load 57
buckling stability 57
buffer 81
buffer blasting 139
buffer tuber 131
building density 152
building envelope 155
building line 152
building model 152
building sun shading 156
building thermal shading 156
building thermotics 155
built environment 44
built-in beam 56
built-in camber 81
built-in travel detector 94
built-up back pressure 85
built-up member 154
bulb support 76
bulb tubular unit 72
bulb type hydrogenerator 104
bulging 144
bulk density 23, 139
bulkhead gate 89
bulky grain 135
bulldozer 148
bump-free switch-over 79
buoyancy force 53
burglar alarm system 131
buried channel 9
buried drain 69
buried fault 14
buried terrace 9
burst strength 146
busbar 104
busbar cavern 132
busduct 113
bushing 110
business tax 164
bus tie breaker 113
bus type current transformer 110
busway 113
butterfly valve 83
buttress dam 61
butt splices 56
butt welding 147
buzzer 121
by-pass air duct 158
bypassing leakage 22
bypass tunnel 139

C

cab 92
cable 112
cable bearer 87
cable branch box 130
cable core 131
cable crane 150
cable distribution head 130
cable drag scraper 148
cable drum 93
cable duct bank 113
cable hoist 92
cable rack 113
cable routing 113
cable shaft 113
cable-stay pile 66
cable terminal 113
cable tray 87, 113
caboperated crane 80
cage induction motor 114
calcareous cement 11
calcareous concretion 11
calc-dolomite marble 11
calcite 12
calcite infilled 15
calcite vein 12
calcium coated 14
calc-sinter 16
calibration 52, 94
caliche nodule 11
camber 60
cam relationship 75
canal for water release 39
canal structure 69
candela 115
canned motor pump 82
canopy 154
cant column 154
canteen 153
cantilever 55
cantilever crane 150
cantilevered extension 62
cantilever form 147
cantilever retaining wall 69
cantilever sheet-pile 66
cap 112
capable fault 7
capacitance 98
capacitive reactance 98
capacitive susceptance 98
capacitive transducer 68
capacitor 107
capacity benefit 38
capillary rise 21
capillary water 15
capital cost 41
capital farmland 46
capital fund 166
capitalized value 41
carbonaceous shale 11
carbonation of concrete 57
carbonatite formation 7
carbonic acid corrosion 16
carbon sequestration 43
carbon sink 43
carbon steel 91
card reader 132
carriage 171

carryover regulation 36
cartridge 138
cartridge valve 93
cascade development 3
casement window 154
cash flow 40
cash inflow 40
cash outflow 40
casing cover 82
casing liner side plate 83
casing placement 27
casing ring 83
cast iron pipe 137
cast steel 137
casualty accident 171
cataclasite 11
cataclastic breccia 11
cataclastic structure 15
catch drain 144
catchment area 32
catchpit 87
catenary 111
catenary constant 111
caterpillar excavator 148
caterpillar gate 89
cathode 129
cathodic protection 92
cation 16
caulk 136
cavern complex 22
caving 18
cavitation 63，73
cavitation damage 63
cavitation erosion 73
cavitation margin 73
cavitation pitting guarantee 73
cavitation resistance 63
cell 129
cell can 129
cell lid 129
cement 11
cement cartridge rock bolt 142
cement cream 146
cemented masonry cofferdam 139
cementitious material 136
cement silo 149
center bulb seal 90
center drift tunneling 140
center rotating disconnector 107
centralization lubrication 93
centralized control 123
centralized load 111
centrifugal fan 157
centrifugal pump 81
ceramic-coated rod 94
ceramic coating 94
certificate of origin 171
certified discharge capacity 85
chainage 29
chain hoist 92
chain plate 90
chalcopyrite 12
chamber freeboard 95
chamber levelness 95
chamber retaining wall 69
change order 169
channel busbar 113
channel diversion 138
channelized waterway 38
char 86
characteristic angle 127
characteristic curver 82
characteristic impedance 100
characteristic period 58
charge efficiency 130
charge factor 130
charge for one interval 139
charge for trouble 164
charger sizing 130
chatter 85
check 25
check damper 157
check flood level 4
check torque 92
check valve 83
chemical alteration 17
chemical bonding 147
chemical composition 91
chemical grouting 141
chemical oxygen demand 45
chemical weathering 17
chevron drain 69
Chezy formula 70
chilled water 157
chimney drain 61
chimney effect 86
chlorite 12
chopped lightning impulse
  test 116
chronic toxicity 86
churn and grabbing drilling 26
chute 62
chute block 63
chute spillway 62
circle diagram 98
circuit-breaker 107
circuit-breaker failure
  protection 126
circuit crest working
  off-state voltage 119
circuit crest working
  reverse voltage 119
circuit element 98
circular runout 94
circular slide 18
circulating medium 26
circulating water pump 82
circulation flush 83
circulation pipe 156
circumferential stress 21
civil architecture 152
clamp 112
clamp seal 90
clamshell shovel 148
clarification 168
clarification of welds 91
classified catalogue 173
classified highway 135
classified indexing 172
classified level 173
class of cube strength 145
clastic flow mechanism 22
clastic sediment 10
clastic texture 13
clay-bound macadam

pavement 135
clayey soil 10
clay infilled 15
clay seam 13
clay shale 11
clay stone 11
clearance between poles 108
clearance to earth 109
clearance to obstacles 112
clear water 26
cleavage 12，15
Cleveland open cup 25
clevis 112
cliff 9
climbing (jacked) form 147
clinker 86
clinker cement 136
clinograph 26
clogging 22
cloister 153
closed-circuit grouting 141
closed fold 13
closed impeller 82
closed-loop control 121
closed position 109
closed traverse 29
close-open operation 109
close-range photogrammetry 29
closing force 54，91
closing overvoltage 118
closing speed 109
closing time 109
closure dike 138
closure gap 138
closure of arch 60
closure temperature 60
coagulation sedimentation 159
coal seam 10
coarse aggregate 135
coarse grained texture 13
coarse gravel 10
coarsely-crystalline texture 13
coarse sand 10
coarse sediment concentration 64
coating adhesion 92
coaxiality 87
cobble 10
cobble soil 10
coefficient of collapsibility 19
coefficient of curvature 23
coefficient of measure
conversion 139
coefficient of subgrade
reaction 54
cohesion 20
cohesionless granular
structure 15
cohesionless sealer 146
cohesionless soil 10
cohesive soil 10
coil condenser 158
cold accumulation 158
cold bend 147
cold gagging 147
cold joint 146
cold standby reserve 103
cold wave 33
cold weld 91
cold working pressure 85
collage belt 6
collapse 18
collapsible loess 10
collar beam 154
collecting sump 144
collector 144
collimating point 29
collimator 30
colloidal grout 141
colonnade 154
colorimeter 115
column 18
columnar joint 14
column pipe 82
combination curve 74
combination fault 126
combined energy dissipation 63
combined sewer overflow 159
combustibility 85
commissioning 170
common alarm 123
common battery switch 130
common defect 170
common storage 36
common winding 109
communication 130
communication module 122
communication satellite 131
communication security 70
community dial office 130
commutating voltage 119
commutation 119
commutation capacitor 119
commutation circuit 119
commutation reactor 119
commuting allowance 163
compaction criterion 143
compaction density 19
compact limestone 11
compactness 19，143
compactor 149
compact shelving 173
companion specimen 25
compartment 113
compatibility 70
compensating joint 65
compensation for land
requisition 165
compensation subsidy 165
compensatory tension 67
complete at the first attempt 88
complete characteristics 82
complete evacuation 161
complete freezing 59
completely weathered 17
complete shutdown 125
complex assembly 91
component 91
composite flexural member 56
composite floor system 154
composite geomembrane 136
composite geotextile 136
composite insulator 112
composite roof truss 154
comprehensive archives 172
comprehensive breaking

strength 111
comprehensive cycle efficiency 68
comprehensive processing plant 134
compressed seal 90
compressibility 79
compressible stratum 17
compressible-type filler 136
compression bearing capacity 56
compression-concentration tendon 67
compression-dispersion tendon 67
compression modulus 20
compression wave 7
compression zone 54
compressive strength 19
compressive structural plane 15
computer aided design 70
computer tomography 27
concave bank 9
concave slope 9
concavo-convex 9
concealed beam 56
concert hall 153
conchoidal fracture 13
concrete age 145
concrete air content 146
concrete blockyard 149
concrete bucket 149
concrete face rockfill dam 60
concrete face slab 61
concrete hollow gravity dam 59
concrete mixing plant 149
concrete pavement 135
concrete precooling system 149
concrete preheating system 150
concrete preparation 149
concrete remover 147
concrete slotted gravity dam 59
concrete socket 60
concrete solid gravity dam 59
concrete sprayer 148
concrete spreader 149
concrete tetrahedron 138
concrete transport skip 149
concrete vibrator 149
concreting in lifts 145
condensate flow rate 158
condensate line 158
condensate pump 82
condenser 107
condenser bushing 110
condenser mode 107
condensing mode 125
conductance 98
conducting direction 119
conduction through 119
conductivity 98
conductor 98
conductor bundle 112
conductor vibration 111
conduit 64
conduit type hydropower station 3
cone crusher 148
cone deposit 35
confidential secret 173
confined flow 16
confined water 15
confining overlying bed 16
confining pressure 54
confining stratum 15
confining underlying bed 15
conformity 13
conglomerate 10
conglomeratic sandstone 11
conical cowl 157
conical gear actuator 85
coniferous forest 44
conjugate beam 56
conjugated fault 14
connecting bolt 137
connecting rod 76
connecting traverse 29
connection box 130
connection line load 103
connectivity 4
connector 129
consequent bedding structure 22
consequent landslide 18
consequent slope 22
conservation 173
consignment 171
consistency 145
consolidated drained shear test 24
consolidated drained triaxial test 24
consolidated undrained shear test 24
consolidated undrained triaxial test 24
consolidation grouting 141
consolidation of soil 19
consolidation pressure 24
consolidation settlement 22
constant-angle arch dam 60
constant-center arch dam 60
constitution of power sources 37
constitutive relation 28
construction camp 134
construction control network 29
construction delay 171
construction detailed design stage 25
construction duration 150
construction intensity 150
construction joint 56
construction mapping 26
construction mechanization 148
construction preparation 134
construction quality 170
construction scheme 134
construction sequence 150
construction supervision 4
construction survey 29
construction torque 92
construction transportation 134
construction traverse 29
consumer surplus 40
consumer tariff 41
contact grouting 141
contacting travel 108

contact potential difference 98
contact scouring 22
contact tension fitting 112
contained fire 160
container 171
containerized transport 135
containment grouting 141
context database 173
continent 8
continental facies 10
continental platform 7
continuous beam 154
continuous concreting 145
continuous gradation 143
continuous running duty 114
continuous sampling 26
continuous scouring sand basin 70
continuous seam 154
contorted fold 13
contour 29
contour ditch 144
contour interval 29
contour living hedgerow 47
contour tillage 47
contract agreement 168，169
contraction joint 146
contractor 170
contractor camp 134
contractor's all risk insurance 165
contractor's personnel 170
contractor's representative 170
contractor's superintendence 170
contract price 171
contract time limit 169
contract variation 169
control board 124
control cable 113
control function 79
controllable set 100
controlled blasting 139
controlled station 123
controlled variable 78
controller 114
control network 28
control observation 67
control panel and console 121
control panel room 132
control point 29
control switch 108，114
control system dead time 79
control system interfaces 124
control systems component 80
control systems proper 80
convection heating 155
conventional automation 125
conventional concrete 144
convergence measurement 29
convergent deformation 67
converse guide 90
convertible shovel 148
convex bank 9
conveyance system 64
conveyor case 82
conveyor vane 82
convey tunnel 64
cooling load 155
cooling medium 120
cooling system 105
coordinate 29
copper loss 105
corbel 57
core box 27
core disking 21
core drilling (boring) 26
core extraction 26
core fault 106
core logging 27
core loosening 105
core record spacer 27
core type transformer 109
core wall 61
coring 145
corner seal 90
corona bar 112
corona discharge 117
corona interference 111
corona loss 117
corona shielding 105
corridor 153
corrosion allowance 83
corrosion prevention 92
corrosion proof measure 65
corrosion resistance 66
corrosive carbon dioxide 16
corrugated sheet 137
corundum 12
cost 164
cost effectiveness ratio 39
cost engineer 163
cost engineering technician 163
cost index 163
cost plus award-fee contract 168
cost plus fixed-fee contract 168
cost recovery 39
cost reimbursement contract 168
counterbalance valve 93
counter drain 69
counter-electromotive force 97
counterfort retaining wall 69
counter relay 126
counter thrust bearing 76
counterweight 95
coupled continuous charging 139
coupling 92，97，156
coupling bolts 76，87
coupling capacitor 131
coupling filter 131
coupling flange 76
coupling guard 83
courtyard 153
covered karst 18
covering cultivation 47
cowl 157
crab 81
crab rock loader 148
crack 91
crack aperture 14
crack filling 14
crack healing 14
crack resistance 55
crane capacity load 147
crane rail 90
crane rock beam 150

crawler crane 150
creep 18
creepage distance 118
creep deformation 20
creeping slide 18
creep rupture 20
creep sliding-tensile fracture 22
crenulation cleavage 15
crescent-rid reinforced bifurcation 65
crescent stiffener 65
crest overflow 62
crest spillway 62
crib dam 61
critical cultivation slope 47
critical damping 79
critical flow 63
critical path 150
critical schedule 150
critical sequence 150
critical span 111
crop raiding 44
cross beam 154
cross bedding 13
cross blast interrupter 108
cross brace (X-brace) 154
cross country fault 126
cross head 76
crosshole resistivity probe 27
crosshole seismic probe 27
crosshole tomography 27
cross joint 14
cross main 161
cross product 100
cross-section 5
cross subsidization 40
cross vault 154
crowbar circuit 121
crown 64
crown cantilever 60
crown collapse 23
crucible 25
crumbling soil 23
crumple 14
crushed 21
crushed ice 146
crushed rock 136
crushed stone 135
crusher 148
cryogenic valve 84
cryptocrystalline texture 13
crystalline corrosion 16
crystalline limestone 11
crystalline substance 12
crystalloblastic texture 13
cultivated land 50
culturaland historic relic 50
cultural palace 152
cultural relics site 4
culvert diversion 138
cumulative frequency 33
cumulative percent passing 143
cumulative sediment load 34
curb stones 61
curing agent 137
curing mat 146
current carrying capacity 111
current circuit 127
current component 126
current error 111
current injection circuit 109
current-limiting fuse 114
current-limiting reactor 110
current livelihood 48
current-meter 75
current relay 126
current resonance 99
current transducer 122
current transformer 110
current zero 108
curtain grouting 141
curve fitting 33
cushion block 142
custody 173
customs clearance 171
customs declaration 171
customs duty 164
cut and fill slope 66
cut-and-try method 28
cut hole 139
cut-in deflector 77
cut-off collar 65
cut-off current 108
cut-off dike 69
cutoff trench 61
cutoff wall 66
cut slope 66
cutting 91
cutting head and loading at foot 66
cutting ring 24
cutting shoe 26
cutting slope and unloading 66
cycle stress 57
cyclic action 54
cylinder 94
cylindrical gate 89
cylindrical gear actuator 85
cylindrical surge chamber 64
cylindrical ventilator 157

D

dacite 11
dado 153
daily load 103
daily regulating reservoir 68
dam block 59
dambreak flood 34
dam crest 60
dam foundation 22
dam-gap diversion 138
dam heel 59
dammed lake 18
damp 58
damped oscillation 99
damper 59
damping type governor 77
damping winding 105
damp-proof course 156
dam shell 61
dam slope 60
dam toe 59
dangerous goods 135
dangerous voltage 115
Darcy's law 70

dark brownish yellow 12
dark mineral 12
dash-bond coat 154
dashpot 78
data acquisition software 122
data base 70
data communication 123
data server 122
data transfer 70
datum plane 29
datum point 29
daywork labor 163
dead band 79
dead-end clamp 112
dead load 54
dead short 101
dead tank circuit-breaker 107
dead time 109
dead water level 4
dead zone 127
debounce 122
debris 23
debris flow 18
decay 99
deck beam 154
deck charging 139
declassification 173
decomposing corrosion 16
decomposition weathering 17
decoration 154
decorative plant 44
deenergize-to-shutdown 125
deep cut in hillside 23
deep flexural member 55
deep-focus earthquake 7
deep footing 54
deep slot type rotor 114
de-excitation by inversion 121
defect elimination 101
defective soil aeration 46
defects liability period 172
defects notification period 172
definite time relay 126
deflected current 63
deflecting bucket 63
deflection angle 65
deflector 77
deflector position sensor 124
deflector servomotor 77
defoaming agent 137
deformation modulus 20
deformation resistance 20
deforming slope 22
degradation 57
degree of freedom 58
degree of saturation 19
dehumidification 173
dehumidizer 93
de-hydrogenation treatment 91
delay blasting 139
delay detonator 138
delayed automatic reclosing 129
delayed toxicity 86
delivery pipe 158
delivery tunnel 64
delta configuration 112
delta connection 98
delta deposit 35
delta-wye conversion 99
deluge system 161
demagnetization 97
demand guarantee 168
demobilization 170
demolition blasting 139
demonstration farm 50
dense-graded aggregate 136
densely jointed belt 14
density 19
density current 35
density relay 109
density sensor 109
dental concrete 146
denudation 9
denudation plain 9
denudation plane 9
departure curve 27
dependable flow 37
depletion premium 40
depreciation 41
depression 9，144
depression air valves 124
depression basin 7
depression cone 17
depression spring 16
depth of seasonal freezing 59
derivative action time 78
derrick 148
descending stage grouting 141
descriptor 173
desert 9
desertification 46
design anchoring 66
design dependability 36
design flood 34
design flood level 4
design frequency 33
design ground motion 58
design head 36
design load 111
design radial load 83
design rainfall pattern 33
design rainstorm 33
design reference period 52
design response spectra 58
design safety criteria 28
design seismic acceleration 58
design seismic intensity 7
design service life 52
design situations 53
design station 35
design threshold 68
design time horizon 36
desilting 3
desilting channel 70
desilting sluice 69
destruction 172
destructive earthquake 7
detached column 154
detached spillway 62
detail survey 29
detection- extinguishing system 125
detention dam 69
deterministic analysis 28
detonation velocity 139

detonator 137
developing fault 102
dewatering screen 149
dewpoint 33
dew-point temperature 155
diabase 12
diagenesis 9
diagonal 90
diagonal turbine 72
diamond 12
diaphragm 61，90
diaphragm valve 83
dielectric 97
dielectric constant 97
diesel generator 113
differential circuit 100
differential gain 78
differential relay 126
differential resistive transducer 68
differential settlement 22
differential surge chamber 64
difficult-flammable 85
diffused luminaire 115
diffuser casing 82
diffuser pump 81
diffuser vane 82
diffusion bucket 63
digital filter 127
digital signature 173
digital substation 103
digital-to-analog converter 127
digital topographic map 30
digital variable 122
dike 12
dilatancy 57
dilute bitumen 136
diluting agent 136
diluvial fan 9
diluvium 10
diorite 11
dip 14
dip slope 22
direct-acting reducing valve 85
direct-axis transient reactance 106
direct-burial cable 113
direct current 98
direct glare 155
directional blasting 139
directional component 127
directional control valve 93
directional jet grouting 141
directional protection 126
directional relay 126
direction angle 14
direct lightning strike 117
direct-loaded safety valve 84
directly affected household 48
direct return system 156
disassembly 87
disbursement 169
disc brake 93
discharge 129
discharge area 85
discharge capacity 64
discharge casing 82
discharge chamber ring 76
discharge current 129
discharge elbow 82
discharge per unit width 138
discharge pressure 82
discharge rate 129
disc holder 85
disconformity 13
discontinuity 15
discontinuity strength 20
discount rate 41
discrepancy 169
discrete element method 70
disc screen 149
dise guide 85
disengaging ratio 127
disengaging value 127
disintegration 17
disintegration weathering 17
disk brake 81
disk-shaped rock core 21
disk type thyristors 121
dislocated water system 8
dislocation 91
dismantling flange 77
dispatching command 123
dispersive clay 10
dispersive soil 23
displacement 94
displacement angle 98
displacement ductility ratio 58
displacement fault 14
displacement sensor 67
displacement ventilation 157
disposable income 48
dispute 169
disruptive discharge 117
dissolved air 75
dissolved oxygen 45
dissolved solid matter 16
dissolving corrosion 16
distorted type flip bucket 63
distortion 91
distributed circuit 98
distributing valve 78
distribution 171
distribution box 113
distribution rebar 56
district allowance 163
divergent oscillation 80
diversion 138
diversion bottom outlet 139
diversion channel 139
diversion discharge criterion 138
diverter 75
divide dike 69
dividing valve 84
division 25
divisional works 169
dogging device 90
dolerite 12
dolerite 11
doline 18
dolomite 12
dolomitic limestone 11
domestic waste 45
domestic water 38
dominant frequency 75

door frame 154
dot product 100
double-block-and-bleed valve 84
double casement window 154
double-chamber surge shaft 64
double circuit line 111
double-curvature arch dam 60
double deck screen 149
double donut 161
double girder 90
double-handling freight 164
double-leaf gate 89
double split barrel 26
double volute casing 82
down conductor 118
downgrade 173
down-hand welding 147
down pipe 153
downstream flange 90
downstream flow connection 63
downstream hazard potential 59
downstream level switch 124
downstream level transducer 124
downstream rockfill zone 61
downstream slope 66
down-the-hole drill 148
downthrown fault 14
draft 172
draft tube 76
draft-tube gate 89
drag fold 13
drag movement 7
drainage area 32
drainage blanket 143
drainage density 32
drainage gallery 64
drain curtain 144
drawdown 17, 36
drawer-type formwork 147
drawn-in 67
drencher system 161
drift ice 21
drillability 26
drill-blast tunneling method 140
drill hole grouting pile 142
drilling exploration 26
drilling log 26
drip 21
drip pipe 158
driving pile 142
drop 32
drop and pull transport 135
drop energy dissipator 63
drop-inlet spillway 62
drop-out fuse 113
drop out of relay 127
drop structure 69
drowned nappe 63
drum brake 81
drum gate 89
drum roller 149
dry curing 146
dry density 19
drying shrinkage 145
dry jet mixing pile 141
dry-laid rubble masonry wall 69
dry mortar 141
dry season 36
dry shiplift 95
dry shotcrete 142
dry to seep 21
dry type reactor 110
dry type transformer 109
dry unit weight 61
dual control 79
dual-drum tamping roller 149
dual-hot-wedge 146
duct heater 158
ductile failure 57
duct resistance 158
due payment 172
dumping site 134
dunes 35
duplex apartment 153
duplex safety valve 84
duplicate feeders 113
duration curve 34
duration series 68
dust concentration 140
dust control 140
dust-proof luminaire 115
dust-tight 94
duty cycle 114
duty free 165
duty type 114
dwelling unit type 153
dyke rock 12
dynamic action 54
dynamical storage 36
dynamic braking 114
dynamic viscosity 19
dynamite 137

E

early screw 92
early-stage diversion 138
early-strength admixture 136
earth clearance 112
earth core rockfill dam 60
earth dam 60
earthed voltage transformer 111
earth fault 126
earth fault current 115
earth fault relay 126
earth flash density 116
earthing brush 105
earthing conductor 118
earthing electrode 118
earthing grid 118
earthing terminal 110
earthing transformer 109
earth leakage current 115
earth plate 130
earth pressure 54
earthquake area 7
earthquake directional components 58
earthquake intensity 7
earthquake magnitude 7
earthquake subsidence 8
earth-rock excavation 139
earth rockfill cofferdam 139
earth-rock fill dam with clay core 60
earth-rock fill dam with

sloping clay core 60
earth station 131
earth wire peak 111
ease of ignition 85
eaves 154
eccentricity 57
echelon structure 7
ecological balance 43
ecological degradation 44
economic appraisal 40
economic benefit 40
economic current density 111
economic impact assessment 43
economic life 40
economic net present value 40
ecosystem service 44
eddy current 97
educational background 48
effective alkali content 136
effective rainfall 32
effective storage 36
effective stress path 23
effluent weir 62
eigenperiod of response spectrum 8
elastic deformation 20
elastic modulus 55
elastic moment 55
elastic-plastic deformation 20
electric actuator 85
electrical accident 118
electrical braking 105
electrical braking torque 114
electrical damper module 78
electrical distance 118
electrical energy 98
electrical fault 118
electrical interference source 80
electrically protective barrier 118
electrically protective screen 118
electrical opening limiting module 78
electrical profiling 27
electrical prospecting 27
electrical safety 118
electrical sounding 27
electric braking 114
electric circuit 98
electric contact 108
electric field 97
electric-field probe 116
electric hoist 92
electric-hydraulic governor 77
electricity 97
electric motor 92
electric potential 97
electric power consumption 98
electric power grid 100
electric power system 100
electric shock 118，148
electric strikes 131
electrode 79，137
electro-erosion 105
electro-hydraulic converter 78
electro-hydraulic proportional valve 78
electro-hydraulic servo-valve 78
electrolyte 129
electrolyte level indicator 130
electrolyte level switch 124
electromagnet 98
electromagnetic actuator 85
electromagnetic braking 114
electromagnetic compatibility 80
electromagnetic emission 122
electromagnetic energy 97
electromagnetic field 97
electromagnetic induction 97
electromagnetic lock 132
electromagnetic radiation 98
electromagnetic relay 127
electromagnetic screen 98
electromagnetic wave 97
electro-mechanical converter 77
electromotive force 97
electronic ballast 115
electronic load controller 77
electrostatic field 97
electrostatic induction 97
elementary frequency 119
elementary period 119
elephant trunk 149
elevated river 32
eligibility 168
elongated and flaky particle 135
elongation 67
eluvium 10
embankment 61
embankment dam 60
embankment escape area 61
embankment material 135
embankment rolling criteria 61
embankment zoning 60
emergency gate 89
emergency lamp 115
emergency lane 135
emergency lighting 87，115
emergency logistics 171
emergency preparedness plan 171
emergency rescue rehearsals 171
emergency reserve 103
emergency response 45
emergency shutdown 125
emergency spillway 62
emergency stair case 160
emergency switchgear 162
emergency ventilation 157
emersed gate 89
emission and dust control 44
emissivity 160
employer 170
employer camp 134
employment rate 48
employment status 48
emptying tunnel 63
emulsified bitumen 136
encoder 92
encryption 123
endangered species 44
end cell 130
end cover 94
end-dump closure 138
endemic disease 45
endemic species 44
end-flared pier 63

end frame 90
endogenetic agent 10
endoscope 75
end sill 63
end stop 81
end support 90
en echelon fault 14
energy benefit 38
energy dissipator 79
energy meter 122
energy output 37
energy-saving design 152
energy storage mechanism 79
engineered slope 66
engineering investigation 3
engineering measure 47
engineer workstation 121
enlarge boring 26
enlargement 140
enterprise overhead 164
entrance 131
entrance controller 132
entry 172
environmental awareness 43
environmental carrying capacity 49
environmental constraints 43
environmental covenant 43
environmental degradation 44
environmental deterioration 43
environmental flow 45
environmental illumination 132
environmental impact assessment 43
environmental integrity 43
environmental monitoring 43
environmental quality 43
environmental sensitive object 44
environment capacity 43
environment ecology 43
environment-friendly 43
eolian deposit 10
eolian landform 9
eolian monadnock 9
epicenter 7
epidote 12
epidote vein 12
epilimnion 45
epimetamorphic rock 11
epoxy filler 90
epoxy insulated busbar 113
epoxy resin grout 141
epoxy resin mortar 141
epsilon-type structure 7
equalization charge 130
equalizer pulley (sheave) 93
equilibrium temperature 120
equipment procurement 4
equipotential bonding 118
equi-pressure surface 53
equity capital 40
equivalent intensity 8
equivalent network 99
equivalent uniform live load 54
equopotential bonding 129
erect fold 13
erection all risk insurance 165
erection bay 65, 87
erection tolerance 87
erection welding 91
Erlenmeyer flask 25
erosional terrace 9
erosion force 46
erosion plain 9
erosion-resisting performance 66
erratic boulder 10
escalator 153
escape gradient 61
escape hatch 161
escape lighting 115
escape shaft 160
escape sign luminaire 115
escape stair 154
estuary 32
eternal tensile load 67
ethernet 122
ethnic awareness 48
ethnic discrimination 48
ethnic group 48
ethnic minorities 48
eutrophication 45
evacuation route 160
evaporation 32
evaporite 10
evapotranspiration 33
evenness 140
event log 122
event recording 124
everyday tension 111
evidential sampling 170
excavated slope 142
excavation gradient 66
excavation in layers 140
excavation in open-cut 139
exceeding probability 52
excellent surrounding rock 20
excessive exhaustion of groundwater 38
excess pore pressure 17
exchange loss or gain 166
excitation 99
excitation transformer 120
exciter 120
execute component 127
exfoliation 18
exhaust 148
exhaustion cone of groundwater 38
exhaust shaft 157
exhaust valve 156
existing structure 68
exit access 161
exit door 160
exit gradient 17
exit sign luminaire 115
exit stairway 161
exogenic agent 10
expanded waterstop 137
expanding cement 136
expansion 19
expansion joint 56
expansive agent 137
expansive concrete 145
expansive soil 10
expected output 37

expenditure of idleness 172
experimental heat release 86
expiry date 169
exploitable potential 36
exploitable reserve 139
exploration adit 27
exploration borehole 26
explosion crater 139
explosive magazine 134
export duty 164
exposed penstock 65
exposure dose 86
exposure fire 161
ex-situ conservation 44
extended rating current 111
extension 130
extensometer 67
exterior finishing 154
external energy interrupter 108
external fault 126
externality 40
extinguishing agent 87
extrados 60
extra-high voltage 100
extraordinary rainstorm 33
extrapolation 75
extreme temperature refractory 162
extrusion concrete side wall 144
exudation 17
eye bolt 137
eye observation 67

F

face joint 61
face runout 83
face shovel 148
facies transformation 13
factorage 165
factory acceptance 92
factory lumber 137
Fahrenheit temperature 32
fail-safe 122
failure envelope 20
failure load 111
fail year 50
fair face concrete 144
fair surrounding rock 20
fall and fall fishway 69
fallouts 23
false alarm 86
false firing 119
false set 146
false trip 127
fan delivery 157
fan for primary air 158
fan outlet 157
fascine works 69
fatigue strength 57
fault 14
fault-block mountain 7
fault clay 11
fault current 101
faulted basin 7
faulted bedding plane 13
fault-folded mountain 7
fault fractured zone 14
fault gouge 11
fault impedance 101
fault phase 118
fault ride through 103
fault scarp 8
fault signal 86
fault striae 14
fault striation 14
fault-tolerance 122
fault triangular facet 8
fault trough valley 8
fauna 44
feasibility study 4
feasibility study stage 25
feather joint 14
feedback control 4
feedback path 79
feed chute 149
feeder 101
feeder breaker 113
feeder disconnector 107
feeder screen 149
feeding ground 45
feeding regime 50
feed-in tariff 41
feed main 161
feeler gauge 68
feldspathic sandstone 11
fence time 50
fender 95
ferromagnetic substance 97
ferromagnetism 97
ferroresonance 99
ferruginous imbueing 14
ferruginous sandstone 11
ferry 134
fertility rate 48
fibre dispersion 131
fibre optic cable 131
fibre strain 131
field breaker 121
fieldbus 122
field current transducer 121
field discharge resistor 121
field inquiry 48
field reconnaissance 26
field subsidy 163
field voltage transducer 121
field winding 120
field work 29
filing 172
filled slope 142
filler 136
fillet 60，61
fillet welding 147
filling and emptying 69
filling pile 142
filling valve 90
fill moisture content 143
film stress 65
filter 17，107，125
filter capacity 94
filter criteria 61
filter efficiency 157
filter fineness 94
filter obstruction 125
final exit 160
final gap-closing 138

final payment 169
final rolling 144
final screw 92
final set 146
final statement 169
final value 41
financial analysis 39
financial benefit 39
financial charges 164
financial internal rate of return 39
financial net present value 39
financial opportunity cost of capital 41
financial price 39
financial subsidy 39
financial sustainability 39
financing 4
fine aggregate 135
fine gravel 10
fineness modulus 23
fines 135
fine sand 10
fine steel 137
fingerling 50
finish coat 154
finished aggregate 136
finisher 149
finish screening 143
fire activation risk 159
fire alarm system 159
fire break 160
fire bund 87
fire compartment 160
fire coupling 87
fire curtain 160
fire dike 160
fire door 160
fire effluent 86
fire escape 160
fire extinguisher 87
fire fighting access 160
fire forcible entry tool 160
fire hydrant 161
fire ladder 160
fire lift 87
fire lookout 86
fire pool 160
fire prevention 157
fire resistance 159
fire resistant barrier 160
fire resisting damper 87
fire resisting duct 87
fire resisting partition 86
fire retardance 86
fire retardant paint 137
fire separation distance 160
fire-smoke detection 162
fire stop 162
firestop collar 159
fire wall 87
firewood forest 50
firing failure 119
firm energy 37
firm output 37
first critical speed 83
first demand guarantee 168
first impound period 67
first order leveling 28
first order traverse 28
first-stage facing 61
fishery 50
fish hatchery site 45
fish migration 45
fish rearing 50
fishway 69
fissure 14
fissure artesian groundwater 16
fissure filling 14
fitting 112
fixed ball valve 84
fixed cone valve 89
fixed contact 108
fixed end beam 56
fixed hoist 92
fixed unit-price contract 168
fixed wheel gate 89
flake ice 146
flake-shaped particle 23
flame detector 124
flame front 86
flame retardant material 160
flame spread 86
flaming 86
flammability 85
flap gate 89
flashboard 89
flashing 153
flash-over tendency 139
flash point 86
flat 10
flat beater 149
flatbed trailer 150
flatness 87
flat-plate vibrator 149
flat plinth 61
flat rate tariff 104
flat slab dam 61
flat slope 9
flat-topped weir 62
flat type seal 90
flat welding 147
flexible busbar 113
flexible filler 61
flexible gate disc 84
flexible hose 94
flexible slip form 147
flexible support 140
flexural bearing capacity 56
flight lock 69
flip bucket 63
floating ball valve 84
floating boom 89
floating charge 130
floating gate 89
floating ring seal 83
float switch 124
flocculation 45
flood 34
flood control 3
flood detention area 38
flood dyke 38
flood frequency 34
flood hydrograph 34
flood peak 34

floodplain 9
flood regulation storage 36
flood routing 33
flood storage and reclamation 38
flood surcharge 34
floor slab 153
flora 44
flow cleavage 15
flow control valve 93
flow-duration curve 34
flowing concrete 144
flow transducer 124
flow valve 93
flow velocity 53
fluctuating backwater zone 35
fluctuating pressure 53
fluidal texture 13
flume 69
fluorescent lamp 115
fluorescent tracer 17
fluorite 12
fluoroelastomer 94
flush bolt 137
flush gate 89
flushing channel 69
flushing tunnel 62
flutter 85
fluvial facies 10
flux 137
fly ash 135
flyover type powerhouse 65
flysch formation 7
fly-wheel effect 79
flywheel moment 107
foam extinguishing agent 87
focal mechanism 7
fog flashover 112
fold 13
fold axis 13
fold basement 7
fold bundle 7
folded-plate structure 152
folder 172
fold hinge 13
fold limb 13
foliated metamorphic rock 11
follower 142
follow-up control 79
follow-up tube 142
fonds 172
foot valve 84
forage crops 50
force (weight) density 19
forced air change 158
forced circulation system 159
forced convection 161
forced exciting 120
forced oscillation 99
forced ventilation 161
force majeure 169
forebay 64
forecast lead time 35
foreign exchange premium 40
forepoling 140
forepoling bolt 66
forepoling pipe-shed 140
foreshock 7
forest coverage rate 44
forest land 50
fore toe 69
fork angle 65
forklift 150
form removal 146
form waling 147
formwork jumbo 148
forward blocking state 119
forward breakdown 119
forward path 79
foundation 22
foundation gallery 64
foundation pit dewatering 138
foundation restraint crack 146
foundation ring 76
foundation settlement 22
foundation upheaval 22
foundation upthrow 22
fountain 16
Fourier transformation 79
fractile 25
fraction 19
fracture 12
fracture cleavage 15
fracturing grouting 141
fragmental structure 15
frame beam 56
frame column 56
frame crane 150
framed revetment 142
framed tube structure 152
frame structure 152
franchise 169
Francis turbine 72
free beam 55
freeboard 59
free calcium oxide 16
free cantilever 55
free carbon dioxide 16
free draining 69
free face 15
free fall weir 62
free-flow tunnel 63
free nappe 63
free on board 164
free on board destination 165
free oscillation 75
free porosity 38
free roller gate 89
free-standing intake structure 64
freewheel diodes 120
freeze-thaw erosion 46
freezing and thawing cycle 145
freight costs 163
freight traffic intensity 135
freqency division filter 131
frequency 58
frequency band 99
frequency converter 119
frequency module 77
frequency regulating capacity 103
frequency relay 126
frequency response 79
frequency setting module 78
frequency spectrum 100
frequency transducer 122
fresh 17

fresh concrete 144
friction anchor bolt 142
friction surface 92
friction torque 73, 75
frieze 65
frog hammer 149
front shroud 83
front wall 154
frost boiling 9
frost damage 59
frost-free period 50
frost heaving force 54
frost upheaval 9
frozen soil 10
fry 50
full charge 130
full conduit flow 64
full depth one stage method 141
full exemption 165
full face driving method 140
full lift safety valve 84
full lightning impulse 117
full-load shutdown 79
full-penetration weld 91
full water spout 160
full width rising closure 138
function generator 79
fundamental component 103
fundamental frequency 99, 120
fund flow 166
funnel 18
furrow-ridge tillage 47
fuse 113
fuse-disconnector 113
fuse monitoring 120
fuse-plug spillway 62
fusion welding 147

G

gabbro 12
gabion dam 61
gable 154
gaged basin 34
gallery 59
gallery ventilation 140
gangway 147
gantry 92
gantry crane 92
gap gradation 143
gap-graded aggregate 136
gapless metal oxide arrester 116
gas burst 23
gas dielectric breakdown 117
gas discharge 117
gasify 86
gas insulated line 111
gas metal welding 147
gasoline 137
gated spillway 62
gated weir 62
gate gallery 64
gate groove 90
gate position indicator 92
gate slot 90
gate valve 83
gathering-arm rock loader 148
gating 138
gauge-datum 67
gauge pressure 72
gauging station 35
gear box 76
gearmotor 93
gear pump 93
general exhaust ventilation 157
general-shear failure 58
generation interconnection 101
generator access hatch 76
generator air brake 125
generator efficiency 105
generator fault 106
generator floor 65
generator hall 65
generator-motor 107
generator pit 87
gentle slope 9
geodetic datum 29
geographical information system 28
geogrid 136
geoid 29
geological disaster 18
geological mapping 26
geologic hazards inquiry 26
geologic park 44
geo-mechanical model 28
geomembrane 136
geometric similarity 75
geomorphy 8
geostatic stress field 21
geosyncline 7
geotectonics 6
geotectonic system 6
geotextile 136
geothermal anomaly 27
geyser 16
girder 154
glacial cirque 9
glacial deposit 10
glacial drift 10
glacial landform 9
glaciers lake 33
glacier valley 9
gland cover 82
glare 155
glass insulator 112
global positioning system 28
globe diaphragm valve 84
globe valve 83
glory hole spillway 62
gneiss 11
gneissic structure 13
gobi 9
GPS fitting vertical survey 29
grabbing crane 80
graben 7
grab excavator 148
gradient 9
grading ring 112
grain composition 25
grain crops 50
grain modulus 23
grain output 50
grain size analysis 34
granite 11
granite-gneiss 11

granite-pegmatite 11
granite-porphyry 11
granodiorite 11
granular cementation 25
granularity 19
granular limestone 11
granular texture 13
granulated blast-furnace slag 135
granulite 11
granulitic texture 13
graphical solution 28
graphic progress 150
graphite 12
grass and crop rotation 47
grassland degradation 46
gravel 10
gravelly soil 10
gravitational erosion 46
gravitational field 54
gravitational prospecting 27
gravity anomaly 27
gravity dam 59
gravity drain 69
gravity irrigation 39
gravity spring 16
gray water 159
grazer 44
greasy luster 12
green concrete 144
greenhouse gas emission 44
green lighting 155
green space rate 152
greywacke 10
grille lamp 115
grill footing 54
groin 63
groin sill 70
groove 91
ground crack 8
ground gate 89
ground motion 58
ground penetrating radar 27
ground sill 62
ground station 131
ground stress field 21
ground subsidence 18
ground surface exploration 26
groundwater exhaustion 38
groundwater regime 38
groundwater table 15
groutability 141
grouted anchor bolt 142
grouted rock bolt 143
grout emitting 142
grout enriched vibratable concrete (GEV) 145
grouting 66
grouting gallery 64
grouting pressure 141
grout interconnection 142
grout take 141
guaranteed efficiency 82
guaranteed torque 80
guarantee for retention bond 169
guarantor 168
guard rail 61
guide bearing 76
guide bearing housing 77
guide bearing journal 77
guide bed 13
guide fossil 13
guide hole 23
guide pile 29
guide roller 81
guide sleeve (bush) 94
guide structure 69
guide vane 76
guide vane bearing 76
guide vane opening 74
guide vane pulsation 73
guide vane servomotor 78
guide vane stem 76
guide vane thrust bearing 76
guide vane torque 75
guide wall 63，140
gully 9
gully density 46
gully erosion 46
gunite car 148
gusset plate 90
gutter 153
guttering 161
guy 112
gymnasium 153
gypsum 12
gyratory crusher 148

H

habitable space 153
habitat suitability 43
halite 12
hammer crusher 149
hammer down the hole 26
hand drill 148
hand-dug pile 142
handing over inspection 170
handling carrying 171
hand-over 171
hanger 90
hanger rebar 147
hanging valley 9
hardener 136
hardness 12
hard rock 20
harmful gas 23
harmonic component 103
harmonic frequency 120
harmonic number 99
harmonic order 99
harmonic resonance 114
harmonic source 103
hatch 154
hatching 45
haul ramp 135
haunch 140
hazard identification 171
hazardous environment 171
hazardous substance 171
head dependability curve 37
head gate 89
heading and bench method 140
heading and cut method 140
heading face 140
head loss 64
head office overhead 170

headrace canal 64
headrace tunnel 64
headward erosion 46
head ward scouring 22
headway 154
heat barrier 161
heat bonding 146
heater 94
heat fire detector 160
heating 155
heating capacity 156
heating load 156
heating medium 155
heating riser 156
heat insulating layer 157
heat load 155
heat of hydration 146
heat release rate 86
heat sealing 146
heat transfer 156
heat-transfer fluid 157
heavy crawler dozer 148
heavy duty scaffold 147
heavy-outsized pieces 135
heavy rainstorm 33
heavy wall drive barrel 26
heavyweight aggregate 136
hedonic method 40
helical stairs 154
hematite 12
hermetically sealed cell 129
hermetic separation 80
heterogeneous jointer 137
heterogeneous texture 13
hex bolt 137
hiatal texture 13
hierarchical control of power system 123
high-flow period 36
high frequency oscillator 131
high frequency patching bay 131
highland 8
high-level earthquake 58
high-lift pump 148
high-low tracks 95
high-low wheels 95
highly corrosive 16
highly relaxed 18
highly weathered 17
high-profile layout 132
high-range water-reducing admixture 137
high response excitation 102
high-rise structure 152
high-set differential protection 126
high-strength bolts 92
high temperature retarding admixture 137
high-velocity flow 63
high voltage oscilloscope 116
high voltage rectifier 116
high voltage switchgear 108
high voltage winding 109
hill 8
hillside 8
hinged bent column 57
hinged crest gate 89
hinge pin 84
historical and cultural heritage 44
historic culture site 21
hoist 81
hoist building 132
hoisting capacity 94
hoisting mechanism 81
holdings 172
hold-off interval 119
hole-through tendon 143
holing through survey 29
hollow 9
hollow gravity arch dam 60
hollow jet valve 89
hollow spot 146
homogeneous deformation 20
homogeneous earth dam 60
honeycomb 146
honeycomb weathering 17
hook 93
hook bolt 137
hook crane 80
hook-release device 92
hook-release lifting beam 92
hopper loader 148
hopper wagon 150
horizontal acceleration 8
horizontal bedding 13
horizontal control survey 28
horizontal drainage 69
horizontal filter 61
horizontal frost heaving force 54
horizontal slope 22
horizontal stage 47
horizontal stationary screen 149
hornblende 12
hornfels 11
horseshoe section 64
horst 7
host rock 12
hot standby 94
household head 48
HSE construction 134
humidification 173
hump-shaped weir 62
hurdle cut-off rate 40
hydraulic actuator 85
hydraulic aerial cage 150
hydraulic amplifier stage 80
hydraulic brake 75
hydraulic cement 136
hydraulic connectivity test 17
hydraulic drop 63
hydraulic fill dam 62
hydraulic fracturing method 23
hydraulic gauge 17
hydraulic gradient (slope) 53
hydraulic gun 149
hydraulic hoist 92
hydraulic jack 95
hydraulic jump 63
hydraulic motor 94
hydraulic oil 93
hydraulic performance 75
hydraulic similitude 75
hydraulic steel structure 3

hydraulic structure 3
hydraulic thrust 74
hydraulic transient 79
hydraulic tunnel 62
hydro-chemical analysis 16
hydrodynamic bearing 83
hydrodynamic pressure 53
hydroenergy computation 36
hydrogenerator 104
hydrogeological test 17
hydrograph ascending limb 33
hydrographic net 32
hydrograph recession limb 33
hydrological computation 33
hydrological series 34
hydrological station 35
hydrological telemetry system 36
hydrometry 36
hydromica 12
hydrophone 68
hydropower 36
hydropower plant 4
hydropower potential 36
hydropower station 3
hydropower station at dam-toe 3
hydropower station in dam 3
hydrostatic head 54
hydrostatic pressure intensity 53
hydrothermal alteration 17
hydrothion 16
hypolimnion 45
hysteresis 75

I

ice age 6
ice and floating debris release 70
ice avalanche 18
ice break forecast 36
ice cover 59
ice dam 59
ice load 111
ice regime 59
ideal synchronization 102
ideal transformer 100
ideal voltage source 98
idle capacity 37
idling 114
idling operation 78
igneous rock 11
ignite 86
ignition source 86
illite glimmerton 12
illuminance 115
image compression 173
image processing 30
imbricate fault 14
immaterial assets 168
immediate settlement 22
impact crusher 149
impact hand wheel 84
impact statement 40
impact-type energy dissipator 63
impedance 98
impedance converter 100
impedance matching 98
impedance relay 126
impeded drainage 39
impeller 82
impeller blade 83
impeller cap 83
impeller chamber 83
impeller cone 83
impeller crown 83
impeller hub 83
impeller ring 83
impeller skirt 83
impeller vane 83
impervious blanket 66
impervious core 66
impervious curtain 66
impervious layer 60
import surtax 164
imposed deformation 57
impounding 21
impounding in stages 138
impoundment 21
impressed current 92
impulse current 117
impulse current generator 116
impulse current shunts 116
impulse earthing resistance 118
impulse sparkover voltage 117
impulse transducer 77
impulse turbine 72
impulse voltage divider 116
incapacitation 86
incense coil hoist 92
incident wave 100
incipient fire 161
incipient tractive force 35
incipient velocity 35
inclination transducer 67
inclined boring 26
inclined disc butterfly valve 84
inclined intake structure 64
inclinedjet turbine 72
inclined joint 59
inclined shaft 26
inclined ship lift 95
inclined weir 62
inclusion texture 13
income-in-kind 48
incoming breaker 113
incoming transformer 120
incoming trunk 130
incomplete fusion 91
increased-modulus zone 61
incremental command 123
independent accounting 165
independent coordinate 28
Independent expense 165
independent laboratory 75
independent third party 168
index bed 13
index fossil 13
index of thermal inertia 155
indigenous culture 48
indigenous inhabitant 48
indigenous population 48
indirect cost 164
indirect lightning strike 117
indissolved solid matter 16
individual member 91
indoor damp environment 57
induced overvoltage 118

induced polarization 27
induced stress 21
induced voltage 97
inducer 83
inducing joint 146
inductance 98
inductive ballast 115
inductive reactance 98
inductive susceptance 98
inductive transducer 68
inductor 98
industrial architecture 152
industrial television 131
inert gas 80
inert gas arc welding 147
inerting 86
infant mortality rate 48
infiltration 16
infiltration gallery 158
inflatable dam 61
inflatable rubber seal 125
inflowing flood 34
inflowing sediment 35
inflow runner 158
influence radius 17
influent seepage 16
infrared detector 131
infrared rays operated crane 80
infrastructure 4
infrequent flooded zone 49
inherent braking torque 114
inherent frequency 99
inhibit reclosing 129
initial attack 161
initial burning 161
initial charge 130
initial compression curve 19
initial dewatering 69
initial impounding 138
initial reading 68
initial rolling 144
initial settlement 22
initial speed 72
initial stress 21
initial support 140
initiation 139
injection flush 83
injection rate 141
injector 77
inland transportation 134
inlet duct 157
inlet-raised discharge tunnel 62
inner guide ring 76
inner jacket 131
inorganic modified grout 92
in quadrature 98
inrush current 126
inrush of water 23
inselberg 9
insensitivity 79
insequent landslide 18
insequent valley 9
in-situ conservation 44
inspection gallery 64
installation accuracy 87
installation error 87
installed capacity 37
installment payment 172
instantaneous angular speed 80
instantaneous detonator 138
instantaneous power 99
instantaneous release 114
instantaneous unit hydrograph 33
instantaneous value 98
instruction command 123
instrumentation 67
instrument pier 68
insulated paint 137
insulating coating 160
insulating stick 118
insulation 159
insulation breakdown 118
insulation fault 101
insulation shielding 113
insulator 112
insulator set 112
insulator string 112
insurance and freight (CIF) 164
insurance certificate 169
insurance period 165
intact rock 15
intact specimen 25
intake 64
intake/outlet 68
intake gate 89
intake with gate shaft 64
intangible assets 40
intangible flood damage 38
integral action time 78
integral gain 78
integral structure 15
integrated travel detector 94
integrated wastewater discharge 4
integrating circuit 100
integrity 173
integrity of rock mass 21
interbedded layers 13
intercalation 13
intercepting well 144
interconnection 98
interface wave 27
intergranular space 25
interharmonic component 120
interharmonic frequency 120
interim acceptance 170
interior wall 154
interlaced channel 32
interlayer 13
interlayer shear zone 14
interlock 80
interlocked structure 15
intermediate coat 92
intermediate rock 11
intermediate shaft 83
intermediate support 111
intermed pier 62
intermittent flow 35
intermittent gauging 36
intermittent irrigation 39
intermittent mixer 149
intermittent periodic duty 114
intermittent river 32
intermontane basin 9
internal cause 10

internal erosion 61
internal fault 126
internal insulation 118
internal overvoltage 118
internal rate of return 40
internal scour 22
internal thermal insulation 156
international competitive bidding 168
international settlements 164
interphase short-cicuit 102
interpolation 75
interruption 103
intersection line 65
interstratified weathering 17
inter-unit gallery 153
interval terrace 47
intrados 60
intraformational joint 14
in-transit depot 134
intrusion detection 123，132
inundated cultivated land 49
inundation inventory survey 49
invar tape 68
invasive species 44
inverse Laplace transform 100
inverse plummet 68
inverse time relay 126
inversion 119
inverted siphon 69
inverter 119
investigation stage 25
investment per kilowatt 166
involuntary relocation 48
involuntary resettler 48
inwash 22
iron disseminated 14
iron-manganese disseminated 14
irregular intake structure 64
irreversible deformation 20
irreversible serviceability limit states 53
irrigated land 50
irrigation 3
irrigation duty 39
island arc belt 6
isochrone 16
iso-efficiency curve 82
isohyet chart 33
isolated operation 78
isolated-phase bus 113
isolated power plant 104
isolating transformer 109
isolation 123
isopiestic line 16
isoseismals 8
isotopic age 6
isotropy 19

J

jacket 113
jacket cover 82
jaw crusher 148
jet 157
jet grouted cutoff wall 140
jib crane 92
jib loader 148
job opportunity 48
joint 147
joint 14
joint filling 15
joint frequency 14
joint gauge 68
joint grouting 141
joint load var control 126
joint persistence ratio 14
joint set 14
jumper flag 112
jumper lug 112

K

kaolinitic 12
Kaplan turbine 72
karren 18
karst 18
karst aquifer 18
karst cave 18
karst channel 18
karst depression 18
karst erosion 18
karst landscape 18
karst topography 18
katametamorphic rock 11
kerogen shale（oil shale） 11
keyed contraction joint 146
key trench 140
kick-off meeting 170
kinematic viscosity 19
knee point voltage 111

L

labor 163
labor consumption norm 164
labor force 48
labor insurance 48
labor safety and industrial hygiene 4
labyrinth seal 76
labyrinth spillway 62
laccolith 12
lacing 144
lacustrine deposit 10
lacustrine facies 10
lacustrine plain 9
lag compressor 125
lagoon facies 10
laitance 146
lake facies 10
laminar flow 17
laminated load 34
laminated wood 137
lamprophyre 11
lamprophyre vein 12
land 49
land bearing capacity 49
land circulation 49
land desertification 45，46
land disturbance 45
landform 8
land reclamation 49
land requisition 49
land revenue 50
landscape design 152
landscape forest 44
landslide 18

landslide bulge 18
landslide crown 18
landslide dam 18
landslide deposit 10
landslide flank 18
landslide fracture 18
landslide graben 18
landslide lake 18
landslide main scarp 18
landslide terrance 18
land subsidence 23
land use right 49
lane 135
lapped splices 56
lap welding 147
lap winding 104
large perturbation 79
large-scale agricultural production 49
large well drilling 26
laser alignment system 68
lateral compression 7
lateral contraction 57
lateral deflection 57
lateral movement gate 89
lateral pressure 21
lateral stability 57
lateral strain 57
laterite 10
later stage diversion 138
latter loss 33
leaching erosion 46
lead-lag control 125
leakage 16
leakage around dam abutment 22
leakage flux 99
leakage loss 75
leakage stopping 138
lean concrete 145
learn about 25
lease 163
lebus groove drum 93
left lateral strike-slip activity 14
left lateral wrench 14
left-over land 50
legend 5
length scale ratio 74
lengthways overturn 95
lens 13
lenticle 11
leucite 12
level 29
level ditch 47
leveling adjustment 29
leveling coat 144
leveling point 29
levelness 87
level net 29
level switch 124
level transducer 124
lever reducing valve 85
lid sealing compound 129
life insurance 48
lifting beam 92
lifting eye 90
lifting force 54
lifting-lie gate 89
lifting pipe 82
lift joint 145
lift thickness 143
lift yoke 147
light attenuation 131
light concrete structure 155
light grayish green 12
lightning arrester 115
lightning current 116
lightning outage rate 117
lightning overvoltage 118
lightning rod 117
lightning withstand level 117
limestone 11
limit deviation 87
limiter stage 173
limit states 53
limit switch 124
limonite 12
linear circuit 98
linear pneumatic actuator 85
linear pressure 144
linear running water 21
linear strain 57
linear transducer 79
lineation 15
line charging 121
line current 99
line drop compensation 110
lined tunnel 140
line fault 101
line telephone 130
line trap 131
line-type fire detector 86
lining 140
lining 66
link 131
link box 113
lintle 90
liquid column manometer 75
liquid limit 19
liquidometer 93
liquid state 19
lithite 11
lithofacies 10
lithology 9
litter layer 46
littoral facies 10
livelihood restoration 49
live line detection 118
live load 54
livestock husbandry 50
livestock population 50
livestock shed 50
live storage 36
live tank circuit-breaker 107
living quarter 49
load acceleration constant 78
load break switch 108
load bus 101
load case 54
load center 103
load commutation 119
load details 130
load distribution 54
loader 148
load estimation 113
load flow 101

load indicator 92
load inertial 79
loading and unloading 171
loading charge 163
load limiter 92
load loss 110
load moment 80
load rejection 78
load reserve 103
load stability 102
load transfer 101
loam 10
lobby 153
local compression bearing capacity 56
local control 121
local control level 122
local control unit 122
local currency 166
local damage 56
local equipotential bonding 118
local instability 22
localized fire 160
local labor 163
local materials 163
local purchase 163
local terminal 121
local ventilation 157
lock bolt 137
lock chamber 69
lock gate 89
lockout reclosing 129
lock washer 92
loess 10
loess plateau 9
loess ridge 9
loess terrace 9
logical circuit 127
logic volume 173
logistics 171
lonestone 10
long crested weir 62
longitudinal cofferdam 139
longitudinal joint 14，59
longitudinal slope 135
longitudinal valley 9
longitudinal vertical bracing 155
longitudinal wave 7
loop pipe network 158
loose granular structure 15
loosened rock mass 18
loosened zone 21
loosening blasting 139
loose-rock dam 61
loudspeaker 130
Love wave 7
low-angle dip 14
low-carbon lifestyle 43
lower bound solution 28
lower bracket 104
lower limit earthquake 7
lower pit 76
lower reaches 32
lower reservoir 68
lower shaft 83，105
low-flow period 36
low flow regulation 37
low heat expansive cement 136
low-heat micro- expansion concrete 145
low lift safety valve 84
low-rise apartment 153
low voltage apparatus 113
low voltage ride through 103
lubricated plug valve 84
lubricating oil 137
lubrication system 95
Lugeon test 17
lumber 137
lumen 115
luminaire 115
luminance 115
luminous flux 115
lumped circuit 98
lump-sum contract 168
lux 115

M

macadam pavement 135
machine hall 65
machine room 92
machining allowance tolerance 87
magma 11
magmatic activity 7
magnesium sulfate corrosion 16
magnet 97
magnetic anomaly 27
magnetic ballast 115
magnetic circuit 99
magnetic door contact 131
magnetic field 97
magnetic flux 97
magnetic hysteresis 97
magnetic induction 97
magnetic pole 97
magnetic release 114
magnetic remanence 97
magnetic saturation 97
magnetic screen 97
magnetic sensors 124
magnetic starter 114
magnetic susceptibility 97
magnetic yoke 105
magnetite 12
magnetization 97
magnitude-frequency relation 8
magnitude interval 7
main frequency 143
main grid 100
main powerhouse 65
main rockfill zone 61
main shock 7
main stream 32
main structure 4
maintainability 104
maintained command 123
main tectonic movement 6
maintenance-free battery 130
maintenance free bearing 90
maintenance interval 103
main terminals 105
main transformer 109
main waterway 38
major cycling way 141
major flood 34

major hazard 171
major inundation impact 49
make-break time 109
making 108
mammal species 44
mandrel driven pile 142
manifold 77
manifold penstock 65
manual bypass switch 130
manuscripts 172
map crack 146
mapping control point 29
marble 11
marine facies 10
marine ingression 6
marine premium 165
marine regression 7
mark 29
marker bed 13
marl 11
marshalling kiosk 113
Marshall stability 144
masonry-concrete structure 152
masonry dam 61
masonry member 153
masonry-timber structure 152
mass concrete 144
massive metamorphic rock 11
massive structure 15
master switch 114
material consumption norm 164
matrix 12
matrix approach 40
mattress 69
maturity degree of concrete 145
maximum credible earthquake 58
maximum design earthquake 58
maximum principal stress 21
mean annual runoff 34
mean annual temperature 32
meandering channel 32
mean flow 34
mean sediment discharge 34
mean sediment load 34
measuring component 127
measuring instrument transformer 110
measuring weir 68
mechanical air discharge 148
mechanical air supply 157
mechanical brake 75
mechanical butt splicing 56
mechanical floor 152
mechanical hydraulic governor 77
mechanical linkage 78
mechanical opening limiter 78
mechanical power 72
mechanical property 91
mechanical seal 83
mechanical smoke control 161
mechanical smoke exhaust system 157
mechanical synchronization 95
mechanical ventilation 156
mechanism of soil erosion 46
mechanization level 50
medical insurance 164
medium dense 19
meizoseismal area 8
melt drip 86
membrane structure 152
Mercalli intensity 7
meridian 29
meridian velocity 82
meridional contour 75
metaconglomerate 11
metadata 173
metal-clad switchgear 113
metal foil capacitor 107
metalized capacitor 107
metallic luster 12
metallic sheath 87
metal oxide arrester 116
metamorphic rock 11
metapelite 11
metasandstone 11
meteorological station 32
method of block limit equilibrium 28
method of non-equilibrium thrust 28
mica schist 11
microclimate 44
microfissure 14
microlitic texture 13
microquake 8
micro-seismic wave survey 27
micro-seismograph 8
middle outlet 62
middle reaches 32
midfeed system 156
migmatite 11
migmatization 11
migratory population 48
migratory species 44
mild channel 32
mild region 33
millisecond blasting 139
mimic board 123
mineral admixture 135
mineralization 16
mineralization of groundwater 37
mineral water 16
miniature circuit breaker 114
minimum operating level 4
minimum principal stress 21
minimum service head 158
minor cycling way 141
miscellaneous fill 139
Mise-a-la-masse method 27
misoperation 118
miter block 90
miter end 90
miter gate 89
miter guide 90
mitigation measure 45
mixed blast interrupter 108
mixed erosion 46
mix-in-place pile 142
mobile channel 32
mobile crane 92
mobile fire extinguisher 87
mobile monitoring 67
mobile terminal 121
mobilization 170

mobilization of construction machinery 148
model to prototype conversion 75
model turbine 74
moderate erosive environment 57
moderately corrosive 16
moderately developed 21
moderately hard rock 20
moderately jointed rock 15
moderately pervious 17
moderately thick layer structure 15
moderately weathered 17
moderate temperature valve 84
mode-superposition method 58
modifier 136
moduled case circuit breaker 114
modulus of sediment yield 34
moist unit weight 143
moisture content 19
molasse formation 7
moment of inertia 79
monetary indemnity (compensation) 49
monolith 59
monorail crane 92
monorail hoist 92
mono-stable relay 126
monsoon rain 33
montmorillonite 12
monumental architecture 152
monzonite 11
mooring force 54
moraine soil 10
mortar leveling 154
mortar material 141
mosaic structure 15
most confidential level 173
motion limiter 81
motor operating mechanism 108
mountain foot 8
mountain pass 8
mountain range 8
mountain slope 8
movable block 93
movable concentrated load 54
movable partition 153
movable target 68
moving contact 108
moving flat bed 35
moving screen method 75
moving uniform load 155
mucking 140
muddy clay 10
mud eruption 23
mudflow terrace 18
mudstone 11
multicore cable 113
multi-disc swing check valve 84
multi-element transducer 122
multi-leaf gate 89
multi-level intake 64
multi-mode optical fibre 131
multiphase system 99
multiple boom jumbo 148
multiple circuit line 111
multiple cropping 50
multiple cyclicities 6
multiplexing equipment 130
multipoint displacement meter 67
multipole switch 125
multi-rate energy meter 122
multi-reservoir regulation 37
multispeed motor 114
multi-stage lock 69
multi-stage pump-turbine 72
multi-stories apartment 153
multistory building 152
multi-turn actuator 85
muscovite 12
mushroom stone 9
mutual induction 97
mylonite 11
mylonitic texture 13

N

naked flame 160
narrow gradation 143
narrow slot method 23
natatorium hall 153
national control survey net 28
national geodetic net 28
natural arching 27
natural density 19
natural disaster 44
natural earthing substance 118
natural erosion 46
natural frequency 58, 75
natural ground line 5
natural growth rate 48
natural habitat 44
natural heritage 44
natural landscape 21, 44
natural organic polymer 26
natural period of vibration 58
natural runoff plot 47
natural slope 22
natural water content 61
nature reserve 4
navigation capacity 38
navigation clearance 38
navigation density 38
navigation dependability 38
navigation lock 69
necessary accessories 163
needle 77
needle deflector link 77
needle rod 77
needle servomotor 77
needle-shaped particle 23
needle stroke 74
needle valve 83
negative plate 129
negative pressure 53
negative-pressure chute 149
negative sequence component 99
negative sequence impedance 102
negative sequence network 102
negative terminal 129
neotectonics 8
neritic facies 10
net head 36
net heat flux 160

net positive suction head 73
network attack 123
network configuration 100
neutral conductor 99
neutral point 99
nickel plating 94
nitrogen 80
node 98
no-deviation regulation 77
no-difference regulation 77
no-fines concrete 144
noise 45
noise filter 127
noise level 45
noise pollution 45
no-load current 110
no-load field current 120
no-load field voltage 120
no-load operation 78
no load test 88
nominated subcontractor 170
non-armoured cable 112
nonbrittle impervious zone 143
non-clogging impeller 83
non collapsible loess 10
non-combustibility 85
non-commercial forest 50
noncompetitive bid 168
non-contact switch 114
non-control zone 123
non-destructive testing 91
non-dispersible agent 137
non-dispersible underwater concrete 145
nonflammable material 162
non-graded road 135
non-homogeneous earth dam 60
nonlinear circuit 98
non-linear discharge resistor 121
non-member system 56
non-overflow cofferdam 139
non-overflow section 59
non-point source pollution 45
non-pressure pipeline 159
nonrefundable payment 172
non-regime channel 32
non-renewable resources 44
non-return valve 83
non-saturated sediment transport 35
non-scouring velocity 35
non-self-regulating canal 64
non-self restoring insulation 118
non-silting velocity 35
non-structural measure 38
non-traded goods 39
non-uniform sediment 34
non-unit protection 127
non-vibrating compaction downward 144
nonwoven geotextile 136
noricite 11
normal continuous 73
normal erosion 46
normal fault 14
normal force 57
normal load 111
normally consolidated soil 19
normal moveout correction 27
normal pool level 4
normal shutdown 125
normal stress 57
nose of waterstop 137
nose vane 76
no-slump concrete 144
"no-stress" strain meter 67
nozzle 77
nuclie 75
nuisance trip 127
numeraire 40
nursery garden 50
nursery stock 50

O

oblique bedding structure 22
oblique distance 15
observation series 67
observation target 29
occupational disease 4
occupational health and safety 4
odd-shaped 64
off-circuit tap-changer 110
off-dam reference 68
off-line filing 173
off-peak tariff 104
offset 146
offset of jet to runner 75
offshore structure 152
off state 119
ogee spillway 62
oil-bath lubrication 93
oil conservator 110
oil filter 93
oil forced air forced cooling 110
oil forced water forced cooling 110
oil immersed transformer 109
oil leakage sump 110
oil level gage 93
oil pressure tank 79
oil suction pipe 94
old channel 9
old landslide 9
olivinite 12
on/off variable 122
once through system 158
one percent chance flood 34
one side welding 87
one-way continuous slab 56
one-way reinforcement 147
one-way slab with multispans 56
on-grid energy 41
on-grid tariff 163
on-load tap-changer 110
on-load tap-changing transformer 109
on-off command 125
on-site access 134
onsite trunk road 135
oolitic limestone 11
opal 12
open caisson 142
open channel flow 64
open-close time 109
open earth-rock excavation 139

open gearing 93
open-graded aggregate 136
open ground 9
open impeller 82
opening 109
opening limiter 79
opening loop gain 78
open inlet 64
open-loop control 121
open phase operation 102
open spillway 62
open switchyard 66
open terrain 9
open weir 62
open well 158
operating basis earthquake 58
operating mode transition 79
operating stroke 94
operational amplifier 100
operation circuit 127
operation dispatching 171
operation platform 147
operation schedule 171
operation time limit 127
operator intervention 125
ophiolite belt 6
opportunity cost 40
opposed-blade damper 157
optical cable 131
optical fibre cord 131
optical fibre splice 131
optical phase conductor 131
optical receiver 131
optical repeater 131
optical sensors 124
optimum moisture content 143
optimum water content 19
orange peel excavator 148
organic fertilizer 50
organic soil 10
orientation 14
oriented boring 26
oriented core 26
orifice plate 75
original phreatic surface 61
ornament 154
orogenic belt 6
orogenic cycle 6
orogeny 6
orthoclase 12
orthometamorphite 11
ortho-rock 11
oscillating jet grouting 141
Ottawa sand 25
outage 103
out-dip structural plane 155
outdoor substation 104
outer casing 82
outer corner 65
outer fixed end 142
outflow discharge 62
outflowing sediment 35
outgoing transformer 120
outgoing trunk 131
outlet bucket 63
outlet channel 62
outlet gate 89
outlet through dam 62
out-of-step operation 102
out-of-step protection 128
output 37
output cable 129
output impedance 98
outside fire escape 160
outwash deposit 10
oven-dried aggregate 146
overall stability 20
overcharge 130
over-chute 69
over-consolidation ratio 19
over-current discrimination 127
over-current release 114
over-excavation 139
overexcitation limiter 120
overflow cofferdam 139
overflow dam section 62
overflow spillway 62
overflow surge chamber 64
overflow type powerhouse 65
overhang 60
overhang crane 150
overhanging eaves 153
overhanging platform 147
overhead building 153
overhead crane 80
overhead earthing wire 117
overhead line 100
overhead traveling crane 150
overhead welding 147
overlapping placing 145
overlay welding 91
overload capacity of transmission line 103
overload release 114
overreach pilot protection 128
over-reinforced 147
overshoot 80
oversize aggregate 135
oversize and weight cargo 135
over-standard flood 138
overstocking 44
overthrust fault 14
overtime allowance 163
overtravel 108
overturned anticline 13
overturned fold 13
overturning moment 59
overvoltage 118
overyear regulation 36
ox-bow lake 9
ozone layer 44

P

package deal contract 168
packaged refrigerating unit 157
packaging recycling 163
packed soil 154
packer 142
packing gland 142
packing seat 84
packing sleeve 83
paddle mixer 149
pad foot vibratory roller 149
painting 92
paleochannel 9

paleo-earthquake 7
paleogeography 9
palimpsest texture 13
palletized transport 135
panel point 111
pan tilt 132
pantograph disconnector 107
pantry 153
paper capacitor 107
parallax 29
parallel-blade damper 157
parallel dike 69
parallel double gate 84
parallel gate valve 84
paralleling operation 105
parallelism 87
parallel measure 25
parallel operation 78
parallel radial arm 90
parallel resonance 99
parametamorphite 11
parapet 59
parapet 153
parent company guarantee 168
parking lot 153
parlor 153
Parshall flume 68
partial cutoff 140
partial discharge detector 116
partially penetrating well 17
partial safety factor 52
partial shutdown 125
particle shape 10
particulate grout 141
particulate matter 44
partition wall 153
part-turn actuator 85
pass at the first attempt 88
pass band 99
passing bay 135
passive bus 101
passive earth pressure 54
passive filter 107
pasted plate 129
patent 169
patio 153
pattern anchoring 66
pattern bolting 66
pavement 135
pay as you earn 163
payment due 169
peak acceleration 8
peak intensity 150
peak load power plant 104
peak-load tariff 104
peak making current 109
peak-ripple factor 100
peak shaving and valley filling 103
peak withstand current 108
pearly luster 12
pea stone 135
pebble 135
pebble 10
pedestal pile 142
pedestrian bridge 147
pegmatite 11
pegmatite vein 12
pelitic texture 13
Pelton turbine 72
penalty 172
pendant-operated crane 80
pendulum 80
peneplain 9
penstock expansion joint 65
penstock roundness tolerance 65
percentage differential relay 128
percent passing 143
perch groundwater 15
percolation 16
percusive reverse circulation drill 148
percussion drilling 26
perennial base flow 37
perennial river 32
performance certificate 169
performance guarantee 168
peridotite 12
perimeter grouting 141
perimeter precaution 132
periodic duty 114
periodic logging 121
peripheral hole 139
peripheral velocity 75
periphery joint 61
permanent action (load) 53
permanent deformation 57
permanent droop 78
permanent fault 101
permanent gauge 35
permanent joint 146
permanent land requisition 134
permanent leakage 22
permanent magnet 97
permanent magnet generators 124
permanent resident 49
permanent structure 4
permanent support 140
permeability 16，99
permeable dike 69
permeable layer 15
permeation resistance class 145
permissive pilot protection 128
permittivity 97
perpendicularity 87
persistent command 123
persistent situation 53
personal accident insurance 164
personal protective equipment 148
personnel fall prevention system 148
phanerocrystalline texture 13
phase 98
phase angle 73
phase comparison relay 126
phase controlled charger 130
phase difference 98
phase lag 98
phase lead 98
phase reversal switch 125
phase segregated protection 127
phase sequence 99
phase terminals 105

phase-to-earth clearance 111
phase-to-phase clearance 111
phase winding 109
phasor 98
phenocryst texture 13
phonolite 12
phosphating 94
phosphorous slag powder 135
photoelectric emission 98
photograph interpretation 30
photometer 115
phreatic line 61
phreatic water 15
phyllite 11
phyllonite 11
physical alteration 17
physical contingency 166
physical protection 132
physical weathering 17
phytoplankton 45
pier 76
pier-head powerhouse 65
pier nose 63
piezoelectric effect 98
piezometer 67
piezoresistive transducer 68
pigtail 131
pile extension 142
pile foundation 66
piling 66
pillar drain 61
pilot distributing valve 78
pilot drift tunnel 27
pilot heading 23
pilot hole 142
pilot servomotor 78
pilot servo-positioner 78
pilot valve 125
pilot wire protection 128
pin gate 89
pinger 93
pinnate shear joint 14
pinnate tension joint 14
pintal 90
pintle assembly 90
pintle shoe 90
pintle socket 90
pipe fittings 94
pipe segment (section) 65
pipe shaft 154
piping 22, 94
piston 94
piston accumulator 79
piston reducing valve 85
piston rod 94
pit 76
pit (trench) exploration 26
pit liner 76
pit-run aggregate 136
pitted surface 146
placement lift 145
placing and spreading 143
plagioclase 12
plagio-gneiss 11
plain 8
plain bar 137
plain concrete 144
plain gate 89
plain of subaerial denudation 9
plain sedimentation 159
planar 15
planar slide 18
planation surface 9
plane angle 74
plane survey 29
plane-table 30
planned exploitative reserve 139
planning stage 25
plantation agriculture 50
plant waterway 64
plastic blende drain 137
plastic deformation 20
plastic flow-tensile fracture 22
plasticizer 136
plastic limit 19
plastic moment 55
plastic primacord tube 138
plastic sealer 137
plastic squeezing out 23
plateau 8
plate bearing test 24
plate collision 6
plate compactor 149
plate geotectonic setting 6
plate suture zone 6
platform type formation 7
platy limestone 11
platy structure 13
plinth 61, 153
plot ratio 152
plough horizon 21
plug 64
plug valve 83
plug weld 91
plumbline well 68
plume 157
plunge 14
plunge pool 63, 70
plunger coil 78
plunging anticline 13
plunging fold 13
plutonic 12
pneumatic actuator 85
pneumatically operated valve 84
pneumatic drill 148
podium 153
point bearing pile 142
point discharge 117
point precipitation 33
point source pollution 45
polarization current 97
polarization mode dispersion 131
polarized relay 126
pollutant 44
pollution flashover 112
polymer-cement ratio 141
polymer concrete 145
polymer latex 141
polyphase system 99
polyurea 137
polyurethane 94
polyurethane grout 141
pond 39
pondage power station 3
pontoon causeway 147

pool scouring 22
population influx 48
porcelain bushing 110
porcelain insulator 112
porch 153
pore groundwater 15
pore pressure 17
pore water 15
porosity 19, 91
porphyrite 11
porphyritic texture 13
porphyry 12
port 98
portable luminaire 115
port-adhesive grouting 141
portal collapse 23
port charge 165
positive displacement 94
positive sequence component 99
positive sequence impedance 102
positive terminal 129
post arc current 108
post-cast strip 146
post-resettlement support 49
post-tensioned anchor 142
post-tensioned
prestressed concrete structure 55
post-trip logging 121
potable water 16
potential drop 97
potential energy 53, 72
potential failure surface 59
potentiometric transducer 68
pot hole blasting 139
pouring orifice 93
powder extinguishing agent 87
power capacitor 107
power conversion 119
power dispatching 123
power factor 74
power factor controller 121
power frequency 100
powerhouse at dam toe 65
powerhouse on river bank 65
powerhouse within dam 65
power line carrier 128
power-off brake 114
power output governor 79
power pulsation 73
power quality 103
power relay 126
power setting module 78
power spectral density 73
power supply 100
power supply failure 124
power supply module 122
power system abnormality 126
power system collapse 102
power system oscillation 102
power system stabilizer 80
power transformation 100
power transformer 109
power transmission 100
power tunnel 64
pozzolan 135
pozzolana cement 136
pre-action system 161
precast beam 154
precast concrete 144
precautionary earthquake 58
precipitation 32
precipitation-runoff relation 34
precooled aggregate 146
pre-feasibility study 4
preferential duty 164
preferential tax policy 49
pre-flood drawdown 38
preheat 91
preheated aggregate 146
preliminarily identify 25
preliminary crusher 149
preliminary investigation 23
premonitory symptom 7
prepackaged concrete 145
preparatory period 150
prepayment 169
preplaced-aggregate concrete 145
pre-preparatory period 150
prepressed anchor bolt 142
preshrinking mortar 141
presplit blasting 139
pressure-actuated seal 90
pressure balancing pipe 76
pressure energy 72
pressure fluctuation 80
pressure gage 94
pressure gradient 53
pressure inlet 64
pressure meter switch 94
pressure pipeline 159
pressure pulsation 73
pressure relay 94
pressure relief device 85, 156
pressure relief valve 83, 110
pressure seal 84
pressure sensor 93
pressure sustaining valve 85
pressure switch 124
pressure tap 75
pressure transducer 124
pressure tunnel 63
pressure weighbeam 75
prestress 57
prestressed anchorage 66
prestressed concrete 145
prestressing tendon 67
pre-stretching 94
pretension 67
pretension force 92
pre-torque 96
prevailing wind 33
price contingency 166
price distortion 40
price inflation 48
price level 163
price variance 164
primacord 138
primary air 158
primary breaker 149
primary control network 28
primary grout hole 141
primary joint 14
primary load 111
primary return air 158
primary traverse 29

primary triangulation point 28
primary winding 109
prime paint 137
primer coat 92
priming 82
principal stress 21
principal tapping 110
prism drainage 69
private automatic exchange 130
private branch exchange 130
private financing 40
probability distribution 52
probability mass 52
probability of failure 52
probability of survival 52
probability of withstand 118
probable maximum flood 34
probable maximum precipitation 32
process control level 121
process control mode 169
Proctor optimum density 143
productivity 150
profit 164
programmable logic controller 124
progressive collapse 57
progressive deposition 35
progressive failure 22
progress payment 169
project-affected community 48
project affected persons 48
project contracting 168
project development mode 3
project development purposes 3
project employer 168
project entity 4
Project features 5
project framework 40
project insurance premium 165
project investment 3
project layout 5
project legal person 4
project option 40
project scale 3
project schedule control 171
proluvial fan 9
proof test 24
propagation velocity of wave 74
property line 49，152
proportional gain 78
proportional valve 78
prospective current 108
protecting wire mesh 66
protection against piping 61
protection panel room 132
protective appliance 147
protective coating 154
protective conductor 118
protective earthing 118
protective fitting 112
protective membrane 146
protective resistor 116
protective spark gap 117
protocol 122
prototype observation 67
prototype turbine 74
proximity effect 97
pseudo-classic architecture 152
public acceptance 48
public building 152
public health 48
public participation 48
public support 48
pulley 81
pulling apart plane 15
pull-out force 67
pull-out resistance 67
pulsating direct current 130
pulse 98
pulse amplifier 121
pulse command 123
pulse pressure gauge 67
pulse transformers 121
pumpability 145
pump concrete 145
pump control valve 85
pump diffuser 83
pump diversion 138
pumped storage power station 4
pumping admixture 137
pumping direction 125
pumping irrigation 39
pump mode 107
pump shaft 83
pump station 148
pump-turbine 72
punching shear 22
punch shear bearing capacity 56
puncture resistance 146
purified drinking water system 159
pycnometer 25
pyrite 12
pyroclastic rock 12
pyrophyllite 12
pyroxenite 12

Q

quakeproof 94
quality assessment 170
quality assurance 170
quality control 170
quality defect 170
quality inspection 170
quality of appearance 170
quality of consumption 103
quality of supply 103
quantile 25
quarantine 171
quarry 23
quarry-run rock 136
quartz 12
quartz andesite 11
quartzite 11
quartz sandstone 10
quartz vein 12
quasi-equity 41
quasi-permanent combinations 53
queen-post supporting pier 56
quenched and tempered 91
quenching 119
quick-acting shutoff gate 89
quick set 146
quick shutdown 125

quincuncial pile 142
quoin block 90
quoin post 90

R

rack 90
radial arm 90
radial bearing 90
radial feeder 101
radial gate 89
radial load pulsation 73
radial thickness of ice 111
radiator 110
radioactive tracer 17
radioactivity log 27
radioactivity survey 28
radioindicator 17
raft footing 54
rail clamping device 93
railing 154
rail track 81
rainfall 32
rainstorm 33
raise boring method 140
raise climber 140
raised head bolt 137
raising cage 140
rammed-soil pile 142
ramp 154
randomfill zone 60
rapid closure 125
rapid drawdown period of water level 138
rapid flow 63
rapid unloading 125
rare earthquake 58
rare species 44
rated accuracy limit primary current 111
rated capacity 81, 129
rated capacity limiter 81
rated fire-resistance period 160
rated lift 85
rated tensile strength 111
rate of karst filling 18
rate of karstification 18
rate of reservoir sedimentation 34
rate of residual value 163
rating adjustment for joint orientation 21
rating of discontinuity conditions 21
rating of rock mass integrity 21
ratio of reinforcement 147
ratio of thickness to height 60
ravelling 18
raw water 158
RCC dam 59
reach 32
reactance 98
reaction turbine 72
reactive aggregate 136
reactive energy 103
reactive load 103
reactive power 99
reactive power compensation 107
reactor 98
readiness for automatic start 126
reading station 68
ready-mix plant 149
rebar mechanical splicing 147
receiver 130
recessed rock anchor 143
recharge by groundwater 37
reciprocating feeder 149
reciprocating screen 149
reclaimed water 159
reclosing overvoltage 118
reclosing pressure relief 85
reconnaissance 25
recording gauge 35
records retention schedule 172
recovery of loss 172
recovery voltage 108
recreation 3
rectangular busbar 113
rectification 119
rectifier 119
rectifier diode 119
rectifying relay 126
recumbent fold 13
recurrence period (interval) 33
reducing coupling 156
redundant system 162
reeving system 81
reference mortar 136
reference node 101
reference plane 82
reference point 27
reference sand 145
reference sighting target 68
reflectance 115
reflected wave 100
reflectivity 115
reflux distillation 25
reflux valve 156
refracted wave 100
refraction survey 27
refractory clay block 160
refractory concrete 145
refractory insulation 160
refractory integrity 160
refractory shield 160
refresh time 123
refrigeration 157
refuge storey 153
refusal criterion 141
regenerative braking 114
regime channel 32
regime sediment charge 35
regional economic integration 4
regolith 10
regular dewatering 69
regulated flow 36
regulating guarantee 74
regulating pond 64
regulating ring 76
rehabilitation and upgrading 4
reignition 108
reimbursable expenditures 169
reinforcement 66
reinforcement bar 137
reinforcement concrete 145
reinforcing-bar truss 140

reinforcing mesh 66
rejection 170
relational database 70
relative compactness 143
relative density 19
relative efficiency 73
relatively crushed 21
relatively integral 21
relative non-scalable loss 74
relative watertight layer 15
relaxation zone 23
relaxed rock mass 18
relay 126
release 108
release agent 25
reliability 52
relief tunnel 62
relief well 61
relieving pressure 85
relocation border 49
reluctance 99
remanent magnetization 97
remedial action scheme 129
remedying defect 170
remote backup protection 126
remote control 121
remote sensing 27
remote terminal 121
remuneration rate 169
renewable resources 44
repeated action 54
repeated grouting 141
repetition survey 29
repetitive load 23
replacement 66
representative station 35
reproducibility test 25
required rate of return 40
reseating pressure 85
reserve busbar 104
reserve capacity 103
reserved rock sill 139
reservoir-affected area 49
reservoir backwater 35
reservoir bank-collapse 22
reservoir bank ruin 22
reservoir flood routing 37
reservoir immersion 21
reservoir impoundment 138
reservoir induced earthquake 22
reservoir inflow 34
reservoir leakage 21
reservoir outflow 34
reservoir regulation 37
reservoir silting 35
reset of relay 127
reset time 127
resettlement area 49
resettler 48
reset value 127
resident population 48
residual capacity 129
residual hole rate 139
residual remanence 77
residual soil 17
residual strength 20
residual stress 21
residual uplift pressure 53
resilient seats 85
resin anchor bolt 142
resistance 98
resistance of heat transfer 156
resistance of river channel 35
resistance reducing agent 118
resistance thermometer 68
resisting moment 59
resistivity 98
resistor 98
resolution 68
resonance 58, 99
resonance overvoltage 118
response 99
response time index 161
restorability 58
restoration plan 45
restrained deformation 57
restricted capacity 37
restrike 108
resynchronization 102
retaining concrete plug 66
retarding admixture 137
retention 166
retention money guarantee 168
retention period 172
reticulated shell 153
retraction range 67
retrieval 172
retrogressive scouring 22
return flow zone 157
return on equity 41
return on investment 41
return pipe 156
return ring 76
return ring vane 76, 83
return valve 156
reuse of return flows 38
revealed preference 40
reverberation sound 155
reversal slope 22
reverse bedding structure 22
reverse blocking state 119
reverse blocking valve device 119
reverse breakdown 119
reverse circulation drilling 26
reversed return system 156
reverse power protection 128
reverse rotary rig 148
reverse slope 142
reverse tainter gate 89
reversible machine 107
reversible serviceability limit states 53
reversible turbine 72
reversing drum mixer 149
review 25
revolving screen 149
rework 170
rheological behavior 20
rheology 20
rhyolite 11
rhyolite-gneiss 11
rhyotaxitic structure 13
ribbed bar 137
ribbed slab 56
ridge beam 154

riding wheel 95
rift valley 7
rig for model test 75
right lateral slip offset 14
right plummet 68
rigid busbar 113
rigid gate 84
rigid structural plane 15
rim 76
rim-generator tubular unit 72
rim seal 77
ring 122
ring beam 154
ring bus configuration 104
ring closing 102
ring-follower gate 89
ring gate 89
ring infiltrometer 67
ring manifold 75
riparian woodland 47
riprap 143
riprap slope protection 66
rise 57
rising limb of inflow hydrograph 33
risk of flashback 117
river-bank spillway 62
river basin 32
river basin divide 32
riverbed intake structure 158
river bend 9
river channel 32
river closure 138
river control works 69
river eco-restoration 44
river sediment 34
riverside intake structure 158
river system 32
river terrace 32
river training 47
road class 135
road transport 134
roadway across the dam 135
rock bolting 66
rock burst 23
rock core 26
rock-crib dam 61
rock debris dam 60
rocker arm 84
rocker shovel 148
rockfall 18
rockfill dam 60
rockfill slope protection 66
rockfill zone 60
rocking ring girder support 65
rock mass integrity 21
rock mass rating 20
rock mechanics 19
rock plug blasting 139
rock quality designation 20
rod 90
rod chamber 93
rod drop 27
rod insulator 112
rodless chamber 93
rod mill 149
rolled earth-rock fill dam 60
roller 90, 95
roller chain 90
roller compacted concrete 144
roller compaction efficiency 144
roller gate 89
roller race 90
rolling 143
rolling bearing 90
rolling face 90
roof board 153
roof bracing 153
roof gate 89
roof purlin 153
roof smoke screen 87
roof truss 153
roof vent 87
roof waterproofing 153
rotary dehumidifier 158
rotary pneumatic actuator 85
rotary rig 148
rotating exciter 120
rotating jet grouting 141
rotating seal ring 76
rotational frequency 73
rotational inertia 74
rotational shear structure 7
rotational transducer 79
rotodynamic pump 81
rotor 104
rotor fan 105
rotor hub 105
rotor rim 105
rotor winding 105
rotor yoke 105
rough 15
roughness 14
rough surface 145
round crested weir 62
rounded fragment 9
route survey 29
routine observation 67
row house 153
rubber dam 61
rubber waterstop 137
rubble 136
rubble drain 144
rubble masonry dam 61
runaway speed of turbine 72
runner 76
runner band 76
runner blade 76
runner blade trunnion 76
runner cart 77
runner chamber 76
runner cone 76
runner crown 76
runner hub 76
runner torque 73
runner transport door 77
runoff 33
runoff modulus 33
runoff regulation 37
runoff yield 33
run-of-river hydropower station 3
rust protection 94

S

sacrificial anode 92

saddle back 8
saddle support 65
safe evacuation 160
safe reservoir yield 36
safety current 115
safety curtain 87
safety helmet 148
safety island 8
safety margin 53
safety net 148
safety rail 148
safety shutdown 80
sag 111
salient pole machine 104
saline-alkali swale 50
salinity 16
salinization 45
salty fog acting zone 57
same-floor drainage 159
sampling 25
sanctuary 50
sand air current 46
sand and gravel 23
sand boiling 22
sand drain 144
sand erosion 73
sand liquefaction 8
sand ripple 35
sand-shifting control forest 50
sandstone 10
sandy clay 10
sandy conglomerate 10
sandy loam 10
sandy soil 10
sanitary fixtures 159
sanitation facility 45
sash window 154
satellite photograph 30
saturated sediment transport 35
saturated soil 17
saturated unit weight 143
saturation swelling stress 19
sawn timber 137
scaffold 147
scalar product 100
scaling 23
scarifying 143
scarp 9
scenic resort 44
schedule of payment 169
schist 11
schistose structure 13
scorch 86
scour 34
scour beneath dam 22
scouring 22
scouring funnel 35
scour resistance 66
scraper ring 94
screening curve 143
screen size gradation 143
screw 93
screw conveyor 149
screwed plug 156
screw stem hoist 92
scrotiform weathering 17
sealant 138
seal cage 83
seal coat 144
sealed cell 130
sealed sprinkler head 161
sealing face 84
sealing ring 94
seal plate 90
seal welds 91
seasonal energy 37
seasonal flood 38
seasonally frozen ground 59
seasonal watercourse 32
seat angle 85
seat area 85
seawater spraying zone 57
secondary beam 154
secondary circuit 127
secondary consolidation 19
secondary crusher 149
secondary dam (saddle dam) 62
secondary stress 57
secondary structure 4
secondary winding 109
second harmonic restrain 126
second-stage facing 61
sectional internal force 55
sectional modulus 57
section circuit-breaker 113
sector gate 89
security alarm system 159
security of the site 170
sediment accumulation 34
sedimentary formation 7
sedimentary rock 10
sedimentation 34
sedimentation basin 47
sediment delivery ratio 34
sediment detention reservoir 35
sediment detention weir 35
sediment erosion and deposition balance 34
sediment flushing flow 46
sediment free flow 35
sediment-laden flow 35
sediment runoff 34
sediment transport 34
sediment trapping dam 47
seepage 16
seepage area 16
seepage deformation 17
seepage failure 61
seepage path 16
seepage pressure 16
segregation 146
seismic acceleration 8
seismic action 54
seismic appraisal 59
seismic area 7
seismic belt 7
seismic combination 53
seismic earth pressure 58
seismic focus 7
seismic fortification intensity 58
seismic inertia force 58
seismic intensity 7
seismic isolation 59
seismic network 8
seismic precautionary zone 58

seismic response spectrum 8
seismic source 7
seismic strengthening 59
seismo-active fault 7
seismogenic fault 7
seismogenic mechanism 7
seismogenic structure 7
seismogenic tectonics 7
seismostation 8
selector 79
self-balancing type 82
self-circulation venting 159
self-commutation 119
self-condensing 144
self extinguishbility 86
self-hardening slurry 141
self-leveling 144
self-lubricating bearing 90
self-prestressed cement 136
self-priming 158
self-regulating canal 64
self restoring insulation 118
self-stressing grout 141
self-supporting scaffold 147
self-sustained oscillation 102
self weight collapsible loess 10
self-weight stress 21
semiconductor converter 119
semi-fixed penstock 65
semi-gravity retaining wall 69
semi-outdoor powerhouse 65
semi-solid state 19
semitrailer 150
semi-umbrella hydrogenerator 104
sensitive object 49
sensitive soil 10
sensitivity shift 68
separate network operation 102
separation 82
separator 129
sequence control system 121
sequence interlocking 124
sequence placement 145
sequential reclosing 129
series compensation 107
series connection 98
series extension 34
series interpolation 34
series reactor 110
series representativeness 34
series resonance 99
series structure 79
serpenite 12
serpentine 12
serviceability limit states 53
service building 65
service business 49
service drain 144
service gate 89
service life 94
service reliability 104
service seal 87
service spillway 62
service water 125
servomotor 78, 115
servomotor cushioning time 80
servomotor stroke 78
servo valve 93
set-point command 123
setting 127
setting elevation 73
setting value 127
setting verify 88
settlement allowances 165
settlement joint 56
settlement of soil mass 19
severe bum 161
severe erosive environment 57
sewage disposal charge 164
sewerage system 159
SF_6 circuit breaker 113
shaft current 105
shaft deflection 83
shaft exploration 26
shaft jumbo 148
shaft log 26
shaft orientation survey 29
shaft runout 83
shaft-sinking method 140
shaft spillway 62
shaft torque 73
shaking feeder 149
shaking screen 149
shale 11
shallow flat ravine 9
shallow-focus earthquake 7
shallow footing 54
shallow sea facies 10
shaped composite form 147
shared storage 36
sharp-crested weir 62
sharp transition 20
shear deformation 22
shear-friction strength 20
shear joint 14
shear modulus 20
shear resistance 57
shear-rupture strength 20
shear wave 7
sheath 113
sheave 81
sheep-foot roller 149
sheet-metal fire door 160
sheet pile cell wall 141
sheet-pile cofferdam 139
sheet-pile retaining wall 69
sheet piling 142
shelf retaining wall 69
shell 84
shell type branch pipe 65
shell type transformer 109
shield 7
shielded full face rock TBM 148
shielding ring 112
shifted formwork 147
shifting channel 32
ship breasting force 54
ship chamber 95
ship fleet 38
ship lift 69
ship lock 69
shoal training 69
shock current 115
shoe brake 81

shop fabricated runner 87
shop pre-assembled 92
shored with cribbing 26
short-circuit 101
short-circuit current 101
short-circuit impedance 110
short-circuit making capacity 108
short-delay blasting 139
shotcrete 66
shotcrete robot 148
shoulder 135
shrinkage limit 19
shunt 121
shunt capacitor 107
shunt compensation 107
shunt reactor 110
shunt release 108
shut off head 82
shut out 171
shutter dam 62
shuttle cableway 150
shuttle conveyor 149
side beam 90
side channel spillway 62
side column 90
side cover type pump 81
side dump truck 150
side guide 90
side shoring 26
side shovel 95
sidewalk 147
side wall 23
sieving (screening) 143
sight distance 135
sighting point 29
sign 29
signal circuit 127
signal recording time 74
signal relay 126
signal-to-noise ratio 131
signal valve 161
signboard 118
silicagel breather 110
siliceous cement 11
sill 12
silt 10
siltation volume 34
silt content 136
silt flow 35
silt grain size 34
silt mudstone 11
silt sill 70
siltstone 11
silt training sill 70
similarity 28
simulated rainfall experiment 47
simultaneous blasting 139
simultaneous observation 67
single-centered arch 60
single circuit line 111
single conductor 112
single core cable 113
single gate 84
single girder 89
single-girder bridge 81
single leaf swing gate 89
single liquid grouting 141
single-mode optical fibre 131
single phase auto-reclosing 129
single phase transformer 109
single pole disconnector 107
single-purpose project 38
single shot reclosing 129
single-stage pump 81
single stage pump-turbine 72
single-suction pump 81
sinidine 12
sinkhole 9
sinter 18
sinusoidal change 79
site 170
site acceptance 170
site access 134
site accommodation 134
site agent 170
site assembly 92
site category 58
site design 152
site fabricated runner 87
site leveling 134
siting 49
ski-jump energy dissipation 63
ski-jump spillway 62
skin disease 45
skin effect 97
skin fire 162
skin plate 89
skin weld 91
skip loader 148
skipped vibration 145
skylight 153
slab footing 54
slack bus 101
slag 91
slate 11
slaty cleavage 15
slave operation 120
slave station 123
sleeve expansion joint 65
sliced rock 11
slickenside 14
slickenside 14
slide block 90
slide cutting bedding plane 18
slide damper 157
slide escape 87
slide-resisting pile 66
sliding-buckling 22
sliding formwork 147
sliding gate 89
sliding resistance 20
sliding ring girder support 65
sliding support 90
slightly developed 21
slightly mineralized water 16
slightly open 14
slightly pervious 17
slightly weathered 17
slip 114
slip cleavage 15
slip cliff 18
slip coefficient 92
slip foundation 18
slip mass 18
slip resistance analysis 59

slipring 105
slipring cover 105
slip-ring induction motor 114
slip tongue 18
slit-type flip bucket 63
slope 9，22
slope collapse 22，46
slope cracking 22
slope cutting 142
slope dressing 142
slope face drainage 69
slope failure 22
slope feeding vehicle 149
slope flow 144
slope of the droop graph 78
slope protection 66
slope protection works 47
slope stabilization project 47
slope trimming 142
slope wash 10
slope water works 47
sloping spreader 149
sloping stationary screen 149
sloping terrace 47
slopping placement 144
slot-and-wedge anchor bolt 142
slot outlet 157
slotted spillway bucket 63
slot-tube anchor bolt 143
slow shear test 24
sludge pump 82
sluice board 62
sluice chamber 62
sluice gate 89
sluice tunnel 62
sluice weir 62
sluicing-siltation earth dam 47
slump 145
slumping 18
slurry 26
slurry cutoff 140
slurry wall stabilizing 140
slush ice 146
smart card 132
smart grid 100
smart substation 103
smoke barrier 161
smoke bay 160
smoke damper 87
smoke detector 124
smoke extraction 157
smoke flue 154
smoke proof staircase 160
smoke removal 140
smoke shaft 161
smooth 15
smooth blasting 139
smooth drum vibrating roller 149
smoothing reactor 120
smouldering 86
snowmelt flood 138
snowmelt runoff 33
snubber 119
snubber circuit 121
socket 112
sodium silicate grout 141
soft (weak) structural plane 15
soft cushion 65
soft mineral 12
soft particle 23
soft rock 20
soft starting 114
soil and water conservation 46
soil auger 26
soil-cement slope 142
soil dowel 143
soil erosion 46
soil erosion zoning 46
soil evaporation 39
soil flow 22
soil-forming rock 9
soil gleization 45
soil impervious zone 61
soil mechanics 19
soil moisture content 39
soil nailing 66
soil resistivity 118
soil-rock mixture 10
soil-structure interaction 54
soil tedding 143
solar fraction 159
solar irradiation 159
solarization 146
solar lamp 115
solenoid valve 124
solid content passing through hydroturbine 73
solid grain size 35
solid-head buttress dam 61
solidification slurry 141
solid mineral composition 35
solid state 19
solid waste pollution control 45
solution crack 18
sonic echo exploration 28
sorting 171
sound absorption 155
sound insulation 155
soundness 146
source voltage 97
sown area 50
space truss 153
spalling 18
spandrel 64
spanning member 56
spark blasting cap 138
sparkover 117
spark tester 116
spatial behavior 57
spatial truss 153
spawner 45
spawning ground 45
special load 111
special operation personnel 163
special-shape flip bucket 63
special tool 94
specific creepage distance 118
specific energy 72
specific enthalpy 158
specific expansion 19
specific frictional resistance 156
specific grain surface 25
specific gravity 19

specific penetration resistance 24
specific surface area 19
specific toxicity 86
specific yield 17
specified time relay 126
specimen 25
speed adjustment 114
speed constant 82
speed deviation 80
speed governor 79
speed increaser 76
speed reducer 93
speed sensor amplification 78
speed signal generator 125
speed signal setter 79
sphere gap 116
spherical branch pipe 65
spherical plain bearing 90
spheroidal weathering 17
spider 105
spider hub 105
spillway tunnel 62
spindle 84
spinning reserve 103
spiral case 76
spiral casing floor 65
spiral rebar 137
spiral stairs 154
splice 147
splice reinforcement 56
split flange 87
split-phase transformer 109
split ring 83
split spacing grouting 141
splitting and peeling off 23
spontaneous ignition 86
spooling of a wire rope in multilayer 94
spool valve 93
spot bolting 66
spot welding 147
spray chamber 158
spray cooling 146
sprayer 149
spray lubrication 93
spreader 149
spread footing 154
spreading and leveling 145
spreading gradation 143
sprig-up 161
spring 16
springer 64
spring flood 138
spring plate 85
spring washer 92
SPT blow count 24
spur dike 69
squeezed blasting 139
squeezing rock 15
squirrel cage rotor 114
stability against overturning 59
stability against sliding 22
stability margin 102
stabilized power supply 120
stable temperature field 146
stacker 148
stacker reclaimer 148
stadia survey 29
stadium 153
staff gage 35
stage acceptance 170
stage casing 82
staged design flood 34
stage-discharge relationship 35
stage grouting 141
stage loading 23
staggered joint 59
staggered piling 142
stair 154
stair riser 154
stair step 154
stairway 154
stakeholder engagement 48
stalactite 18
stalacto-stalagmite 18
stalagmite 18
stall 82
standard deviation 25
standard impulse current 117
standard insulation level 117
standard penetration test 24
standard runoff plot 47
standard sphere gap 116
standard sprinkler 161
standard switching impulse 117
standby capacity 37
stand-by duration 104
stand-by power supply 113
standing wave 100
stand post hydrant 161
standstill seal 87
stand-up time 140
star 122
star connection 98
starter 115
starting bus 125
starting sequence dropout 124
start torque 114
statement on completion 169
state variable 99
state vector 99
static action 54
static bypass switch 130
static exciter 120
static frequency convertor 107
static investment 166
static leveling system 67
static phase shifter 107
static suction head 73
static synchronous compensator 107
static var compensator 107
stationary electric contact 108
stationary flat bed 35
stationary seal ring 76
station service 113
station service switchgear 113
station service transformer 113
statistical switching overvoltage 118
statistics 173
stator 104
stator coil 104
stator core 104
stator core fault detection 106

stator frame 105
stator winding 104
stay 112
stay cone 77
stay ring 76
stay vane 76
steady non-uniform flow 64
steady state 78
steady-state heat transfer 156
steady-state runaway speed 72
steady state short-circuit current 101
steady water level 17
steam jacket type valve 84
steam trap 84
steel casting 91
steel-fiber shotcrete 142
steel fibre concrete 145
steel forging 91
steel gabion 138
steel pallet 148
steel sheet-pile cellular cofferdam 139
steel strand 137
steel-tape armoured cable 112
steeping field 39
steep slope 9
stem 84
step 154
step change 79
step displacement input 79
step-down substation 104
step-down transformer 109
stepped column 154
stepped spillway 62
stepped structural plane 15
step-up substation 104
step-up transformer 109
step voltage 115
stereogram 5
stereo-model 70
stiffener 90
stiffener ring 65
stilling basin 63
stirrup 147
stochastic variable 52
stock 12
stockyard 134
stone desert 9
stone masonry 143
stop gate 89
stoplog gate 89
stop valve 83
storage allocation 36
storage battery 129
storm erosion 46
storm water runoff 33
straightness 87
strainer 125
strain hardening 20
strain pulsation 73
strain-slip cleavage 15
strain softening 20
strand 112
stranded conductor 112
stratified rock 15
stratified structure 15
stratum dip 13
stratum gap 13
stratum orientation 13
stratum strike 13
stream channel 32
stream field 53
stream line 61
stress concentration 21
stress contour 60
stress control 56
stress envelope 20
stress history 21
stress path 23
stress pulsation 73
stress redistribution 21
stress relaxation 21
stress release (relief) 18
stress release fissure 18
stress-strain monitoring 67
strike 14
strike-slip fault 14
string transducer 68
strip intercropping 47
stripping layer 23
stripping top soil 139
stroboscopic effect 115
stroboscopic light 75
strongly pervious 17
strong-motion seismograph 8
structural basin 14
structural dome 14
structural integrity 57
structural measure 47
structural member 52
structural plane 15
structural reliability 3
structural resistance 52
structural vibration control 59
structural wall 153
structure durability 58
strut bracing 90
stud bolt 137
stuffing box 84
subangular 10
sub-blocky structure 15
sub-conductor 112
subcritical flow 63
subgrade 135
subgrade 22
sub-massive structure 15
submerged arc welding 147
submerged bearing 83
submerged density 19
submerged dike 69
submerged energy dissipator 63
submerged gate 89
submerged trashrack 64
submerged unit weight 143
submersible taiter gate 89
submission 168
subrounded 10
subsciber's set 131
subsidiary drain 144
subsidy 39
subsidy policy 49
substation 103
substation gantry 132
substitution approach 39

substructure 65
subsurface drainage 69
subsurface pipe drain 69
subsurface waterlogging 39
subsystem debagging 96
subterranean natural resource 44
successively zigzag spread 8
successor 168
suction bell 82
suction casing 82
suction cover 82
suction elbow 82
suction tube 83
sudden change relay 126
sudden-expand gate slot 90
sulfates corrosion 16
sulphate-resisting cement 136
sulphur-sinter 16
summer flood 138
summing point 79
sump tank 79
sunk bolt 137
sunk hydrant 161
superconductor 98
supercritical flow 63
superelevation 135
superheat 156
superimposed back pressure 85
superimposed stress 21
superposed beam 154
superposition theorem 99
superstructure 65
supervision level 121
supporting 66
supporting insulator 112
supporting pier 65
support span 91
suppositious seam 154
surcharge preloading consolidation 142
surcharge storage 36
surface burn 86
surface cleanness 92
surface coat 92
surface erosion 46
surface feature 29
surface flame spread 162
surface flash 86
surface geophysical prospecting 27
surface outlet 62
surface ponding 159
surface preparation 92
surface roughening 145
surface runoff 33
surface tension 74
surface waterlogging 39
surge arrester 115
surge shaft 64
surging 82
surrounding rock 20
surveillance area 132
survey data 29
survey mark 29
survival rate of afforestation 46
susceptance 98
susceptible area of debris flow 47
suspended hydrogenerator 104
suspended load 34
suspended structure 152
suspension bridge 135
suspension clamp 112
suspension index 35
suspension insulator set 112
suspension work 35
suture belt 6
swale 50
swamping 45
Swedish slice method 70
Swedish slipcircle method 70
swelling 19
swelling out of surrounding rock 23
swelling rock 15
swelling soil 10
swelling test 24
swing check valve 84
swing seal device 90
swirl diffuser 157
switch 122
switch 108
switch cubicle 108
switch-disconnector 113
switched valve device 119
switch-fuse-disconnector 113
switching 119
switching value 127
switchyard 66
syenite (sinaite) 11
symbol 5
symmetrical short-cicuit 101
symmetry 87
synchro check 129
synchro check relay 126
synchronization 129
synchronizing shaft 95
synchronous clock 123
synchronous condenser 104
synchronous motor 114
synchronous swing 102
synchronous time 102
syncline 13
synclinorium 13
synthetic organic polymer 26
synthetic paint 137
synthetic storm 33
systematic anchoring 66
system impedance 101
system pipe 161
system software 122

T

table feeder 149
table mountain 7
tabular structure 13
tacheometric polygon 29
tacheometric survey 29
tachogenerator 77
tack coat 144
tafone 9
tail gate 89
tailing hold structure 47
tailrace channel 64
tailrace surge chamber 64
tailrace tunnel 64

tainter gate 89
taking-over 171
talc 12
talc schist 11
tally 171
talus fan 9
tamper 149
tamper alarm 131
tamping 143
tandem roller 149
tangential frost heaving force 54
tangential principal stress 27
tangential strain 57
tangential stress 57
tantile 52
tapered deposit 35
taper foot roller 149
tapped line 101
tapping 110
tax avoidance 165
tax reduction 165
tearing strength 147
tearing strength 94
tectonically stable zone 8
tectonic breccia 11
tectonic framework 7
tectonic joint 14
tectonic line 7
tectonic stress 21
tectonite 11
teed line 101
teleadjusting 123
teleindication 123
telemetry 36
telemetry station 36
telemonitoring 123
telephone harmonic factor 105
telesignalization 123
teleswitching 123
temperature action 54
temperature gradient 146
temperature protection 127
tempering 91
template 75
temporary casing 27
temporary droop 78
temporary land occupation 49
temporary leakage 22
temporary support 140
tender design stage 25
tender security 168
tender sum limit 168
tendon 137
tensile strength 20
tensile structural plane 15
tension anchor bolt 142
tension clamp 112
tension crack 14
tensioned grout tendon 67
tension fault 14
tensioning 67
tension joint 14
tension support 111
terminal block 127
terminal box 113
terrace 9, 47
terrace house 153
terrain roughness 54
terrain slope break 8
terrestrial species 44
test fluid 85
test on completion 171
texture 13
thalweg 32
theodolite 29
theoretical potential 36
theoretic delivery 145
thermal action 54
thermal conductivity 55
thermal diffusivity 55
thermal efficiency 156
thermal equilibrium 114
thermal insulation 157
thermal overload release 114
thermal radiation 86
thermal relay 126
thermal spray 92
thermal stability 156
thermocouples 68
thermocouple transducer 68
thermostatic water bath 25
thin-film oven test 24
thin-plate weir 75
thin-slab structure 152
thin-webbed beam 56
third-order leveling 28
threaded sector 137
three-column disconnector 107
three-dimensional geonet 142
three-girder reinforced bifurcation 65
three phase auto-reclosing 129
three phase short-cicuit 101
three pole disconnector 107
three-way reinforcement 147
three-way valve 156
three winding transformer 109
threshold 68
throat bush 83
throat ring 76
throttle bush 83
throttle valve 83, 93
through beam 154
through bolt 137
through capacity 95
through crack 61
through funnel vortex 64
through impedance 110
through switchboard 131
through transportation 171
through way type valve 84
thrust bearing 105
thrust-bearing bracket 105
thrust bearing housing 77
thrust block 60
thrust collar 65, 105
thrust fault 14
thrust pad 77
thrust runner 105
thunderstorm day 116
thyristor bridge 121
thyristor rectifier 120
tidal hydropower station 4
tie bolt 83
tie-line load 103

tight blasting 139
tilting uplift 8
timber 137
time attendance 131
time-average pressure 53
time coordination 127
time-current characteristic 108
time delay for stopping auxiliaries 125
time-dependent effect 53
time-domain analysis 75
time preference rate 40
time relay 126
time tagging 123
tipping drum mixer 149
toe drain 61
toe slab 61
toe support fills 144
toe wall 61
tolerance 91
tongue 112
toothed disk 77
toothed wheel 124
top anchorage assembly 90
topaz 12
top-down excavation 139
top drift method 140
top filter valve 110
topographical constrain 25
topography 8
topology 80
toppling 18
top soil 10
torque counterweight 95
torque deviation 80
torque fluctuation 80
torque wrench 92
torsioned flip bucket 63
torsion member 56
torsion stirrup 56
torsion structural plane 15
total life cycle 53
total quality control 170
total salinity 16
total station 29
total stress method 28
total suspended solids 45
total volume of voids 143
touch voltage 115
tour gauging 36
tourism 3
tourmaline 12
toxicant 86
toxicant concentration 86
toxic hazard 86
toxic potency 86
traceability 75
tracer isotope 17
trachyte 11
track 90
track connection station 134
track crane 150
track haulage 140
trackless haulage 140
track-mounted drill 148
traffic capacity 135
traffic density 135
traffic switchboard 131
trailer 150
trailing conveyor 149
training alignment 69
training dike 69
trajectory nappe 63
tranquil flow 63
trans-basin water transfer 38
transfer busbar 104
transfer chute 149
transfer function 79
transfer impedance 98
transfer station 134
transfer story 154
transfer switch 114
transient 74
transient component 99
transient fault 101
transient overvoltage 118
transient response 79
transient stability of power system 102
transition current 120
transition joint 113
transition layer 144
transition zone 60
trans-lithospheric fault 7
transmissibility 17
transmission capacity 100
transmission line corridor 111
transmission loss 103
transmissivity 16
transmitter 131
transport in assembly 135
transport intensity 135
transport of segmentation 135
transposition 112
transposition storm 33
transposition support 111
transversal cofferdam 139
transversal section 5
transverse bearing 90
transverse joint 59
transverse load 111
transverse valley 9
transverse wave 7
trash boom 89
trashrack 64
traveling hoist 150
traveling wave 100
travelling cavitation 75
traverse leg 29
traverse point 29
traverse survey 29
traversing 80
tremie pipe 149
tremie placing 145
trench drain 69
trenching by cutting 142
trench installation 140
trenchless installation 140
trend 14
trial-and-error method 28
trial-load method 60
trial running without load 88
triangulation 29
triangulation net adjustment 28
triangulation network 28

tribrach 30
tributary 32
trickle charge 130
trigger advance angle 119
trigger delay angle 119
triggering 119
trigonometric leveling 29
trilateration 29
trip for out of step 129
tri-phosphor 115
triple split barrel 26
trip speed 83
trolley 81, 92
trough 9
truck 92
truck crane 150
trunk channel 39
trunking 113
trunnion 90
trunnion beam 90
trunnion girder 90
trunnion hub 90
trunnion yoke 90
truss arch 56
tube anchor bolt 143
tubular turbine 72
tubular valve 89
tuff 12
tuff powder 135
tunnel 22
tunnel boring machine 148
tunnel diversion 138
tunnel mapping 26
tunnel reaming method 140
tunnel spillway 63
turbine blade position 125
turbine discharge 36
turbine floor 65
turbine inlet water passage 77
turbine input power 72
turbine mode 107
turbine nozzle position 125
turbine output power 72
turbine pier 65
turbine pit 87
turbine vent valve 125
turbine walkway 87
turbulence 75
turbulent flow 17
turnaround loop 135
turn buckle 112
turnkey contract 169
turn-off arm 119
turnover form 147
twin seat bidirectional valve 84
twisted structure 54
twisting channel 32
twist outlet 157
two-centered arch 60
two-lagoon one-way tidal power station 69
two-way extensometer group 67
two-way flat slab 56
two-way joint meter 68
two winding transformer 109
typical floor 153
typical rainstorm 33

U

ultra-basic rock 12
ultra-high voltage 100
ultra-rapid fuses 121
umbrella hydrogenerator 104
unavailable layer 23
unbonded circular anchored tendon 142
unconfined compressive strength 19
unconformity 13
unconsolidated undrained shear test 24
uncultivated land 50
underconsolidated soil 19
undercut 91
undercut slope 22
undercutting 60
under-excavation 139
underexcitation limiter 120
underexcitation operation 106
underfrequency protection 128
underground cavern 23
underground complex 22
underground corrosion 22
underground diaphragm wall 66
underground powerhouse 65
underground river 18
underlapped 80
underpower protection 128
underreach 126
underreach pilot protection 128
under-reinforced 147
undersize aggregate 135
undervoltage protection 128
undervoltage release 114
under water lamp 115
underwater topography 27
undissolved air 75
undulating 15
undulating structural plane 15
unevenness 140
unfactored load 53
ungaged basin 34
ungated spillway 62
ungated weir 62
ungraded aggregate 136
ungrouted contraction joint 146
uniaxial compressive strength 19
uniaxial tensile strength 20
uniclinal structure 14
unidirectional valve 84
uniform cross-section beam 154
uniform sediment 34
union 156
unit bay 65
unit works 169
universal fire truck 87
unkeyed contraction joint 146
unlined tunnel 140
unmanned hydropower plant 121
unsaturated soil 17
unsawn timber 137
unset condition 132
unstable 20
unstable slope 142
unsteadiness 75

unsuccessful reclosing 129
unsymmetrical load 103
unsymmetrical short-cicuit 101
unuseful layer 23
unwanted operation 127
upcurved spillway bucket 63
updating time 123
up duration 104
uplift 7
uplift pile 142
upper bound solution 28
upper reaches 32
upper reservoir 68
upright fold 13
upstream cofferdam 138
upstream end of backwater 35
upstream flange 90
upstream impervious blanket 61
upstream rockfill zone 61
upstream slope 66
upthrown fault 14
upturned deflector 63
urban expansion 48
urban fire protection 161

V

vacuity 53
vacuum breakers 159
vacuum drain 159
vacuum extinction chamber 108
vacuum fuse 114
vacuum grouting 141
vacuum mat 147
vacuum relief valve 85
vadose region 16
valley 8
valley flat 32
valley load 103
valve 129
valve-actuating mechanism 85
valve spool stroke 78
valve type arrester 115
valve voltage drop 119
vane pump 93
vapour pressure 72
var 100
var-hour meter 122
variable action 53
variable centre arch dam 60
variable frequency starting 107
variable radius arch dam 60
variance 25
varmeter 122
vector-borne disease 45
vector product 100
vegetable 50
vegetable land 50
vegetation clearance 45
vegetation cover ratio 46
vegetation measure 47
veiling reflection 155
vein 12
vein rock 12
veinwater 15
velocity energy 72
velocity field 63
velocity fluctuation 63
velocity head 72
vent cap 130
ventilating duct 154
ventilation 156
ventilation by exhaust 140
vermiculite 12
vertical boring 26
vertical extension 161
vertical girder 89
vertical joint 61
vertical-lift gate 89
vertical shaft 26
very slightly pervious 17
very strongly pervious 17
vesicular structure 13
vibo-treated composite soil 142
vibrating compaction upward 144
vibrating screen 149
vibration 58
vibration damper 112
vibration displacement 73
vibration probe 124
vibration transducer 124
vibration velocity 73
vibratory roller 149
vibroflotation 66
vibro-replacement 142
video alarm verification 132
video compressed format 132
video controller 132
video detection 131
video switcher 132
video transmission 132
villa 153
virgin compression curve 19
viscoelastic deformation 20
viscoelastic modulus 20
visco-plastic deformation 20
viscosity 145
viscous debris flow 18
visual inspection 91
vitreous luster 12
vitreous texture 13
void ratio 19
volcanic breccia 12
volcanic rock 12
voltage build up from zero 103
voltage circuit 127
voltage component 127
voltage controlled bus 101
voltage dip 103
voltage drop 97
voltage dropping diode 130
voltage fluctuation 103
voltage forming circuit 127
voltage injection circuit 109
voltage matching transformer 111
voltage recovery 103
voltage regulator 120
voltage relay 126
voltage resonance 99
voltage sag 103
voltage source inverter 119
voltage transducer 122
voltage transformer 111
voltage-var control 126

voltmeter 122
volumetric reinforcement ratio 55
volute casing 82
volute pump 81
vortex elimination 64
vortices 75
voussoir 60
vugular pore 13

W

wafer type butterfly valve 84
wall-slab structure 152
wall-stabilizing soil 136
washer 92
washing screen 149
wash load 34
waste disposal 45
water-absorbing ratio 143
water aeration 45
water and soil conservation 4
waterborne disease 45
water bursting 23
water-cementitious material ratio 145
water-cement ratio 145
water column pressure 91
water compressibility 80
water content 19
water conveyance canal 39
water conveyance system 3
water-cooled aggregate 146
water-cooled condenser 158
water curtain for fire compartment 161
water discharge density 161
water distribution canal 39
water diversion irrigation 39
water divide 9
water environment capacity 45
water expansion anchor bolt 143
water filling by gate split 91
water filling by pass pipe 91
water gushing 23
water hammer 74
water influx 23
water level amplitude 95
water level fluctuating zone 57
water main 158
water pollution control 45
waterpower 36
water pressure test 17
waterproof 94
waterproof coating 136
waterproof explosive 137
water proof lamp 115
waterproof membrane 136
waterproof paint 137
waterproof sealant 136
water quality 16
water reducing admixture 137
water release structure 3
water resources allocation 37
water resources crises 37
water resources regionalization 37
water retaining section 59
water retaining structure 3
water reuse system 158
water riser pipe 158
water seal cage 83
water sealing 82
water seeps 23
watershed 9，32
watershed dam 46
watershed sediment yield 46
water spraying and sand emitting 8
waterstop 137
water storage irrigation 39
water storage pool 47
water supply 3，38
water supply fittings 159
water supply pipe 158
water table 15
water-taking at different levels 38
watertight bottom layer 144
watertight diaphragm 66
watertight surface layer 144
waterway transport 134
waterway tunnel 63
wave bedding 13
wave erosion 46
waveform distortion 103
waveguide 98
wave runup 34
wave setup 34
wave wall 59
wave winding 104
waviness 75
wax-sealed sample 25
weak interlayer 13
weakly corrosive 16
weakly pervious 17
weakly relaxed 18
weathered soil 23
weathered zone 17
weathering 17
weathering and alteration 17
weathering crust 17
weathering fissure 14
web 90
wedge 84
wedge double gate 84
wedge equilibrium method 28
wedge gate valve 84
wedge out 13
wedge plate 144
wedge slide 18
wedge storage 36
weekly regulation 36
weighing hopper 149
weighted average head 36
weighted cover zone 61
weighted graded filter 60
weight span 111
weir gauge 68
weld 91
weld fittings 94
welding gun 150
welding joint 147
weldment 91
welfare withdrawal 164
well-distributed gradation 61
well-graded aggregate 136

wet curing 146
wetland 44
wet pipe system 161
wet shotcrete 142
wetted section 64
wharf 134
wheel 93
whole year diversion 138
wicket gate 89
widening 135
wildlife conservation 44
wildlife habitat 44
wildlife migration 44
wildlife species 44
winch 95
windage loss 76
wind deflection 111
wind erosion 46
wind-induced vibration 54
window sill 154
window spandrel 154
wind setup 34
wind wave 34
wing trench 140
wing wall 69
winning bid 168
wintering ground 45
winterized concrete plant 149
wiper seal 94
wire alignment transducer 68
wire box 138
wired alarming 86
wire mesh with shotcrete 142
wire rope 93
wire rope hoist 92
wire rope hoist 95
wire strain gauge 67
withdrawable circuit breaker 113
withholding tax 164
withstand voltage 117
workability 145
work activity 150
workmanship procedure 91
worm 93
worm gear 93
wormkalk 11
wound rotor 114
written discharge 169

X

xenolith 12

Y

year-end dividend 166
yield deformation 20
yield limit 20
yield strength 20
yoke 84
yoke nut 84
yoke plate 112

Z

zero-lapped 80
zero sequence component 99
zero sequence impedance 102
zero sequence network 102
zero shift 68
zigzag fold 13
zoned earth dam 60
zone of capillarity 15
zoning of water conservancy 37
zooplankton 44